SCIENCE OF SUPERSTRONG FIELD INTERACTIONS

Other Related Titles from AIP Conference Proceedings

To learn more about these titles, or the AIP Conference Proceedings Series, please visit the webpage **http://proceedings.aip.org**

SCIENCE OF SUPERSTRONG FIELD INTERACTIONS

Seventh International Symposium of
The Graduate University for Advanced
Studies on Science of Superstrong
Field Interactions

Shonan Village, Hayama, Japan   13–15 March 2002

EDITORS
Kazuhisa Nakajima
The Graduate University for Advanced Studies
Shonan Village, Hayama, Japan
and
High Energy Accelerator Research Organization (KEK)
Ibaraki, Japan

Masayuki Deguchi
The Graduate University for Advanced Studies
Shonan Village, Hayama, Japan

SPONSORING ORGANIZATION
The Graduate University for Advanced Studies, Japan

◉ **CD-ROM INCLUDED**

AMERICAN INSTITUTE OF PHYSICS

Melville, New York, 2002
AIP CONFERENCE PROCEEDINGS ■ VOLUME 634

Editors:

Kazuhisa Nakajima
High Energy Accelerator Research Organization (KEK)
Accelerator Laboratory
Oho 1-1, Tsukuba
Ibaraki, 305-0801
JAPAN
E-mail: nakajima@post.kek.jp

Masayuki Deguchi
The Graduate University for Advanced Studies
(SOKEN-DAI)
Shonan Village, Hayama
Kanagawa, 240-0193
JAPAN
E-mail: deguchi@soken.ac.jp

L.C. Catalog Card No. 2002112234
ISBN 0-7354-0089-X
ISSN 0094-243X
Printed in the United States of America

CONTENTS

PLENARY TALKS

POSTER PRESENTATIONS

SHONAN LECTURES

Preface

The recent evolution of ultra-intense lasers makes it possible to open frontiers of science on super-strong field interactions, which cover explorations in a broad area ranging from atom and molecular science to nuclear and high energy particle physics. In this decade, high intensity lasers that achieved the interaction energy range of MeV have created novel scientific phenomena in atomic and molecular dynamics in strong fields: high field chemistry, ionization dynamics, higher harmonic generation, coherent X-ray radiation, nonlinear plasma dynamics, energetic photon and particle generation, and particle acceleration. In the next decade, the strong field frontier is anticipated to reach the ultra-relativistic regime of GeV to TeV, in which a new paradigm shift will take place toward nuclear physics, high energy physics and even astrophysics and cosmology in a quest for the origin of matter and the universe.

The seventh international symposium of the Graduate University for Advanced Studies (SOKEN-DAI) was held in order to shed light on this fast evolving field to enhance its rapid growth in a truly international and interdisciplinary forum. It was the first officially organized conference on strong field science in Japan. Its purpose was to convene specialized scientists to discuss the most recent results on various aspects of the strong field interactions with matter and to foster new ideas that enrich scientific exploration in this field. The main symposium HAYAMA Session was held on March 13-15, 2002 at the Graduate University for Advanced Studies in Hayama. The public lectures took place on March 16 at the Cross Tower Hall, Shibuya, Tokyo, with introductory science lectures oriented to a public audience as well. On March 11-12 preceding the specialized symposium, the SHONAN lectures took place to give a series of tutorial lectures that covered an introductory overview followed by the most recent advances on several sub-fields. In consequence, students and young scientists participating in the SHONAN lectures offered unique opportunities to follow a number of invited lectures and to introduce them to the subsequent international symposium for the new frontier science as well as allowing the graduate students the opportunity to obtain credits from the university.

The symposium and tutorial lectures were convened with enthusiasm for a new frontier of science in the friendly air of the Shonan area, where all participants could take pleasure in exciting talks and discussions as well as an enjoyable banquet and a lovely scenery of Mt. Fuji, viewed from the SOKEN-DAI campus. The plenary talks by the outstanding speakers reviewed recent advances of a wide-range of world-wide research on strong field science. A number of excellent poster presentations beyond our expectations revealed growing interest and export in this field especially among young scientists around Japan. The results of the symposium are briefly summarized below.

The ionization dynamics, high field chemistry, atom and molecular dynamics in strong fields initiated strong field science. The strong field interactions with precise coherent phase control could manipulate a fast non-thermal chemical process to create a new state of matter. The high harmonic generation and a short pulse X-ray generation by ultrashort pulses have been typical important subjects of strong field science, which have made successful achievements. In this conference the achievement of very high brightness coherent radiation, exceeding a large scale synchrotron light source is reported. High quality (coherence and very short pulse) radiation will open up a wide range of applications. On the nonlinear plasma dynamics and particle generation, many new exciting results have been discussed in plenary talks and poster presentations, such as soliton observation, very high quality particle beams which would be impossible to produce by conventional technology. The applications to fast ignition and particle accelerator technologies will be expected in the near future. The field of nuclear physics matured in a sense of plentiful experiments and possible applications, such as neutron source and isotope creation through fusion and photo-nuclear reactions. The applications to high energy physics, astrophysics and cosmology are really a forefront of super-strong field science in a sense that we are searching for what we can do with high field physics and where we should go. The useful stimulating talks and encouraging discussions have been presented at this symposium, such as a super high intensity field generation and super high energy acceleration. Now we are really convinced that the forefront of high field science will quickly advance forward with links of a wide range of science, such as atomic and molecular science, plasma physics, nuclear and high energy physics, and astrophysics.

This proceedings is being published to record the results of the symposium in super-strong field interactions as well as the SHONAN lectures. It is a great pleasure to serve the readers with an informative reference work and to contribute a new step to advances in strong field science.

Kazuhisa Nakajima

Chair of the Organizing Committee
The Graduate University for Advanced Studies
High Energy Accelerator Research Organization

The Organizing Committee for the Seventh International Symposium of The Graduate University for Advanced Studies (SOKENDAI)

Kazuhisa Nakajima (Chairman: Dept. of Photoscience)
Koji Hirata (Coordination Center for Research and Education)
Tetsuyuki Yukawa (Coordination Center for Research and Education)
Masayuki Deguchi (Coordination Center for Research and Education)
Toshio Kobayashi (Dept. of Cyber Society and Culture)
Yoshiaki Itoh (Dept. of Statistical Science)
Ken Takayama (Dept. of Accelerator Science)
Yoshiro Azuma (Dept. of Materials Structure Science)
Hiroyuki Iwasaki (Dept. of Particle and Nuclear Physics)
Toshinori Suzuki (Dept. of Structural Molecular Science)
Toshitaka Kajino (Dept. of Astronomical Science)
Takako Kato (Dept. of Fusion Science)
Yoko Satta (Dept. of Biosystems Science)
Yoshiyasu Matsumoto (Dept. of Photoscience)
Humikazu Shibazaki (Coordination Center for Research and Education)
Keiichi Kodaira (President of the university)
Naoyuki Takahata (Vice-president of the university)

The International Symposium Secretariat

Masayuki Hoshi
Yoshiko Takeda
Kenichi Fujii
Tuneyuki Nakajima

Acknowledgments

The symposium and publication of the proceedings were fully supported by the Graduate University for Advanced Studies.

Plenary Talks

High-Brightness Coherent Soft X-Ray Generation by High-Order Harmonics

Katsumi Midorikawa, Eiji Takahashi, and Yauso Nabekawa

Laser Technology Laboratory, RIKEN
Hirosawa 2-1, Wako-shi, Saitama 351-0198, Japan

Abstract. We investigate the energy scaling of high-order harmonics in Ar using a loosely focused beam under the phase-matched condition. By adjusting the argon gas density and the pump laser focusing condition, a total output harmonic energy as high as 0.7 μJ is obtained in the spectral region of 34.8 to 25.8 nm (the corresponding order of the 23rd to 31st harmonic), while the 27th order harmonic (29.6 nm) energy attained is as high as 0.33 μJ with almost perfect spatial profile. Both high efficiency and good spatial quality are achieved simultaneously.

INTRODUCTION

High-order harmonic generation (HHG)[1, 2] can produce ultrashort coherent extremely ultraviolet (XUV) pulses from a compact laser system. Compared to other XUV light sources, such as synchrotron orbit radiation (SOR), free electron lasers (FEL), X-ray lasers (XRL), and laser produced plasma, high-order harmonics (HH) have many advantages such as ultrashort pulse duration, high peak intensity and brightness, wide tunability, and low equipment cost. This emission source is useful for XUV nonlinear optics, attosecond physics, and XUV spectroscopy. Recently, several interesting applications using a HH have been demonstrated in solid-state and plasma physics[3] as well as atomic and molecular spectroscopy. Also, some attempts were made in nonlinear physics to produce two-photon ionization of rare gases in the XUV.[4, 5] Moreover, Hentschel *et al.*[6] recently demonstrated the measurement of attosecond pulse duration using a HH.

For further development of a variety of applications of HH, one of the most important issues is its energy scaling. The high-energy HH is expected to boost new physics in the soft X-ray region. For example, Bandrauk *et al.*[7] reported the advantages of high intensity short wavelength radiations to Coulomb explosion imaging, and Ishikawa and Midorikawa[8, 9] pointed out the possibility of a nonlinear optical effect in the soft X-ray region with a high-energy harmonic pulse.

Experimental and theoretical optimization of conversion efficiency to increase an output energy of HH were reported with a few mJ pump laser energy.[10, 11] However, the energy scaling under such an optimized condition has not been studied experimentally, although we can utilize much larger pump energy. Furthermore, the conditions that give the maximum efficiency and output did not always generate high harmonics with good spatial quality in previous works.[12]

CP634, *Science of Superstrong Field Interactions*, edited by K. Nakajima and M. Deguchi
© 2002 American Institute of Physics 0-7354-0089-X/02/$19.00

Here, we report the energy scaling of HH in Ar under the optimized phase-matched condition. Our scaling method demonstrated a linear increase of harmonic energy with respect to the geometrical focusing area of the pump pulse, while keeping an almost perfect spatial profile of the harmonic output. Moreover, we performed the direct measurement of HH output energy using an XUV photodiode. The high output energy achieved in our scheme allows the direct energy measurement with a single shot.

ENERGY SCALING PROCEDURE

Most of the high-order harmonics generation (HHG) was performed by using a pulsed gas jet, where an intense laser pulse is focused. Phase-matching is dominated by a Gouy phase shift of the laser field and the medium's dispersion as it passes through the focus. Ditmire et al.[13] performed the direct measurement of the conversion efficiencies of HH using an X-ray charge coupled device (CCD) camera. Although they had generated individual harmonics with energies as high as 60 nJ at wavelengths as short as 20 nm, they had not commented on their beam quality.

Recently, by using a self-guided beam or a capillary waveguide in a static long gas-cell, conversion efficiencies were improved.[10, 14-18] In those cases, phase-matching was performed by balancing dispersions due to neutral atoms, free electrons, and focusing geometry. As pointed out by Constant et al.,[11] when the medium length is comparable to or longer than the absorption length, it becomes important to consider the effect of the absorption term of the target medium. Therefore, the photon number, N_q of the qth harmonic on axis per unit of time and of area is given by

$$N_q = N_0^2 |d(q\omega_0)|^2 \frac{4(L_{abs}L_{coh})^2}{L_{coh}^2 + (2\pi L_{abs})^2}\left[1 + exp\left(-\frac{L_{med}}{L_{abs}}\right) - 2cos\left(\frac{\pi L_{med}}{L_{coh}}\right)exp\left(-\frac{L_{med}}{2L_{abs}}\right)\right], \quad (1)$$

where $d(q\omega_0)$ is the atomic dipole moment induced by the laser field, and N_0 and L_{med} are neutral gas density and medium length, respectively. Coherence length L_{coh} and absorption length L_{abs} correspond to $\pi/\Delta k$ and $1/2\alpha$, respectively. Here, α is the absorption coefficient for the qth harmonic. To estimate Δk, one needs to consider the geometrical phase advance and the atomic dispersion for both fundamental and XUV light. As is found in Eq. (1), when the coherence length is comparable to the absorption length, N_q saturates as soon as the medium length becomes longer than the absorption length. Therefore, in order to increase the HH output, the coherence length should extend much longer than the absorption length. When the coherence length is much longer than both the absorption length and medium length, the net HH energy yield is proportional to the following relationship,[13]

$$N_q \propto A_{spot}\left(PL_{med}\right)^2 \quad\quad\quad (2)$$

where A_{spot} is the spot area of the pump pulse at the focusing point, which corresponds to $\pi\omega_0^2$, ω_0 corresponds to the spot size, and P is the target gas pressure.

As described above, the photon number of HH is proportional to the square of the medium length and the target gas pressure. It is also proportional to the square of the dipole strength at the given harmonic order. The dipole strength is assumed to follow $d(q\omega_0) \sim (1-\eta)I_0^5$,[19] where η and I_0 correspond to the ionization probability and the laser intensity at the focus, respectively. The focused laser intensity is limited by the ionization threshold of the target gas medium, because the negative dispersion due to the free electrons breaks phase-matching and ionization decreases the target atomic gas density. In the nonionized condition of the target medium, the gas pressure must be adjusted to cancel the geometrical phase shift of the pump laser pulse.

To extend the interaction length between the laser pulse and the target medium, the Rayleigh length $z_0 = \pi\omega_0^2/\lambda_0$ must be extended. Since a Gouy phase is given by $\Delta k_{gouy} = q/(z_0 +z^2/z_0^2)$, the increasing of the Rayleigh length results in the decreasing of the Gouy phase shift. Therefore, to satisfy the phase-matched condition, the target gas density must be decreased in inverse proportion to the medium length. In other words, the PL_{med} product must be constant under the optimized phase-matched condition in neutral atoms. Consequently, as is derived from Eq. (2), to enhance the output energy of HH under the phase-matched condition, the spot size of the pump pulse at the focus should be increased. This leads to the increase of input laser energy to maintain a fixed pump intensity.

Designed harmonic energy scaling parameters are shown in Table 1. Optimized parameters for f =500 mm pumping were obtained from the previous experiment.[10] For newly designed parameters, the laser focusing condition and the pump energy are scaled up approximately 10-fold.

TABLE 1. Designed High Harmonic Energy Scaling Conditions.

Focusing Length [mm]	500	2500	5000
Pump Energy [mJ]	5	25	50
Sopt Size: ω_0 [μm]	60	130	200
Medium Length: L_{med} [mm]	10	50	100
Gas Pressure: P_{Ar} [Torr]	20	4	2

EXPERIMENT

Experimental studies were carried out with a 10 Hz Ti:sapphire laser system based on a chirped pulse amplification (CPA). This system produced an output of 200 mJ with a pulse width of 35 fs. The wavelength was centered at 800 nm. The pump pulse was loosely focused with a fused silica lens, and delivered into the target chamber through a CaF$_2$ window. We set the focus at the entrance pinhole of the interaction cell. The interaction cell had two pinholes on each end surface of the bellow arms. These pinholes isolated the vacuum and gas-filled regions. The interaction length was variable from 0 to 150 mm in the interaction cell. Argon gas was statically filled in the interaction cell. The generated harmonics illuminated a 120 μm (H) x 15 mm (V) slit of the spectrometer. A flat-field grating (1200 grooves/mm) relayed the image of the slit to a microchannel plate (MCP), while it was spectrally resolved along the horizontal axis. CCD camera detected two-dimensional fluorescence from a phosphor screen placed behind the MCP. Therefore, we measured the spectrally resolved far-field profiles of HH. The spectral response of the measurement equipment was assumed to be flat, because the measured spectral region was sufficiently narrow. The absolute energy of HH was measured directly with an unbiased silicon XUV photodiode (XUV-100).[20] This detector has a wide range of sensitivity from 200 nm to 0.07 nm (6 eV to 17.6 keV). The sensitivity of this photodiode follows the simple linear law $N_e = E_{ph}/3.63$ eV, where N_e is the number of photoelectrons created by a single XUV photon of energy E_{ph}. The values of spectral sensitivity and quantum efficiency of the XUV photodiode were referred to in Ref. 21, and calibrated with 266 nm Q-switched YAG laser pulses in this experiment.

RESULTS AND DISCUSSION

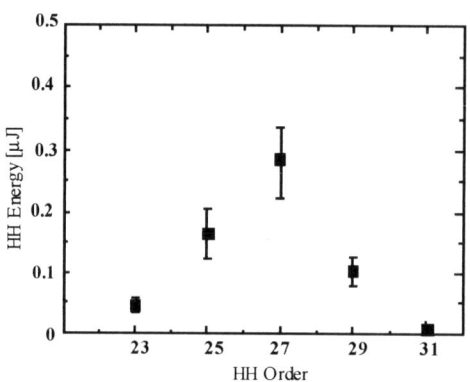

Figure 1. Experimentally obtained energy yield of the 27th harmonic for a 10 cm long medium with 1.8 Torr Ar.

The absolute output energy of HH was measured using an XUV photodiode. This photodiode was inserted in front of the spectrometer, and the output signal was recorded directly on an oscilloscope. A thin aluminum filter (0.2 μm) was placed between the sources and the detector to eliminate the pump pulse and pass only the 11th through 45th harmonics (17 to 70 eV). In practice, low order harmonics (<21st) are not phase-matched in our experimental condition and are strongly absorbed in argon gas.

Therefore, the XUV photodiode detected only the 23rd to 31st order harmonic energies. Figure 1 shows the HH energy yield for the 10 cm interaction. The measured total energy of HH was ~0.7 μJ. The individual harmonic energies were estimated from the spectral distribution of HH. From the relative harmonic strength distribution, the 27th harmonic energy was estimated to be ~0.3 μJ. The error bars resulted from the fluctuation of the pump laser energy. The maximal output of 330 nJ was recorded with a pump energy of 20 mJ.

We also measured the pump spot area dependence of the 27th harmonic energy by changing the pump focusing length. Figure 2 shows the experimentally obtained energy yield of the 27th harmonic as a function of the spot area of the pump pulse at the focus. Note that the spot area is calculated from the focusing geometry in the vacuum. As is shown in Table 1, we initially planned to extend the input laser energy to ~50 mJ. However, since the frequency spectrum of the pump laser pulse was modulated by the self-phase modulation (SPM) at the entrance window (CaF$_2$), we could use only 20 mJ in this experimental setup. When we increased the input energy above 20 mJ, we could obtain neither good spatial quality nor high energy within the observed spectral region. Therefore, the pump energy and the truncated diameter for an f = 5000 mm pumping geometry were set at 20 mJ and 18 mm, respectively. The spot size of the pump pulse at the focus was measured to be approximately 200 μm, which was nearly Gaussian. The output energy of the 27th harmonic shows linear dependence on the spot size of the pump pulse. However, the input laser energy was not scaled up along the designed scaling parameters (see Table.1). For an f = 500 mm pumping, the energy and the diameter of the pump pulse were 5 mJ and 7 mm, respectively. In the middle range of our optimized scaling condition, we used an f = 2500 mm focusing lens. In this condition, the energy and the diameter of the pump pulse were 12 mJ and 13 mm, respectively, while the argon gas pressure and the medium length were 4 Torr and 5 cm,

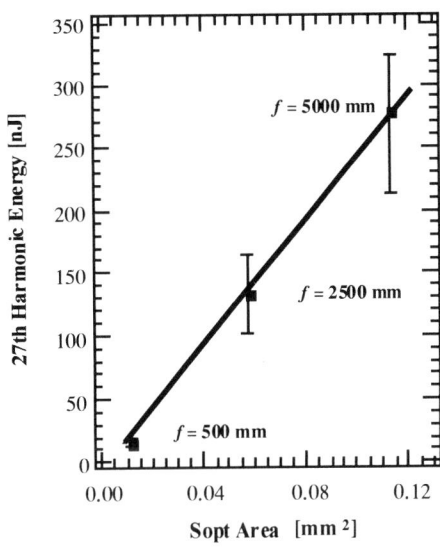

Figure 2. Experimental energy yield of the 27th harmonic as a function of the geometrical spot area of the pump pulse at the focus

respectively. The $P_{Ar}L_{med}$ product was also kept constant for the f = 2500 mm pumping. When the interaction length was increased to 10 cm in the 1.8 Torr Ar gas, the output energy of the 27th harmonic was enhanced more than 10 times, compared with the 20 Torr, 1 cm interaction, while the pump energy was increased by only a factor of four. Therefore, the conversion efficiency was improved by a factor of two, and attained a value of 1.5×10^{-5}.

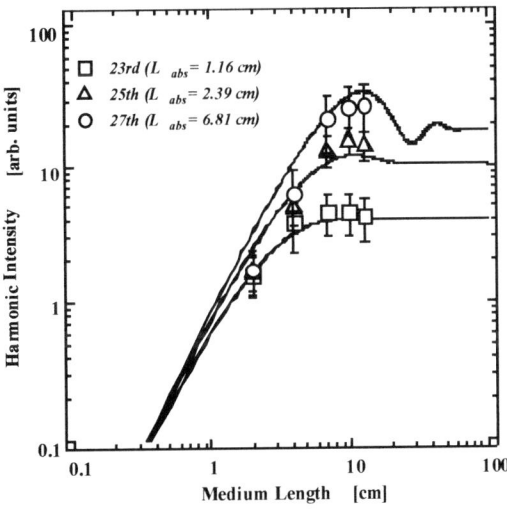

Figure 3. Harmonic intensity on axis as a function of medium length. The solid line corresponds to the calculated photon number for $L_{coh} = 15$ cm with 1.87 Torr Ar.

The evolution of HH intensities in the spectral region from the 23rd to 27th order harmonics measured as a function of medium length is shown in Fig.3. Based on Eq. (1), we estimated the coherence length L_{coh} in the experiment. The absorption length L_{abs} for 1.8 Torr argon gas was calculated to be 1.16 cm for the 23rd, 2.39 cm for the 25th, and

6.81 cm for the 27th harmonic from Ref. 24. The solid line shows theoretically fitted intensities for the 23rd, 25th and 27th harmonics. The coherence length was estimated to be ~15 cm by fitting the theoretical curves. As is pointed out by Constant *et al.*,[11] the optimizing conditions of medium, coherence and absorption lengths are given by $L_{med} >3 L_{abs}$, $L_{coh} >5 L_{abs}$. The 23rd and 25th order harmonics satisfied the optimized condition for this relation. Therefore, those orders were saturated under our experimental conditions. On the other hand, the 27th order harmonic did not satisfy the above conditions yet, because of low absorption.

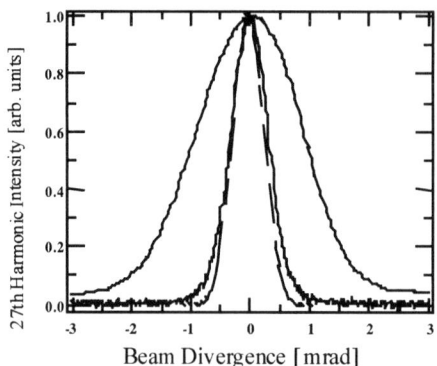

Figure 4. Normalized 1D spatial distribution of the 27th harmonic. Bold and thin lines correspond to P=1.8 Torr, 10 cm medium length and P=20 Torr, 1 cm medium length, respectively. The dashed line shows a numerical 1D spatial distribution of the 27th hamonic under the 10 cm medium length condition.

Figure 4 shows the single-shot far-field spatial profiles of the 27th harmonic. The profile was integrated with respect to wavelength. The abscissa represents the emission angle, and the ordinate represents the normalized intensity. Emission angles (mrad) can be altered to the spatial profile of the harmonic by considering the distance between the harmonic source and the spectrometer slit. No data smoothing procedures were made for the experimental results described below. The thin line corresponds to the experimental result from a 20 Torr, 1 cm interaction under the optimized phase-matched condition. The output beam divergence was ~2 mrad (FWHM). The bolid line corresponds to the 27th harmonic from a 1.8 Torr, 10 cm interaction. To extend the interaction length under the phase-matched condition, the pump pulse was loosely focused with an f =5000 mm planoconvex lens. As a result of phase-matching, the spatial quality of the 27th harmonic also showed a Gaussian-like profile, and the beam divergence decreased by a factor of three. The measured beam divergence was 0.7 mrad (FWHM).

The beam emittance of HH under the phase-matched condition was improved with the increase of the medium length. The beam divergence of HH is related to a noncollinear angle of HH. This type of phase-matching is called the Cerenkov phase matching.[17] The Cerenkov phase matching is similar to the Gouy factor in that the phase of the pump pulse is independent of the gas density and the laser intensity. In the off-axis direction, the phase-mismatch is given by $\Delta k = k_q - q k_0 - \Delta k_{cere}$, $\Delta k_{cere} = q\pi\theta/\lambda_0$, where θ is the noncollinear angle of HH. By using Eq. (1), we performed the calculation of the 1D spatial distribution of the 27th harmonic including the Cerenkov phase matching. The dashed line in Fig. 4 shows a calculation result in the f =5000 mm experimental condition. In that calculation, the phase-mismatch factors were assumed to be zero, except for the Cerenkov factor Δk_{cere}. The spatial intensity distribution of the pump pulse at the focal plane can be described by $I(r) = I_0 exp(-2r^2/w_0^2)$. The emitted photon number was defined to follow $I(r)^{10}$, and the

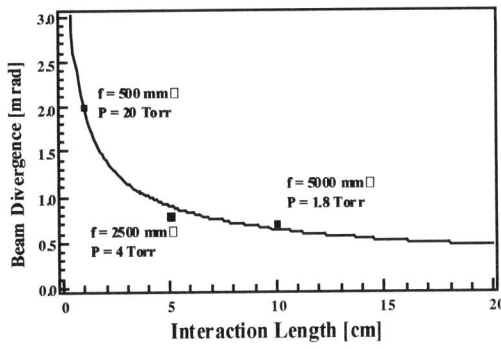

Figure 5. Experimentally obtained beam divergence of the 27th harmonic under optimized phase-matched condition. The solid line correspond to the fitting curve $(1/(L_{med})^{1/2})$ for experimental results.

interaction laser intensity was assumed to be constant along the medium length. Calculated and experimental spatial distributions in the phase-matched condition of f = 5000 mm are compared in Fig. 4. Good agreement in profile and divergence are clearly recognized.

Figure 5 shows beam divergence as a function of medium length under the optimized phase-matched condition. The solid line corresponds to the fitting curve for the experimental results, which was inversely proportional to the medium length with a power of 0.5. The beam divergence angle of HH under the phase-matched condition $(k_q - qk_0 \sim 0)$ is given by $\theta \propto 1/(L_{med})^{1/2}$. Therefore, the measured beam divergence of the 27th harmonic was proportional to $1/(L_{med})^{1/2}$. This result is in good agreement with the theoretical prediction including the Cerenkov phase matching. It is clearly demonstrated that to generate the low-emittance harmonic beam, the long medium length is more useful than a thin gas jet.

CONCLUSION

When the interaction length was increased to 10 cm in 1.8 Torr argon gas, the total output energy in the spectral region of 34.8 to 25.8 nm (the corresponding order of 23rd to 31st harmonics) attained was 0.7 μJ, and the maximal 27th harmonic energy was estimated to be 0.33 μJ. The output energy of the 27th harmonic (29.6 nm) was enhanced by more than 10-fold, compared with 20 Torr, 1 cm interaction, while the pump energy was increased by only a factor of four. Therefore, the conversion efficiency was improved by a factor of two, and attained a value of 1 .5 x10^{-5}. From the measured beam parameters for the 27th harmonic, the peak brightness of this coherent soft X-ray was estimated to be 8 x10^{26} photon/mm^2/mrad^2/s, assuming a beam diameter of 60 μm at the exit of the gas cell. This is the maximum brightness achieved to date, compared to the values produced by various coherent soft X-ray sources.[22, 23]

REFERENCES

1. McPherson, A., Gibson, G., Jara, H., Johann, U., Luk, T. S., McInstyre, I. A., Boyer, K., and Rhodes, C. K., J. Opt. Soc. Am. B **4**, 595-601 (1987).
2. Ferray, M., L'Huillier, A., Li, X. F., Lompré, L. A., Mainfray, G., and Manus, C., J. Phys. B: At. Mol. Opt. Phys. **21**, L31-L35 (1988).

3. Salieres, P., Dero, L. Le., Auguste, T., Monot, P., d'Oliveira, P., Campo, D., Hergott, J.-F., Merdi, H., and Carre, B. , *Phys. Rev. Lett.* **83**, 5483-5486 (1999).
4. Kobayashi, Y., Sekikawa, T., Nabekawa, Y., and Watanabe, S., *Opt. Lett.* **23**, 64-66 (1998).
5. Descamps, D., Roos, L., Delfin, C., L'Huillier, A., and Wählstrom, C.-G., *Phys. Rev.* **64**, 031404 (2001).
6. Hentschel, M., Kienberger, R., Spielmann, Ch., Reider, G. A., Milosevic, N., Brabec, T., Corkum, P., Heinzmanns, U., Dreschers, M., Krausz, F., *Nature* **414**, 509-513 (2001).
7. Bandrauk, A. D., and Chelkowski, S., *Phys. Rev. Lett.* **82**, 273004 (2001).
8. Ishikawa, K., and Midorikawa, K., *Phys. Rev.* **65**, 043405 (2002).
9. Ishikawa, K., and Midorikawa, K., *Phys. Rev.* **65**, 031403 (2002).
10. Tamaki, Y., Itatani, J., Obara, M., and Midorikawa, K., *Phys. Rev.* **62**, 063802 (2000).
11. Constant, E., Garzella, D., Breger, P., Mevel, M., Dorrer, Ch., Blanc, C. Le, Salin, F., and Agostini, P., *Phys. Rev. Lett.* **82**, 1668-1671 (1999).
12. Nisoli, M., Priori, E., Sansone, G., Stagira, S., Cerullo, G., and Silvestri, S. De, *Phys. Rev. Lett.* **88**, 033902 (2002).
13. Ditmire, T., Crane, J. K., Nguyen, H., DaSilva, L. B., and Perry, M. D., *Phys. Rev.* **51**, R902-R905 (1995).
14. Rundquist, A., Durfee III, C. G., Chabg, Z., Herne, C., Backus, S., Murnane, M. M., and Kapteyn, H. C., *Science* **280**, 1412-1415 (1998).
15. Tamaki, Y., Nagata, Y., Obara, M., and Midorikawa, K., *Phys. Rev.* **59**, 4041-4044 (1999).
16. Tamaki, Y., Itatani, J., Nagata, Y., Obara, M., and Midorikawa, K., *Phys. Rev. Lett.* **82**, 1422-1425 (1999).
17. Durfee III,. C. G., Rundquist, R., Backus, S., Herne, C., Murnane, M. M., and Kapteyn, H. C., *Phys. Rev. Lett.* **83**, 2187-2190 (1999).
18. Takahashi, E., Nabekawa, Y., Otuka, T., Obara, M., and Midorikawa, K., *CLEO/Pacific Rim 2001, Conf. Digest (IEEE), Postdeadline Paper,* 4-5 (2001).
19. Krause, J. L., Schafer, K. J., and Kulander, K. C., *Phys. Rev.* **68**, 3535-3538 (1992)
20. United Detector Technologies Inc..
21. Krumrey, M., Tegeler, E., Goebel, R., and Kohler, R., *Rev. Sci. Instrum.* **66**, 4736-4737 (1995).

Ultrafast time-resolved X-ray diffraction

K. Sokolowski-Tinten[1]. C. Blome[1], J. Blums[1], A. Cavalleri[2], C. Dietrich[1],
A. Tarasevitch[1], D. von der Linde[1]

[1]*Institute for Laser- and Plasmaphysics, University of Essen, 45117 Essen, Germany*
[2]*Material Science Division, Lawrence Berekely Nat. Lab., Berkeley, CA 94720, USA*

Abstract. Femtosecond laser-generated plasmas emit ultrashort X-ray pulses in the multi-keV
range, which allow the extension of X-ray spectroscopy into the ultrafast time-domain. We
report here on the generation of such short X-ray pulses and their application for time-resolved
diffraction as a means to directly study ultrafast structural dynamics in laser-excited solids.

INTRODUCTION

Most of our knowledge on the atomic structure of matter is due to X-ray
spectroscopies, because the short wavelength of X-rays gives direct access to the
spatial scale of atomic arrangements. However, standard X-ray experiments (using for
example synchrotron radiation) provide essentially a *static* picture because the time
resolution of such experiments is rather limited. Most fundamental processes in nature,
such as chemical reactions and phase transitions involve *dynamic* changes of the
atomic arrangements, which occur typically on time-scales comparable with the
natural oscillation periods of atoms and molecules, that is femtoseconds to
picoseconds. Therefore, great efforts have been made during the past few years to
extend X-ray spectroscopy into the ultrafast time-domain.

One of the most successful approaches has been made possible by the recent
progress in ultrafast laser technology, namely *chirped pulse amplification*. Multi-
Terawatt, ultrafast table-top scale laser systems are now available in an increasing
number of laboratories all around the world. When focused onto solid targets, such
intense light pulses generate high-density plasmas emitting short bursts of hard X-rays
up to the MeV region [1]. This new kind of high brightness ultrashort pulsed X-ray
sources allow an extension of the established time-resolved techniques of ultrafast
optical spectroscopy into the hard X-ray range, thus providing simultaneously atomic
scale *temporal* and *spatial* resolution.

This contribution discusses the generation of femtosecond, multi-keV X-ray pulses
and their application for time-resolved diffraction as a means to directly study ultrafast
structural dynamics in laser-excited solids.

CP634, *Science of Superstrong Field Interactions*, edited by K. Nakajima and M. Deguchi
© 2002 American Institute of Physics 0-7354-0089-X/02/$19.00

SHORT X-RAY PULSES FROM LASER-PRODUCED PLASMAS

As has been first demonstrated by Kühlke et al. [2], the high-density micro-plasmas, generated through the interaction of very intense light pulses with solid targets, represent an efficient source of hard X-rays. Particularly interesting is the strong characteristic line emission (for example K_α-emission) of these plasmas. This line radiation is believed to result from the interaction of energetic electrons with the non-excited target material underneath the thin plasma layer [3], a process quite similar to an ordinary X-ray tube. Because those energetic electrons are produced by direct acceleration in the strong laser field, a very short duration of the X-ray pulses, of the order of the laser pulse duration, can be expected.

Femtosecond laser-plasma-driven X-ray sources are quite attractive, in particular for small, university-scale laboratories, because they combine simplicity with low cost, as compared to other, accelerator-based approaches [4]. A convenient implementation of such a source uses a thin metallic wire, which is continuously moved through the focus of the laser, as a target [5]. The left part of Fig. 1 shows a photograph of the wire-target assembly used at our set-up for time-resolved diffraction.

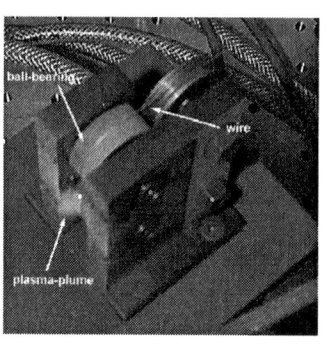 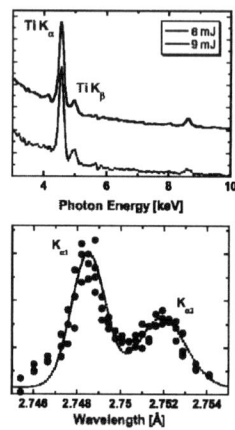

FIGURE 1. Left: photograph of the Titanium wire-target assembly. Right: X-ray emission spectra of the laser excited Titanium target (top: overview spectrum obtained by photon counting/pulse height analysis with a X-ray CCD; bottom: higher resolution spectrum of the spin-orbit split K_α-lines obtained with a crystal spectrometer).

It should be kept in mind that due strong laser-induced ablation the target has to be moved between two consecutive laser pulses in order to provide a fresh surface area for each individual pulse. As a consequence the wire-target offers two main advantages compared to other target schemes: (i) it allows a very compact design, which simplifies shielding issues, and (ii) virtually infinite measurement times are possible simply by providing a sufficiently long spool of wire. Our source, which runs at 10 Hz repetition rate, can operate for nearly 70 h with a 500 m long spool of wire and a pulling velocity of typically 200 μm per pulse!

We have chosen titanium as target material. The Ti-K_α-emission (4.51 keV) of our source allows Bragg diffraction on a wide variety of materials and overcomes the inherent limitations with respect to lattice parameters and/or diffraction orders of sources operating at longer wavelengths. The left part of Fig. 1 displays spectra of our source in the keV-range. The top viewgraph represents an overview spectrum obtained by operating an X-ray CCD-detector (thinned, back-illuminated) in the single photon counting mode with pulse height analysis (to assure single-photon detection we had to reduce the laser energy). The dominant features are the titanium K_α- and K_β-lines at 4.51 keV and 4.93 keV, respectively. The bottom viewgraph shows a high resolution spectrum of the spin-orbit split $K_{\alpha1}$ and $K_{\alpha2}$-lines, which was obtained with a crystal spectrometer.

As has been mentioned above, the X-ray-tube-like K_α-emission from laser-produced plasmas is caused by fast electrons accelerated in the intense laser field. The efficiency of this process depends on the energy of the electrons and is therefore strongly influenced by the details of the laser-plasma interaction. For example, it is common experience [6,7] and also supported by theoretical calculations [8] that for given laser- and material parameters the highest laser intensities not necessarily lead to highest K_α-yield. This is demonstrated by the data shown in the left part of Fig. 2, where we measured the relative yield of our titanium K_α-source (dots) as a function of the position of the focusing optics (f = 150 mm) relative to the target.

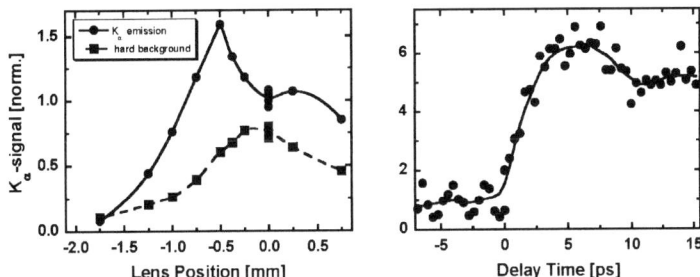

FIGURE 2. Left: normalized X-ray signal as a function of the lens position (relative to the focus; dots: Ti- K_α-emission; squares: hard background). Right: Ti-K_α-yield in the two-pulse excitation scheme as a function of the delay time between the plasma generating pre-pulse and the main pulse.

The K_α-signal has been normalized to the yield with the titanium target exactly in the focal plane of the optics (zero position), corresponding to the highest laser intensity. The highest K_α-flux is observed 0.5 mm away from the focal point, which corresponds to approximately two times the Rayleigh length. At the same time the background signal detected by the CCD (squares; arbitrarily normalized to fit into the plot window of Fig. 2), which is due to hard X-rays and secondary radiation, is significantly reduced. Therefore, optimizing the focusing conditions does not only increase the K_α-flux, but allows at the same time a substantial improvement of the signal-to-noise ratio.

A second way to tailor the plasma properties in order generate exactly those electrons which have the highest efficiency in the production of K-shell holes is the use of a double-pulse excitation scheme. This scheme separates the steps of plasma production and electron acceleration. A first, medium intensity laser pulse is used for plasma generation. After a certain delay, the main high intensity pulse interacts with the pre-formed plasma to accelerate the electrons. The right part of Fig. 2 shows the K_α-emission as a function of time-delay between the pre-pulse ($I \approx 10^{15}$ W/cm^2) and the main pulse ($I \approx 5 \times 10^{16}$ W/cm^2), which produces the fast electrons.

In agreement with results obtained at a silicon K_α-source at 1.8 keV [9] we observe an enhancement of almost an order of magnitude at a delay time of just a few picoseconds. In [9] this enhancement has been attributed to resonance absorption in the slightly expanded plasma (scale length $\approx \lambda$), which leads to a very efficient production of electrons in the optimum energy range (note that the cross section for ionization of K-shell electrons peaks for most materials at approximately 3 – 4 times the K_α-energy).

TIME-RESOLVED X-RAY DIFFRACTION

It is an important advantage of laser-plasma driven X-ray sources that the laser driving X-ray generation provides at the same time an absolutely synchronous source for optical excitation. Therefore, the basic experimental concept of ultrafast time-resolved measurements, the so-called *pump-probe* scheme, which is well established in the optical domain, can be directly extended to the hard X-ray regime: an optical *pump*-pulse is used for excitation while an ultrashort X-ray pulse serves as a *probe* to monitor the transient dynamics induced by the pump.

Because laser-plasma driven X-ray sources are essentially monochromatic, they are particularly suited for time-resolved Bragg-diffraction experiments. A schematic of such an experiment is shown in Fig. 3.

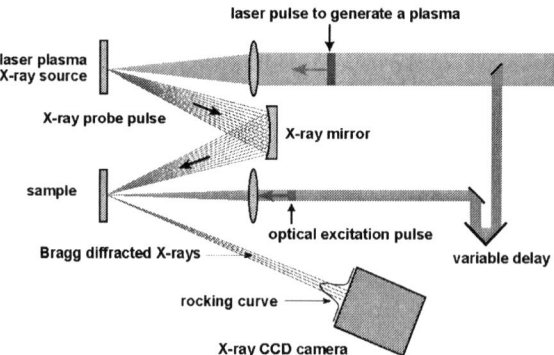

FIGURE 3. Schematic of an optical pump X-ray probe experiment for time-resolved diffraction.

In our set-up near-infrared laser pulses from a 10 Hz amplified Ti:sapphire laser system with pulse energies of about 150 mJ and a pulse duration of 120 fs are focused onto the surface of the moving titanium wire. A small fraction is split off the main laser pulse and (after passing through an optical delay line) is used for sample excitation.

The K_α-radiation from the plasma is emitted incoherently into the full solid angle. Efficient use of the produced X-rays requires, therefore, re-collection and focusing of the X-rays onto the surface of the sample under investigation with a spot size comparable or smaller than the area excited by the optical pump. Focusing of the K_α-radiation is accomplished with the help of a toroidally bent crystal. As has been discussed in detail by Misalla et al. [10], monochromatic point-to-point imaging of the plasma source can be achieved in this way. The experimental geometry is shown in the left part of Fig. 4.

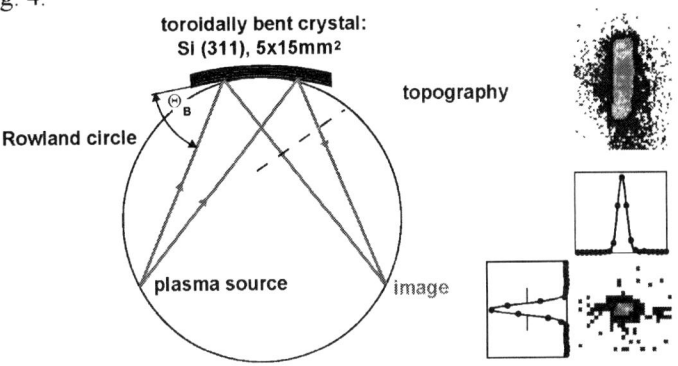

FIGURE 4. Focusing of the keV-plasma emission. Left: experimental geometry. Right top: *topography* of the bent crystal mirror; Right bottom: 1:1 image of the plasma source (FWHM: 85 μm).

For a given crystallographic orientation of the mirror the horizontal and vertical bending radii are determined by the imaging geometry on the Rowland-circle and the requirement that the Bragg-condition has to be fulfilled for the chosen wavelength.

The results of an imaging experiment at our Ti-K_α-source using a bent Silicon crystal with (311) surface orientation are shown in Fig. 4. In the lower right the spatial distribution of the focused X-rays, as detected with our X-ray CCD (pixel size 27 μm), is displayed. The spot is nearly circular and exhibits a FWHM of approximately 85 μm. This focus typically contains 30.000 *detected* K_α-photons per pulse.

The upper right part of Fig. 4 shows the local reflection characteristic of the mirror, which is obtained by placing the CCD-detector away from the image plane close to the mirror. This reflection characteristic provides direct information on the surface topography of the mirror because of its very narrow *rocking curve* (angular dependence of the diffraction efficiency for a fixed X-ray wavelength). Any spatial deformation of the mirror surface leads to a deviation of the local incidence angle from the Bragg-angle and thus to a reduction of the diffraction efficiency. The Silicon-crystal used in our set-up shows an excellent topography with a peak-to-peak non-uniformity of the reflectivity of only 10 %.

In the diffraction experiment the crystalline sample is placed in the image plane under the appropriate Bragg-angle. The optical pump spot and the X-ray focus must overlap spatially on the surface of the sample, and the diffracted X-rays are recorded by an X-ray sensitive area detector. Please note, that the X-rays, incident from the X-ray mirror onto the sample, cover an angular range much larger than the width of the *rocking curve*, which can be, therefore, recorded in a single exposure without any angle-scanning of the sample.

A major problem in optical pump, X-ray probe diffraction experiments results from the different penetration depths of optical radiation and multi-keV X-rays. In semiconductors and metals the optical absorption depth is usually below one micrometer, becoming even shorter at high levels of excitation due to strong nonlinear contributions to the overall absorption. Therefore, only a very thin layer of the order of 100 nm near the surface can be optically excited. X-rays, on the other hand, have typically a penetration depth of a few microns or more.

To overcome this mismatch between pump- and probe depths we used thin crystalline films grown on silicon substrates. Using surfactant mediated growth techniques [11] highly perfect single-crystalline films can be obtained on large size substrate (4" wafers). Most important, these films are stress free and grow with their natural lattice constant. Therefore, it is easily possible in a diffraction experiment to separate the contributions of the thin film from the bulk substrate if the difference in the lattice constants and the corresponding difference in the Bragg-angles is sufficiently large.

This is the case for the thin Germanium films grown on Silicon, which we used in the diffraction experiment discussed in the next chapter. An example of the diffraction pattern of such a Ge/Si-heterostructure is shown in Fig. 5.

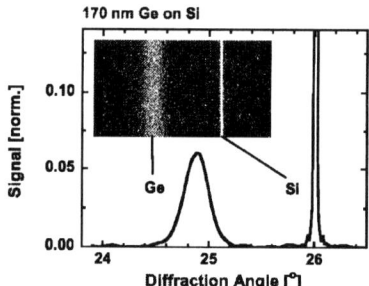

FIGURE 5. Bragg-diffraction profile (*rocking curve*, 111-reflection) from a 170 nm thick, single-crystalline Germanium film grown on Silicon using the laser-driven Ti-K_α X-ray source. Insert: image directly recorded on the X-ray CCD-detector.

The data shown in Fig. 5 represent the angular diffraction profile (*rocking curve*) of the 111-Bragg reflection of a 170 nm thick Germanium film on Silicon obtained from a two-minutes integration on the CCD (the insert shows the CCD-image). The Bragg-peaks of the Germanium film ($\theta_B = 24.88^0$) and the Silicon substrate ($\theta_B = 26^0$) are well separated. Note, that the significantly larger width of the Germanium-peak is related to the small thickness of the overlayer and *not* to structural imperfections.

X-RAY PULSE DURATION

As in an ordinary, all-optical pump/probe experiment the temporal resolution of an optical pump/X-ray probe experiment is determined by the duration of the probe pulse.

While there are reliable and very sophisticated methods to accurately measure the pulse shape of optical pulses (for example FROG), comparable methods for the multi-keV X-ray range are not yet available.

Nevertheless some information on the X-ray pulse duration can be obtained from a time-resolved diffraction experiment. In the most general case the observed transient changes of the diffraction signal represent the convolution of the actual material response with the X-ray probe pulse. If the material response is sufficiently fast the measured transients directly provide the X-ray pulse duration.

A process, which should give such a very fast material response, is femtosecond laser-induced melting of semiconductors. It has been investigated for nearly 20 years with ultrafast optical techniques [12-14] and there is strong evidence that a transition from the ordered solid phase to the disordered liquid state is possible within just a few hundred femtoseconds. Therefore, this process has become some kind of a *test-case* for ultrafast X-ray techniques [15-19], in particular diffraction, and a number of recent studies [18-20] have clearly demonstrated sub-picosecond X-ray response.

Results from our own work [19] are depicted in the left part of Fig. 6. It shows the angular integrated X-ray reflectivity of the 111-reflection of a 170 nm Germanium film as function of pump-probe time delay for two different pump fluences.

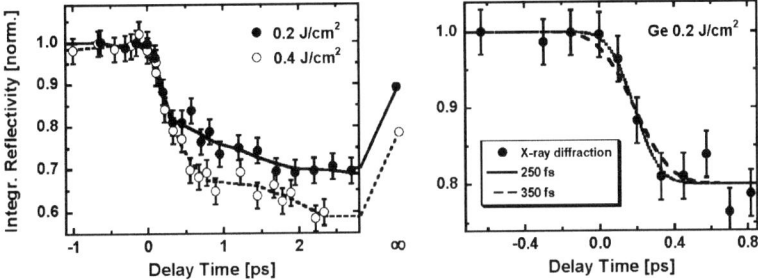

FIGURE 6. Left: angular integrated X-ray reflectivity of a 170 nm Germanium film (111-reflection) as a function of the time delay between the optical pump pulse and the X-ray probe for two different pump fluences. Right: fits of the measured diffraction data (0.2 J/cm^2) assuming a step-like material response and Gaussian-shaped X-ray pulses.

The most prominent feature is the rapid initial drop of the integrated diffraction efficiency within a few hundred femtoseconds, a clear indication for a very fast loss of order over a depth of about 40 nm. Here we do not discuss further the details of ultrafast melting and the interesting material behavior observed on longer time-scales (for this the reader is referred to [19]), but want to focus on the X-ray pulse duration.

To get an estimate on the pulse duration we assume the limiting case of a completely *instantaneous*, step-like material response (which obviously *over-estimates* the X-ray pulse duration!). In the right part of Fig. 6 the initial decrease of the diffraction signal measured for a pump fluence of 0.2 J/cm^2 has been fitted by a convolution of a step-like material response with Gaussian-shaped X-ray pulses of different duration. Satisfactory fits are obtained with X-ray pulse widths between 250 fs and 350 fs. As a result we can give an upper limit for the X-ray pulse duration of about (300 ± 50) fs. To our knowledge, these are the shortest X-ray pulses in the multi-keV range reported so far!

ACKNOWLEDGMENTS

The authors are indebted to I. Uschmann and E. Förster for providing the X-ray-mirror, and to M. Kammler and M. Horn-von-Hoegen for preparation of the heterostructure samples. Financial support by the *Deutsche Forschungsgemeinschaft*, the *European Community*, the *German Academic Exchange Service*, and the *National Science Foundation* is gratefully acknowledged.

REFERENCES

1. Kmetec, J.D., et al., *Phys. Rev. Letters* **68**, 1527 (1992).
2. Kühlke, D., Herpers, U., and von der Linde, D., *Appl. Phys. Letters* **50**, 1785 (1987).
3. Rousse, A., Audebert, P., Geindre, J.P., Fallies, F., Gauthier, J.C., Mysyrowicz, A., Grillon, A.G., and Antonetti, A, *Phys. Rev. E* **50**, 2200 (1994).
4. Schoenlein, R.W., Leemans, W.P., Chin, A.H., Volfbeyn, P., Glover, T.E., Balling, P., Zolotorev, M., Kim, K.-J., Chattopadhyay, S., and Shank, C.V., *Science* **274**, 236 (1996); Schoenlein, R.W., Chattopadhyay, S., Chong, H.H.W., Glover, T.E., Heimann, P.A., Shank, C.V., Zholents, A.A., and Zolotorev, M., *Science* **287**, 2237 (2000).
5. Rose-Petruck, C., Jimenez, R., Guo, T., Cavalleri, A., Siders, C.W., Raksi, F., Squier, J.A., Walker, B.C., Wilson, K.R., and Barty, C.P.J., *Nature* **398**, 310 (1999).
6. Eder, D.C., *Appl. Phys. A.* **70**, 211 (2000)
7. Fill, E., Bayerl, J, and Tommasini, R., *Rev. Sci. Instr.* **73**, 2190 (2002).
8. Reich, C., Gibbon, P., Uschmann, I., and Förster, E., *Phys. Rev. Letters* **84**, 4846 (2000).
9. Bastiani, S., Rousse, A., Geindre, J.P., Audebert, P., Quoix, C., Harminaux, G., Antonetti, A, and Gauthier, J.C., *Phys. Rev. E* **56**, 7179 (1997).
10. Missalla, T., Uschmann, I., Förster, E., Jenke, G., and von der Linde, D., *Rev. Sci. Instr.* **70**, 1288 (2000).
11. Horn-von-Hoegen, M., *Appl. Phys. A* **59**, 503 (1994).
12. Shank, C.V., Yen, R., and Hirlimann, C., *Phys. Rev. Letters* **50**, 454 (1993).
13. Saeta, P., Wang, J.K., Siegal, Y.N., Bloembergen, N., and Mazur, E., *Phys. Rev. Letters* **67**, 1023 (1991).
14. Sokolowski-Tinten, K., Bialkowski, J., and von der Linde, D., *Phys. Rev. B* **51**, 14186 (1995).
15. Chin, A.H., Schoenlein, R.W., Glover, T.E., Balling, P., Leemans, W.P., and Shank, C.V., *Phys. Rev. Letters* **83**, 336 (1999).
16. Siders, C.W., Cavalleri, A., Sokolowski-Tinten, K., Toth, C., Guo, T., Kammler, M., Horn-von-Hoegen, M., Wilson, K.R., von der Linde, D., and Barty, C.P.J., *Science* **286**, 1340 (1999).
17. Lindenberg, A.M., Kang, I., Johnson, S.L., Missalla, T., Heimann, P.A., Chnag, Z., Larsson, J., Bucksbaum, P.H., Kapteyn, H.C., Padmore, H.A., Lee, R.W., Wark, J.S., and Falcone, R.W., *Phys. Rev. Letters* **84**, 111 (2000).
18. Rousse, A., Rischel, C., Fourmaux, S., Uschmann, I., Sebban, S., Grillon, G., Balcou, P., Förster, E., Geindre, J.P., Audebert, P., Gauthier, J.C., and Hulin, D., *Nature* **410**, 65 (2001).
19. Sokolowski-Tinten, K., Blome, C., Dietrich, C., Tarasevitch, A., von der Linde, D., Horn-von-Hoegen, M., Cavalleri, A., Squier, J.A., and Kammler, M., *Phys. Rev. Letters* **87**, 225701 (2001).
20. Feurer, T., Morak, A., Uschmann, I., Ziener, C., Schwoerer, H., Reich, C., Gibbon, P., Förster, E., Sauerbrey, R., Ortner, K., and Becker, C.R., *Phys. Rev. E* **65**, 16412 (2002).

Interference Stabilization in Atoms and Molecules

M.V. Fedorov

General Physics Institute, Russian Academy of Sciences
38 Vavilov St., Moscow, 119111, Russia

Abstract. The existing theoretical and experimental works on interference stabilization of Rydberg atoms are overviewed. Physical origin of the phenomenon, its main features and the conditions of existence, as well as the main theoretical approaches used for its description are discussed. These ideas are used to describe interference stabilization of molecules with respect to photodissociation. Specific calculations are carried out for the hydrogen molecular ion H_2^+. The main features of strong-field photodissociation are described. The conditions of stabilization and destabilization of a molecule are found and related to features of the initial-state molecular vibrational wave function.

INTRODUCTION

In the most general form, stabilization of a decaying quantum system means that its decay can be slowed down under the influence of some external fields or forces. The decay of atoms driven by a strong laser field is associated usually with the irreversible process of one-photon or multiphoton ionization. In molecules the decay process to be slowed down by a strong field can be related to either ionization or dissociation. In this report two such phenomena will be discussed: strong-field photoionization of Rydberg atoms and photodissociation of molecules. In both cases the mechanism of stabilization is assumed to be related to the strong-field Raman-type transitions between close bound levels: Rydberg levels in atoms and vibrational levels of the ground electronic state in molecules. This mechanism is known as the interference stabilization (IS) [1-3]. Specific manifestations of IS in atoms and molecules can differ rather significantly because of big differences in features of atomic and molecular spectra and transitions. Discussion of differences and similarity between the atomic and molecular IS is very instructive and useful for understanding the physics of the phenomena. Both rather well established and new results on IS in atoms and molecules are described.

INREFERENCE STABILIZATION OF RYDBERG ATOMS

The effect of IS in Rydberg atoms was described and discussed for the first time in 1988 [1]. Many features and manifestations of this phenomenon are summarized in Ref. [2]. As mentioned above, IS in Rydberg atoms arises owing to strong-field

CP634, *Science of Superstrong Field Interactions*, edited by K. Nakajima and M. Deguchi
© 2002 American Institute of Physics 0-7354-0089-X/02/$19.00

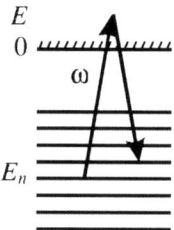

E

0

ω

E_n

FIGURE 1. A scheme of Raman-type transitions between Rydberg levels.

Raman-type transitions between close Rydberg levels E_n, with the main contribution to the corresponding two-photon matrix elements given by the atomic continuum (Λ-type transitions, Fig. 1). Subsequent transitions to the continuum from coherently re-populated levels E_n interfere with each other, and this interference suppresses the process of photoionization and gives rise to IS. The existence criterion of IS is equivalent to the criterion of efficient Λ-type transitions, and it has the form $\Gamma_i \geq E_{n+1}$ – E_n, where $\Gamma_i = (\pi/2)\varepsilon_0^2 |d_{nE}|^2$ is the Fermi-Golden-Rule (FGR) ionization width of the Rydberg level E_n, d is the dipole moment, and ε_0 is the light field-strength amplitude. Near by the ionization threshold, $n \gg 1$ and $E \ll 1$ (in atomic units) the dipole matrix elements d_{nE} can be approximated by their quasiclassical expression [3, 4] $d_{nE} \sim n^{-3/2}\omega^{-5/3}$, where ω is the light frequency, $\omega \ll 1$, and $E \approx E_n + \omega$. This reduces the IS criterion to the form $\varepsilon_0 \geq \omega^{5/3}$.

In the theory of IS, the wave function $\Psi(t)$ of an atom in a light field $\varepsilon(t)=\varepsilon_0 \cos(\omega t)$ is expanded in a series of the field-free atomic wave functions

$$\Psi = \sum_n C_n(t)\psi_n + \int dE\ C_E(t)\psi_E \qquad (1)$$

with unknown time-dependent probability amplitudes $C_n(t)$ and $C_E(t)$. Equations for these probability amplitudes follow from the Schrödinger equation, and these equations can be significantly simplified owing to a series of beautiful features of Rydberg atoms. Not dwelling upon the details, let us mention only the main steps of such simplifications. The approximations which can be used are RWA and adiabatic elimination of the continuum. The latter is related to the flat-continuum approximation, which means that the dipole matrix elements d_{nE} sufficiently slowly depend on the energy in the continuum E. With the procedure of adiabatic elimination performed, the set of equation for $C_n(t)$ and $C_E(t)$ is reduced to the set of differential equations for the discrete-level probability amplitudes $C_n(t)$ only. Connection with continuum in these equations is characterized by the only decay constant $\Gamma \sim \varepsilon_0^2 / n^3 \omega^{10/3}$ identical for all the initial and final levels E_n and $E_{n'}$.

The simplest model in which IS exists is the model of two discrete levels (denoted as, e.g., E_1 and E_2) plus a continuum. In such a model, equations for $C_1(t)$ and $C_2(t)$ take the form

$$i\dot{C}_1(t) - E_1 C_1(t) = -\frac{i}{2}\Gamma\left(C_1(t) + C_2(t)\right),$$

$$i\dot{C}_2(t) - E_2 C_2(t) = -\frac{i}{2}\Gamma\left(C_1(t) + C_2(t)\right). \tag{2}$$

Both in this simple model and in a general case of a multilevel Rydberg atom one can either solve the initial value problem or look for the stationary solution of equations like Eqs. (2). In the last case, one gets the quasienergy solutions $C_n(t) = \exp(-i\gamma t) a_n$ and complex quasienrgies of the system under consideration. For the simplest two-level system the arising complex quasienergies are easily found to be given by

$$\gamma_{\pm} \equiv \frac{1}{2}\left\{E_1 + E_2 - i\Gamma \pm \sqrt{(E_1 - E_2)^2 - \Gamma^2}\right\}. \tag{3}$$

These complex quasienergies are plotted in Fig. 2 in the form of quasienergy zones, the boundaries of which are determined as $Re(\gamma_{\pm}) + |Im(\gamma_{\pm})|$ and $Re(\gamma_{\pm}) - |Im(\gamma_{\pm})|$. In dependence on $\Gamma \propto \varepsilon_0^2$ the quasienergy zones at first broaden, than they have a branching point at $\Gamma = E_2 - E_1$, and at $\Gamma > E_2 - E_1$ one of the two quasienergy levels continues to broaden whereas the other one narrows. Formation of a narrow long-living qausienergy level in the strong-field limit indicates a possibility of stabilization of the system. At $\Gamma \geq E_2 - E_1$ both narrow and wide quasienergy levels are seen to be centered at $Re(\gamma_{\pm}) = (E_2 - E_1)/2$, i.e., exactly between the field-free levels E_1 and E_2.

In a multilevel system of Rydberg levels E_n decaying to the continuum, all the quasienergy levels behave similar to the narrowing quasienergy level of the two-level system: they narrow with a growing fild-strength amplitude ε_0, and they appear to be localized between the field-fee Rydberg levels, at $(E_{n+1} - E_n)/2$ [1, 2].

To see explicitly the effect of stabilization, one has to solve the initial-value problem to get, for example, the total probability of ionization per pulse in its dependence on the peak light intensity. Such calculations were done [6] for the pulse envelope of the form $\varepsilon_0(t) = \varepsilon_0 \sin^2(\pi t/\tau)$. Both multilevel structure of Rydberg levels and degeneracy in the angular momentum l were taken into account. One of the results

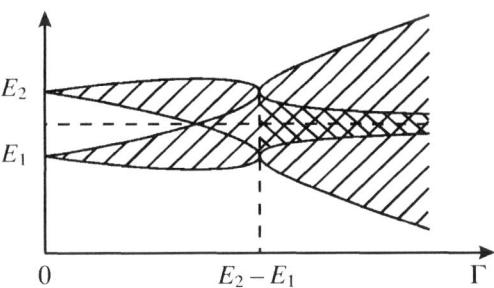

FIGURE 2. Quasienergy zones of the model decaying two-level system characterized by Eqs. (2), (3).

21

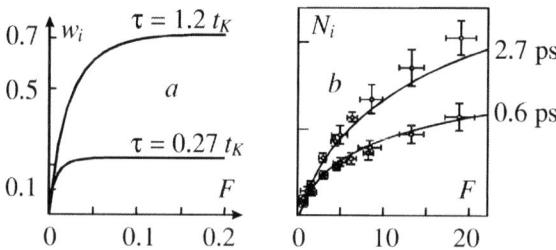

FIGURE 3. (*a*) The calculated [6] total probability of ionization per pulse w_i vs. fluence $F=I\times\tau$ (F in arbitrary units), and (*b*) experimentally measured [7] ion yield $N_i(F)$, F in J/cm^2, in both cases $n = 27$; $t_K=2\pi\, n^3$ is the Kepler period.

is shown in Fig. 3*a*. The probability of ionization is plotted in dependence on fluence $F=I\times\tau$ for two different pulse durations. The curves of Fig. 3*b* show the experimentally measured (under similar conditions) ion yield vs. fluence F [7]. In both cases the curves corresponding to shorter pulse duration and, hemnce, higher intensity are located lower than the curves corresponding to longer pulses and lower intensity. This is a direct confirmation of stabilization, i.e., decrease of the yield with a growing light intensity. Theoretical and experimental results are seen to close to each other.

To conclude this Section, it's reasonable to mention an approach to the theory of IS alternative to that discussed above. This approach involves an attempt to use quasiclassical (WKB) description for solving directly the nonstationary Schrödinger equation [8]. The result is given by the following very simple analytical expression for the probability of ionization from a Rydberg level

$$w_i = \int_0^1 dx \int_{-\infty}^{\infty}\frac{dt}{2t_K}\left\{1-J_0^2\left[x\frac{2^{2/3}\,3^{1/6}\,\Gamma\!\left(\frac{2}{3}\right)\varepsilon_0(t)}{\omega^{5/3}}\right]\right\}, \qquad (4)$$

where Γ and J_0 are the gamma- and zero-order Bessel functions. The dependence $w_i(I)$ calculated with the help of this formula for a trapezoidal pulse envelope is plotted in Fugure 4 together with the results of exact numerical solution of the Schrödinger equation [9]. The coincidence looks rather impressive.

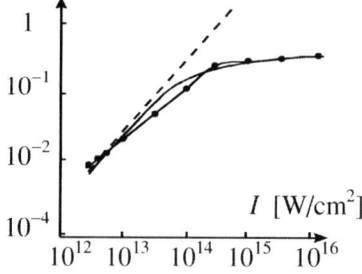

FIGURE 4. The total probability of ionization per pulse vs. the peak light intensity: numerical (dots) and analytical (4) (solid line) solutions; $n = 5$ and trapezoidal envelope $\varepsilon_0(t)$ with the switch-on/off and plateau periods equal to, respectively, 2 and 10 optical cycles [9]; dashed line – perturbation theory.

RAMAN-TYPE TRANSITIONS IN MOLECULES

A scheme of Raman-type transitions to be taken into account is shown in Fig. 5. The frequency w is assumed to be high enough to provide one-photon transition between the ground and first excited, unstable, electronic state. The levels to be re-populated via Λ-type transitions are the vivbrational levels E_v of the ground electronic state. In analogy with Eq. (1), let us expand the wave function of a molecule in a series of field-free wave functions

$$\Psi = \psi_0(\mathbf{r}, R) \sum_v C_v(t) \exp(-iE_v t) \chi_v(R) + \psi_1(\mathbf{r}, R) \int dE\, C_E(t) \exp(-iEt) \chi_E(R), \quad (5)$$

where R and $\mathbf{r}$ are the internuclear distance and the electron position vector, $\chi_v(R)$ and $\chi_E(R)$ are nuclear wave functions in the electronic states U_0 and U_1, v is the vibrational quantum number and E is the energy of a molecule in the unstable state U_1, $\psi_0(\mathbf{r}, R)$ and $\psi_1(\mathbf{r}, R)$ are the electronic wave functions of a molecule in the states U_0 and U_1. The main difference with the case of Rydberg atoms concerns the dipole matrix elements d_{vE}. They were calculated for the molecular ion H_2^+ [3], and the dependence of d_{vE} on E was shown to have an oscillating character. This means that the continuum of nuclear motion in the excited electronic state is not flat. As a consequence, the procedure of adiabatic elimination of the continuum appears to be inapplicable. However, a kind of a generalized procedure of semi-adiabatic elimination (explained below) was shown to be valid. In the framework of this procedure of semi-adiabatic elimination equations for the ground-state vibrational probability amplitudes $C_v(t)$ have the form

$$\dot{C}_v(t) = -\frac{\varepsilon_0^2(t)}{4} \sum_{v'} Q_{v,v'}(\omega) C_{v'}(t) \exp\{i(E_v - E_{v'})t\} \quad (6)$$

where $Q_{v,v'}(\omega)$ are the Raman-type two-photon matrix elements

$$Q_{v,v'}(\omega) = \int_{-\infty}^{0} dt'\, R_{v,v'}(t') \exp\{-i(\omega + E_{v'})t'\} \quad (7)$$

and the function $R_{v,v'}(t)$ are given by

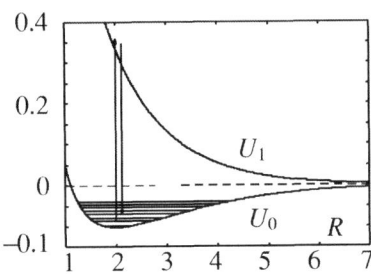

FIGURE 5. Potential curves $U_{0,1}(R)$ (in atomic units) and Raman type transitions.

23

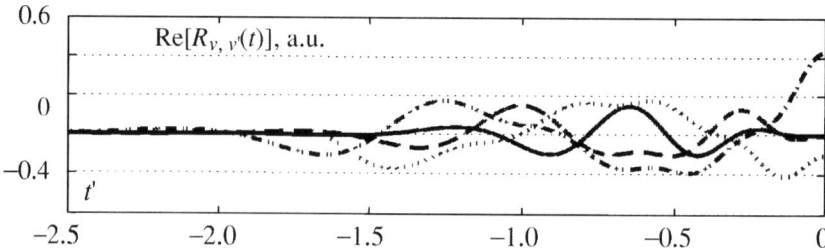

FIGURE 6. Real parts of the functions $R_{v,v'}(t')$ (8); solid line – $v=2$ and $v'=0$, dashed line - $v=2$ and $v'=1$, dash-dotted line - $v=2$ and $v'=2$, dotted line - $v=2$ and $v'=3$.

$$R_{v,v'}(t') = \int_0^\infty dE\, d_{vE} d_{Ev'}\, \exp(iEt').$$ (8)

In a pure form adiabatic elimination of the continuum is valid when $R_{v,v'}(t')$ can be approximated by the delta-functions. However, for molecules this approximation does not hold good. Calculated explicitly for H_2^+, real parts of several functions $R_{v,v'}(t')$ are shown in Fig. 6. These functions are seen to be localized in a time interval $\delta t' \sim 2$ fs around $t'=0$. For pulse duration of 70 fs $\delta t' \ll \tau$ and $1/\omega_e \sim 10$ fs, where $\omega_e \sim 0.01$ a.u. is the vibrational frequency of H_2^+. For this reason, the slow functions $\varepsilon_0(t'+t)$ and $C_{v'}(t'+t)$ can be taken out of the integrals over t' at $t'=t$ to give Eqs. (7). On the other hand, the function $\exp(-i\omega t')$ is not slow compared to $R_{v,v'}(t')$, and must be retained under the symbol of integral in Eq. (7). This is why such an approximation is defined as the semi-adiabatic elimination of the continuum. The factor $\exp(-i\omega t')$ under the symbol of integral in Eq. (7) determines the difference between the pure adiabatic and semi-adiabatic elimination of the continuum, as well as the dependence of the Raman-type matrix elements $Q_{v,v'}$ on the light frequency ω.

With the help of transformation $C_v(t)=\exp(-iE_v t)b_v(t)$ Eqs. (6) can be reduced to the form of equations with constant coefficients, which have stationary, or quasienergy, solutions. The corresponding eigenvalues are the complex quasienergies γ, to be found from the equation

$$Det\left\| (\gamma - E_v)\delta_{v,v'} + \frac{\varepsilon_0^2}{4} Q_{v,v'}(\omega) \right\| = 0.$$ (9)

Two quasienergy zones (starting form $v = 2$ and 3) are shown in Fig. 7 in the dependence on the light intensity I for $\omega = 0.338$. The boundaries of zones are determined as $Re(\gamma) + |Im(\gamma)|$ and $Re(\gamma) - |Im(\gamma)|$. Spacing between the boundaries equals the width of quasienergy zones (levels) $|Im(\gamma)|$. It's seen that the two zones show in Fig. 7 behave very similar to the two-level zones in Rydberg atoms (Fig. 2). Again, these zones at first broaden, then overlap, and then a narrowing zone is formed on the background of the broadening one. Narrowing of a quasienergy zone is a manifestation of stabilization.

24

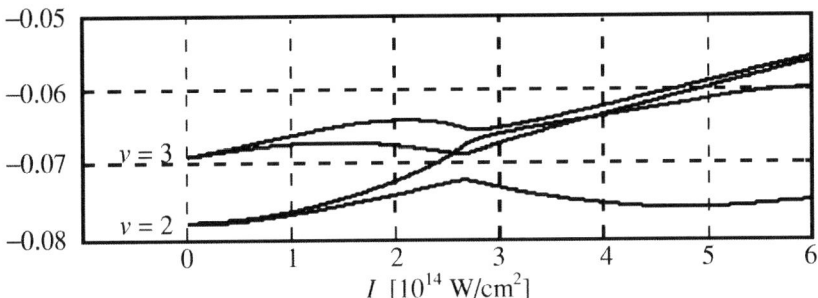

FIGURE 7. Quasienergy zones; $\omega = 0.338$ [all in a.u.]

Alternatively, Eqs. (6) can be solved directly to determine the time-dependent probability amplitudes $C_v(t)$, populations of vibrational levels $|C_v(t)|^2$, probability of dissociation $w_D(t) = 1 - \sum_v |C_v(t)|^2$, and total probability of dissociation per pulse $w_{D\,tot} = w_D(\tau)$. To estimate the role of Raman-type transitions, the time-dependent and total probabilities of dissociation will be compared with the FGR with saturation probabilities

$$w_D^{FGR}(t) = 1 - \exp\left(-\frac{\pi}{2}|d_{vE}|^2\Big|_{E=E_{v_0}+\omega}\int_{-\infty}^{t}dt'\,\varepsilon_0^2(t')\right)$$ (10)

and $w_{D\,tot}^{FGR} = w_D^{FGR}(\tau)$. The calculated dependencies $w_{D\,tot}(I)$ and $w_{D\,tot}^{FGR}(I)$ are shown in Fig. 8 (correspondingly, by the solid and dotted lines). The exact probability of

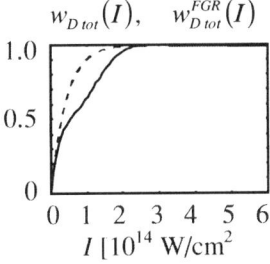

FIGURE 8. Total probability if dissociation per pulse vs. the peak light intensity: exact calculation (solid line) and FGR with saturation formula (10) (dashed line); $\omega = 0.338$ a.u., $v_0 = 2$.

dissociation is seen to have a knee-structure, which indicates in this case a partial stabilization of a molecule with respect to photodissociation.

An informative cvharacteristics of the degree of stabilization is the difference between the FGR and exact probabilities, $X(\omega, I) \equiv w_{D\,tot}^{FGR} - w_{D\,tot}$. Stabilization and destabilization regions correspond to $X(\omega, I) > 0$ and $X(\omega, I) < 0$, respectively. The calculated total dissociation probabilities and the function $X(\omega, I)$ are plotted in

Fig. 9. in their dependence on frequency ω at a given peak intensity I.

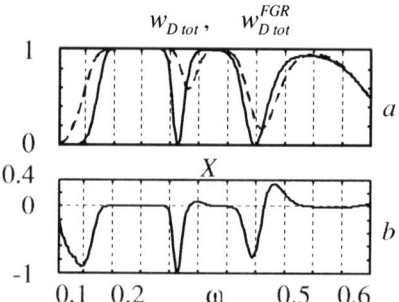

FIGURE 9. Exact (dashed line) and FGR with saturation (solid line) total probabilities of dissociation per pulse (a) and their difference X (b) vs. frequency ω (in atomic units); $I = 2 \times 10^{14}$ W/cm^2, $v_0 = 2$.

The deep hollows of $w_{D\,tot}^{FGR}(\omega)$, in accordance with the Franck-Condon rule, correspond to almost zero matrix elements for constant-R transitions from such intrenuclear distances where the initial vibrational wave function turns zero. The strong field smoothes these deep hollows and shifts them mainly to higher-frequency regions. For this reason the regions where $w_{D\,tot}(\omega) > w_{D\,tot}^{FGR}(\omega)$ and $w_{D\,tot}(\omega) < w_{D\,tot}^{FGR}(\omega)$ arise around every hollow of the curve $w_{D\,tot}^{FGR}(\omega)$, and these are, correspondingly, destabilization and stabilization regions. Owing to the Franck-Condon principle this frequency-picture can be converted to the internuclear-distance picture, as it's shown in Fig. 10. Stabilization and destabilization regions correspond to the left- and

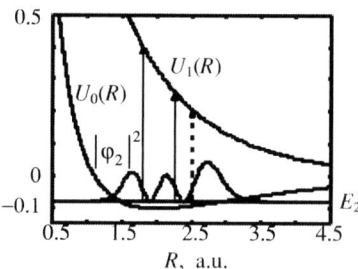

FIGURE 10. Stabilization/destabilization frequencies and a structure of the initial-state vibrational wave function.

right-hand sides of the hollows of the initial squared vibrational wave function $|\varphi(R)|^2$. The corresponding frequencies can be found as distances along the vertical lines from the initial vibrational level E_2 to the potential curve $U_1(R)$. They are shown by solid lines with arrows for stabilization and dashed line for destabilization regions.

REFERENCES

1. Fedorov M. V., and Movsesian, A. M., *J Phys. B* **21**, L155-L158 (1988).
2. Fedorov, M.V., *Atomic and free electrons in a strong light field*, Singapore-London-New York: World Scientific, 1997.
3. Sukharev M.E. and Fedorov M.V., *Phys. Rev. A* **65**, 033419(12) (2002).
4. I.Ja. Bersos, *JOSA B* **7**(5), 617-621 (1990).
5. Delone, N.B., Goreslavsky, S.P., and Krainov V.P., *J. Phys. B* **27**, 4403-4419 (1994).
6. Fedorov, M.V., Tehranchi, M.-M., and Fedorov, S.M., *J. Phys. B.*, **29**, 2907-2924 (1996).
7. Hoogenraad, J., Vrijen, R.B., and Noordam, L.D., *Phys. Rev. A* **50**(6), 4133-4138 (1994).
8. Fedorov, M.V., and Tikhonova, O.V., *Phys. Rev. A*, **58**, 1322-1334 (1998).
9. Popov, A.M., Volkova, E.A., and Tikhonova, O.V., *Sov. Phys. JETP* **86**, 328 (1998).

Generation and Transport of Fast Electrons in Laser Irradiated Targets at Relativistic Intensities

F. Amiranoff[1], S.D. Baton[1], L. Gremillet[1], O. Guilbaud[1], M. Koenig[1],
E. Martinolli[1], J.J. Santos[1], M. Rabec Le Gloahec[2], C. Rousseaux[2],
T. Hall[3], D. Batani[4], A. Bernardinello[4], G. Greison[4], E. Perelli[4],
F. Scianitti[4], M.H. Key[5], J.A. Koch[5], A.J. Mackinnon[5], R.R. Freeman[6],
R.A. Snavely[6], C. Andersen[6], T.E. Cowan[7], R.B. Stephens[7],Y. Aglistkiy[8]

[1] *Laboratoire pour l'Utilisation des Lasers Intenses*
UMR 7605 CNRS-CEA-Ecole Polytechnique-Paris VI, Palaiseau, France
[2] *Commissariat à l'Energie Atomique, Bruyères-le-Châtel, France*
[3] *Department of Physics, University of Essex, Colchester, UK*
[4] *Dipartimento di Fisica "G.Occhialini" and INFM, Università di Milano-Bicocca, Italy*
[5] *Lawrence Livermore National Laboratory, Livermore, CA, USA*
[6] *Department of Applied Sciences, University of California Davis, CA, USA*
[7] *General Atomics, San Diego, CA, USA*
[8] *Science Applications International Corporation, McLean, VA, USA*

Abstract. The transport of relativistic electrons in solid targets irradiated by a short laser pulse at relativistic intensities has been studied both experimentally and numerically. A Monte-Carlo collision code takes into account individual collisions with the ions and electrons in the target. A 3D-hybrid code takes into account these collisions as well as the generation of electric and magnetic fields and the self-consistent motion of the electrons in these fields. It predicts a magnetic guiding of a fraction of the fast electron current over long distances and a localized heating of the material along the propagation axis.
In experiments performed at LULI on the 100 TW laser facility, several diagnostics have been implemented to diagnose the geometry of the fast electron transport and the target heating. The typical conditions were: $E_l \leq 20$ J, $\lambda = 1$ μm, $\tau \approx 300$ fs, $I \approx 10^{18}$-5.10^{19}W/cm^2. The results indicate a modest heating of the target (typically 20-40 eV over 20 μm to 50 μm), consistent with an acceleration of the electrons inside a wide aperture cone along the laser axis.

INTRODUCTION

The fast-igniter scheme for the inertial laser fusion relies on the heating of a compressed DT core by a beam of high-energy electrons accelerated by an intense laser beam in the plasma surrounding the target [1-3]. The main physical issues are: (i) the transfer of the laser energy to fast electrons with a well adapted energy spectrum and angular distribution; (ii) the transport of these fast electrons from the acceleration region to the compressed core itself; (iii) the heating of the fuel by the fast electrons,

CP634, *Science of Superstrong Field Interactions*, edited by K. Nakajima and M. Deguchi
© 2002 American Institute of Physics 0-7354-0089-X/02/$19.00

and; (iv) the resulting ignition of the fuel and neutron production. All of these aspects have already been addressed in theoretical, numerical and experimental studies [1-12].

The results obtained at the LULI laboratory on electron transport and solid target heating with a 100 TW laser beam are detailed in this paper.

EXPERIMENTAL SET-UP AND RESULTS

The experiments have been performed on the LULI 100 TW laser facility. The 350 fs, 1.057 µm laser pulse with an energy up to 10 J was focused by a $f/3$ off-axis parabola at normal incidence onto flat solid targets, either Al foils or sandwich targets. The laser focal spot was 15 µm to 20 µm in diameter corresponding to a maximum intensity of $5 \ 10^{19}$ W/cm^2. Several optical as well as x-ray diagnostics have been implemented to diagnose the geometry and the efficiency of the target heating.

Visible Self Emission

The target rear side was imaged with a $f/2$ optical system on both a CCD camera and a streak camera. The spectral sensitivity domain extends from 350 nm to 880 nm with a maximum at 600 nm, except for a 60 nm region around 530 nm, the second harmonic of the laser light. A typical time-integrated image consists of two features. A large and smooth halo, due to thermal emission of the cooling and expanding plasma lasts for nanoseconds. On the contrary, a smaller spot in the center is emitted in less than 10 ps. It is due either to the electrons crossing the back surface and emitting optical transition radiation (OTR) [13], or to the same electrons emitting bremmsstrahlung radiation while reflecting in the Debye sheath [14]. The diameter of this prompt feature (shown in Fig.1) increases with Al target thickness (Fig.2), showing an electron spread in the target with a typical half angle of $\theta/2 \sim 17 \,^\circ$. At the same time the corresponding integrated signal drops rapidly at small thicknesses before decreasing more smoothly.

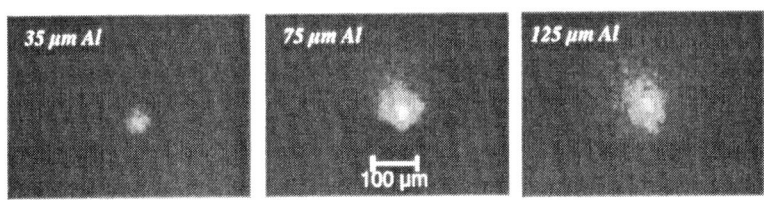

FIGURE 1. Time-resolved images for three Al thicknesses and $I \approx 10^{19} W/cm^2$.

To estimate the light intensity we used the code PâRIS [12]. This code models both collision effects (multiple scattering and slowing down) and self-generated electric and magnetic fields. Simulation conditions for the initial electronic population were chosen in order to be as close as possible to experimental conditions. The initial hot electron temperature is given by the scaling law [Beg] $T_h(r,t) = 100 \, (I_{17}(r,t))^{1/3}$ [keV] where $I_{17}(r,t)$ is the laser intensity in units of 10^{17} W/cm^2. Injecting an electron population with a 1.6 J total energy, a pulse duration of 300 fs, over a 20 μm focal spot into a 75 μm thickness Al target, PâRIS gives 5.2 10^{12} outgoing electrons with a total energy of 0.46 J. From this, the estimates of the OTR signal and the synchrotron radiation are both close to 4.10^{-13} J, in good agreement with the measured value.

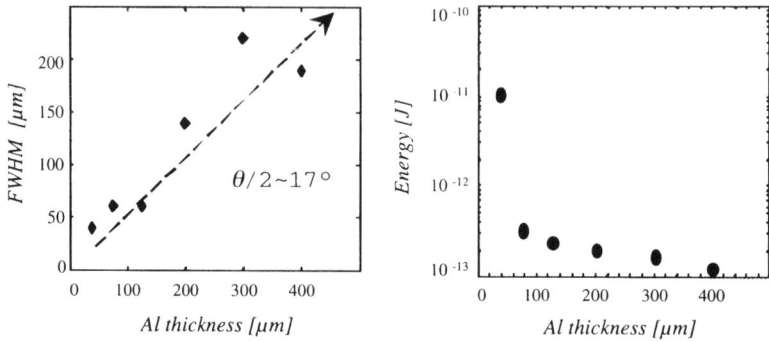

FIGURE 2. FWHM of the time-resolved images (left), integral of the signal (right)

Reflectometry in the Visible Domain

A probe beam at $\lambda = 0.53$ μm was incident at 45° on the back side of the target. A second optical system measured the reflectivity of the target as function of space and time. Two modes of operation have been used. In the first mode, the probe beam was compressed to 300 fs and the image was measured on a CCD camera for different time delays between the interaction beam and the probe beam. In the second mode, the probe beam was frequency chirped to about 100 ps, and the image was sent through a spectrometer. The time-frequency relation then gives a time resolution of 2-3 ps over the whole duration of the probe beam [15].

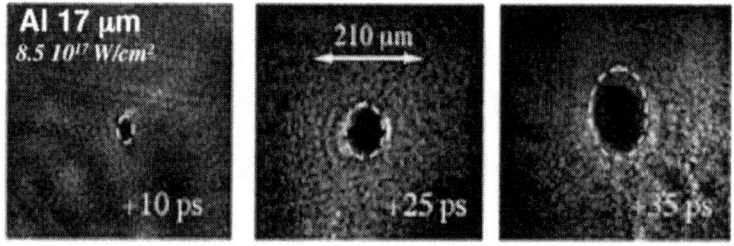

FIGURE 3. Rear side Reflectometry images; 17 μm Al target, 8.5 10^{17} W/cm^2.

The images presented in Fig.3 show a low reflectivity spot, which expands at an average radial speed of 2-3 10^6 m/s. The reflectivity in the center drops to values below 0.3. The chirped pulse data show that the change in reflectivity starts a few ps after the main pulse, expanding at a very high initial radial velocity of 5-7 10^6 m/s before dropping to 10^6 m/s after 30 ps. For thin targets (35 μm), the initial diameter of the perturbation is only slightly larger that the focal spot while it increases for larger thicknesses.

The values of the reflectivity suggest that the back side temperature is between a few eV and 100 eV. Assuming that the radial expansion corresponds to the thermal electron velocity, we infer a temperature starting at $T \approx 120$ eV then decreasing to about 20 eV for the average expansion velocity.

X-ray K_α Spectroscopy

A conically bent KDP crystal x-ray spectrometer was used to measure the spectrum between 7 Å and 8.5 Å. This domain includes the cold Al K_α line as well as the shifted K_α line of heated and ionized Al. A typical spectrum is shown in Fig.4 for a 28 μm Al target and an incident laser energy of 10 joules. It shows a broad cold K_α emission spot (8.339 Å) and a shifted K_α line at 8.275 Å. This K_α line is relatively weak and is only visible up to 50 μm targets. The shift corresponds to an ionization stage (Z^*) of 5. From the model of Lee and More [16], the estimated temperature is of the order of 40 eV. This result is preliminary and a more accurate model will be developed to calculate the ion populations.

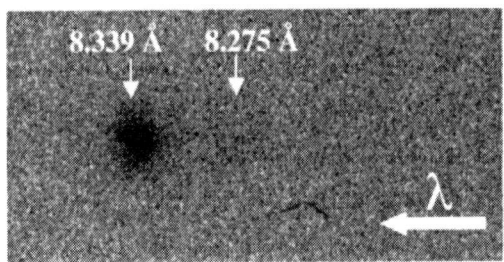

FIGURE 4. Spectrum for a 28 μm Al target: cold K_α line (8.339 Å) and ionized K_α line (8.275 Å).

Rear Side XUV Imaging

The back side of the target was imaged in the XUV domain at 180 Å with a 5 μm spatial resolution using a multidieletric spherical mirror and an additional flat mirror with both a 50% reflectivity. Images obtained for 2.2 μm and 27 μm Al targets are shown in Fig.5. For the thinnest targets, the diameter at half-max intensity was 67 μm, much larger than the laser focal spot. For larger thicknesses, the increase in diameter suggests propagation in a $\approx 50°$-$60°$ cone angle.

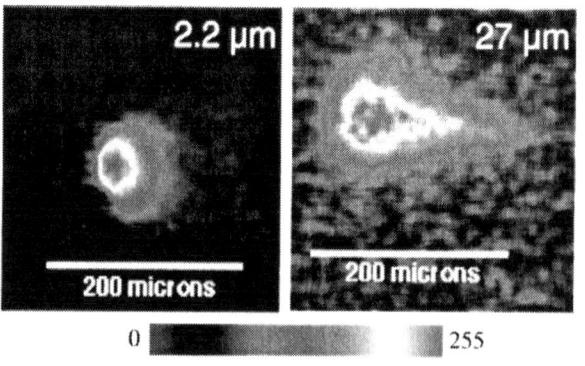

FIGURE 5. XUV pictures obtained for two Al target thicknesses

The XUV signal is mainly due to thermal radiation of the cooling and expanding plasma at the back of the target. From the absolute intensity, and using 1D-LASNEX simulations for the evolution of the heated plasma, one can infer the initial temperature of the emitting region. At maximum laser energy, and for thicknesses up to 30 μm, T ≈ 20 eV. For larger thicknesses, the temperature decreases and the diameter increases so that the signal is too low to be detected.

X-ray K_α Imaging of Fluor Layers

In order to diagnose the propagation of the fast electrons in the bulk of the target, the Ti K_α line at 4.5 keV of embedded 20 μm thick Ti layers in Al/Ti/Al sandwich targets was imaged using a spherical Bragg crystal with a resolution of 10 μm. Two examples are shown in Fig.6. In agreement with the XUV images, the data show a minimum radius of 34 μm and a cone angle of 54°.

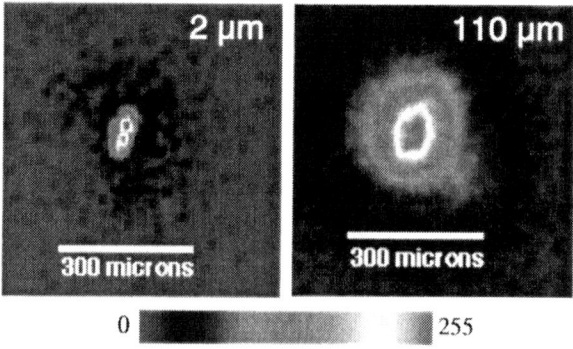

FIGURE 6. Ti K_α pictures for a 2 μm Ti target and a sandwich target with 110 μm of Al in the front.

PÂRIS SIMULATIONS

The experimental data indicate a temperature of the order of 20 eV up to Al thicknesses of 30 µm. Results obtained from PâRIS simulations are shown in Fig.7. When the electrons are injected at normal incidence, most of them are magnetically guided along the axis. The final temperature is of the order of a few hundred eV (and even a few keV near the front side of the target) over a radius of a few microns only, in disagreement with the experimental data. On the contrary, if the electrons are injected inside a wide aperture cone, only part of them are magnetically guided and the diameter of the heated region is much wider. The final temperature is of the order of a few 10 eV. These two features are in much better agreement with the measurements.

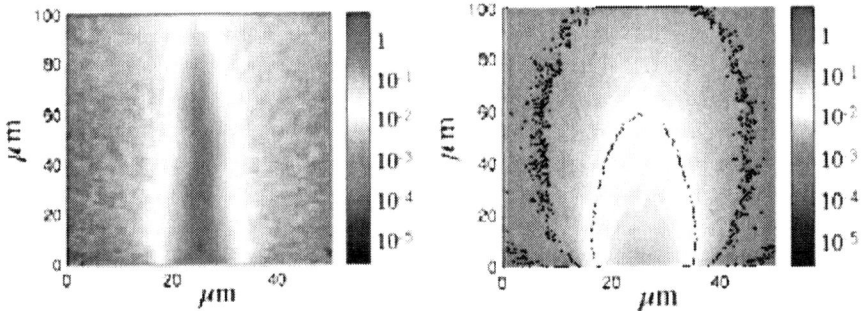

FIGURE 7. Temperature profiles in keV obtained from PâRIS simulations. Electrons are injected from the bottom. Left: 1.6 J of electrons injected at normal incidence inside a 6 µm FWHM focal spot in 300 fs with $T_h = f(I)$ and $I_{max} = 3 \ 10^{19}$ W/cm^2. Right: same conditions except 1.2 J injected in a ±40° cone angle and $I_{max} = 2 \ 10^{19}$ W/cm^2.

CONCLUSION

Experiments on the generation and transport of fast electrons and on the resulting heating of the target have been conducted on the 100 TW LULI laser facility at a maximum intensity of $\approx 3 \ 10^{19}$ W/cm^2 and a maximum laser energy of 10 J. The results show a modest heating of the target, of the order of 20-40 eV up to 50 µm of Al. This is consistent with about 10% to 30% of the laser energy being transferred to fast electrons inside a large angle cone, typically ±20° to ±40°.

ACKNOWLEDGMENTS

The authors would like to thank the technical staff of LULI for their help in running these experiments. Part of this work has been supported by the European contracts: LULI ACCESS HPRI-1999-CT 00052 and TMR ERBFMGE-CT95-0044 and by Grant E1127 from Région Ile-de-France. One of us (J.J.S.) was financed by MCT (Portugal) under the contract PRAXIS XXI BD/18108/98. Part of this work has been performed under the auspices of the U.S. Department of Energy by the LLNL under contract No. W-7405-ENG-48.

REFERENCES

1. Tabak, M., *et al.*, *Phys. Plasmas* **1**, 1626 (1994).
2. Atzeni, S., *Phys. Plasmas* **6**, 3316 (1999).
3. Meyer-ter-Vehn, J., *Plasma Physics and Controlled Fusion* **43**, A113 (2001).
4. Beg, F.N., *et al.*, *Phys. Plasmas* **4**, 447 (1997).
5. Wharton, K.B, *et al.*, *Phys. Rev. Lett.* **81**, 822 (1998).
6. Tatarakis, M., *et al.*, *Phys. Rev. Lett.* **81**, 999 (1998).
7. Gremillet, L., *et al.*, *Phys. Rev. Lett.* **83**, 5015 (1999).
8. Borghesi, M., *et al.*, *Phys. Rev. Lett.* **83**, 4309 (1999).
9. Zepf, M., *et al.*, *Phys. Plasmas* **8**, 2323 (2001).
10. Koch, J.A., *et al.*, *Phys. Rev. E* **65**, 016410 (2001).
11. Kodama, R., *et al.*, *Nature* **412**, 798 (2001).
12. Gremillet, L., *et al.*, *Phys. Plasmas* **9**, 941 (2002).
13. Frank, I., and Ginzburg, V., *J. Phys. USSR* **9**, 353 (1945).
14. Jackson, J.D., *Classical Electrodynamics*, John Wiley & Sons, 1975.
15. Benuzzi-Mounaix, A., *et al.*, *Phys. Review* **60**, R2488 (1999).
16. Lee, Y.T., *et al.*, *Phys. Fluids* **27**, 273 (1984).

Short Pulse Laser Driven Ion Beams – Experiments and Applications

Markus Roth[*], Matthew Allen[+], Patrick Audebert[#], Abel Blazevic[*], Erik Brambrink[*], Thomas E. Cowan[+], Julien Fuchs[#], Jean-Claude Gauthier[#], Matthias Geißel[*], Manuel Hegelich[~], Stefan Karsch[~], Jürgen Meyer-ter-Vehn[~], Hartmut Ruhl[+], Theodor Schlegel[*]

[*]Gesellschaft für Schwerionenforschung mbH, Planckstr.1, 64291 Darmstadt, Germany
[+]General Atomics, P.O. Box 85608, San Diego, California 92186-5608
[#]Laboratoire pour l'Utilisation des Lasers Intenses, 91128 Palaiseau, France
[~]Max-Planck-Institut für Quantenoptik, Garching, Germany

Abstract. We present the results of a study on the acceleration of intense ion beams from solid targets irradiated with laser intensities up to 5×10^{19} W/cm^2. A strong dependence of the ion beam parameters on the conditions on the target conditions and laser parameter was found. The ion beam characteristic revealed a highly laminar acceleration and an excellent beam quality superior to that from conventional accelerators. We succeeded in shaping the ion beam by the appropriate tailoring of the target geometry and we performed a characterization of the ion beam quality. The production of a heavy ion beam could be achieved by suppressing the amount of protons at the target surfaces. Finally, we demonstrated the use of short pulse laser driven ion beams for radiography of thick samples with high resolution.

INTRODUCTION

The acceleration of intense ion beams from relativistic laser plasma interaction is a new and rapidly growing field of research[1,2,3,4]. Recently, new experiments, using ultra-intense short pulse lasers[5] with intensities exceeding 10^{19} W/cm^2 have shown collimated beams of protons that have a very low emittance, while reaching energies of up to 50 MeV[6], which is understood as rear surface emission accelerated by the TNSA (Target Normal Sheath Acceleration) mechanism[7]. This ion beam generation is attributed to electrostatic fields produced by hot electrons acting on protons from adsorbed water vapor and hydrocarbons[8]. Relativistic electrons generated from the laser-plasma interaction, having an average temperature of several MeV, envelope the target foil and form an electron plasma sheath on the rear, non-irradiated surface. The electric field in the sheath ($E_{stat} \sim kT_{hot}/e\lambda_D$, $\lambda_D = (\varepsilon_0 kT_{hot}/e^2 n_{e,hot})^{1/2}$) can reach $>10^{12}$ V/m, which field-ionizes atoms on the surface and accelerates the ions very rapidly normal to the rear surface. Protons, having the largest charge-to-mass ratio, are preferentially accelerated in favor of heavier ions over a distance of a few microns, and up to tens of MeV. This forms a collimated beam with an approximately exponential energy distribution with 5-6 MeV. Because of the dependence of the ion beam on the formation of the sheath, this process should reveal information about the electron transport through the target. We expect details of the ion acceleration will also depend on the target material and surface conditions. Therefore we carried out

CP634, *Science of Superstrong Field Interactions*, edited by K. Nakajima and M. Deguchi
© 2002 American Institute of Physics 0-7354-0089-X/02/$19.00

experiments to investigate the influence of these target parameters on the ion beam production.

EXPERIMENTS

The experiments were performed at the Laboratoire pour l'Utilisation des Lasers Intenses (LULI). Pulses of up to 30 J at 300 fs pulse duration at λ=1.05 µm were focused with an f/3 off-axis parabolic mirror onto free standing target foils at normal incidence, at intensities up to 5×10^{19} W/cm^2. The focal spot (FWHM) measured in vacuum was about 8µm. Amplified spontaneous emission (ASE) occurred 2ns before the main pulse at a level of 10^{-7} of the main pulse energy and preformed a plasma.

The diagnostic setup is depicted in Fig.1. The free standing target was probed by a frequency doubled laser beam parallel to the surface to determine the plasma conditions on the front and rear surface. A stack of radiochromic film (RCF) was positioned a few cm behind the target to measure the spatial beam profile. Due to the pronounced energy loss of ions at the end of their range (Bragg-peak) different layers of the RC film pack allow the imaging of the ion beam at different energies. Details about the RCF are given in[6]. A slot in the center of the RCF allowed a free line of sight for the charged particle spectrometers fielded at 0^0, 6^0 and 13 degree to provide the energy distribution of the emitted electrons and ions.

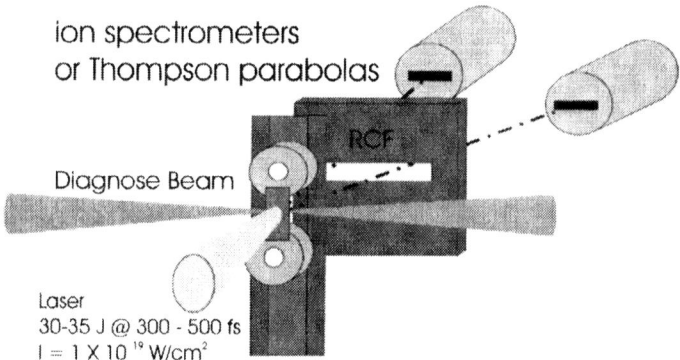

FIGURE 1. experimental setup. The free standing target is irradiated at normal incidence. A slit in the radiochromic film gives a line of sight for the particle spectrometer.

Two absolutely calibrated, permanent magnetic ion spectrometers were mounted at a distance of about 1m from the target covering a solid angle of 5×10^{-6} sr. The protons were recorded in nuclear emulsion track detectors which allow single particle detection without being overwhelmed by the blinding x-ray flash from the laser plasma. A light tight paper in front of the emulsion stopped protons below 1.8 MeV. As a measure of the total yield of protons we used a titanium catcher foil. The ^{48}Ti is transmuted by a (p,n) reaction by protons above a sharp reaction threshold at ~5 MeV to an excited state of the ^{48}V isotope. We observed the gamma deexcitation lines of the

^{48}Ti(p,n)^{48}V reaction, which provided the total activation and therefore the yield of protons above the reaction threshold of ~5 MeV.

To detect heavy ions we used two high resolution Thompson parabolas. The parallel electric and magnetic fields in the Thompson parabolas discriminated ions with respect to their momentum and charge-to-mass ratio, at the plane of the CR-39 track detectors. By etching the CR-39 in sodium hydroxide, the material damage caused by the impacts of ions above a threshold of a few hundred keV become visible. Microscopic scanning provides position as well as the size of the impact, which is proportional to the atomic number, Z of the ion.

RESULTS

Hydrodynamic Target Stability

For the effective acceleration of the ions an undisturbed back surface of the target is crucial to provide a sharp ion density gradient as the accelerating field strength is proportional to T_{hot}/el_0, where T_{hot} is the temperature of the hot electrons and l_0 is the larger of either the hot-electron Debye length, or the ion scale length of the plasma. The preceding ASE launches a shock wave into the target which causes a destruction of the acceleration sheath. Therefore the target thickness was chosen to guarantee an undisturbed rear surface based on calculations using the hydrocode MULTI[9]. The result of the simulation is shown on the left side of Fig.2. The inwards propagating shock wave reaches the rear surface at about 8 ns after the onset of the prepulse. Thus, the targets should maintain an undisturbed back surface for a 5 ns prepulse. When we applied a prepulse at a contrast ratio of 10^{-7} of the main pulse 10 ns before the main pulse the maximum energy of the protons dropped to 2 MeV from the typical 10-20 MeV range typical of low-prepulse shots.

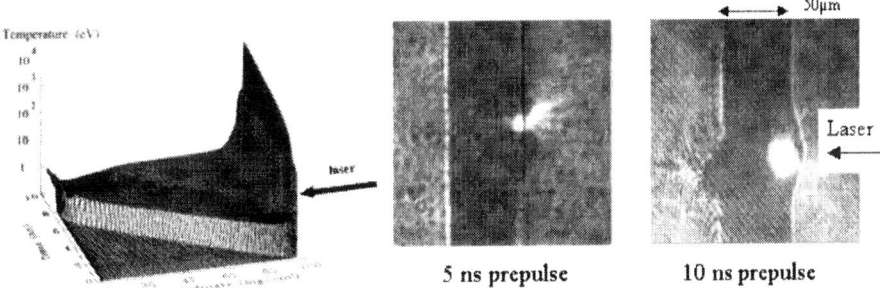

FIGURE 2: left: simulation of the shock wave launched by the prepulse. right: Images of the plasma conditions on the front and rear target surface. right part: perturbation of the rear surface due to prepulse induced shock wave breakout. No protons were detected.

In 10 ns a shock wave launched by the prepulse penetrates the target and causes a rarefaction wave that diminishes the density gradient on the back and therefore drastically reduces the accelerating field. Fig. 2 also shows interferometric measurements of the target surface with and without the additionally applied prepulse. The front surface always shows the blowoff plasma, extending up to about 200μm, caused by the ASE. In absence of a prepulse (left image) the rear surface is unperturbed and a high-energy proton signal could be detected on the RCF. When we observed the presence of an extended plasma at the rear surface due to the applied pre-pulse, no protons above the detection threshold of our RCF could be measured. This result is also in excellent agreement with recent experiments using a second laser to generate a plasma at the rear target surface[10].

Angular Dependence

The angular dependence of the energy distribution of the proton beam was measured with two ion spectrometers positioned at an angle of 0° and 13° respectively. The measured spatial distributions of protons on the dispersion plane were deconvoluted (with respect to the entrance aperture)[11] and corrected for the spectrometer dispersion. The energy of the protons emitted normal to the target rear surface extended up to 25 MeV. The maximum energy of the protons dropped to about 13 MeV at an angle of 13^0, consistent with a 2-D model of the sheath acceleration process. The spectral shape of each proton energy distribution is generally continuous up to the cut-off energy. The best fit to the spectrum obtained by the ion spectrometers, as well as to the spectral information extracted from the stacked RCF packages was obtained by using a two component exponential distribution with 2 and 6 MeV respectively.

Yield, Surface Dependence

In previous experiments[6] using metal targets the origin of the protons was found to be contaminant layers of water vapor and hydrocarbons. The total yield of protons could be increased significantly using plastic targets, due to acceleration of protons from the bulk material, however the laminarity of the beam was largely disrupted, with the spatial pattern of the accelerated protons exhibiting a large degree of filamentary-like structure. To investigate the influence of the target conditions on the creation of the ion beam, we varied the target composition and structure of the rear surface.

We used thin (48μm) targets of gold with either a flat or structured rear surface. The results showed a clear dependence of the spatial uniformity of the proton beam on the structure of the back surface. In contrast to the homogenous, collimated beam from the gold target, protons emitted from the structured gold rear surface showed filaments. To discriminate between conductivity and surface quality effects, we next used ~100 micron plastic and glass targets. While the flat surfaces of glass and plastic yielded a strong, but filamented proton beam, there were no protons detected above 1 MeV from the roughened targets. The similar beam patterns obtained from plastic and glass targets exclude the origin (surface or bulk) of the protons to be the reason for the onset of the filamentary structures. In contrast, due to the strong coupling of the ion acceleration mechanism to the electron distribution at the rear surface of the target the smooth, laminar beam quality from metal targets indicates a rather homogenous electron transport through the target. Insulating material seem to disrupt the electron transport, which causes filamentation of the electron distribution and therefore also a non-homogenous ion acceleration.

Structuring the gold surface maintained a smooth surface with hills and valleys. The surface of the plastic and glass targets was largely destroyed by numerous cracks. When the material on the rear surface is exposed to the strong electric field generated by the electron plasma sheath, it is field ionized instantaneously. A shallow, wavelike surface, such as for the roughened gold targets, is expected to lead to a microlensing phenomenon, consistent with the observed filamentation of the accelerated protons as been calculated for the case of a single concave depression of the surface[7]. In the case of a destroyed surface, the cracks and defects on the plastic and glass create many sharp excursions. The ion plasma created by the field is therefore extended over a larger scale length normal to the surface. We expect this to partially compensate the charge separation sheath, and therefore strongly suppress the ion acceleration.

A well known technique to determine the total yield of fast protons is to use nuclear reactions in a catcher material. For our proton beam, we used the ^{48}Ti(p,n)^{48}Va reaction that provided a sharp threshold at proton energies of ~5 MeV. The total yield of ^{48}Va activations produced in a typical shot was 10^7. We from that deduce a total flux of 10^{11} laser-accelerated protons, assuming an energy distribution with a temperature of 2 MeV. This represents a total conversion efficiency of about 1% of the laser energy to accelerated protons.

Proton Beam Shaping

An important question to be addressed for any future application of laser-accelerated protons and ions is the possibility of tailoring the proton beam, either collimating or focusing it, by changing the geometry of the target surface. We first attempted to defocus the beam in one dimension, by using a convex target. Using a 60 µm diameter Au wire as a target basically constituted such a one-dimensional de-focusing lens, and we observed a line as shown in Fig. 3. Tilting the wire also changed the orientation of the line, which results from the radial, fan-shaped expansion of the protons normal to the wire.

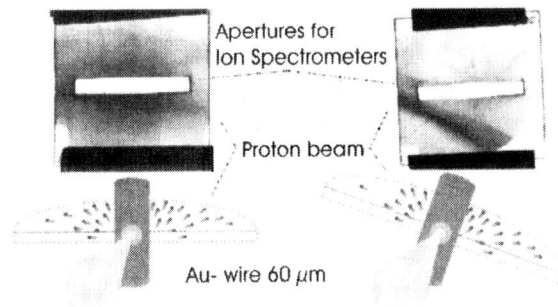

Figure 3. Experimental setup and RCF images of experiments with 60 µm gold wires. The convex rear surface constitutes a de-collimating cylinder-lens. Accordingly the proton beam was formed into a line.

We then attempted to focus the protons by modifying the curvature (concave) of the target foil. Focusing laser generated protons is essential for many applications like ion-induced material damage research, proton driven fast ignition[12], proton radiography, and the use as next generation ion sources. Due to the gaussian-like shape of the hot electron Debye sheath that causes the acceleration, there is an energy dependent angle of divergence that has to be compensated to focus the ions in the energy range of interest. Therefore the effective focal length of a curved target rear surface is longer and is dependent of the proton energy. The results, that will be published elsewhere show a strong reduction in the divergence of the central core of the proton beam representing ballistic collimating of laser produced proton beams.

Heavy Ion Beam Production

We next attempted to control the accelerated ion species, and in particular selectively accelerate either protons or heavy ions. Due to their larger charge-to-mass ratio, which causes the protons to outrun the other ion species during the ambipolar expansion, protons are accelerated faster taking most of the energy from the electrostatic sheath. Therefore the amount of protons had to be reduced significantly. For targets of solid metals (gold, aluminum) the majority of the protons is due to water vapor and hydrocarbons at the target surface. We reduced these impurities by resistively heating the targets.

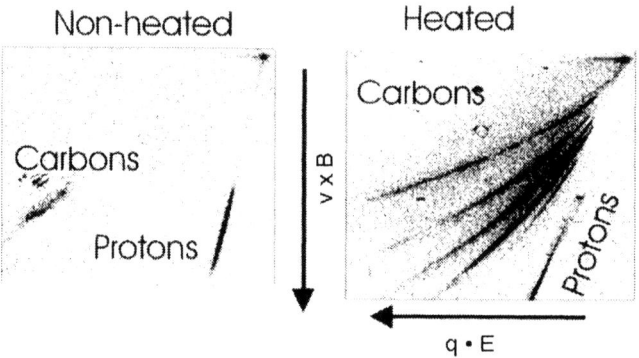

Figure 4. Heavy ion beam production. In contrast to the strong proton signal (left), removing the hydrocarbons from the target rear surface results in a strong heavy ion (carbon) signal (right).

The targets consisted of thin foils of 50 μm Al coated with 1 μm of carbon. To detect the heavy ions with respect to their momentum and charge-state distribution, we substituted the ion spectrometers with two Thompson parabolas at an angle of 0^0 and 13^0. The ions were recorded in CR-39 plastic track detectors. We compared the yield for heated and non-heated targets, as shown in Fig. 4. As expected, for the non-heated targets (left) a strong proton signal was observed together with a weak signal of carbon ions. The result changed dramatically for the heated targets as shown in the right part of Fig. 4. A sharply reduced proton signal was detected in these experiments together with a much more intense heavy ion signal (carbon and aluminum ions). We observed a higher yield, much higher ion energies and ions at higher charge states.

40

Proton Beam Emittance

For most of the future applications of laser generated ion beams the beam quality is the most important characteristic. The radiochromic film data suggest that protons or other light ions accelerated by the TNSA mechanism may have a usefully small emittance in the sense of an actual ion beam.

To precisely estimate our emittance, we used penumbral imaging of edges at different distances from the target with the magnetic spectrometers, to directly measure the core emittance of the proton beam. We determine the normalized emittance of protons from flat gold foils to be ~0.2 pi mm-mrad, and factor of at least two smaller than the resolution limited measurements in[6]. The results of this analysis and subsequent modeling, developing a 2-D extension of the model in[13], suggest that we observe a rather cold proton beam, which is smoothly diverging and highly laminar. From these data, we deduce that the proton temperature is less than ~1 keV.

Radiography Using Laser Accelerated Proton Beams

The excellent beam quality of the ion beam is ideally matched to the requirements for imaging techniques. One scheme of particular interest is the use of laser accelerated protons to radiograph samples to study their properties. Due to the different interaction mechanism protons can provide complementary information to techniques like x-ray backlighting. Because of the copious amounts of protons accelerated in a very short time, laser accelerated protons provide a new diagnostic quality in research of transient phenomena. We performed a first set of experiments to demonstrate the feasibility of these laser accelerated proton beams for radiography applications. In contrast to experiments for object imaging[14], where the target was exposed to electrons closely behind the target, we have chosen a different geometry.

FIGURE 5. Radiography of a compound target. Details of the target are given in the text.

We used a distance of 5 cm between the proton source and the target in order to minimize any charging of the target and placed the detector (RCF) close to the object to reduce deflection effects and a compound target of different materials to be imaged by the protons. It consisted of an 1 mm thick epoxy ring structure, several copper wires of 250 μm diameter, a hollow cylinder with 300 μm steel walls, several Ti sheets of 100 μm thickness and a glass hemisphere of 900 μm diameter and 20 μm wall thickness. The protons were recorded in multiple layers of RCF to detect the image at different proton energies. Fig.5 shows the radiography of the target for final proton energies of 7.5 MeV. The image constitutes a negative image of the areal density of the target. The names of the collaborating institutes have been engraved on the epoxy ring, which results in a reduced thickness and therefore a higher energy deposition of the protons in the respective layer. The areal density variation of the hollow cylinder, including a hole in the wall on the right hand side can be seen as well as a thin metal rod placed inside the cylinder. The results show a clear dependence on the areal density rather than residual charging effects, in contrast to the experimental technique used in[14]. The time of exposure in this experiment was estimated to be in the order of tens of picoseconds, based of the initial proton beam pulse duration and the energy dependent dispersion of the pulse from the source to the target.

Conclusion

We have presented a detailed investigation of the target conditions on the proton and ion beam production from intense laser solid interactions. The observed strong dependence on the rear surface conditions in agreement with the target normal sheath acceleration mechanism. The target conductivity appears to have a major influence on the quality of the ion beam, and the quality of the surface finish of the target is very important for maintaining a high gradient sheath and a laminar beam. It has been shown that tailoring the ion beam (yield, shape, composition, homogeneity) by means of target shape and composition is possible, and we present first observations of laser-accelerated ion beam shaping. Finally the successful generation of an heavy ion beam (carbon, aluminum) further encourages speculation that laser-accelerated ion beams may become a useful tool in a variety of future applications.

This work was supported by the EU, Contract No. HPRI CT 1999-0052

REFERENCES

[1] A. P. Fews, et al., Phys. Rev. Lett. **73**, p. 1801 (1994).
[2] K. Krushelnick, et al., Phys. Rev. Lett. **83**, p. 737 (1999).
[3] A. Maksimchuk, et al., Phys. Rev. Lett. **84**, p. 4108 (2000).
[4] M. Zepf, et al., Phys. Plasmas **8**, p. 2323 (2001).
[5] M. Perry and G. Mourou, Science **264**, p. 917 (1994).
[6] R. Snavely et al., Phys. Rev. Lett. **85**, p. 2945 (2000).
[7] S.C. Wilks, et al., Phys. Plasmas **8**, p. 542 (2001).
[8] S.J. Gitomer, et al., Phys. Fluids **29**, p. 2679 (1986).
[9] R. Ramis, R. Schmalz and J. Meyer-ter-Vehn, Comp. Physics Communic. **49**, 475 (1988)
[10] A.J. MacKinnon, et al., Phys. Rev. Lett. **86**, p. 1769 (2001).
[11] LB. Lucy, Astron. J. **79**, 745 (1974)
[12] M. Roth, et al., Phys. Rev. Lett. **3**, Vol. 86, 436 (2001)
[13] L.M. Wickens and J.E. Allen, Phys. Fluids 24, 1984 (1981).
[14] M. Borghesi, et al., Plas. Phys. and Contr. Fusion **43**, p. A267 (2001).

Lower Phase Velocity of Focused Laser Beam and Vacuum Laser Acceleration

N. Cao, Y.K. Ho [a], Q. Kong, J. Pang, L. Shao, Y.J. Xie

Institute of Modern Physics, Fudan University, Shanghai 200433, China
a) Author to whom correspondence should be addressed.
FAX: +86-21-65643815. Electronic address: hoyk@fudan.ac.cn

Abstract. An acceleration channel has been found in the field of a focused laser beam propagating in vacuum, which shows similar characteristics to that of a wave guide tube of conventional accelerators: a subluminous wave phase velocity in conjunction with a strong longitudinal electric field component. Relativistic electrons injected into this channel can remain synchronous with the accelerating phase for sufficiently long time and receive considerable energy from the field. We call this acceleration scheme CAS (capture and acceleration scenario). The basic conditions for CAS to occur are examined and the output properties of electrons accelerated by this scheme are also presented in this paper.

Advances in laser technology have resulted in a new class of compact, ultrashort pulsed lasers with extremely high intensity [1-3], and high power lasers continue to spur new concepts for advanced laser-driven accelerators [4-9]. Among them, the far-field laser acceleration of free electrons in vacuum has received wide attention [10] because it avoids a number of potential problems in other accelerating scheme. Nevertheless, vacuum laser acceleration has a foundational and long-standing question that has not been answered totally till now. Can a free electron have net energy exchange with a laser beam in far-field where the interaction length is unlimited [11]?

According to the Lawsoon-Woodward theorem [12], a light beam in the far-field is the superposition of propagating plane waves, thus its effect on a particle is the sum of plane wave effects. Since a particle moves at velocity less than c and a plane wave moves at c in a vacuum, the particle must slip in the phase of the wave, and because of phase slippage, the effect of light on a particle averages to zero if the interaction is unbounded.

More recently it was shown [10], via experiments and simulations that a focused laser pulse interacting with low energy electrons (at rest or near at rest) could be ponderomotively scattered and receive net energy gain by interacting with the nonlinear ponderomotive laser force.

In our studies, we found that, for a focused laser beam, there exists a region characterized by lower phase velocity ($V_\varphi < c$). If the amplitude of the longitudinal electric field in the region is strong enough, and if one can inject fast electrons into it, then these electrons can remain synchronous with the accelerating phase for sufficiently long time such that they receive considerable energy from the field. We named this acceleration scheme CAS (the capture and acceleration scenario)[13]. Simulations indicated that, at the regime of the CAS, the electron's trajectory is perturbed significantly by the laser field as it enters the high intensity

CP634, *Science of Superstrong Field Interactions*, edited by K. Nakajima and M. Deguchi

region, which in effect limits the interaction region. Hence, large energy gains are obtained without limiting the interaction distance by the use of additional optics. For presently available lasers (>100TW), substantial energy gains (>100MeV) can be obtained.

This paper is organized as follows. The subluminous phase velocity region in the laser field is derived firstly. Secondly, the acceleration channel is described. Then, the dynamics behaviors of CAS electrons are discussed and the output properties of electrons interacting with the laser field are given lastly.

Numerical simulation methods used here are similar to those we used previously [13]. A four- dimensional energy-momentum configuration (γ, p_x, p_y, p_z) is used to specify an electron state, where γ is the Lorentz factor and the momentum p is normalized in the units of $m_e c$, the angle $\theta = \tan^{-1}(p_x/p_z)$ is the electron injection angle with respect to the z axis, the laser propagation direction. The electron dynamics is governed by the relativistic Newton-Lorentz equation. For simplicity, throughout this paper, time and length are normalized by $1/\omega$ and $1/k$, respectively. Without losing generality, we assume that, initially electrons move freely without the influence of laser field, the center of the laser pulse and that of the injected electron bunch reach the point $x = y = z = 0$ simultaneously at $t = 0$, Δt_d is used to specify the relative delay between the laser pulse and an electron.

For a laser beam of Hermite-Gaussian(0,0) mode polarized in x-direction and propagating along z-axis, the phase of the Gaussian beam field is given by

$$\varphi(x,y,z) = kz - \omega t - \varphi(z) - \varphi_0 + \frac{k(x^2 + y^2)}{2R(z)} \tag{1}$$

where $\varphi(z) = \tan^{-1}(z/Z_R)$ is the Grouy phase shift, $Z_R = k w_0^2/2$ is the Rayleigh length and $R(z) = z(1 + Z_R^2/z^2)$ is the radius of the curvature. The phase velocity $(V_\varphi)_l$ of the wave along a particle trajectory can be calculated using the equation

$$\frac{\partial \varphi}{\partial t} + (V_\varphi)_l (\nabla \varphi)_l = 0$$

(2)

where $(\nabla \varphi)_l$ is the gradient of the phase field along the trajectory. From it, we can get the phase velocity along the direction normal to the equal-phase face (The minimum phase velocity) $V_{\varphi m}$ as follows

$$V_{\varphi m} = \frac{ck}{|\nabla \varphi|} = \frac{ck}{\sqrt{\left(\frac{2\rho l}{w_0(1+l^2)}\right)^2 + \left(k - \frac{1-f_\varphi}{Z_R(1+l^2)}\right)^2}} \tag{3}$$

with

$$f_\varphi = \frac{r^2(1 - z^2/Z_R^2)}{w_0^2(1 + z^2/Z_R^2)} \tag{4}$$

and $\rho = \sqrt{x^2 + y^2}/w_0$, $l = z/Z_R$, $r = \sqrt{x^2 + y^2}$. From Eq.(3), it is straight forward to find that the subluminous phase velocity region: $V_{\varphi m} < c$ occurs approximately in

44

the region $r>w(z)$. In this region, the magnitude of the minimum phase velocity is of the order $V_{\varphi m} \sim c[1-1/(kw_0)^2]$, with an angle relative to z-axis $\sim 1/(kw_0)$. The distribution of $V_{\varphi m}$ on the plane $z=0$ is shown in Fig 1. Another feature of Fig.1 is that at $z=0$, the subluminous phase velocity region occurs only for $r>w_0$.

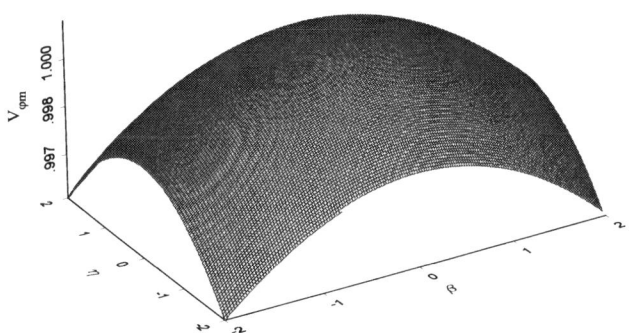

FIGURE 1. The distribution of the minimum phase velocity $v_{\varphi m}/c$ versus $\beta = x/w_0$ and $\eta = y/w_0$ in the $z=0$ plane of the focused laser beam with $kw_0=60$.

For accelerating particles, in addition to the subluminous phase velocity, the field strength, i.e., the amplitude of the longitudinal electric field, is also an important factor. We thereby introduce a quantity Q that combines these two factors together to represent the ability of the laser field for accelerating charged particles. We call it acceleration quality factor, which is defined by

$$Q = Q_0(1-V_{\varphi m}/c)[x/w(z)]\exp[-(x^2+y^2)/w(z)^2]$$

for $V_{\varphi m} \leq c$, and Q=0 for $V_{\varphi m}>c$ (5)

Where Q_0 is a normalization constant chosen to make Q in the order of unit. In Eq.(5), $1-V_{\varphi m}/c$ represent the contribution from the phase velocity, and the remaining term is proportional to the amplitude of the longitudinal electric field, it describes the effects of laser intensity on electron's behavior.

The distribution of the acceleration quality factor Q on the plane $z=0$ for a focused laser beam with $kw_0=60$ is given in Fig.2.

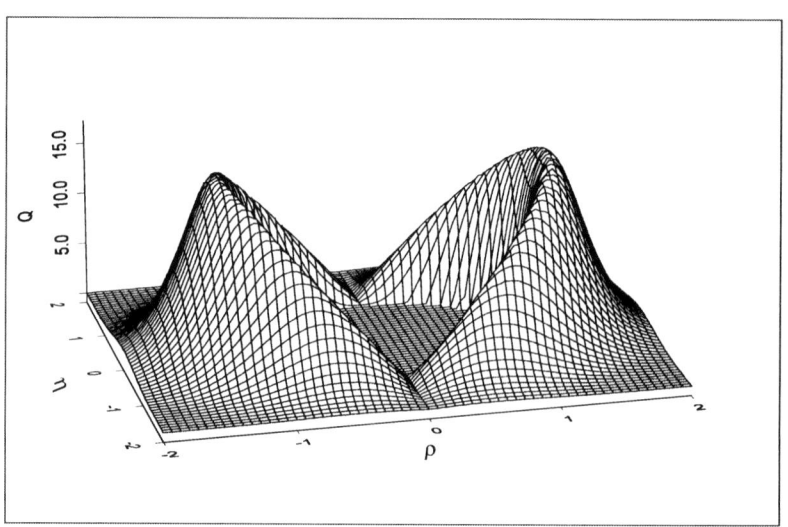

FIGURE 2. The acceleration quality factor Q versus $\rho = x/w_0$ and $\eta = y/w_0$ in the $z=0$ plane of a focused laser beam with $kw_0=60$.

It can be seen from Fig.2 that significant values of Q emerge just beyond the beam width and concentrated in the areas near the polarization plane. This is because the lower phase velocity region locates outside the beam width and the large amplitudes of the axial electric field distribute near the polarization plane. This is the favorable region for accelerating electrons and we call it the acceleration channel of the laser beam propagation in vacuum. It is also of interest to note that in the region near the beam axis, the factor Q=0 because $V_{\varphi m}>c$ there. It means the region near beam axis is not good for accelerating charged particles. This is because the phase velocity here is the highest, and the amplitude of the axial electric field is small. Along the diffraction angle (~ $1/kw_0$), exceeding several Rayleigh lengths, the intensity of the laser field become much weaker and therefore, the effect of acceleration become inconspicuous. From the above discussions, we may say that there exists an acceleration channel in the field of the focused laser beam propagating in vacuum, which shows similar characteristics to that of a wave guide tube of conventional accelerators: subluminous phase velocity in conjunction with a strong longitudinal electric field component. Consequently, if one can inject fast electrons into this channel, then some of them may remain synchronous with the accelerating phase for sufficiently long time such that they receive considerable energy from the field.

Now we turn to study the conditions under which CAS phenomenon can occur. Simulation results show that there exists a threshold value of laser intensity $(a_0)_{th}$ which is sensitive to the beam width w_0. One can observe the CAS phenomenon only as $a_0 \geq (a_0)_{th}$. Fig.3 shows the electron maximum final energy γ_{fm} vs. the laser intensities a_0 at different beam width, γ_{fm} is the maximum value of γ_f as φ_0 is varied over the range $(0, 2\pi)$.

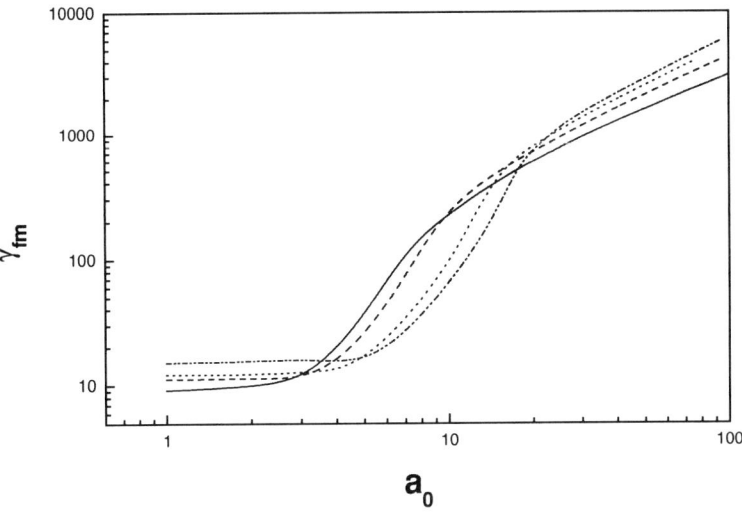

FIGURE 3. The maximum output energy γ_{fm} as a function of the laser intensity a_0 at different laser beam width. Solid line for $w_0=60$, $P_x=0.9$, $P_z=9$; dashed line for $w_0=80$, $P_x=1.1$, $P_z=11$;dotted line for $w_0=100$, $P_x=1.2$, $P_z=12$; and dash dot dot line for $w_0=120$, $P_x=1.4$, $P_z=14$; momentums in y-direction in all cases equal zero, we terminate the calculation at $\omega t=400000$.

Fig.3 illustrates that, (i) $(a_0)_{th}$ is strongly dependent on the laser beam width w_0 , it increases quickly with w_0 . (ii) The scaling law for net energy gain of electrons from the laser field is approximately as $\gamma_{fm} \propto a_0{}^n$ with $n{\sim}1$ as $kw_0{<}150$. This feature is consistent with the mechanism underlying CAS, namely the acceleration occurs primarily in the acceleration channel by the longitudinal electric field. (iii) With the increasing of beam width w_0, the maximum output energy γ_{fm} increase also. This is due to the fact that the work (ΔW) done by the laser field on the electron in the longitudinal direction is proportional to the beam width and can be approximately estimated by $\Delta W \propto E_z \cdot Z_R \propto \dfrac{E_0}{kw_0} \cdot \dfrac{kw_0{}^2}{2} \propto E_0 w_0$, where E_0 is the transverse electric field amplitude at focus, and Z_R the Rayleigh length.

In addition to the laser intensity, the electron initial incident momentum and incident angle are also important factors for CAS to emerge.

Roughly speaking, for CAS to occur, the laser intensity should be large enough as $a_0{\geq}5$, and electrons should be injected into the laser field in a small angle (typically $\tan\theta=0.1$) with injection energy ranging 5-15MeV.

The output properties of the electrons interacting with the laser pulse in vacuum are shown in Fig.4.

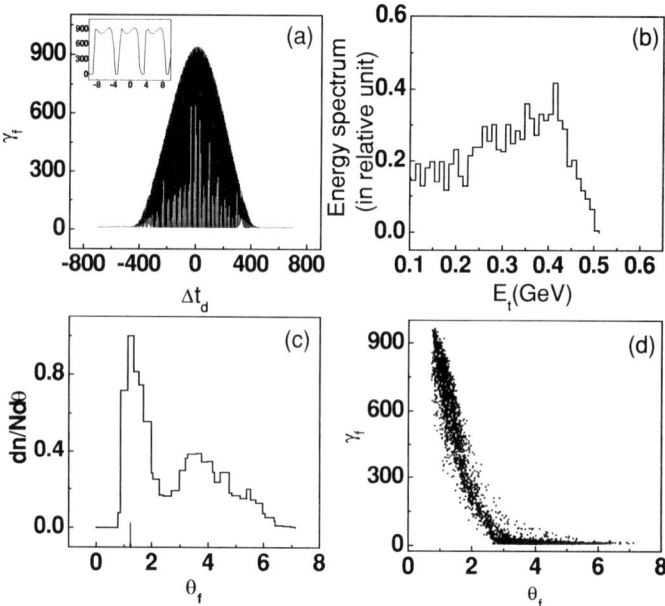

FIGURE 4. The output properties of electron bunch interacting with laser pulse of the same size. The laser parameters used are $a_0=30$, $w_0=60$, $\tau=300$. The electron bunch impinges on laser pulse with $\tan\theta=0.1$, $E= 4.9$MeV, and space emittance $0.1\pi\cdot$mm$\cdot$mrad. (a) Dependence of the electron final energy γ_f on the relative delay time Δt_d. Only those electrons in the x-z plane are considered. (b) Energy spectrum, (c) Angular spectrum, of the output electrons; (d) Correlation of the distribution of electrons final energy γ_f and scattering angle θ_f.

Our study shows that the output property of the interaction is a high-energy electron macro-pulse consisted of many micro-pulses whose duration corresponds to the periodicity of the laser field (Fig.4(a)). Due to the fact that the incident electrons encounter all phases of each laser period, energies of CAS electrons spread widely. For the specific example of Fig.4, the output energy range from 245MeV to 491MeV, with average energy about 356MeV and the number of CAS electrons is about 34% of the total incident electrons.

The output electrons can be roughly divided into two groups: CAS electrons concentrated in the small scattering angle region and inelastic scattering (IS) electrons correspond to the broad peaks in the larger angle region (Fig.4(c)). The simulation results of the energy-angular correlation are roughly consistent with the equation derived from the classical Hamilton-Jacobi theory [14] (Fig.4(d)).

$$\theta_f = \tan^{-1}\left(\sqrt{p_{\perp i}^2 + 2(\gamma_i - p_{zi})(\gamma_f - \gamma_i)}\Big/(p_{zi} + \gamma_f - \gamma_i)\right) \qquad (6)$$

The deviations are due to that we use Gaussian beam to describe the laser field, and in CAS case, the longitudinal component plays a very important role [13]. Another reason is that we set the emittance of the initial states of the electrons $0.1\pi\cdot$mm$\cdot$mrad in the calculations.

In conclusion, we find that for a focused laser beam propagating in free-space, there exists, surrounding the laser beam axis, a subluminous wave phase velocity region. Relativistic electrons injected into this region can be trapped in acceleration phase and remain in phase with the laser field for sufficiently long time, thereby receiving considerable energy from the field. The basic conditions for CAS to occur are the laser intensity should be large enough as $a_0 \geq 5$, and electrons should be injected into the laser field in a small angle (typically $\tan\theta=0.1$) with injection energy range be in 5-15MeV. The output electrons can be divided into two groups: the CAS electrons with higher energy concentrated in small scattering angle region and the IS electrons with lower energy gains scattered widely in space. The output electrons can be run through a magnetic spectrometer to select a near-monoenergetic electron micro-pulse train. Studies show that CAS is an attractive laser acceleration scheme.

ACKNOWLEDGMENTS

The authors would like to thank A.M. Sessler and E.H.Esarey for enlightened discussions. This work is supported partly by the National Natural Science Foundation of China under contracts No.19984001 and 10076002, National High-Tech ICP Committee in China, Engineering-Physics Research Institute Foundation of China, and National Key Basic Research Special Foundation (NKBRSF) under Grant No.1999075200.

REFERENCES

1. Mourou, G. A., Barty, C. P. J., et al., *Phys. Today* **51**(1), 22-28(1998).
2. Barty, C. P. J., *Laser Focus World* **93**,1996.
3. Perry, M. D., and Mourou, G., *Science* **264**, 917-923(1994).
4. *Advanced Accelerator Concepts*, edited by P. Schoessow, AIP Conference Proceedings 335, New York: American Institute of Physics, 1995.
5. Sprangle, P., Esarey, E., et al., *Phys. Plasmas*, **3**, 2183(1996).
6. Huang, Y. C., Zhang, D.,et al., *Appl. Phys. Letters* **68**, 753(1996).
7. Hafizi, B., Ting, A., et al., *Phys. Rev.E* **55**, 5924-5933(1997).
8. Nakajima, K. Fisher, D., et al., *Phys. Rev. Letters* **74**, 4659(1995).
9. Umstadter, D. Chen, S. Y. et al., *Science* **273**, 472-475(1996); Modena, A., Najmudin, Z., et al., *Nature* **337**,606-608(1995).
10. Monot, P., Auguste, T., et al., *Phys. Rev. Letters*. **70**, 1232-1235(1993); Moore, C. I., Knauer, J. P., et al., *ibid*, **74**, 2439-2442(1995); Malka, G., Lefebvre, E., et al., *ibid*,**78**, 3314-3317(1997); Stupakov, G. V., and Zolotorev, M. S., *ibid*. **86**, 5274(2001); Salamin,Y. I., and Keitel, C. H., *ibid*, **88**, 950-954(2002).
11. Sessler, A. M. *Am.J.Phys* **54**,505 (1986); *Phys.Today* **1**,26 (1988).
12. Woodward, P. M., *J. IEEE* **93**, 1554(1947); Lawson, J. D., *IEEE Tran. Nucl. Sci.* NS-**26**, 4217(1979); Esarey, E., Sprangle, P., et al., *Phys. Rev. E* **52**, 5443(1995); Sprangle, P., Esarey, E., et al., *Phys. Plasmas* **3**, 2183(1996).
13. Wang, P. X., Ho, Y. K., et al., *Appl. Phys. Letter* **78**, 2253-2255(2001); Wang, P. X., Ho, Y. K., et al., *J. Appl. Phys* **91**,856-866(2002); Wang, J. X., Ho, Y. K., et al., *Phys. Rev. E* **58**, 6575(1998).
14. Salamin, Y. I., Faisal, F. H. M., *Phys. Rev. A* **55**,3678-3683(1997); Hartemann, F. V., Meter, J. R. V., et al., *Phys.Rev.E* **58**, 5001-5012(1998).

Acceleration of Proton Beams by Relativistically Self-focused Intense Laser Pulses

James Koga*, Kazuhisa Nakajima†, Mitsuru Yamagiwa* and Alexei Zhidkov‡

*Advanced Photon Research Center, JAERI, Kyoto-fu, 619-0215, Japan
†High Energy Accelerator Research Organization, Tsukuba, 305-0801, Japan
**Nuclear Engineering Research Laboratory, Graduate School of Engineering, The University of Tokyo, 22-2 Shirane-shirakata,Tokai, Naka, Ibaraki, 319-1188, Japan

Abstract. Two dimensional particle-in-cell simulations are performed which show the formation of an extremely large electrostatic field near the front of a relativisitically self-focused laser pulse propagating in an underdense plasma. The size of the field is found to reach a maximum of ~ 6.5 TV/m for a 100TW laser pulse propagating over a distance of about 1mm in a plasma at about 3% of the critical density. We use this field generated by relativistically self-focused ultraintense laser pulses to accelerate injected ions.

INTRODUCTION

Ultraintense laser interactions with matter can generate enormous numbers of highly energetic electrons, photons, and ions. Recent laser-matter interaction experiments have revealed acceleration of $\sim 10^{10}$ electrons with the maximum energy up to ~ 100 MeV and production of $\sim 10^{12}$ ions up to several tens of MeV[1].

In two-dimensional PIC simulations for propagation of ultraintense laser pulses of the order of 10^{20}W/cm^2 in underdense plasmas of $\sim 10^{20}$cm^{-3}, we have found that large amplitude of positive electrostatic fields of the order of a few TV/m are generated over a ~ 1 mm scale in the front of a relativistically self-focused laser pulse. These enormous fields imply the capability of accelerating protons. The solitary structure of the electrostatic field with a spatial and temporal size of the order of μm can produce a femtosecond ion pulse. We propose a new acceleration mechanism for ions due to enormous accelerating fields generated by relativistically self-focused laser pulses in plasmas. This mechanism may open up a new regime of ultraintense laser-matter interactions and new fields of high energy particle physics.

SIMULATION PARAMETERS

To study the self-focusing of a high intensity short pulse laser in a plasma we use the code PCUBE (Progressive Parallel Plasma Code) which is a 2 dimensional fully relativistic particle-in-cell(PIC) code[2] running on the Compaq ES40 227 node parallel computer. The simulation box is 672μm (23000 cells) by 37.4μm (1280 cells) in the x

CP634, *Science of Superstrong Field Interactions*, edited by K. Nakajima and M. Deguchi

and y directions respectively. The boundary conditions are periodic in the y direction and outgoing in the x direction. There is a vacuum region at one end of the simulation box of length 21.9μm. The plasma density is chosen to be 5.3×10^{19}cm^{-3} which corresponds roughly to doubly ionized Helium gas at atmospheric pressure. There are 8 electrons and 8 ions in each simulation cell with an ion to electron mass ratio of 1836. The linearly s-polarized laser pulse (E_z, B_y) starts in the vacuum region and propagates into the plasma. The parameters of the laser are of the 100 TW Ti:sapphire laser at the Japan Atomic Energy Research Institute [3]. The pulse length is 19 fs with a spot size of 10μm. The wavelength is 0.8μm. The corresponding unitless laser strength parameter $a_0 = 7.4$ where $a_0 = eE_0/m_0\omega_0c$, E_0 is the peak electric field, m_0 is the electron mass, and ω_0 is the laser frequency.

The critical power P_{cr} for the relativistic self-focusing of a Gaussian laser pulse is given by [4], $P_{cr}[\text{GW}] = 17(\frac{\omega_0}{\omega_p})^2$, where ω_0 is the laser frequency and ω_p is the plasma frequency. With the current density $P/P_{cr} = 179$, where P is the laser power. Thus, the laser pulse should relativistically self-focus in the plasma. Also given the condition for length of the laser pulse L_t necessary for the optimum generation of a wake field [5]: $L_t = \pi c/\omega_p$, we find that $L_t = 2.3\mu$m whereas the laser pulse length is 4.9μm. Under these conditions a large wake field behind the pulse should not occur.

SIMULATION RESULTS

Figure 1(a) shows the laser pulse after it has propagated 256μm ($340c/\omega_p$). The laser pulse has relativistically self-focused, has filamented, and the central portion has narrowed. From the line profile taken down the center of the pulse (Figure 1(b)) we can see that the front of the laser pulse has steepened. This is due to the fact that the front of the laser pulse has been depleted compared with the initial Gaussian profile. The amplitude is approximately a factor of 2 higher than the initial laser pulse amplitude. Figure 2 (a) shows the electron density at the same propagation distance. In the central portion of the pulse the electrons have been completely ejected. An electron cavity from the main part of the pulse has formed which is $8c/\omega_p$ wide and $6c/\omega_p$ long. Electrons have built up in front of this cavity. Figure 2 (b) shows a profile of the electron density down the center of the evacuated region. In the front of the laser pulse the density is 25 times the initial background plasma density which is just below the critical density at 32.9 times the background density.

Due to this large buildup of electrons at the front of the pulse, there is a large positive electrostatic field in the propagation direction of the laser pulse created there. Figure 3 shows the structure of the electric field after the same propagation distance. Figure 3(b) shows the line profile of the electric field down the center of the pulse. The electric field rises rapidly to a maximum of 6.5 TeV/m and gradually drops off until it becomes negative behind the laser pulse. We can compare this maximum field to the wave-breaking limit field. Using a linear group velocity for the laser pulse the wave-breaking limit field becomes $E_{WB} = 2.17$ TV/m [6]. We see that the field generated in the front of the laser pulse is greater than the wave-breaking limit field.

The source of the large density created at the front of the laser pulse can be determined

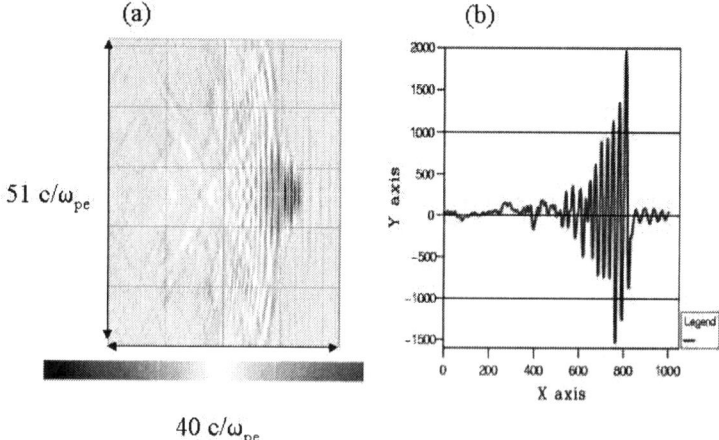

FIGURE 1. (a) E_z field of the self-focusing laser pulse, (b) profile down the center of the pulse after the laser has propagated $340c/\omega_p$.

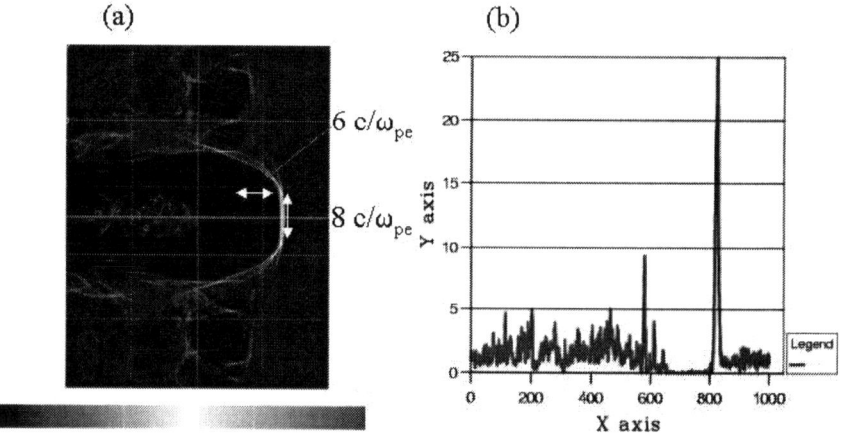

FIGURE 2. (a) Electron density of the background plasma after the laser has propagated $340c/\omega_p$. (b) Profile of the electron density down the center of the simulation box.

from the equation of continuity: $\frac{\partial n_e}{\partial t} + \frac{\partial(n_e v_x)}{\partial x} = 0$ where n_e is the electron density, x is the propagation direction of the laser pulse, and v_x is the velocity of electrons in the propagation direction. In the frame comoving with the laser pulse we get[7]:

$$n_e = \frac{n_{e0}}{1 - \beta_x} \qquad (1)$$

where n_{e0} is the initial electron density and β_x is v_x/c. From this equation we see that where the background electron velocity is high the density is high. We can determine the

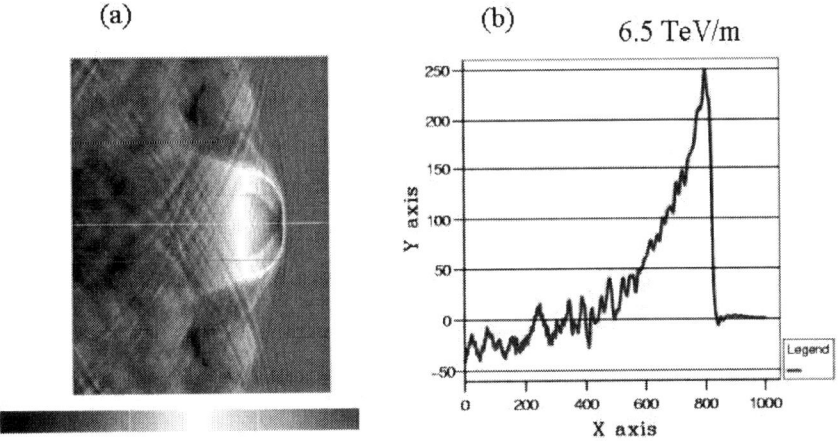

(a)   (b)   6.5 TeV/m

FIGURE 3. (a) The structure of the E_x electrostatic field after propagating $256\mu m(340c/\omega_p m)$. (b) Profile of the electron density down the center of the simulation box.

peak of this electron density relative to the laser field by using the equations of motion of an electron in a plane wave. For a plane wave of the form $E(\phi) = a_0 \sin(\phi)$ where $\phi = \omega[t - \frac{x(t)}{c}]$ is the phase, a_0 is the normalized amplitude, and ω is the frequency of the laser pulse, the velocity of the electron β_x is given by [8]: $\beta_x = 1 - \frac{2}{2+a_0^2(\cos(\phi)-1)^2}$. Using Equation 1 we get:

$$\frac{n_e}{n_{e0}} = \frac{2+a_0^2(\cos(\phi)-1)^2}{2}. \tag{2}$$

The maximum in the electron density occurs at $\phi = \pi$. Compared to the simulation results in Figures 1(b) and 2(b), this corresponds to the same position in the laser wave for the maximum density. Using the value of $a_0 = 5.57$ corresponding to the field at the same time as Figure 1(b) from Equation 2 we get $n_e/n_{e0} \approx 63$ for the maximum density. This is about 2.5 times larger than the maximum value of the density from the simulation.

The maximum density can also be estimated from the electrostatic potential between the front and back of the laser pulse occurring from the buildup of electrons at the front of the pulse and the resulting evacuation at the back The maximum attainable potential, $\Delta\Phi$, can be estimated by: $e|\Delta\Phi| \approx 2\pi n_e d^2 = mc^2 a_0^2/2$ where n_e is the electron density, d is the width of the electron bunch, and the last term represents the maximum energy that electrons can attain in a finite time duration plane wave. Rewriting this equation we get $n_e = \frac{a_0^2}{4\pi d^2 r_e^2}$ where r_e is the classical electron radius. Using $a_0 = 5.57$ and assuming $d = \lambda_0$ we get $n_e/n_{e0} \approx 26$. This is in close agreement with the simulation results.

Figure 4 shows the maximum normalized amplitude of the laser pulse E_z and the corresponding electrostatic field E_x generated by the pulse as a function of propagation distance of the laser pulse. The laser pulse relativistically self-focuses to a peak nor-

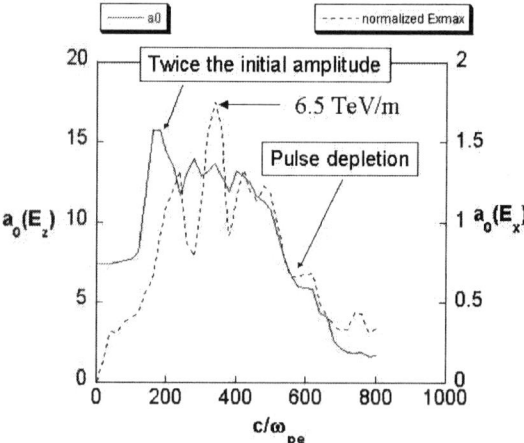

FIGURE 4. (Normalized E_z field of the self-focusing laser pulse (left axis, solid line) and electrostatic field E_x (right axis, dotted line) as a function of the laser pulse propagation distance.

malized amplitude of $a_0 = 15.7$ after propagating $160c/\omega_p$. This is more than twice the initial amplitude. After the initial peak in self-focusing amplitude, the pulse begins to deplete. This depletion is due to the absorption of the laser pulse by the plasma. By the end of the simulation $\approx 89\%$ of the laser energy has been deposited in the plasma. The laser is propagating in a near vacuum at the speed of light while the electron cavity created by the laser pulse propagates at near the plasma group velocity. Theoretically in one dimension the depletion distance is [9]: $l_{pulse}\frac{\omega_0}{\omega_p} \approx 257\frac{c}{\omega_p} = 188\mu m$. This value is smaller than that seen from the simulation results. One factor may be the effect of the relativistic factor γ of the background electrons on the plasma frequency ω_p or two dimensional effects.

As shown in Figure 4 the electrostatic field peaks after the laser field peaks. The maximum electrostatic field is 6.5 TeV/m ($a_0 = 1.75$). Throughout most of the laser propagation the electrostatic field amplitude is about 10% of the laser field amplitude. Also, in conjunction with the laser pulse depletion the electrostatic field decreases.

ION ACCELERATION SCHEME

We propose to use the large field created at the front of the pulse to accelerate injected protons to higher energies. In order to determine the minimum energy of injection of the protons we can use the theory developed for determining the minimum energy of injection for electrons in a wakefield [10]:

$$\gamma_{max/min} = \gamma_p(1 \pm \beta_p\sqrt{2\gamma_p\Delta\phi}) \tag{3}$$

assuming $\gamma_p\Delta\phi \ll 1$ where $\gamma_{max/min}$ refers to the maximum and minimum energy of electrons which can be trapped in a potential well of a normalized potential difference

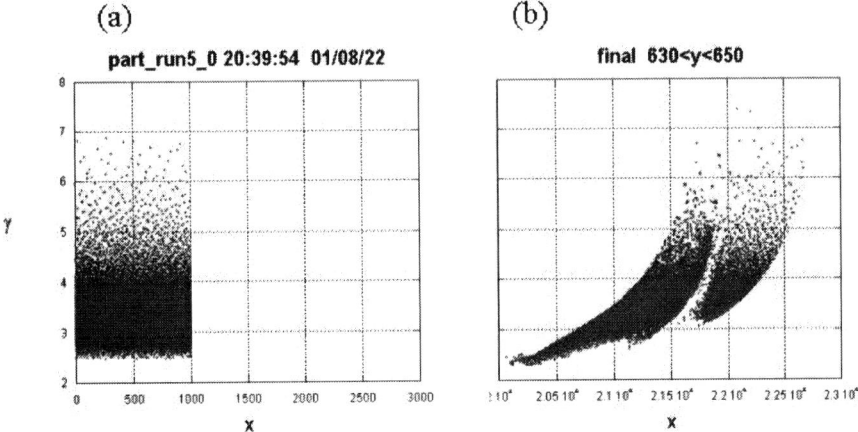

(a) part_run5_0 20:39:54 01/08/22

(b) final 630<y<650

FIGURE 5. (a) Initial and (b) final $\gamma - x$ phase space of the central portion of the injected proton beam at $630\Delta \leq y \leq 650\Delta$ where the center of the simulation is at $y = 640\Delta$

$\Delta\phi$. The potential difference is:

$$\Delta\phi = \phi_{max} - \phi_{min} = a_0 \frac{m_e}{m_i} \tag{4}$$

where a_0 is the normalized amplitude of the electrostatic field generated by the laser pulse. Using the maximum electrostatic field $a_0 = 1.75$ we get $\Delta\phi = 9.53 \times 10^{-4}$ and with $\gamma_p = 5.73$ resulting in $\gamma_p\Delta\phi = 5.46 \times 10^{-3}$ satisfying the condition for the validity of Equation 3. Using these values we get: $\gamma_{max/min} = 6.18/5.3$ or correspondingly, $T_{max/min} = 4.72/3.92\text{GeV}$ This implies that if we inject protons at the minimum energy and can accelerate them to the maximum energy we can get an increase of up to 0.8 GeV. Due to the fact that the actual group velocity of the self-focused laser pulse differs from the one predicted by linear considerations we measured the velocity of the pulse from the simulations. The values are $\beta_{sim} = 0.952$ and $\gamma_{sim} = 3.27$. Using these values we get $T_{sim} = 2.06\text{GeV}$ giving a lower injection velocity. In order to insure that some protons are injected in front of the electrostatic field a proton beam with an initial energy spread was input into the simulation. The beam was uniformly distributed in the y direction and placed in the same initial position as the laser pulse with a length of $29.2\mu\text{m}$ (1000Δ). The proton beam had a very low density so that the protons are only pushed by the electrostatic field generated by the laser pulse. Figure 5 shows the (a) initial and (b) final $\gamma - x$ phase space of the central portion of the injected proton beam at $630\Delta \leq y \leq 650\Delta$ where the center of the simulation is at $y = 640\Delta$. In Figure 5(a) is shown the initial proton beam with a spread in γ between 2.47 and 6.85. After the proton beam has propagated with the laser over $642\mu\text{m}$ ($880c/\omega_p$) the proton beam has spread due to the initial spread in energies (Figure 5(b)). There is a gap between the front portion of the beam and the back part of the beam. This may be due to the electrostatic field generated by the laser pulse. It can be seen in the plot that there are now protons which have

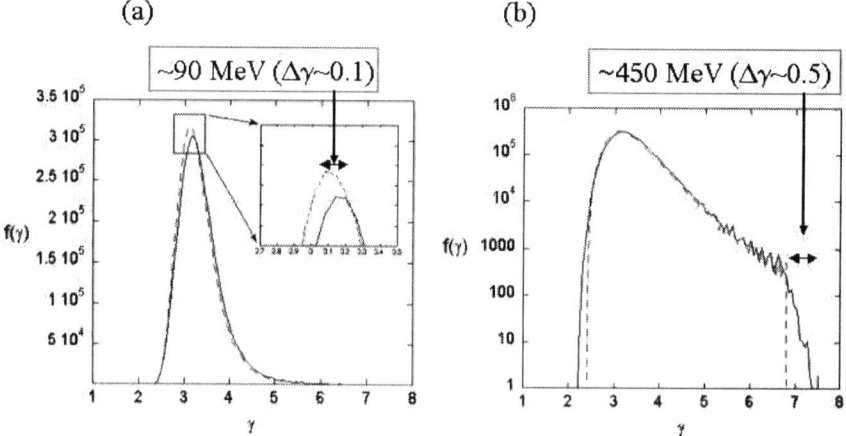

FIGURE 6. Initial (dotted lines) and final (solid lines) distribution $f(\gamma)$ of the injected proton beam. (a) linear scale, close-up of distribution (inset) and (b) log scale.

energies greater than $\gamma = 7$. In Figure 6 is shown the energy distribution of the injected proton beam. In Figure 6(a) one can see that the final distribution (solid line) is shifted from the initial distribution (dotted line). In a close-up of the peak in Figure 6(a)(inset) the shift corresponds to $\Delta\gamma \approx 0.1$ or an equivalent acceleration of 90 MeV. In a log plot of the distribution in Figure 6(b) one can see that there is an increase in the high energy tail of the proton beam. One can see that this increase corresponds to $\Delta\gamma \approx 0.5$ or an equivalent acceleration of 450 MeV. The 90 MeV and 450 MeV increases in energy correspond to average acceleration gradients of 0.14 and 0.7 TeV/m, respectively. These gradients are reasonable in comparison to the maximum electrostatic field gradient of 6.5 TeV/m.

The optimal energy and position of the injected protons still need to be determined. The phase space area occupied by the beam increased in the transverse momentum by about a factor of 2 from the initial spread. For this paper the main point was to show the possibility of acceleration of protons by the large electrostatic field generated at the front of the laser pulse.

There are several factors which affect or limit the acceleration of ions. One is the depletion of the laser pulse. In Figure 4 we showed the maximum electric field as a function of propagation distance. The maximum field was found to decrease in accordance with the depletion of the laser pulse. Thus with the depletion of the laser pulse the acceleration efficiency also drops. Since the depletion is slower at lower densities according to theory [9], it is better to accelerate protons using low density plasmas. At lower densities, however, the initial injection energy of the protons needs to be higher. Another factor affecting the acceleration is the "snaking" instability which causes the laser pulse to deviate from it's original propagation direction [11]. All these factors need to be considered for the optimal acceleration of protons.

CONCLUSION

In this paper we have proposed using the large electrostatic field created in the front of a relativistically self-focused laser pulse to accelerate ions. We have shown from 2 dimensional PIC simulations that an accelerating field of the order of 6.5 TeV/m can be excited by a 100 TW laser pulse propagating in a plasma at a density of 5.3×10^{19} cm^{-3}. The large field is found to be generated by the buildup of electrons at the front of the relativistically self-focused laser pulse. The buildup of electrons can be accounted for by using the continuity equation. Protons injected in the front of such a self-focused pulse can be accelerated in coincidence with the electric field which moves at a velocity nearly equal to the laser group velocity in plasmas. The maximum kinetic energy gain for injected protons with an average energy of 2 GeV is found to be 450 MeV for a laser intensity of the order of 10^{20}cm^{-3} with a pulse duration of ~ 20 fs.. The acceleration will be limited by the depletion of the laser pulse. The next stage of this investigation will be the acceleration of stationary protons using an inhomogeneous plasma with a density gradient. These type of laser-matter interactions will open up a new regime in high energy beam science.

ACKNOWLEDGMENTS

J. K. wishes to thank Y. Ueshima, T. Tajima, B. Bulanov, and Y. Kato for useful discussions. Also, J. K. wishes to thank various COMPAQ computer personnel for aid in running the simulations.

REFERENCES

1. Cowan, T. E., Roth, M., Johnson, J., Brown, C., Christl, M., Fountain, W., Hatchett, S., Henry, E. A., Hunt, A. W., Key, M. H., MacKinnon, A., Parnell, T., Pennington, D. M., Perry, M. D., Phillips, T. W., Sangster, T. C., Singh, M., Snavely, R., Stoyer, M., Takahashi, Y., Wilks, S. C., and Yasuike, K., *Nucl. Instr. Meth. Phys. Res. A*, **445**, 130–139 (2000).
2. Ueshima, Y., Sentoku, Y., and Kishimoto, Y., *Nuclear Instruments and Methods in Physics Research A*, **455**, 181–184 (2000).
3. Yamakawa, K., Aoyama, M., Matsuoka, S., Kase, T., Akahane, Y., and Takuma, H., *Opt. Lett.*, **23**, 1468–1470 (1998).
4. Sun, G. Z., Ott, E., Lee, Y. C., and Guzdar, P., *Phys. Fluids*, **30**, 526–532 (1987).
5. Tajima, T., and Dawson, J. M., *Phys. Rev. Lett.*, **43**, 267–270 (1979).
6. Akhiezer, A. I., and Polovin, R. V., *Sov. Phys. JETP*, **3**, 696–705 (1956).
7. Ueshima, Y., Kishimoto, Y., Sasaki, A., and Tajima, T., *Laser and Particle Beams*, **17**, 45–58 (1999).
8. Hartemann, F. V., Fochs, G. P., LeSage, G. P., Luhmann, N. C. J., Woodworth, J. G., Perry, M. D., Chen, Y. J., and Kerman, A. K., *Physical Review E*, **51**, 4833–4843 (1995).
9. Bulanov, S. V., Kirsanov, V. I., Naumova, N. M., Sakharov, A. S., and Shah, H. A., *Physica Scripta*, **47**, 209–213 (1993).
10. Esarey, E., and Pilloff, M., *Physics of Plasmas*, **2**, 1432–1436 (1995).
11. Naumova, N. M., Koga, J., Nakajima, K., Tajima, T., Esirkepov, T. Z., Bulanov, S. V., and Pegoraro, F., *Physics of Plasmas*, **8**, 4149–4155 (2001).

Electronic and Nuclear Wave Packet Dynamics of Molecular Systems in Intense Laser Fields

Hirohiko Kono, Yukio Sato, and Yuichi Fujimura

Department of Chemistry, Graduate School of Science, Tohoku University,
Sendai 980-8578, Japan   E-mail: kono@mcl.chem.tohoku.ac.jp

Abstract. Intramolecular electronic dynamics and tunnel ionization of H_2 in an intense laser field ($I \approx 10^{14} \, Wcm^{-2}$ and $\lambda = 760$ nm) were examined with accurate evaluation of three-dimensional two-electron wave packet dynamics. Based on the results of population analysis of "field-following" adiabatic states, an electrostatic model in which each atom in a molecule is charged by laser-induced electron transfer and ionization proceeds via the most unstable atomic site is proposed. This simple electrostatic approach is applied to investigate the structural deformations of CO_2 cations in an intense field.

INTRODUCTION

Interaction of atomic or molecular systems with high-power laser fields results in nonperturbative electronic dynamics such as above-threshold ionization and tunnel ionization [1]. In a high-intensity and electronically nonresonant long-wavelength regime (intensity $I > 10^{13} W/cm^2$ and wavelength $\lambda > 700nm$), a laser electric field significantly distorts the Coulombic potential that the electrons are placed in. The distorted potential forms a "quasistatic" barrier through which an electron or electrons can tunnel. In the case of molecules, a large part of the electron density is transferred among nuclei within a half optical cycle [2]. Consequently, tunnel ionization proceeds through the most unstable atomic site [3]. Laser-induced intramolecular electron transfer, which is governed by electron-electron repulsion and molecular structure (e.g., bond length), also induces structure deformation of molecules [2]. Resultant structure deformations in turn change the electronic response to the field, e.g., intramolecular electron transfer followed by tunnel ionization [4]. It is known that tunnel ionization is greatly enhanced at a specific internuclear distance that is much longer than the equilibrium internuclear distance R_e (known as enhanced ionization) [5].

To quantitatively understand the ultrafast electronic dynamics in intense fields, it is necessary to solve the time-dependent Schrödinger equation for the electronic degrees of freedom of a molecule. We have developed an efficient grid point method, the dual transformation method [6], for accurate estimation of propagation of an electronic wave packet. In this method, both the wave function and the Hamiltonian are transformed consistently to overcome the numerical difficulties arising from the divergence of the Coulomb potentials. We have applied this method to small molecular systems such as H_2^+ and H_2 [3,4]. To the best of our knowledge, our treatment of H_2 is the first

CP634, *Science of Superstrong Field Interactions*, edited by K. Nakajima and M. Deguchi
© 2002 American Institute of Physics 0-7354-0089-X/02/$19.00

accurate evaluation of two-electron dynamics of a molecule. The vibrational degree of freedom is also incorporated in the calculation of H_2^+ without resorting to the Born-Oppenheimer approximation [3].

In this paper, we present the results of a theoretical investigation of intense-field-induced phenomena of H_2. In a low frequency case ($\lambda > 700$ nm), the localized ionic (bond) states H^+H^- and H^-H^+ appear alternately according to the laser cycle, through which tunnel ionization proceeds. The electron pair is always created around the proton at which the dipole interaction energy with the field becomes lower (descending well). It is expected that the clarified mechanism of enhanced ionization for H_2 would serve as a prototype of tunnel ionization of multi-electron molecules. We have also investigated the electronic dynamics of H_2 in a high-frequency case in which the laser frequency ω is larger than the energy difference ω_{X-B} between the ground state $X^1\Sigma_g^+$ and first excited state $B^1\Sigma_u^+$ ($\lambda_{X-B} \sim 100$ nm). For $\omega > \omega_{X-B}$, the electron pair is formed in the ascending well. This indicates that it is possible to control the ionization probability in various ways.

Levis *et al.* recently reported selective bond dissociation and rearrangement of polyatomic molecules with optically tailored, intense-field laser pulses [7]. Intense laser chemistry has now opened up. The characteristic features of electronic dynamics of H_2 in an intense laser field leads to a simple electrostatic concept that the dynamics of bound electrons and the subsequent ionization process of a multi-electron polyatomic molecule can be clarified in terms of "field-following" adiabatic states. The properties of adiabatic states of multi-electron molecules can be determined by *ab initio* molecular orbital (MO) methods. The calculated time-dependent potentials are used to evaluate the nuclear dynamics (such as bond stretching) until the next ionization process. We apply this approach to a CO_2 molecule in an intense field and show that the two C-O bonds can be symmetrically stretched in the CO_2^{2+} stage.

ELECTRONIC DYNAMICS OF H_2

In this section, we present the results of an investigation of the two-electron wave packet dynamics of H_2 in intense fields. The position of the jth electron is designated by cylindrical coordinates (ρ_j, z_j and φ_j). The z_1 and z_2 axes are parallel to the molecular axis. The internuclear distance R is fixed. We assume that the molecule is aligned parallel to the field polarization direction [8]. Electron transfer in H_2 is then characterized by motion along the molecular axis. To represent the wave packet $\psi(t)$, we therefore employ the reduced density $\bar{P}(z_1,z_2)$ obtained by integrating $|\psi(t)|^2$ over the degrees of freedom other than z_1 and z_2. As an example, $\bar{P}(z_1,z_2)$ of the state X at $R=4$ a.u. is drawn in Fig. 1(a). The reduced density clearly demonstrates that the covalent components around $z_1 = -z_2 = \pm R/2$ are dominant at $R=4$ a.u. The localized ionic components $|H^+H^-\rangle$ and $|H^-H^+\rangle$ contained in $\psi(t=0)$ are both 19%. Here, $|H^+H^-\rangle$ and $|H^-H^+\rangle$ are defined as H^- ions located at $z_1 = z_2 = \pm R/2$.

We first discuss a case in which the laser frequency ω is much smaller than ω_{X-B}. The laser electric field $\mathcal{E}(t)$ that the H_2 interacts with is assumed to be $f(t)\sin(\omega t)$, where $f(t)$ is the pulse envelope at time t. The parameters are as follows: $\omega = 0.06$ a.u. ($\lambda = 760$ nm); $f(t)$ is linearly ramped so that $f(t)$ attains its maximum $f_0 = 0.12$ a.u. after one cycle. The corresponding intensity I is given by $3.5 \times 10^{16} f^2(t)\,\mathrm{Wcm^{-2}}$. The ionic component around the descending potential well, where $z\mathcal{E}(t) < 0$, increases as the field approaches the first local maximum at $t = \pi/2\omega = 26.2$ a.u. $= 0.634$ fs. At $t = \pi/2\omega$, $\mathcal{E}(t) = 0.03$ a.u. and $|\langle\psi|H^+H^-\rangle|^2 = 0.31$. See Fig. 1(b). The laser field forces the two electrons to stay near a nucleus for almost a half cycle. The packet at $t = \pi/\omega$ is nearly identical to the initial one shown in Fig. 1(a), indicating that the electronic response to the field is basically adiabatic. Until then, no ionization current is observed. A quarter cycle later, as shown in Fig. 1(c), the density around $z_1 = z_2 = R/2$ becomes as large as $|\langle\psi|H^+H^-\rangle|^2 = 0.54$ because of the stronger field $\mathcal{E}(t = 3\pi/2\omega) = -0.09$ a.u. The ionic character increases as $|\mathcal{E}(t)|$ increases.

As indicated by the broken line in Fig. 1(c), an electron is ejected from the localized ionic structure H^+H^-. If the two nuclei are far from each other, the ionization potential of the localized ionic structure is considered to be as low as $I_p(H^-) = 0.75$ eV. The

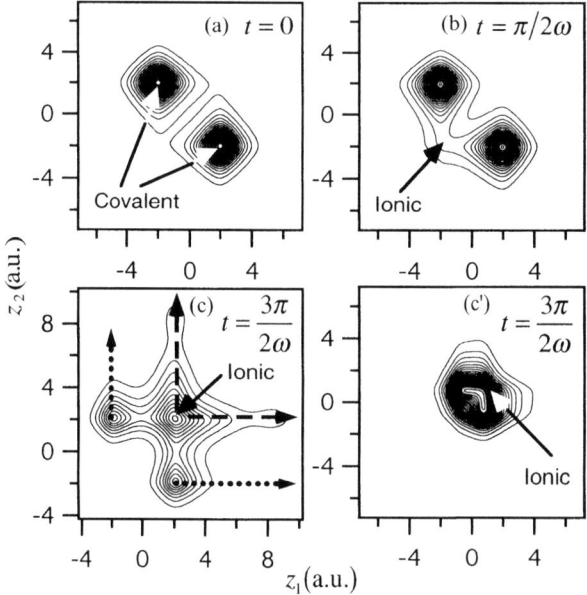

Figure 1. Snapshots of the electronic wave packet dynamics of H_2 in an intense field at (a) $t = 0$, (b) $\pi/2\omega$, and (c) and (c') $3\pi/2\omega$ ($\omega = 0.06$ a.u.). The coordinates z_1 and z_2 for the two electrons are parallel to the molecular axis. The reduced density $\overline{P}(z_1, z_2)$ is represented by a contour map. The instantaneous field strengths are 0.03 a.u. at $t = \pi/2\omega$ and -0.09 a.u. at $t = 3\pi/2\omega$. The snapshots (a)-(c) are those obtained for $R = 4$ a.u. As shown in (a), the covalent components around $z_1 = -z_2 = \pm R/2$ are dominant in the initial ground state at $R = 4$ a.u. As shown in (b) and (c), an ionic component H^-H^+ or H^+H^- is created near the descending well. As indicated by the broken line in (c), an electron is ejected from the structure H^+H^-. The reduced density at $t = 3\pi/2\omega$ for $R = 1.6$ a.u. is drawn in (c').

60

localized ionic structure is hence unstable and is regarded as a doorway state to ionization. The ionization current from the covalent structure, denoted by a dotted line in Fig. 1(c), is relatively small. At $R = 4$ a.u., the rate of ionization from a *pure* localized ionic state is at least five-times greater than that from a *pure* covalent state. We have calculated the wave packet dynamics at different values of R for the same pulse. The reduced density at $t = 3\pi/2\omega$ for $R = 1.6$ a.u. ($\approx R_c$) is shown in Fig. 1(c'). Despite the intense field strength at $t = 3\pi/2\omega$, the ionization current is very small. The change in R from 4 a.u. to 1.6 a.u. reduces the ionization rate by about two orders of magnitude. As R decreases, the rate of ionization from H^-H^+ decreases owing to the more attractive force of the distant nucleus, while the rate of a covalent state is almost independent of R. On the other hand, the population of the H^-H^+ created increases as R decreases. For instance, for $R = 1.6$ a.u., $|\langle\psi|H^+H^-\rangle|^2$ is as large as 0.74 at $t = 3\pi/2\omega$ (at $t = 0$, $|\langle\psi|H^+H^-\rangle|^2 = 0.58$.). As a result, ionization is enhanced at the critical distance $R_c = 4$-6 a.u. The explanation that the created H^-H^+ is the main doorway state to ionization is reinforced by the fact that ionization is suppressed for triplet states (H^-H^+ is not created owing to the Pauli exclusion principle).

We have examined the intramolecular electronic dynamics that governs the ionization process by analyzing the populations of field-following adiabatic states defined as eigenfunctions of the instantaneous electronic Hamiltonian $H_{elec}(t)$ including the interaction with $\mathcal{E}(t)$ [9]. The effective $H_{elec}(t)$ for H_2 is constructed from the three lowest electronic states, X, B, and EF. One of the three resultant adiabatic states is the lowering ionic adiabatic state characterized by H^-H^+ or H^+H^-, which decreases in energy by $-|\mathcal{E}(t)|R$; on the other hand, the energy of the covalent adiabatic state $H \cdot H$ is relatively insensitive to $|\mathcal{E}(t)|$. Therefore, the lowering ionic state and the covalent character-dominated initial state can cross each other in energy. By solving the time-dependent Schrödinger equation for the effective $H_{elec}(t)$, as shown below, we have found that the difference in ionization dynamics between small R and the R cases originates in the character of the level crossing of the lowest two adiabatic states.

The initial ground state is adiabatically connected with the lowering ionic state after the crossing. As $|\mathcal{E}(t)|$ increases, the lowest adiabatic state starting from the X state becomes more ionic. In the case of $R \leq 4$ a.u., the energy gap at the avoided crossing between the two lowest adiabatic states is as large as that at zero field strengths because the transition dipole moment of the B-X transition, which enlarges the gap, increases as $\approx R/\sqrt{2}$ up to $R \approx 3$ a.u. As a result, nonadiabtic transitions to upper adiabatic states hardly occur. Ionization occurs from the lowest adiabatic state characterized by H^-H^+ or H^+H^-. Around $R = 6$ a.u., the ionization from the covalent state competes with that from the localized ionic state. As R increases further, the electron density transferred between the nuclei is suppressed: the main character becomes covalent. The ionization at large R (≥ 8 a.u.) mainly proceeds from the remaining covalent component, which outmeasures the created ionic component.

We have also investigated the electronic dynamics of H_2 in a high-frequency case in which ω is close to ω_{X-B}. Even after the incident pulse has fully decayed, H^+H^- and H^-H^+ appear alternately with a period of $2\pi/\omega_{X-B}$ (~500 attoseconds). Contrary to the long wavelength case, for $\omega > \omega_{X-B}$, the electron pair is formed in the ascending well. We have found that ionization is enhanced or suppressed according to whether ω is just below ω_{X-B} (case I) or just above ω_{X-B} (case II). In case I, the first step is the formation of an electron pair in the descending well. The subsequent ionization process is regarded as a hybrid of tunnel ionization and multiphoton ionization. In case II, the electron pair is formed in the ascending well from which ionization is greatly suppressed. This is in sharp contrast to the low-frequency case in which the ascending well is most unstable to ionization. We have numerically demonstrated that two-electron dynamics and ionization can be controlled by adjusting the time lag between two phase-locked ultrashort pulses.

INTENSE FIELD CHEMISTRY OF CO_2

The characteristic features of electronic dynamics of H_2 in an intense laser field leads to a simple electrostatic view that each atom in a molecule is charged by field-induced electron transfer and that ionization proceeds via the most unstable atomic site. In the low-frequency case ($\lambda > 700$ nm), the dynamics of bound electrons can be described in terms of field-following adiabatic states. The properties of adiabatic states can be determined by *ab initio* MO methods. While the time-dependent adiabatic potentials calculated by an MO method are used to evaluate the nuclear dynamics until the next ionization process, the charge distributions on individual atomic sites are used to estimate the ionization probability. For $\omega_{elec} > \omega$, where ω_{elec} represents the characteristic frequency for the electronic transitions, the vibrational motion is mainly determined by the time-dependent potential of the lowest adiabatic state. Here, we ignore field-induced potential crossing at large internuclear distances.

Ultrafast field-induced deformation of molecular structure has been observed for various molecules such as CO_2. Hishikawa *et al.* [10] have experimentally determined the geometry of CO_2^{3+} in a 1.1 PWcm^{-2}, 100 fs pulse ($\lambda = 795$ nm) just before undergoing Coulomb explosions, namely, that the C-O bond length R_{C-O} is stretched to about 3.0 a.u. from the equilibrium value of CO_2 ($R_e = 2.3$ a.u.), and the amplitude of bending is much larger in comparison with CO_2 (the observed mean amplitude of the deviation angle θ from the linear structure being ~40°). We apply the above approach to investigate the deformation stage of CO_2. The adiabatic potential surfaces and charge distributions of CO_2 and its cations are calculated by using the full-optimized reaction space multiconfiguration self-consistent-field method with the 6-311+G(d) basis set [2]. The total charge is assigned to each atom by Mulliken population analysis.

A linearly polarized laser field can align molecules [8]. To achieve high alignment in the polarization direction by adiabatic following of the molecular rotation to the pulse envelope, the pulse duration must be longer than $2\pi/2B \approx 40$ ps, where $B(\approx 0.39$ cm^{-1}) is the rotational constant of CO_2. Since the pulse duration used in the experiment (≈ 100 fs) is shorter than $2\pi/2B$, the alignment is incomplete. The ionization rate has, however, a maximum at the geometry where the molecular O-O axis is parallel to the polarization direction. This parallel case should be studied as the main spatial configuration. When a molecules is perpendicular to the polarization direction, the electric field does not stretch the bonds, while it may cause bending.

We now apply the above-described electrostatic model to the CO_2 case. Consider a linear molecule $O^{P+}C^{Q+}O^{Z+}$ placed parallel to $\mathcal{E}(t)$. Numerical calculations indicate that the charge on C does not change in an intense field. Thus, the charge transfer ionic state $O^{(P-1)+}C^{Q+}O^{(Z+1)+}$ is expected to be a doorway state to tunnel ionization. The field strength $\mathcal{E}_c$ required for its creation is then given by [3]

$$\mathcal{E}_c = \left[I_p(O^{Z+}) - I_p(O^{(P-1)+}) - (Z+1-P)/R_{O-O}\right]/R_{O-O}, \qquad (1)$$

where R_{O-O} is the $O-O$ bond length. An appreciable amount of charge is transferred between the O atoms when $|\mathcal{E}(t)|$ exceeds the value $\mathcal{E}_c$. The estimated value of $\mathcal{E}_c$ serves to predict the intensity required for tunnel ionization.

(i) Neutral CO_2 The charges on the three atoms in the *lowest adiabatic state* of linear CO_2 are calculated as a function of the field strength. At $R_{C-O} = R_e$, the charges of O and C at zero fields are -0.22 and +0.45, respectively. The charge distribution of the main electronic configuration is expressed as $O^{0+}C^{0+}O^{0+}$. When a field is applied, an appreciable amount of negative charge is transferred from the O atom in the ascending well to the O atom in the descending well. From Eq. (1), $\mathcal{E}_c = 0.05$ a.u. at $R_{C-O} = R_e$. The calculated charge of the O atom in the descending well is -0.83 at $\mathcal{E}(t) = 0.1$ a.u. $>$ $\mathcal{E}_c$. The ionic component O^-CO^+ becomes dominant beyond $\mathcal{E}_c$. We thus expect for CO_2 that tunnel ionization via the ionic structure occurs somewhere around $\mathcal{E}_c$. Experimentally, CO_2^+ appears at an intensity around 3.5×10^{13} Wcm^{-2} ($\mathcal{E} \approx 0.03$ a.u.).

We have calculated the potential surface of the lowest adiabatic state of CO_2 as a function of the two C-O bond distances R_1 and R_2 and the bond angle θ. There are two types of bond stretching: symmetric stretching in which $R_1 = R_2$ and asymmetric stretching in which one C-O bond is longer than the other (e.g., $R_1 > R_2 = R_e$). The adiabatic potential of CO_2 is greatly distorted in an intense field; the dissociation energy for a C-O bond is 7 eV at zero fields and is reduced to 4 eV at $\mathcal{E}(t) = 0.1$ a.u. ($>$ $\mathcal{E}_c$). However, bond stretching hardly occurs for the following reason.

The potential of the lowest adiabatic state, $V_0(R_1, R_2, t)$, can be approximated as

$$V_0(R_1, R_2, t) \approx V_0(R_1, R_2) - \mu(R_1 - R_2)\mathcal{E}(t) - \alpha(R_1 + R_2)\mathcal{E}^2(t)/2, \qquad (2)$$

where $V_0(R_1, R_2)$ is the adiabatic potential at zero fields. Because of the second term, at a moment, the barrier for asymmetric dissociation is greatly reduced as mentioned above. It should, however, be noted that the following inequality in temporal or energy scale holds:

$$\omega_{elec} > \omega > \omega_{vib} > \dot{f}(t)/f(t) \sim 1/T_{pulse}, \qquad (3)$$

where ω_{vib} is the vibrational frequency. The experimental condition $\omega > \omega_{vib}$ means that the change in $\mathcal{E}(t)$ is too fast for the vibrational motion to follow $\mathcal{E}(t)$ adiabatically. Since $\omega > \omega_{vib}$ and $f(t)$ does not change in one optical cycle, i.e., $\omega > \dot{f}(t)/f(t)$, $V_0(R_1, R_2, t)$ can be replaced with the cycle-averaged effective potential $\overline{V}_0(R_1, R_2, t)$:

$$\overline{V}_0(R_1, R_2, t) \approx V_0(R_1, R_2) - \alpha(R_1 + R_2)f^2(t)/4. \qquad (4)$$

Thus, the second term in Eq. (2) disappears by cycle averaging. Since $\omega_{vib} >$

$\dot{f}(t)/f(t)$, the vth vibrational state at zero fields is adiabatically transferred to the vth state of $\overline{V}_0(R_1,R_2,t)$. Up to $f(t) \approx 0.15$ a.u. ($\gg \mathcal{E}_c$), the equilibrium geometry of $\overline{V}_0(R_1,R_2,t)$ is almost equal to that of $V_0(R_1,R_2)$. We thus conclude that the neutral CO_2 in an intense field takes a stable linear structure around the zero-field equilibrium geometry.

In conclusion, in the CO_2 and CO_2^+ stages, ionization occurs before the field intensity becomes high enough to deform the molecule.

(ii) CO_2^{2+} The geometry of CO_2^{2+} just after ionization of CO_2^+ is expected to be nearly equal to that of CO_2. Two positive charges in CO_2^{2+} are nearly equally distributed among the three atoms as OC^+O^+, O^+C^+O, and O^+CO^+. An ionic structure favorable for tunnel ionization is $O^-C^+O^{2+}$ created from OC^+O^+. From Eq.(1), $\mathcal{E}_c = 0.18$ a.u. at $R_1 = R_2 = R_e$. Before the field intensity reaches the large threshold value of 0.18 a.u., structural deformation of CO_2^{2+} can occur before ionization.

We examined the nuclear wave packet dynamics on $V_0(R_1,R_2,t)$ of CO_2^{2+}. A realistic pulse shape is employed: $f(t) = f_0 \sin^2(\pi t / T_{pulse})$ for $0 \le t \le T_{pulse}$, where $f_0 = 0.19$ a.u. and $T_{pulse} = 194$ fs. The frequency ω was set at 0.06 a.u. We simulated a case in which the initial wave packet is created on the lowest adiabatic state of CO_2^{2+} when the field envelope $f(t)$ reaches ~0.03 a.u. ($t = 24$ fs.); vertical transition from the ground vibrational state of CO_2 was assumed. In Fig.2, the position of the initial wave packet prepared at $t = 24$ fs is marked with $\times$, and the instantaneous potential surface at $\mathcal{E}(t = 84.7\,\mathrm{fs}) = 0.18$ a.u. is shown. The dissociation energy at zero fields is ~10.7 eV in the case of symmetric stretching and ca. 1.0 eV in the case of asymmetric stretching.

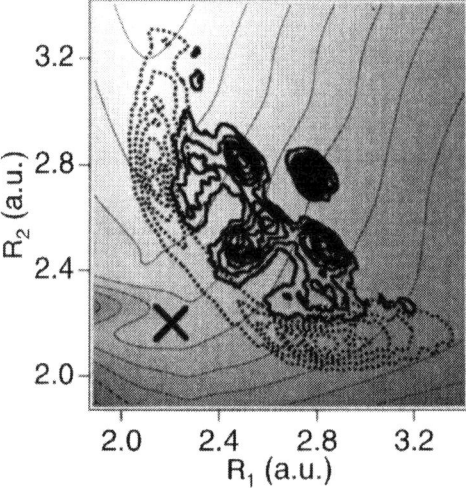

Figure 2. Wave packet dynamics of CO_2^{2+} in an intense laser field. R_1 and R_2 are the two C-O bond lengths. The position of the initial CO_2^{2+} wave packet prepared at $t = 24$ fs is marked with $\times$. The contour plot of the instantaneous potential surface at $t = 84.7\,\mathrm{fs}$ is denoted by thin solid lines (at intervals of 1.4 eV). The contour plot of the wave packet at $t = 84.7$ fs is denoted by bold solid lines and that at $t = 48$ fs is denoted by bold dotted lines.

From the viewpoint of dissociation energy, it is expected that asymmetric streching is predominant. However, the wave packet at $t = 84.7$ fs denoted by bold solid lines shows that symmetric stretching occurs as well as asymmetric stretching. The appearance of simultaneous two-bond stretching can be explained by using the cycle-averaged potential $\overline{V}_0(R_1, R_2, t)$ of CO_2^{2+}. In CO_2^{2+}, the stabilization energy $-\alpha(R_1 + R_2)f^2(t)/4$ due to the induced dipole moment is large in comparison with $V_0(R_1, R_2)$. Consequently, a deep valley along the symmetric stretching coordinate is formed in $\overline{V}_0(R_1, R_2, t)$: the symmetric dissociation becomes barrierless. After bond stretching, ionization to CO_2^{3+} occurs owing to enhanced ionization at larger values of R_{C-O}.

We also propose that the experimentally observed two-body breakup to $CO^+ + O^+$ is due to asymmetric stretching on the adiabatic potential of CO_2^{2+}. The asymmetric stretching originates from the difference in the equilibrium internuclear distance between CO_2^{2+} and CO_2 (or CO_2^+). A part of the wave packet propagates toward the two-body breakup from the Franck-Condon region, as shown by the wave packet at $t = 48$ fs denoted by dotted contour lines in Fig. 2. Although the instantaneous potential is dissociative for the asymmetric channels, asymmetric streching is not enhanced by a field because of the disappearance of $-\mu(R_1 - R_2)\mathcal{E}(t)$ in $\overline{V}_0(R_1, R_2, t)$. As the field intensity is increased, the field-induced symmetric stretching becomes dominant.

The induced dipole of the lowest adiabatic state that shifts the energy downward becomes smaller as the molecule becomes more bent; the curvature of the potential with θ is larger at nonzero fields than at zero fields. Therefore, a large-amplitude bending motion is expected to be hardly induced. However, the curvature of the potential with θ is smaller in CO_2^{2+} than in CO_2. Hence, the wave packet prepared on the lowest adiabatic state of CO_2^{2+} from CO_2 spreads out along θ. The experimentally observed structure of CO_2^{3+} just before Coulomb explosions originates from the field-induced bond stretching of CO_2^{2+} accompanied by a large-amplitude bending motion. The proposed approach is simple but has wide applicability in predicting the *electronic* and *nuclear* dynamics of polyatomic molecules in intense laser fields.

REFERENCES

1. Eberly, J. H., Japanainen, J. , and Rzazewski, K., *Phys. Rep.* **204**, 331- 383(1991); Brabec, T., and Krausz, F., *Rev. Mod. Phys.* **72**, 545-591 (2000).

2. Kono, H., Koseki, S., Shiota, M., and Fujimura, Y., *J. Phys.Chem. A* **105**, 5627-5636 (2001).

3. Harumiya, K., Kawata, I., Kono, H., and Fujimura, Y., *J. Chem. Phys.* **113**, 8953-8960 (2000); *Phys. Rev. A* to be published.

4. Kawata, I., Kono, H., and Fujimura, Y., *J. Chem. Phys.* **110**, 11152-11165 (1999).

5. Bandrauk, A.D., *Comments At. Mol. Phys.* **D1**, 97-115 (1999).

6. Kawata, I., and Kono, H., *J. Chem. Phys.* **111**, 9498-9508 (1999).

7. Levis, R.J., Menkir, G.M., Rabitz, H., *Science* **292**, 709-713(2001).

8. Friedrich, B., and Herschbach, D., *Phys. Rev. Lett.* **74**, 4623-4626 (1995).

9. Kawata, I., Kono, H., and Fujimura, Y., and Bandrauk, A. D., *Phys.Rev.* A**62**, R031401 (2000).

10. A. Hishikawa, A. Iwamae, and K. Yamanouchi, *Phys. Rev. Lett.* **83**, 1127-1130 (1999).

Exotic Behavior of Molecules and Clusters in Intense Laser Light Fields

Kaoru Yamanouchi

Department of Chemistry, School of Science, The University of Tokyo
7-3-1 Hongo, Bunkyo-ku, Tokyo 113-0033, Japan
e-mail: kaoru@chem.s.u-tokyo.ac.jp

Abstract. In order to investigate dynamics of molecules and clusters in intense laser fields, new experimental approaches such as mass-resolved momentum imaging, tandem-type time-of-flight spectrometry, pulsed gas electron diffraction, and coincidence momentum imaging have been developed. From a series of studies using these newly developed techniques, it has been found that mixing among the electronic structures of a molecule induced by intense laser fields plays a crucial role in determining the fate of molecules after ionization.

INTRODUCTION

From recent pioneering studies, it has been revealed that molecules behave in a very characteristic way in intense laser fields $(10^{11} \sim 10^{16}$ W/cm$^2)$ [1], i.e. they significant structural deformation and ionization proceeds almost simultaneously, and the resulting multiply charged molecular ions decompose into atomic fragment ions having a large momentum. The process of this abrupt bond fission generating fragment ions with large kinetic energies is called Coulomb explosion. We focused our attention on the observation that such fragment ions are preferentially ejected along the laser polarization direction, and proposed a new method called mass-resolved momentum imaging (MRMI) [2-4] as a method to evaluate quantitatively the extent of the anisotropy of the fragment ejection. For example, the observed MRMI map for N^{2+} ions ejected from N_2^{z+} $(z = 3 - 5)$ is shown in Fig. 1, from which the existence of the three major explosion pathways contributing to the formation of N_2^+ was clarified.

By the MRMI method, the responses of small polyatomic molecules such as N_2, NO, CO_2, NO_2 and H_2O to intense laser fields were investigated systematically [5-7], and it was found that extensive deformation of their skeletal structure commonly occurs. This characteristic structural deformation of molecules in intense laser fields was interpreted as a result of the formation of so-called dressed states (Fig. 2) in which molecular electronic states are mixed through strong couplings with intense laser fields. The investigation of this type of mixed states of molecules and light fields is considered to be a first step for the realization of control of molecular dynamics in intense laser fields. Because the shape of the potential energy surfaces of the dressed states is varied in response to the temporal change of the magnitudes of the light fields,

CP634, *Science of Superstrong Field Interactions*, edited by K. Nakajima and M. Deguchi
© 2002 American Institute of Physics 0-7354-0089-X/02/$19.00

the nuclear dynamics on the surfaces drawn schematically as an entangled arrow should be governed by the intensity of the light field as well as its wavelength.

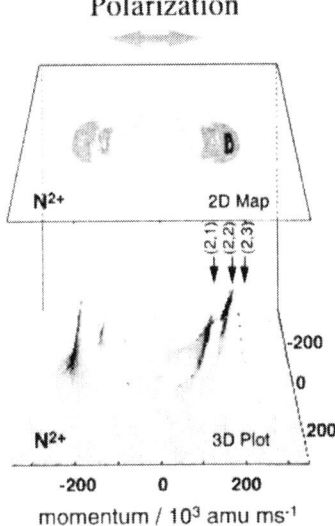

FIGURE 1. MRMI 2D map and 3D plot for N^{2+} ions ejected from N_2^{z+} ($z = 3 - 5$). An arrow indicates the direction of the laser polarization.

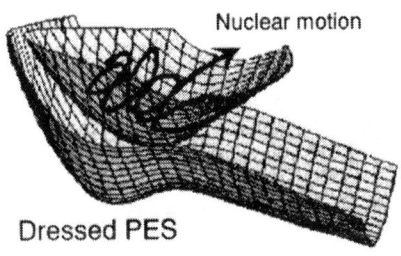

FIGURE 2. Dressed potential energy surface formed by intense laser fields.

TANDEM-TYPE TOF MEASUREMENTS

Only through the measurements of MRMI maps, it is still uncertain whether the dressed state formation, determining the fate of the molecules in intense laser fields, occurs either at a neutral stage, a single-charged stage, or a higher-charged stage. In intense laser fields, in most cases, electrons within a molecule escape from the molecular domain one after another through tunneling ionization, resulting in the formation of multiply charged parent ions. Therefore, it is worthwhile to identify the ion stage at which the major light-induced coupling between molecular electronic states occurs. For this purpose, we developed a tandem-type time-of-flight (TOF) mass spectrometer (Fig. 3) [8,9], and made it possible to identify the fate-determining charge stage by irradiating ions having specific charge and mass numbers with intense short-pulsed laser fields. In the cases of benzene [8] and aniline [9] molecules, it was found that the dressed-state formation occurring at the singly charged ion stage governs their dynamics in intense laser fields. Furthermore, it was clarified that molecular ions are excited instantaneously to the vibrationally highly excited states through the resonant coupling between a pair of electronic states caused by the short-pulsed intense laser fields. From the investigation of aniline-$(NH_3)_n$ ($n = 1$-4) clusters using the tandem-type TOF spectrometer, the responses of the clusters to the light fields were found to be largely dependent on the size of the clusters, i.e. the number of the attached ammonia molecules.

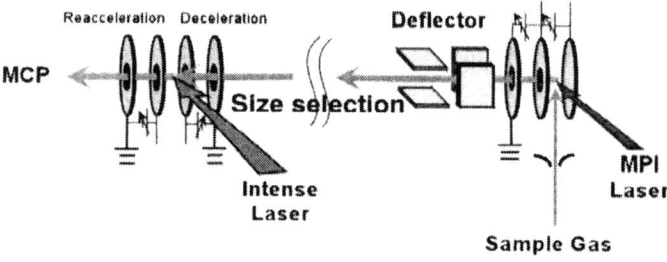

FIGURE 3. Schematic diagram of the tandem-type time-of-flight mass spectrometer.

PULSED GAS ELECTRON DIFFRACTION

It would be wonderful if we could pursue in real time the variation of the geometrical structure of molecules interacting with intense laser fields as well as their alignment process along the laser polarization direction [10]. In order to achieve such a real-time probing, we developed a pulsed gas diffraction apparatus (Fig. 4) [11,12] and attempted to probe molecular responses to intense laser fields directly through an electron diffraction pattern. The short-pulsed electron beam packets were generated by irradiating a photo-cathode of a pulsed electron gun with a short-pulsed laser light and by accelerating the electrons generated through a photoelectric effect to 25 - 40 keV. This pulsed gas electron diffraction method enabled us to record an alignment process of linear polyatomic molecules as a form of electron diffraction images. When ultrashort laser pulses with sub-pico second temporal width are employed for the generation of the electron packets, real time probing of chemical dynamics with ~ 1 ps resolution could be achieved as a series of snap shots of electron diffraction images.

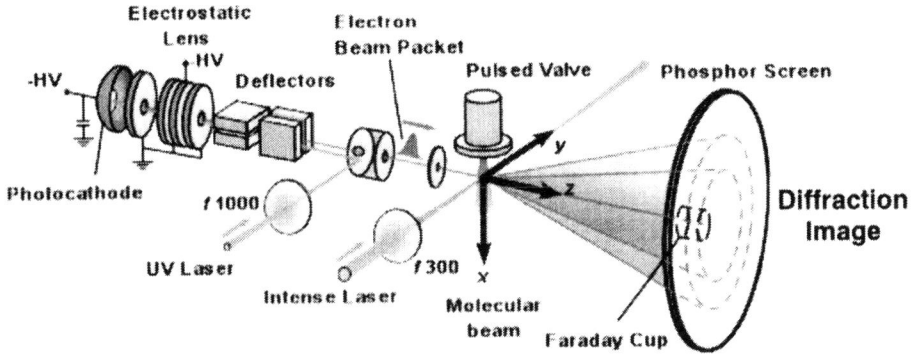

FIGURE 4. Schematic diagram of the pulsed gas electron diffraction apparatus.

COINCIDENCE MOMENTUM IMAGING

We also developed a method called coincidence momentum imaging (CMI) to identify respective events of Coulomb explosion of a single molecule in intense laser fields [13,14]. By using an apparatus shown in Fig.5, which is equipped with a position sensitive detector, we were able to measure momentum vectors of all the fragment ions for respective events of two-body and three-body Coulomb explosion by the double and triple coincidence measurements, respectively. From the CMI measurements, correlations among the nuclear motions within a molecule before the Coulomb explosion were clarified, and furthermore, in the case of the three-body explosion, it became possible to identify separately two types of three-body explosion, i.e. (i) a concerted type in which three atomic fragment ions are ejected almost simultaneously and (ii) a sequential type in which the two-body fragmentation into a diatomic molecular ion and an atomic ion proceeds first followed by the two-body fragmentation of the diatomic molecular ion.

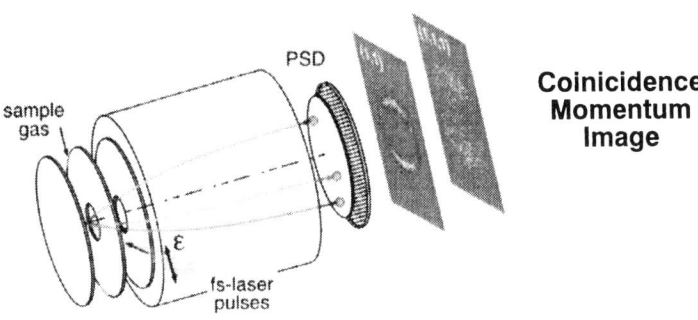

FIGURE 5. Schematic diagram of the coincidence momentum imaging apparatus.

PERSPECTIVE

The recent development of ultrashort pulsed laser technology has afforded us an invaluable opportunity to investigate essential feature of light-matter interaction through phenomena occurring in extremely intense light fields. As described in the present article, when we introduced a new method to investigate molecules and clusters in intense laser fields, we were always encountered by new phenomena, from which we were able to get deeper understanding about the roles of light fields. The sensitive dependence of molecular dynamics on the laser-field intensity as well as on the wavelength of laser light suggests that photochemical processes could be controlled by manipulating properly the properties of short-pulsed laser fields, such as intensity, wavelength, pulse shape, and phase structure. It is certain that interdisciplinary cooperative efforts by researchers in the fields of chemistry, physics, and laser engineering will stimulate further the research on strong-field matter interaction at the forefront.

ACKNOWLEDGMENTS

The results of our studies introduced in the present article were obtained mainly through the CREST project supported by JST (Japan Science and Technology Corporation). I would like to thank my colleagues, Dr. K. Someda, Dr. A. Hishikawa, Dr. K. Hoshina, Dr. A. Iwamae, Dr. M. Kono, Dr. S. Liu, Dr. T. Sako, Dr. Y. Fukuda, Dr. R. Itakura, and Dr. A. Iwasaki, Dr. H. Hasegawa, Mr. I. Maruyama, Ms. J. Watanabe for their efforts in making the project fruitful. Finally, I thank Dr. H. Todokoro, Mr. T. Ohshima, and Mr. Y. Oze for their support in the construction of the gas electron diffraction apparatus.

REFERENCES

1. Yamanouchi, K., Science, **295**, 1659 (2002).
2. Hishikawa, A., Iwamae, A., Hoshina, K., Kono, M., Yamanouchi, K., *Chem. Phys. Lett.* **282**, 283 (1998).
3. Hishikawa, A.; Iwamae, A.; Hoshina, K.; Kono, M.; Yamanouchi, K., *Chem. Phys.* **231**, 315 (1998).
4. Iwamae, A., Hishikawa, A., Yamanouchi, K., *J. Phys. B: At. Mol. Opt.* **33**, 223 (2000).
5. Hishikawa, A., Iwamae, A., Yamanouchi, K., *Phys. Rev. Lett.* **83**, 1127 (1999).
6. Hishikawa, A., Iwamae, A., Yamanouchi, K., *J. Chem. Phys.* **111**, 8871 (1999).
7. Liu, S., Hishikawa, A., Iwamae, A., Yamanouchi, K., *Advances in Multiphoton Processes and Spectroscopy* 13, Y. Fujimura and R.J.Gordon Eds.,World Scientific, p.189 (2000).
8. Itakura, R., Watanabe, J., Hishikawa, A., Yamanouchi, K., *J. Chem. Phys.* **114**, 5598 (2001).
9. Watanabe, J., Itakura, R., Hishikawa, A., Yamanouchi, K., *J. Chem. Phys.* **116**, 9697 (2002).
10. Iwasaki, A., Hishikawa, A., Yamanouchi, K., *Chem. Phys. Lett.* **346**, 379 (2001).
11. Hoshina, K., Yamanouchi, K., Ohshima, T., Ose, Y., Todokoro, H., *Chem. Phys. Lett.* **353**, 27 (2002).
12. Hoshina, K., Yamanouchi, K., Ohshima, T., Ose, Y., Todokoro, H., *Chem. Phys. Lett.* **353**, 33 (2002).
13. Hasegawa, H., Hishikawa, A., and Yamanouchi, K., *Chem. Phys. Lett.* **349**, 57 (2001).
14. Hishikawa, A., Hasegawa, H., and Yamanouchi, K., *Chem. Phys. Lett. in press.*

Single and Crystalized Ions in Ultra-Intense Laser Pulses

Karen Z. Hatsagortsyan[1*], Ulrich D. Jentschura[*] and Christoph H. Keitel[2*]

*Theoretische Quantendynamik, Fakultät für Physik, Universität Freiburg,
Hermann-Herder-Straße 3, D-79104 Freiburg, Germany

Abstract. Relativistic and quantum mechanical effects in the interaction of atomic systems with a superintense laser field are investigated. Various atomic sytems are the object of attention, ranging from highly charged ions to thin crystal layers. A new regime of high-harmonic generation (HHG) based on multiply charged ions is described, where the dynamics of the tunneled wave packet of the electron between birth and recombination is strongly modified by the ionic core, producing a coherent burst of hard X-rays. We also consider relativistic effects in the nonsequential double ionization of helium. The scheme for phase-matched HHG from a laser-driven thin crystal layer in the presence of a high background of free electrons is discussed. For a selective range of parameters, the initial regular structure of the layer is preserved during the strong interaction with the laser pulse due to the suppression of shock waves in the layer. As a result, short hard x-ray pulses may be generated with unprecedented coherence, and thus the possibility of applications in x-ray spectroscopy and holography.

INTRODUCTION

Since its inception, laser radiation has proven to be a powerful tool for both probing and modifying matter. The increasing laser intensity unveils new regimes of laser-matter interaction and opens new perspectives and applications. These include the laser acceleration of particles, the fast ignition of nuclear fusion and the creation of short-wavelength coherent radiation sources. The last substantial progress in laser intensities has been achieved due to the chirped-pulse amplification technique (CPA), which now enables scientists to deliver focused intensities up to the order of 10^{22} W/cm^2 [1]. These laser intensities can impart electrons with tremendous energy. Specifically, in the laser focus, the electrons can be accelerated up to GeV energies [2]. Meanwhile, a dicusssion is in agenda to extend the CPA systems capability up to the order of 10^{28} W/cm^2 using megajoule laser facilities [3].

For high intensities above 10^{15} W/cm^2, the relevant parameter for the description of the nonlinear optical processes is the so-called Keldysh parameter $\gamma = \sqrt{I_p/2U}$, i.e. the ratio of the ionization energy I_p to the electron ponderomotive energy in the laser field U. Here, $U = e^2 E_L^2/4m\omega_L^2$, and e and m are the electron charge and mass,

---

[1] Permanent address: Department of Theoretical Physics, Yerevan State University, A. Manoukian Street 1, 375025 Yerevan, Armenia.

[2] Email address: keitel@uni-freiburg.de

CP634, *Science of Superstrong Field Interactions*, edited by K. Nakajima and M. Deguchi
© 2002 American Institute of Physics 0-7354-0089-X/02/$19.00

respectively. E_L and ω_L represent the laser field amplitude and frequency. In the strong-field regime, $\gamma^{-1} > 1$, the Coulomb barrier is suppressed by the laser field so that the electron can tunnel out through the barrier during the laser period, a process known as tunneling ionization (TI). In the strong-field regime, the relativistic effects, which are the nondipole interaction, the magnetic-field influence and spin effects, begin to play a role when the free-electron oscillation velocity becomes comparable to the velocity of light, i.e. at $\xi := eE_L/mc\omega_L > 1$, where c is the speed of light. The main concern of this report is the investigation of distinct features of relativistic effects in the strong field regime of laser-matter interaction, especially in the HHG process. Various atomic sytems will be the object of attention, particularly, highly charged ions, helium atoms and thin crystal layers.

QUANTUM RELATIVISTIC INTERACTION OF MULTIPLY CHARGED IONS WITH SUPER-INTENSE LASER PULSES

The interaction of atomic systems with high-power laser pulses is a very promising source of high-frequency coherent and short light pulses. The oscillation of small parts of the electronic wavepacket against the ionic core of such laser-driven atoms was shown to lead to the generation of extremely high multiples, harmonics of the applied laser frequency [4]. The shortest highly coherent light generated up to date has the wavelength 2.5 nm [5, 6]. One evident possibility for the increase of the HHG frequency consists in the increase of the driving laser intensity. The immediate ionization of atomic systems exposed to intense laser fields above $10^{15}\,\text{W}/\text{cm}^2$ has been an obstacle to this approach. The use of multiply charged ions circumvents this dilemma, at least to a certain extent. Multiply charged ions of arbitrary charge states as well as their handling can nowadays be carried out as a matter of routine. The remaining electrons are then so strongly bound that even the most powerful laser fields available today can only temporarily remove electrons from the vicinity of the nucleus until further recombination accompanied by

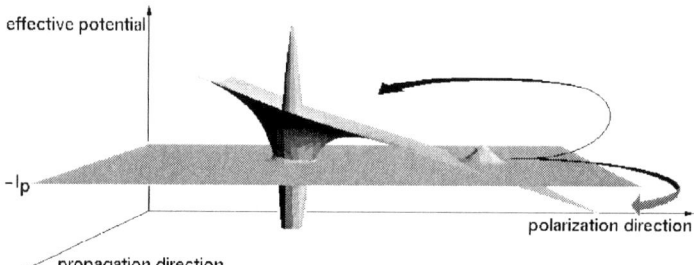

FIGURE 1. Schematic diagram for coherent sub-$\overset{o}{A}$ngstrom hard x-ray generation in multiply charged ions. The effective potential originating from the ionic potential and the laser field at maximal strength, the bound ground state with energy $-I_p$ and the small tunneled part of it are depicted as a function of the spatial coordinates in the laser polarization and propagation directions. Depending on the velocity distribution of the tunneled electron wavepacket the arrows indicate that some fraction of the tunneled wavepacket ionizes in both the laser polarization and propagation directions, while another may return under the strong influence of the deep ionic potential. From [8] with permission.

HHG (see fig. 1).

The dynamics of the electron wavepacket in such intense laser fields are in the relativistic regime. There has been considerable activity recently in this field. From recent results we mention novel means of HHG via slow spin flipping [7]. In Ref. [8], a new regime of HHG is found where the quantum relativistic dynamics of the tunneled electron significantly deviates from the conventional situation. We will give a more detailed description of this result. Atomic Nitrogen in the sixth ionized state is exposed to a superintense laser pulse of the order of $10^{18}\,\mathrm{W/cm^2}$. The applied laser field still corresponds to the weakly relativistic regime, and thus the Dirac equation describing the dynamics of the corresponding wave function may be expanded in powers of the ratio of electron velocity to the speed of light v/c, including up to fourth-order terms. The radiation spectrum (see Fig. 2) shows that high harmonics are produced with wavelengths well below the atomic scale of $1\ \overset{o}{A}$. The very high harmonic part is well separated from the remaining part of the spectrum, in contrast to the "usual" mechanism

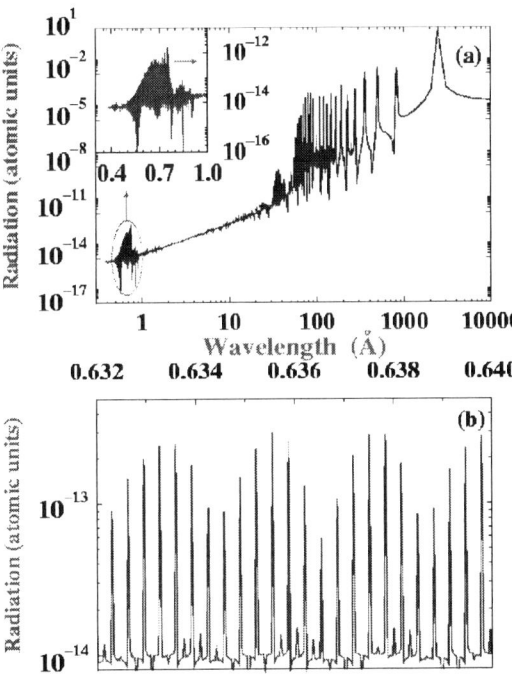

FIGURE 2. The radiation signal with sub-$\overset{o}{A}$ harmonics. Calculated is the x-polarized radiation emitted in the laser magnetic field direction. The laser parameters involve a wavelength of 248nm (KrF laser), an intensity of 1.2×10^{18} W/cm², a 10-cycle linear pulse turn-on and a 10-cycle duration with constant amplitude. The ion is modeled to adapt atomic Nitrogen N^{+6} with all but one electron removed (for the ion model see [9]). In a) the enlargement emphasizes the region of sub-$\overset{o}{A}$ hard x rays. b) It shows the structure for the spectral regime from $0.632 \overset{o}{A}$ to $0.64 \overset{o}{A}$. From [8] with permission.

of HHG. A more intuitive understanding of this new mechanism of HHG can be drawn from the trajectory of the electronic wave packet. As shown in the left side of Fig. 3, the trajectory differs significantly from the conventional situation [4]. The tunneled electron wavepacket between "birth" and recombination is not propagating freely, but its trajectory is rather significantly modified by the strongly attractive potential of the nucleus of the multiply charged ion. The wavepacket starting its return to the nucleus

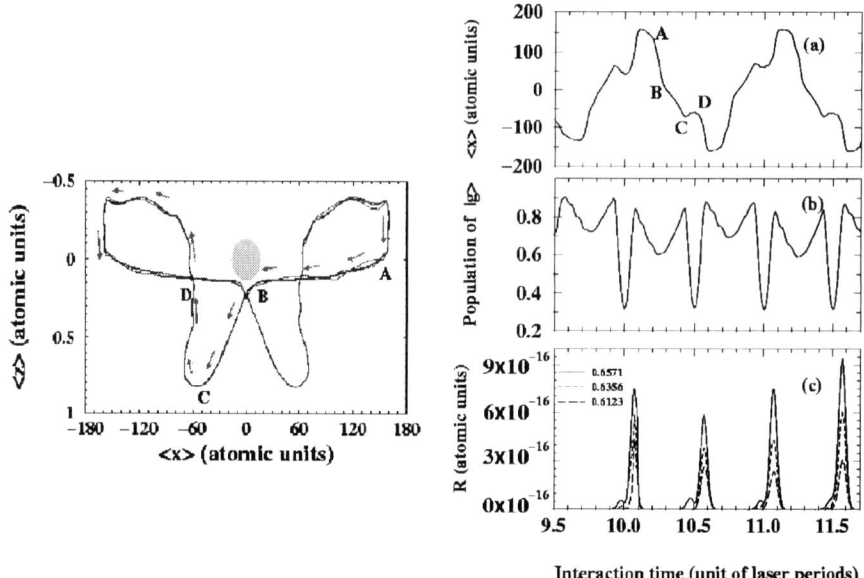

FIGURE 3. Left figure: the weakly relativistic motion of the center-of-mass of the tunneled electronic wavepacket. $\langle x \rangle$ and $\langle z \rangle$ denote the center of mass in the laser polarization direction x and the laser propagation direction z, respectively. Right figure: the physical origin of the sub-$\overset{o}{A}$ harmonics. a) The center-of-mass motion of the tunneled fraction of the electronic wavepacket along the laser polarization direction (x-axis) as a function of the interaction time; b) the corresponding ground state population evolution in time and c) the radiation signal R as a function of time at wavelength $0.6571 \overset{o}{A}$ (solid line), $0.6356 \overset{o}{A}$ (long dashed line), and $0.6123 \overset{o}{A}$ (short dashed line) as given by the reverse Fourier transform of the spectrum in Fig. 2 around the respective frequencies. The parameters are as described in Fig. 2. From [8] with permission.

A is merely scattered elastically at the nucleus B because of its high velocity (see right side Fig. 3: (a) is the mean position of the electron in the laser polarization direction (b) is the ground state population, and (c) is the emitted intensity of the radiation for three selected harmonics with wavelengths below $1 \overset{o}{A}$ as a function of time). Significant interaction with the ionic core takes place during a further recollision, points C–D, involving sufficiently low velocities, as indicated by the strong peak of the ground-state population at this time, see right Fig. 3 (b). The highly excited bound and continuum states relax back to the ground state giving rise to hard x-rays at point D as clearly visible in right Fig. 3 (c). Meanwhile for lower ion charges and laser intensities harmonics

74

are generated around B. Furthermore, we stress that the wavelength of the generated coherent radiation may be further reduced with increasing laser intensity and increasing ionic charge number.

At the end of this section we briefly mention first results concerning relativistic (and magnetic-field) effects on the laser-induced nonsequential double ionization [11]. As is well-known, the yield of the laser-induced double ionization of atoms is many orders of magnitude higher than predicted within a theory using one active electron at a time [10]. This phenomenon has been explained by a recollision of the first ionized electron. In [11], a regime is found beyond the dipole approximation where the probability of double ionization of helium via recollisions is reduced as the recolliding electron may miss the ionic core with the inner electron due to the Lorentz force (generated by the magnetic laser field). This effect confirms indirectly that the recollisions are responsible for the high yield of nonsequential double photoionization. Related effects appear in the angular distribution with enhanced ionization in the laser propagation direction.

X-RAY GENERATION FROM LASER DRIVEN CRYSTALS

A straightforward extension of the atomic HHG scheme into the ultrarelativistic regime of laser intensities is barred by serious problems. First of all, one issue is the ionization and the resulting deterioration of phase matching of the signals of the various ions due to the presence of the electron background. Many efforts were undertaken to improve the phase matching [4] with considerable success except for the hard x-ray regime. The second problem is the radiation pressure in laser propagation direction that prevents the electrons from a recombination. In this situation, it is therefore reasonable to use many-body systems with a regular structure. There has been a lot of activity over the last two decades regarding intense nonrelativistic laser interactions with many-body systems: molecules [12], clusters [13], and solids [14], and laser interactions on the surface of overdense plasmas as formed on laser-driven metallic surfaces [15]. In the latter case, problems arise when approaching the hard x-ray regime due to collective phenomena during the very intense laser-plasma interaction, these collective phenomena substantially reduce the spatial coherence of the harmonic radiation. In [16], HHG via stimulated bremsstrahlung is proposed due to a very short intense laser-crystal interaction, where HHG is limited due to the rather low efficiency of multiple electron scattering.

Usually, strong collective effects during laser-solid interaction arise due to the strong ponderomotive pressure of the laser pulse on the solid surface that launches a shock wave into the bulk of the medium. We propose a setup for strong laser pulse interaction with a thin solid layer where collective effects can be effectively suppressed as the transversal width of the layer, $L_x = L_y = 20\,\mu m$, with respect to the laser beam propagation direction is substantially less than the radius of the laser beam $R_L = 3\,mm$ (see Fig.4). L_i are spatial dimensions of the thin solid layer with separations a_i between the atoms and number of atoms N_i in direction i. The subindices $i \in \{x, y, z\}$ refer to the polarization direction x, magnetic field direction y and propagation direction z of the applied laser pulse, respectively. In this setup, the ponderomotive forces can be estimated to be

ZU/l_L, where $l_L = R_L^2/L_x$, and Z is the effective charge number of the ion. The average ionic displacement is found to be $l_i \approx Z\xi^2 c^2 \tau_L^2 L_x m/(4\,M\,R_L^2)$ during the brief laser-crystal interaction, with M being the mass of the identical ions, and τ_L the interaction time. For a Ti:sapphire laser pulse of 30 fs duration, an intensity of $I = 5 \cdot 10^{21}\,\mathrm{W/cm^2}$ and a potassium crystal ($Z = 17$), we obtain $l_i \approx 0.2\,\overset{o}{A}$ which is far less than the lattice period $a = 4.5\,\overset{o}{A}$. Thus, in the considered setup, the ponderomotive forces are not strong enough to move electrons over an appreciable distance during the interaction. Consequently, no collective instabilities can develop, and the structure of the layer will be preserved during the process.

We assume that HHG at each atom arises due to the tunneling-recollision-mechanism. For a potassium crystal, $Z = 17$ electrons will be ionized on average by the considered laser pulse via over-the-barrier-ionization, and it is the last remaining 18th electron that will generate the harmonics. At these intensities, the medium is highly ionized, and the refractive index for the laser wave n_L is governed by the free electron background. Nevertheless, for the applied parameters the plasma is yet underdense due to dependence of the effective electron mass on the laser intensity, which allows the laser wave to propagate in the plasma, i.e. $\omega_L > \omega_p$. Note, that this setup essentially differs from one considered in Ref. [16], since in the latter case, the transversal amplitude of the electron oscillation exceeds the medium (layer) width, and the laser radiation destroyes the layer: consequently, the above mentioned refractive index becomes irrelevant.

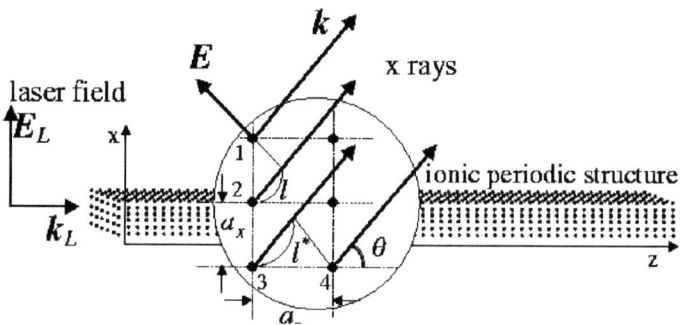

FIGURE 4. Phase matching of HHG from a thin solid layer with periodic ionic structure in an intense and very short laser pulse. Transverse phase matching, i.e. between ion 1 and 2, requires via Eq. (1) $\omega t_1 - kl = 2\pi s_1$ with $l = a_x \sin\theta$, $t_1 = 0$ and integer number s_1. Longitudinal phase matching, i.e. between ions 3 and 4, requires analogously $\omega t_2 - kl^* = 2\pi s_2$ with $l^* = a_z \cos\theta$, time delay $t_2 = a_z n_L/c$ and integer s_2. θ is the angle between the propagation direction of the laser pulse z and that of the harmonic wave $\mathbf{k}$. From [17] with permission of the IOP Publishing Limited.

Maintenance of the periodic structure of the thin crystal layer allows to solve phase matching problem of HHG in the presence of high free electron background. It is achieved by imposing the Bragg condition for phase matching between fundamental and harmonic waves generated from various ions:

$$|\omega t - \mathbf{kl}| = 2\pi s, \; s = 1, 2, ... \tag{1}$$

where $\mathbf{l}$ is the displacement vector of the ions, $\omega, \mathbf{k}$ are frequency and wave vector of the harmonic radiation, respectively, with $k = |\mathbf{k}| = n\omega/c$, refractive index n for the harmonic radiation and t being the time for the laser pulse to propagate the distance $\mathbf{l}$.

With Eq. (1) and Fig. 4 we find the phase matching condition $ka_x \sin\theta = 2\pi s_1$ and $a_z \omega(n_L - n\cos\theta)/c = 2\pi s_2$, with $s_{1,2} = 0, \pm 1, \pm 2, ...$ We consider the two most favorable directions of detection along the laser propagation direction (first scheme: $\theta = 0, s_1 = 0$) and at a nonvanishing angle $\theta = \arccos(n_L/n)$ (second scheme: $\theta \neq 0$, $s_2 = 0$).

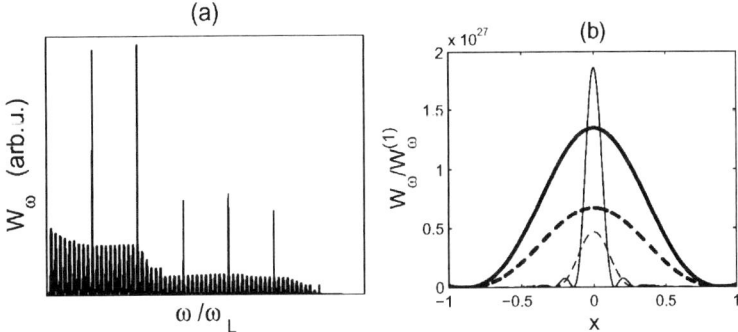

FIGURE 5. a) Sketch of a spectrum with periodic phase matching, b) HHG spectral intensity from a crystal layer $W_\omega = dW/d\omega d\Omega$ scaled by the respective peak value for one ion $W_\omega^{(1)} = (dW^{(1)}/d\omega d\Omega)_{\max}$, where $x = \omega/\omega_L - 10^6$. The narrow features (thin lines) correspond to the coherent HHG from the periodic structure, while the wide features (thick lines) do to the HHG from one ion multiplied by number of ions and magnified by $3 \cdot 10^{14}$, corresponding by intensity to the incoherent HHG from disordered ions . The solid lines correspond to $N_z = 2 \cdot 10^5$ and the dashed lines to $N_z = 10^5$. The parameters employed are: $L_x = L_y = 20\,\mu m$, $\lambda_L = 800\,nm$, $\tau_L = 30\,fs$, $\theta = 0$, $a_x = a_y = a_z = 4.5$ $\overset{o}{A}$, $n - n_L = 0.057$. From [17] with permission of the IOP Publishing Limited.

We calculate the spectral intensity from the regular structure taking advantage of the fact that the electrons responsible for HHG ionize as well as evolve in the laser field in a similar manner. Consequently, their trajectories are congruent apart from translations in space by $\mathbf{l}$ and in time by t. Thus, to a good approximation, the total HHG spectral intensity of the crystal can be expressed via a coherent sum of the HHG spectral intensity arising from single ions (see Fig. 5). The peak spectral intensity of harmonic radiation at the coherence condition (1) is ideally proportional to $(N_x N_y N_z)^2$ and will thus in general significantly exceed the HHG spectral intensity of irregular systems. Reduced enhancements are still possible within the spectral and angular linewidths: $\delta\omega/\omega \approx a_z/sL_z$ and $\delta\theta \approx /\lambda L_x$, with λ being the harmonic wavelength. The efficiency of the proposed x-ray radiation source can be estimated by the spectral brightness of the radiation which we express for convenience in units $photons/(s \cdot mm^2 mrad^2 (0.1\% bandwidth))^{-1})$:

$$B = \frac{dN}{dt\,dA\,d\Omega\,d\omega} = 3 \cdot 10^{15} \frac{N_x N_y N_z n_L \,|\mathbf{n} \times \mathbf{a}_\omega(a.u.)|^2}{a_x a_y a_z(\overset{o}{A})}, \tag{2}$$

where as usual "0.1% bandwidth" means that the spectral variable B is scaled by 0.1% of the considered frequency, $\mathbf{n} = \mathbf{k}/k$ and $\mathbf{a}_\omega = (\omega/e) \int \mathbf{j}_\omega^{(1)} e^{-i\mathbf{kr}} dV$, N is the number of emitted photons, A is the emitter area and Ω is the solid angle in the emission direction.

For a high laser intensity of $I = 5 \cdot 10^{21}\,\mathrm{W/cm^2}$, hard x-ray HHG is possible in principle with unprecedented coherence; e.g., for 1.5 MeV photons, the spectral width becomes $\delta\omega/\omega \approx 1.4 \cdot 10^{-7}$, and the angular divergence of the emitted radiation $\delta\theta \approx 4 \cdot 10^{-8}$. No reliable estimates are currently available for the single-ion-variable $\mathbf{a}_\omega$ entering into the expression for the total spectral brightness in Eq. (2), because of the favorable but complex situation that the tunneled wave packet may recombine with several ions on its journey through the crystal.

The suggested scheme of laser-driven thin crystals is attractive because of its high coherence ($10^{-5}\%$ bandwidth for MeV photons or $10^{-3}\%$ bandwidth for 3 KeV photons). In particular, we hope that our scheme could lead to a highly coherent table-top light source with photon energies above few KeV. Finally the brevity of the hard x-ray pulses may allow for time-resolved x-ray holography and spectroscopy.

ACKNOWLEDGMENTS

This work was supported by Deutsche Forschungsgemeinschaft (Nachwuchsgruppe within SFB 276). KZH is also supported by the Humboldt Foundation.

REFERENCES

1. G. Mourou, C.P.J. Barty, and M.D. Perry, Phys. Today **51**, No.1, 22 031301 (1998); M. D. Perry et al., Opt. Lett. **24**, 160 (1999).
2. Y.I. Salamin and C.H. Keitel, Phys. Rev. Lett. **88**, 095005 (2002).
3. T. Tajima and G. Mourou, Phys. Rev. ST AB **5**, 031301 (2002).
4. C. J. Joachain, M. Dorr, and N. Kylstra, Adv. At. Mol. Opt. Phys. **42**, 225 (2000); C. H. Keitel, Contemp. Phys. **42**, 353 (2001).
5. C. Spielmann, et al., Science **278**, 661 (1997).
6. Z. H. Chang, et al., Phys. Rev. Lett. **79**, 2967 (1997);
7. M. W. Walser and C. H. Keitel, Opt. Comm. **199**, 447 (2001) and Phys. Rev. A **65**, 043410 (2002).
8. C.H. Keitel and S. X. Hu, Appl. Phys. Lett., **80**, 541 (2002).
9. S. X. Hu and C.H. Keitel, Phys. Rev. Lett., **83**, 4709 (1999) and Phys. Rev. A **63**, 053402 (2001).
10. J.B. Watson, et al., Phys. Rev. Lett. **78**, 1884 (1997); T. Weber, et al., Phys. Rev. Lett. **84**, 443 (2000); R. Moshammer, et al., Phys. Rev. Lett. **84**, 447 (2000); A. Becker and F.H.M. Faisal, Phys. Rev. Lett. **84**, 3546 (2000); J.S. Parker, J. Phys. B **34**, L69 (2001).
11. J. Prager and C.H. Keitel, J. Phys. B **35**, L167 (2002).
12. A.D. Bandrauk, Molecules in Laser Fields (Dekker, New York, 1993).
13. See e.g. A. McPherson et al., Nature (London) **370**, 631 (1994); T. Ditmire et al., Nature (London) **398**, 489 (1999); S. X. Hu and Z. Z. Hu, Phys. Rev. Lett **71**, 2605 (1997).
14. L. Plaja and L. Roso-Franco, Phys. Rev. A **45**, 8334 (1992); S. Huller and J. Meyer-ter-Vehn, Phys. Rev. A **48**, 3906 (1993); F.H.M. Faisal and J.Z. Kaminski, Phys. Rev. A **56**, 748 (1997).
15. P.A. Norreys et al, Phys. Rev. Lett. **76**, 1832 (1996); D. von der Linde et al., Phys. Rev. A **52**, R25 (1995); P. Gibbon, IEEE J. Quant. El. **33**, 11 (1997); K. W. D. Ledingham, Phys. Rev. Lett. **84**, 899 (2000).
16. K.Z. Hatsagortsyan and C.H. Keitel, Phys. Rev. Lett. **86**, 2277 (2001).
17. K.Z. Hatsagortsyan and C.H. Keitel, J. Phys. B **35**, L175 (2002).

Some recent developments in nonlinear relativistic plasma dynamics

F. Pegoraro[*], M. Borghesi[†], S.V. Bulanov[**], F. Califano[*], T.Zh. Esirkepov[‡] and A.V. Kuznetsov[§]

[*]*Phys. Dept., University of Pisa and Istituto Nazionale Fisica della Materia, Pisa, Italy*
[†]*Dept. Pure & Applied Physics, The Queen's University of Belfast, Belfast, U.K.*
[**]*General Physics Institute, Russian Academy of Science, Moscow, Russia*
[‡]*Institute of Laser Engineering, Osaka University, Osaka, Japan*
[§]*Moscow Institute of Physics and Technology, Dolgoprudny, Russia*

Abstract. Some recent developments in the analytical and numerical study of the interaction of ultra-intense ultra-short laser pulses with relativistic plasmas are reviewed. Special attention is given to the subject of ion acceleration in view of its applications which range from proton imaging to the acceleration of collimated ion bunches and to the effect of fast magnetic field line reconnection on the evolution of the self-generated magnetic field.

INTRODUCTION

The interaction of ultraintense, ultrashort laser pulses with plasmas provides unique conditions for studying the collective nonlinear dynamics of a macroscopic relativistic system. In these systems the plasma dynamics is characterized by extreme conditions where particles can acquire relativistic energies that correspond to ordered motions such as large amplitude coherent oscillations, electron and ion collimated beams etc. [1]. Such "ordered" energies far exceed the plasma thermal content and are thus a natural source of plasma collective instabilities that can grow to nonlinear amplitudes. In these nonlinear plasma electrodynamic regimes the interplay between particle acceleration and electromagnetic field generation determines how the laser pulse energy is transformed either into plasma energy or into other forms of electromagnetic field energy on the rather short time scale of the interaction between the pulse and the plasma.

ION BEAM GENERATION

One of the most important phenomena presently under investigation in relativistic plasmas is the formation of collimated beams of high energy ions: the effective ion acceleration processes that occur during the laser interaction with matter can lead to important new applications. Emission of collimated beams of highly energetic protons from solid targets irradiated with high intensity laser pulses has been reported by various groups, see Refs.[2]. As an example we can refer to the characteristic properties of proton beams produced at the Rutherford Appleton Laboratory [3] which will be discussed in more

CP634, *Science of Superstrong Field Interactions*, edited by K. Nakajima and M. Deguchi
© 2002 American Institute of Physics 0-7354-0089-X/02/$19.00

detail later in the text. The protons have a broad spectrum, with maximum energies up to a few tens of MeV. The beams have a very high brilliance, if compared to conventional accelerators, and they are very directional, with a divergence of a few degrees at the highest energies. They are emitted from a source with an apparent size of less than $5 - 10\mu m$, for $E > 10\ MeV$, in bursts of ps duration, i.e., 3 orders of magnitude shorter than conventional accelerator bunches.

The generation of fast ions becomes highly effective when the laser radiation reaches the petawatt power range [4]. Analytical estimates and particle in cell simulations [5, 6, 7] show that, by optimizing the laser-target parameters, it becomes possible to accelerate protons up to several hundred MeV.

Advanced ion acceleration scenarios: Energy optimization

For ultra short laser pulses with multiterawatt power in the femtosecond range, the typical time scale of the hydrodynamic expansion of a micron plasma slab is much longer than the laser pulse duration. Under these conditions ions remain at rest. This results in the formation of a positively charged layer of ions. After a time interval $\sim 1/\omega_{pi} = (m_i/4\pi n_0 Z_i^2 e^2)^{1/2}$ the ion layer explodes ("Coulomb explosion"). If we assume that the laser pulse is sufficiently intense so that all free electrons are expelled from the irradiated region of the foil, the electric field near the positively charged layer is given by

$$E_0 = 2\pi n_0 Z_i e l \tag{1}$$

with n_0 the ion density in the foil, $Z_i e$ the ion electric charge, and l the foil thickness. The typical energy of the ions with charge $Z_a e$ accelerated by this electric field is of the order of

$$\mathrm{E}_{max} = 4\pi n_0 Z_a Z_i e^2 l R_\perp \tag{2}$$

where $R_\perp$ is the transverse size of the focal spot which is of the order of the longitudinal size of acceleration region.

Energy spectrum optimization

The "thermal"energy spectra that are obtained in present experiments are not appropriate for applications where energy selection is an important issue, such as in the case where the use of laser accelerated proton beams is proposed for hadrontherapy in oncology [8]. In order to improve the proton beam quality we can use multi-layer targets. In this scheme [8] a foil is used as a target and its rear surface is coated with a thin and transversally narrow hydrogen layer. An ultra short laser pulse irradiates the target: heavy atoms are partly ionized and the resulting free electrons abandon the foil. This leads to a large electric field due to charge separation. Heavy ions with large value of the ratio (μ/Z) remain at rest while the protons in the coating are accelerated. If the proton coating is thin and has a transverse size smaller that the diameter of the focal spot, this scheme provides a controlled acceleration of protons.

Since the proton layer is narrow in the transverse direction we can model the acceleration process by assuming a one-dimensional electric field of the form

$$E(x) = E_0(1 - x/R_0), \quad \text{for} \quad 0 < x < R_0, \quad R_0 = 2R_\perp, \tag{3}$$

At $t = 0$ all particles are at rest with zero velocity. Then. at the exit of the acceleration region, their energy spectrum is given by

$$N(\mathrm{E})d\mathrm{E} = \frac{n_0(x_0)d\mathrm{E}}{|\Omega(2\mathrm{E}m_p)^{1/2}|}, \tag{4}$$

with $\Omega = (eE_0/m_pR_0)^{1/2}$ and $\mathrm{E}(x_0) = m_p[(R_0 - x_0)\Omega]^2/2$, which shows that a small energy spread is obtained when the thickness of the proton layer is narrow. A most important requirement is that the transverse size of the proton layer be smaller than the focal spot so as to reduce the influence of the laser pulse inhomogeneity in the direction perpendicular to its direction of propagation. The pulse inhomogeneity causes an inhomogeneity of the accelerating electric field, which results in an additional energy spread of the ion beam, as seen in experiments. An additional parameter for controlling the properties of the proton beams is provided by deformed targets [6] : protons accelerated from the rear side of a concave foil are focused near the center of the foil curvature.

PROTON IMAGING

The beams of multi-MeV protons generated during the interaction of ultraintense ($I > 10^{19}W/cm^2$) short pulses with thin solid targets can be used as a particle probe in laser-plasma experiments [3, 9]. The beams need not be monochromatic for this application.

Proton imaging employs these proton beams as a diagnostic in a point-projection imaging scheme and provides the possibility of diagnosing electric fields in dense plasmas and laser-irradiated targets. Laser-produced protons also have a great potential as a density diagnostic in thick targets, but in the limit of thin targets (with thickness much smaller than the collisional proton stopping distance) beam perturbations are mainly due to the electromagnetic fields present in or around the targets. As noted in [3], this technique can open up the possibility of accessing the yet unexplored electromagnetic field distributions in indirect drive target assemblies. In a Fast Ignitor context, it can contribute to the study of the electron dynamics and transport following ultraintense laser-plasma interactions, via the detection of the ultralarge electromagnetic fields associated with the electron motion.

Parameters of the proton beams

We refer to the beam parameters described in [3]: the VULCAN laser, operating in the Chirped Pulse Amplification mode (CPA), provides $1.054\mu m$, $1ps$ pulses with energy up to $100J$. The focal spot varied between 8 and $10\mu m$ in diameter at full width at half maximum, containing $30 - 40\%$ of the energy, and giving intensities up to

$5 - 7 \cdot 10^{19} W/cm^2$. The targets used for proton beam production were Al foils, $1 - 2mm$ wide and $3 - 25\mu m$ thick. The proton beams produced under these conditions were bright, typically containing more than 10^{12} protons with energy above $3MeV$ per shot. The energy of the protons typically reached about $25MeV$. The duration of the proton pulse at the source is predicted to be of the order of the pulse duration. Experimental evidence shows that it was below $5ps$.

When a thin object is placed between the source and a detector, there will be a one-to-one correspondence between the points of the object plane and of the detector plane (assuming a purely geometrical propagation for the protons). Distortions to this one-to-one correspondence can be ascribed to deflections undergone by the protons when they cross the object plane, and are ultimately correlated to the electromagnetic fields present in the object plane. Obtaining comparable spatial resolution with proton beams from conventional accelerators requires sophisticated charged particle optics and a *ps* temporal resolution is definitely beyond the possibilities of applications employing such pulses.

The high temporal resolution of this diagnostic makes it ideal in order to study highly transient fields, such as, for example, those arising from the nonlinear plasma dynamics following intense, short pulse interactions. Coherent field structures, such as solitons and vortices are fundamental features of this nonlinear interaction [10].

Electric fields inside plasma bubbles due to postsoliton merging

Analytical and numerical results [11, 12] have shown that slowly propagating, low-frequency, sub-cycle solitons can be generated in the interaction of ultra short ultraintense laser pulses with underdense plasmas. A significant fraction of the laser pulse energy can be trapped in these structures which have a typical size of the order of the collisionless electron skin depth $d_e = c/\omega_{pe}$. The fields inside the solitons consist of synchronously electric and magnetic fields, oscillating at a frequency smaller than the Langmuir frequency ω_{pe} of the surrounding plasma, plus an electrostatic field which arises from the charge separation. Recently, the complex three-dimensional structure of these solitons has been investigated and resolved with the help of sophisticated three-dimensional Particle in Cell simulations [13]

On time scales longer than $(m_i/m_e)^{1/2}\omega_{pe}^{-1}$ the nature of the slow propagating subcycle solitons changes because ions start to expand. As a consequence, a void forms in the ion density and the soliton is changed into a radially expanding postsoliton structure[14] that is largely quasineutral. Due to their expansion, initially separated solitons can merge and form a foam of plasma bubbles [15] consisting of merged postsolitons.

Proton imaging of the bubble fields: test particle approximation

Proton beams as described above can be employed for the detection of such structures. During the time taken by the energetic protons to cross the postsoliton bubble, the oscillating component electromagnetic field changes its sign several times. The deflection of the proton trajectory is thus only due to the electrostatic component of the electric field.

Let us assume that the electric field is localized inside the postsoliton bubble and is directed normally to the walls. The bubble is assumed to have a spherical form: thus the electric field has a radial component only which depends on the coordinates as $E = E_{||}r/R$. We take the postsoliton bubble to be in the $z = 0$ plane while the test proton source is a point source localized at $r = 0$, $z = Z_s$. The bubble radius R in the experiments is much smaller than the proton source distance : $R \ll |Z_s|$. The proton beam reaches the plane $z = 0$ with velocity v_z along z and radial velocity $v_z r_0/|Z_s| \ll v_z$, where r_0 is the radial coordinate at $z = 0$.

If we assume that the electric field in the bubble is relatively weak so that $eE_{||}R/m_p v_z^2 \ll 1$, a proton crossing a bubble with impact parameter r_0 acquires a transverse momentum given by

$$\Delta p_\perp = eE_{||} \frac{2r_0\sqrt{R^2 - r_0^2}}{Rv_z} \tag{5}$$

for $r_0 < R$, and is undeflected, $\Delta p_\perp = 0$, for $r_0 > R$. The density distribution in the x, y plane at the distance Z is equal to $n = n_1$ inside the projection of the bubble on the image plane i.e., inside the circle with radius

$$R_1 = R(1 + Z/|Z_s|), \tag{6}$$

is equal to $n = n_2 > n_1$ at $r = R_1$ and tends to infinity on the radius R_2. If $2eE_{||}|Z_s|/m_p v_z^2 \gg 1$ the position of the maximum density ring is given by

$$R(Z) = \frac{R}{\sqrt{2}}\left(1 + \frac{Z}{|Z_s|}\right) + \frac{eE_{||}RZ}{m_p v_z^2}. \tag{7}$$

As discussed in Ref. [15], in the case the ring radius is only slightly larger than the radius of the bubble image, $R_2 = R_1 + \delta r$ with $\delta r \ll R_1$, we have

$$\delta r = (R/2)\,(eE_{||}|Z_s|/m_p v_z^2)^2. \tag{8}$$

The typical distribution of the proton density at a distant screen exhibits a two ring structure. The radii of the rings depend on the size of the postsoliton, on the postsoliton electric field value, and on the distance to the screen. From the expressions given above we can find both the size of the bubble and the electric field value inside the bubble.

EXPERIMENTAL OBSERVATION OF QUASI-NEUTRAL CAVITATED POSTSOLITONS

In the measurements reported in Ref. [15], the VULCAN CPA pulse was split into two separate 1 ps, 1 μm, 20 J pulses (CPA_1 and CPA_2) which were focussed onto separate targets in a $10 - 15\ \mu m$ FWHM focal spot giving an average intensity of about $10^{19}\ W/cm^2$. The CPA_1 pulse was used as the main interaction pulse and focused into a preformed plasma. The plasma was produced by exploding a thin plastic foil (0.3 μm thick). The CPA_2 pulse was focussed onto a 3 μm Al foil in order to produce a beam of multi-MeV protons, which was used as a transverse particle probe of the interaction region. The main feature observed in the proton images (i.e., the proton beam intensity cross section after propagation through the plasma) was the onset of several bubble structures following the interaction. Proton images of the plasma, obtained with 8 MeV protons and recorded, in different shots, at various delays after the interaction, show 4-5 bubble-like structures at the centre of the plasma corresponding to delays of 20 and 50ps after the pulse plasma interaction. The region where the bubbles are present extend for about 300μm in the transverse direction and for about 150μm in the longitudinal direction. In [15] these observations are shown to be consistent with structures (postsolitons) formed by the expansion and merging of several solitons. For $m_p v_z^2 / 2 = 6\, MeV$, with $Z = 2cm$ and $Z_s = 2mm$, we obtain $E_{\|0} \approx 4 \cdot 10^7\, V/cm, R \approx 50\mu$. This is consistent with the values expected from structures originating from relativistic electron solitons that have been created in the plasma by the laser pulse and that have evolved into larger structures, due to expansion and merging under the effect of the ion dynamics.

ELECTRON BEAMS, MAGNETIC FIELDS AND MAGNETIC RECONNECTION

The generation of beams of relativistic electrons is a ubiquitous feature on the nonlinear interaction of ultra intense laser pulses with underdense and overdense plasmas. The propagation of such beams of relativistic electrons is an issue of fundamental importance for the fast Ignition scheme and their dynamics is closely related to the generation of quasistatic magnetic fields in the plasma [16]. Plasma charge neutrality requires the formation of an electron return current and thus the resulting electron distribution is anisotropic with a velocity spread along the laser beam propagation much larger than the perpendicular spread. The repulsion between the two oppositely directed direct or return currents, or equivalently the effect of the anisotropy of the electron distribution, leads to the spatial separation between the two currents. In a relativistic plasma this is believed to be the most important mechanism of generation of a quasi-steady magnetic field [17]

Magnetic field dynamics

Magnetic field generation from the thermal energy of an anisotropic electron population and the conversion of magnetic into electron energy in the presence of inhomogeneous currents represent two complementary features of the dynamics of a magnetic field in a plasma. Magnetic field generation is related to the mechanism of current separation described above and can be described in terms of the EMCF (electromagnetic current filamentation) instability. This instability is analogous to the well known Weibel instability [18] and occurs in the presence of two populations of counterstreaming electrons.

On the contrary, magnetic field annihilation is best known in the case of the magnetic field line reconnection instabilities studied in laboratory and astrophysical plasmas in the presence of current inhomogeneities. Recently, it has been shown in Ref. [19, 20] that a similar process of "high frequency" magnetic reconnection can affect the 3D spatial distribution and the time evolution of the magnetic field generated by the EMCFI in the laser plasma interaction.

Magnetic field reconnection in laser plasmas

Magnetic field line reconnection allows the restructuring of the magnetic field in a plasma. In a laser plasma the magnetic field generated by the EMCF instability is quasistationary on the times scale of the Langmuir waves and is essentially frozen in the electron fluid. Thus in these plasma regimes the main effect of magnetic field line reconnection consists in allowing the current channels to coalesce, as seen in 3-D PIC simulations in overdense plasmas. To be of interest for laser plasma phenomena this reconnection process must take place on time scales that are not much longer than the electron dynamical time scales and must occur even in the absence of dissipative effects such as resistivity. As in the case of Magnetohydrodynamic (MHD) reconnection, this "high frequency" reconnection which occurs in the frequency range corresponding to the whistler waves [21] leads to the breaking of the field lines of the quasi static magnetic field and to their "reconnection" in a different pattern with the formation of magnetic islands. As in collisionless MHD plasma regimes, this process is made possible by the effect of the electron inertia on the plasma conductivity.

The coalescence of the magnetic channels has important implications for the current dynamics and for the propagation of the energetic electrons in the plasma. Indeed we recall that the magnetic field generated by the EMCF instability under present experimental conditions by a relativistically intense laser pulse is estimated to be of the order of $100MG$. In terms of the associated cyclotron and plasma frequencies we have $\Omega_e/\omega_{pe} \leq 1$: thus in a relativistic laser plasma magnetic fields can affect both the single particle dynamics and the whole collective nonlinear plasma behaviour.

In a laser plasma electron anisotropy is at the basis both of the generation of the magnetic channels and, together with the strong current gradients in these channels, of the development of reconnection and magnetic channel coalescence. The study of the combined development of these competing processes requires a fully 3D description of the plasma [22]. The "minimal" description of the combined magnetic field generation and reconnection is obtained in a 2D-3V configuration where all the vector fields have three components, but are independent of x (direction of the beam propagation).

Numerical 2D-3V kinetic simulations [19] show that the structure and the time evolution of the magnetic field and of the flow pattern, formed after the nonlinear phase of the EMCFI, conform to those produced by magnetic field line reconnection instabilities. The reconnection instability is shown to develop on a fast electron time scale (longer than but comparable to that of the EMCFI) and to lead to a 3-D isotropization of the electron distribution.

CONCLUSIONS

Numerical simulations play a fundamental role in the analysis of these "extreme" regimes which are outside the reach of most analytical developments because of their high dimensionality and because of their fully nonlinear dynamics. As discussed in Ref.[1], these numerical simulations are not only used for validating analytical models, but also play the more vital role of an investigative tool for discovering new phenomena. This simulation analysis must be accompanied by the development of an appropriate terminology, in order to describe the numerical results, and must be guided by an "a priori" understanding of the relevant range of parameters and their scaling. Both the terminology and the parameter estimates can only be obtained from a physical understanding based on the extrapolation of simplified, lower dimensionality, models.

REFERENCES

1. S.V. Bulanov, *et al.* , in: "Reviews of Plasma Physics", **22**, ed. V.D. Shafranov, p. 227, (Kluwer Academic, Plenum Publishers, New York, 2001).
2. E. Clark *et al.* , Phys. Rev. Lett., **84**, 670 (2000); A. Maksimchuk *et al.* , Phys. Rev. Lett., **84**, 4108 (2000); S. P. Hatchett, *et al.* , Phys. Plasmas, **7**, 2076 (2000); R.D. Snavely *et al.*, Phys. Rev. Lett., **85**, 2945 (2000). A. J. Mackinnon, *et al.*, Phys. Rev. Lett., **86**, 1769 (2001).
3. M. Borghesi *et al.* , Phys. Plasmas, **9**, 2214 (2002).
4. T. Zh. Esirkepov, *et al.* , JETP Letters **70**, 82 (1999).
5. S.V. Bulanov, *et al.* , Plasma Phys. Rep. **25** , 701 (1999); F. Pegoraro *et al.*, IEEE Transactions on Plasma Science, **28**, 1226, (2000); Y. Sentoku, *et al.* , Phys. Rev. **E 62**, 7271 (2000).
6. S.V. Bulanov, *et al.* , JETP Lett. **71**, 407 (2000).
7. H. Ruhl, *et al.* , Plasma Phys. Rep. **27**, 411 (2001),
8. S.V. Bulanov, V.S. Khoroshkov, Plasma Phys. Rep. **28**, in press (2002); S.V. Bulanov, *et al.* , Phys. Lett. **A**, in press (2002).
9. M. Borghesi *et al.* , Plasma Phys. Contr. Fus. **43** A. 267 (2001);
10. F. Pegoraro *et. al.* in *Atoms, Solids And Plasmas In Super-Intense Laser Field*, Edt. by D. Batani et al., Kluwer Academic/Plenum Publishers (2001).
11. S. V. Bulanov, *et al.* , Phys. Fluids B **4**, 1935 (1992); S. V. Bulanov, *et al.* , Plasma Phys. Rep. **21**, 600 (1995); T. Zh. Esirkepov, *et al.* , JETP Lett. **68**, 36 (1998).
12. S.V. Bulanov, *et al.* , Phys. Rev. Lett. 82, 3440 (1999); Y. Sentoku, *et al.* , Phys. Rev. Lett. 83, 3434 (1999), F. Pegoraro, *et al.* , Laser and Particle Beams, **18**, 381 (2000); S.V. Bulanov, *et al.* , Physica **D 152-153**, 682 (2001).
13. T. Zh. Esirkepov, *et al.* , submitted to Phys. Rev. Lett., (2002).
14. N. M. Naumova, *et al.* , Phys. Rev. Lett. **87**, 185004 (2001).
15. M. Borghesi, *et al.* , Phys. Rev. Lett. **88**, 135002 (2002).
16. G.A. Askar'yan, *et. al.*, Comm. Plasma Phys. Contr. Fus., 17, 35, (1995).
17. G.A. Askar'yan, *et. al.*, JETP Letters, **60**, 240 (1994);G.A. Askar'yan, *et. al.*, Plasma Physics Reports, **21**, 835 (1995); F. Pegoraro, *et. al.*, Physica Scripta, **T 63**, 262 (1996)/
18. F. Califano *et. al.*, Phys. Rev, **E 56**, 963, (1997); F. Califano*et. al.*, Phys. Rev, **E 57**, 7048 (1998); F.Califano *et. al.*, Phys. Rev, **E 58**, 7837 (1998).
19. F. Califano, *et al.* , Phys. Rev. Lett., **86**, 5293 (2001); F. Califano *et al.* , Physica Scripta, **T 98**, 72, (2002).
20. T. Taguchi, *et al.* , Phys. Rev. Lett., **86**, 5055 (2001).
21. S.V. Bulanov, *et al.* , Phys. Fluids **B 4**, 2499 (1992); K. Avinash,*et al.*, Physics of Plasmas **5**, 2849, (1998), N. Attico *et al.* , Phys. Plasmas, **7**, 2381 (2000); N. Attico *et al.* , Phys. Plasmas, **9**, 458. (2002).
22. Y. Kazimura, *et al.* , Plasma Physics Reports, 27, 1 (2001).

Relativistic Electromagnetic Solitons Produced by Ultrastrong Laser Pulses in Plasmas

Lontano M.[1], Bulanov S.V.[2,3], Califano F.[4], Esirkepov T.Zh.[5], Farina D.[1], Koga J.[3], Liseikina T.V.[6], Mima K.[5], Nakajima K.[7], Naumova N.M.[2], Nishihara K.[5], Passoni M.[1,8], Pegoraro F.[4], Ruhl H.[9], Sentoku Y.[5], Tajima T.[3], and Vshivkov V.A.[6]

[1] *Istituto di Fisica del Plasma, Italian National Research Council C.N.R., Milan, Italy*
[2] *General Physics Institute, Russian Academy of Sciences, Moscow, Russia*
[3] *Advanced Photon Research Center, JAERI, Kyoto-fu, Japan*
[4] *Istituto Nazionale di Fisica della Materia, Universita' di Pisa, Pisa, Italy*
[5] *Institute of Laser Engineering, Osaka University, Osaka, Japan*
[6] *Institute of Computational Technologies, Russian Academy of Sciences, Novosibirsk, Russia*
[7] *High Energy Accelerator Research Organization, Tsukuba, Japan*
[8] *Nuclear Engineering Department, Polytechnic of Milan, Milan, Italy*
[9] *Max-Born Institute, Berlin, Germany*

Abstract. Low frequency, relativistic sub-cycle localised (soliton-like) concentrations of the electromagnetic (em) energy are found in two-dimensional (2D) and in three-dimensional (3D) Particle in Cell simulations of the interaction of ultra-short, high-intensity laser pulses with homogeneous and inhomogeneous plasmas. These solitons consist of electron and ion density depressions and intense em field concentrations with a frequency definitely lower than that of the laser pulse. The downshift of the pulse frequency, due to the depletion of the pulse energy, causes a significant portion of the pulse em energy to become trapped as solitons, slowly propagating inside the plasma. In an earlier phase solitons are formed due to the trapping of the em radiation inside an electron cavity, while ions can be assumed to remain at rest. Later on, after $(m_i/m_e)^{1/2}$ times the laser period, ions start to move and the ion depletion occurs producing a slowly growing hole in the plasma density. In inhomogeneous plasmas the solitons are accelerated toward the plasma vacuum interface where they radiate away their energy in the form of bursts of low frequency em radiation. In the frame of a 1D cold hydrodynamic model for an electron-ion plasma, the existence of multipeaked em solitons has been investigated both analytically and numerically. The analytical expression for a sub-cycle relativistic soliton has been derived for circularly polarized pulses in a cold isotropic plasma, and in the presence of an externally applied magnetic field. Recently, em relativistic solitons in a hot multi-component plasma have been investigated in the frame of an hydrodynamic (adiabatic) model and of a kinetic (isothermal) model.
An overview of the most recent analytical and numerical results on the soliton dynamics is given.

CP634, *Science of Superstrong Field Interactions*, edited by K. Nakajima and M. Deguchi
© 2002 American Institute of Physics 0-7354-0089-X/02/$19.00

INTRODUCTION

The rapid development of the laser technology occurred during the last decade [1] has made available table-top laser sources of femtosecond pulses with intensities up to 10^{21} W/cm^2, which allow one to investigate the radiation-matter interaction under extreme conditions characterised by the electron momentum largely exceeding $m_e c^2$ [2]. Moreover, the exciting perspectives of getting much higher laser intensities in matter in a relatively near future [3], have stimulated a renewed interest towards the physics of ultra-relativistically intense radiation in matter, and in particularly in plasmas. Two-dimensional (2D) and three-dimensional (3D) Particle-in-Cell (PIC) numerical simulation of the interaction of an ultrashort ultraintense laser pulse with a preformed plasma show that a large number of nonlinear processes take place, as for example, laser frequency variation, high-order harmonic generation, the appearance of coherent nonlinear structures (plasma channels, relativistic solitons, vortices), the generation of ultraintense quasistatic electric and magnetic fields, electron and ion acceleration to relativistic energies. Up-to-date reviews of such effects can be found in Refs. 4 and 5. Generally speaking, a non-negligible fraction of the laser pulse energy is involved in these processes, then it is of primary importance to investigate their nature in order to control their occurrence and possibly to take advantage from their "applications".

In this paper we shall focus our attention onto the relativistic electromagnetic solitons (RES), which play an important role in the strongly nonlinear interaction between electromagnetic (EM) radiation and plasmas, and we shall review the most recent theoretical results on this subject. Here, we shall refer to RES as to the localised distributions of the EM field, characterized by a normalized value of the transverse electric field $a = eE/m_e c\omega \geq 1$ (here, e and m_e are the electric charge and the rest mass of an electron, c is the speed of light in vacuum, E and ω are the transverse component of the electric field and the frequency of the EM radiation, respectively), either manifesting a finite drift velocity or being stationary; in both cases, they are long-living EM structures trapped within quasistationary electron density hollows. In a cold plasma, they are the result of the equilibrium of two forces: the ponderomotive force, originated from the nonuniform distribution of the EM energy, and the electrostatic force, associated with the charge separation. We wish to mention that EM soliton-like structures are of concern for astrophysics and cosmology, as well. Indeed, it is believed that spatial density fluctuations in the early Universe (between 10^{-2} and 1 sec after the Big Bang) are at the origin of the galaxies and clusters of galaxies formation [6]. In addition, spatial temperature nonuniformities of the early hot plasma would cause the observed nonuniformities in the distribution of the cosmic microwave background radiation. Therefore the interest for the physics of the RES extends well beyond the laboratory applications and deserves an accurate general analysis.

THE STATUS OF THE RESEARCH BEFORE PIC SIMULATIONS

Let us begin with a short summary of the theoretical studies on RES performed till the end of the nineties. Several analytical investigations dealing with the interaction of the EM radiation of relativistic amplitude with plasmas have been undertaken in the past following the pioneering paper by Akhiezer and Polovin [7]. There, the relativistic nonlinear fluid equations for cold electrons in a uniform positive background, coupled with the Maxwell equations, where used to describe transverse, longitudinal and coupled modes of arbitrary amplitude. In later times, the problem of the propagation of EM radiation of relativistic amplitude in an overdense plasma attracted a lot of attention for its potential use in dense plasma heating [8,9]. Standing wave structures of large amplitude (although, non relativistic) at the boundary with an overdense plasma, accompanied by plasma density bunching, were described by analytical [10] and numerical [11] approaches, allowing movable ions. Exact relativistically intense optical standing waves were considered for the first time in Ref. 12, in the frame of the fluid model developed in Ref. 7. In this paper there are all the ingredients to describe one-dimensional circularly-polarized RES, with fixed ions and the cold electron fluid. One-dimensional (1D) circularly-polarized solitons in an electron-ion plasma have been investigated in Ref. 13, in the quasi-neutral approximation, that is by assuming that the charge unbalance is small if compared with the density of each plasma component. The full problem of 1D circularly-polarized RES in a cold electron-ion plasma has been tackled by Kozlov, Litvak and Suvorov [14]. They have derived a new set of two nonlinear coupled equations for the scalar amplitude of the vector potential and for the electrostatic potential, which allowed to investigate non uniform spatial distributions of the fields (the equations developed in Ref. 7 refer to waves with constant amplitude). Localised solutions in the form of drifting solitons, accompanied by an electron density depression and by an ion density concentration, in both underdense and overdense plasmas, have been obtained by numerical integration of the two relevant second order differential equations. Multi-humped concentrations of the EM energy (either with even or odd number of nodes, p), inside electron density cavities bounded by density "walls", have been found in Refs. 14 and 15 (see also Ref. 16 in the weakly relativistic case with fixed ions). They can be interpreted as due to the trapping of light inside a self-generated electrostatic wave, which is excited at the front of the moving soliton and absorbed at its rear. The asymptotic equilibrium between the relativistic ponderomotive force and the electrostatic field induced by the charge separation leads to a concentration of the ion density in correspondence of the electron depression. It is worth mentioning that in Ref. 14, it has been claimed that i) within the cold fluid approximation, circularly-polarized one-dimensional solitons can exist only if they move at a velocity larger than a critical one, defined as $c(m_e/m_i)^{1/2}$ (m_i is the ion rest mass); ii) the quasi-neutral approximation, which assumes a negligible charge separation to occur, applies to small amplitude drifting solitons only. The nature of localised distributions of EM energy propagating at a speed close to c has been exploited in Refs. 15 and 17, with the aim of exploiting the potentialities to accelerate charged particles to relativistic velocities. In particular, the possibility of using a so-called "triple-soliton" (a combination of two EM and one electrostatic waves in the beat-wave resonance), propagating in a plasma channel, has been investigated in Ref. 17. The main advantage of using a soliton drifting almost at light speed, instead of a laser pulse as a driver for accelerating particles, is that it does not produce any wakefield behind itself, thus overcoming the problem of the pump depletion, so severe in the classical laser-driven accelerator concepts [18].

All the above mentioned investigations dealing with fully relativistic field amplitudes have been based on cold fluid models, with mobile or fixed ions, under quasi-neutral approximation or with the charge separation taken into account. In general it is possible to neglect a finite plasma temperature whenever the velocity acquired by the electrons under the action of the electric field is much larger than the electron thermal velocity. Generally speaking, this condition is well satisfied in laboratory experiments for values of the dimensionless parameter a of the order or larger than unity. However there are a number of reasons urging one to consider RES in warm plasmas. A first example is the following. A cold plasma model leads to density spatial distributions which may become negative for a particular choice of the soliton parameter values. This is due to the fact that if the full plasma cavitation is achieved, in the vacuum region fluid equations are meaningless and one should solve the field equations piece-wise, in the plasma region and in the vacuum region separately, and then match properly at the boundaries (see the recent studies on wave penetration into an overdense plasma in Refs. 19-21). This procedure can be somewhat elaborated and produces consistent density profiles with discontinuous derivatives and very spiky spatial distributions (in these cases their stability can be seriously compromised). The introduction of a finite electron temperature changes the structure of the longitudinal (that is, in the direction of wave propagation) component of the electron motion equation, leading to a non-negative electron density (this can be easily verified if one assumes a Boltzmann equilibrium distribution for the electron density [22]). Moreover, the inclusion of a finite temperature means that the soliton is now the result of the equilibrium of three forces: the ponderomotive force, the electrostatic force, and the thermal force. A second example of the importance of including a finite plasma temperature in the fluid theories of RES is found in cosmology. As we have already mentioned, the formation of RES in the primordial plasma could be an important source of large-scale density and temperature nonuniformities which are at the origin of the galaxies and clusters of galaxies formation [6]. In particular, it is conjectured that in the early epoch of the Universe evolution the matter was in the form of a mixture of electrons, positrons, and photons in thermal equilibrium at a temperature larger than m_ec^2 [6]. It is evident that any problem of propagation of relativistic EM waves in such a hot environment should be formulated in the frame of a hot-plasma model. It is worth noticing that to this aim a proper set of hot-plasma hydrodynamic equations should be derived which treat the thermal motion and the ordered dynamics of the fluid elements consistently from the point of view of the special relativity. Generally speaking, the assumption of a Boltzmann distribution for the particle density in the presence of a ponderomotive and an electrostatic potential [22] represents a model which allows simple analytical calculations, but is devoid of any physical background unless both nonrelativistic field amplitudes and plasma temperatures are considered.

With reference to RES in electron-positron plasmas, we wish to mention few theoretical results. The existence of one-dimensional solitons in a cold magnetised electron-positron plasma has been demonstrated in Ref. 23 and 24. It was argued that in a cold electron-positron plasma no solitons could be produced since no charge separation is expected in the case of equal inertia of the plasma constituents. RES solitons have then been investigated in a cold isotropic electron-positron-ion plasma, where a small fraction of heavy ions causes the appearance of an electrostatic potential distribution which is sufficient to balance the effect of the ponderomotive potential nonuniformity [25]. The case of a hot electron-positron-ion plasma was also studied in Ref. 26, while the hot electron-positron plasma was considered in Refs. 27-29. The above analysis have made use of the slowly varying envelope approximation in order to simplify the Maxwell equations for the EM field trapped in the form of a soliton [24-29]. Interesting discussions on the modelling of a hot electron-positron plasma in the presence of relativistically intense EM radiation can be found in Refs. 30-32.

The above review of the published literature on RES, being by no means exhaustive, shows the deep interest manifested by theoreticians for the subject. In the next Sections we shall describe the most recent analytical and numerical results on RES produced in the wakefield of an ultrastrong laser pulse in a plasma, achieved by the wide international collaboration to which the authors belong.

LOW-FREQUENCY SUB-CYCLE RES IN NUMERICAL SIMULATIONS

Since the early investigations on the laser wakefield acceleration of electrons [33], it was clear that during its propagation through an uniform underdense plasma, the laser pulse undergoes dramatic changes in its characteristics, *i.e.* shape, frequency, and intensity, due to the strongly nonlinear character of the interaction. In particular, it was observed that the laser pulse frequency and amplitude both change during the interaction in such a way as to preserve the total photon number: that is, the process is adiabatic [34]. 1D [34-36] and 2D [37-39] PIC numerical simulations, with mobile electrons and fixed ions, agree on the wave dynamics: during its propagation the laser pulse is strongly depleted both by the electrostatic and magnetostatic wakefield excitation (magnetic vortex formation was studied in Ref. 40; for a relativistic hydrodynamic study of the wakefield excitation see also Ref. 41) and by the stimulated (backward) Raman scattering; therefore its amplitude decreases and, due to the invariance of the total number of photons, its frequency and its group velocity should also decrease. A characteristic propagation length for the nonlinear depletion can be estimated as $\ell_{depl} \approx ct_{depl} \approx \ell_p \, (\omega/\omega_{pe})^2$ for a narrow laser beam, and as $\ell_{depl} \approx a^{-1}\ell_p(\omega/\omega_{pe})^2$ in the case of a broad laser pulse [34,42], where ℓ_p is the pulse length. Therefore, portions of the initially propagating radiation ($\omega > \omega_{pe}$) slow down and become trapped ($\omega < \omega_{pe}$) inside localised density depressions, drifting with a group velocity which is appreciably smaller than the laser pulse speed. Notice that 1D simulations show a finite drift velocity of the soliton (which may be even negative) [35], while 2D simulations predict an almost zero soliton speed in a uniform background plasma [37-39]. The fundamental role played by the lowering of the laser frequency in the soliton formation is supported by the strong spectral broadening [34] and frequency down-shifting (up to a factor 0.25 of the laser frequency) [39] characterising the EM radiation in the regions of the solitons.

PIC simulations have shown that RES are produced in a broad range of physical conditions of the laser-plasma interaction: for linearly and circularly polarized laser radiation, for both s- and p-polarized pulses, for unperturbed plasma densities varying in the range from 0.04 to 2 n_{cr} (where the critical density is defined as $n_{cr} = m_e\omega^2/4\pi e^2$), for initial normalized pulse amplitudes a in the range from 1 to 10. More than 20 % of the laser pulse EM energy is trapped in the form of solitons and does not contribute to the wakefield acceleration process. RES observed in the simulations have no high frequency component, *i.e.* they are sub-cycle solitons, containing half a wavelength of the EM field. As a consequence, they cannot be described in terms of an envelope but the full field structure should be retained in any theoretical approach. Once they have been formed, RES are long-living EM/plasma structures. In particular, numerical simulations of a plasma with mobile electrons and fixed ions show that a single soliton achieves an almost stationary state within a hundred laser radiation cycles. However, in particular conditions solitons are able to interact with the background plasma and between each other in a quite peculiar way. 1D simulations show that, for a proper choice of the laser-plasma interaction conditions, two solitons can be generated which move along trajectories intersecting each other: the two solitons undergo a "collision". After this event, the two solitons emerge presenting the same characteristics they had before their approaching. 2D simulations show that RES are generated with almost zero group velocity in a uniform plasma. If two solitons of opposite phases are produced close to each other, in the course of time their fields tends to oscillate in phase, after that the two solitons merge into a single soliton which preserve the total EM energy [38,43]. A 2D RES can be made to move if it is produced in a nonuniform plasma [39]. In a weakly nonuniform dispersive plasma the soliton moves following the equations of geometric optics: $dr/dt = \partial \aleph/\partial k$, $dk/dt = -\partial \aleph/\partial r$, where the Hamiltonian is the wave frequency $\aleph(r,k) = (k^2 c^2 + \omega_{pe}^2)^{1/2}$. In particular its drift occurs along the direction opposite to the electron density gradient. Then, if a RES is generated in a finite plasma, with its density decreasing towards its boundaries, the soliton is bound to escape from the plasma region and, once at the edge, it leaves the plasma emitting its EM energy in the form of a burst of radiation. The spectrum of the radiation manifests a wide peak well below the laser frequency [39]. The experimental observation of the frequency spectra of the radiation emitted from a plasma used as a target for ultrastrong laser pulses could reveal the onset of physical processes causing the laser frequency down-shift [44]: it could be an indirect proof of the soliton formation.

For dimensionless laser pulse amplitudes $a >> 1$ or for long simulations times (longer than $(m_i/m_e)^{1/2}$ the radiation period) the ion dynamics begins to play an important role in the strong laser-plasma interaction and specifically in the dynamics of RES. This is due to the fact that both the electrostatic field associated with the charge separation where the soliton is localised, and the ponderomotive force act in order to redistribute the ion density. Recent 2D [45-47] and 3D [48-50] PIC simulations with mobile electrons and ions show that the asymptotic soliton dynamics is rich of new physics. The soliton formation, that is the laser frequency down-shift followed by the trapping of part of the EM energy inside an electron cavity, while the ion density remains unperturbed, is similar to what has been observed previously. Indeed, it takes place in a time much shorter than the ion response time ($\approx 2\pi (m_i/m_e)^{1/2} \omega_{pe}^{-1} = 2\pi \omega_{pi}^{-1}$). Later on, the Coulomb repulsion in the electrically non neutral ion core pushes the ions away, in a process which is analogous to the so-called "Coulomb explosion" [51-54], resulting in an ion acceleration to typical energies of the order of $m_e c^2 a_{max}$ [46,47]. The plasma cavity goes on expanding slowly in the radial direction. In the 2D case, the expansion can be modelled within the "snow plow" approximation [55], which for $t \rightarrow \infty$ gives $R \propto t^{1/3}$. A characteristic time scale of the cavity expansion is $\tau = (6\pi R_0 2 n_0 m_i/\langle E_0^2\rangle)^{1/2}$ where $R_0 \approx c/\omega_{pe}$ is the soliton radius [45]. The total EM energy trapped in the cavity is constant and then the field amplitude and the frequency decrease as $E \propto t^{-4/5}$ and $\omega \propto t^{-2/5}$, respectively.

This entity which behaves as a soliton in the early part of its evolution, has been named "post-soliton" in the ion-dominated phase, which is no longer a stationary state.

The soliton formation and their subsequent evolution into a post-soliton have been observed in 3D PIC simulations, as well [48-50]. The EM structure of the 3D soliton is such that the electric field is poloidal and the magnetic field is toroidal. Therefore it has been named TM-soliton. The soliton core manifests an average positive charge, resulting in its Coulomb explosion and in the ion acceleration. Asymptotically, the quasi-neutral plasma cavity is subject to a slow continuous radial expansion, while the soliton amplitude decreases and the ion temperature increases.

ANALYTICAL STUDIES OF LOW-FREQUENCY SUB-CYCLE RES

Several analytical investigations of the RES have been undertaken after it has become clear from multi-dimensional numerical simulations that such coherent EM structures not only play an important role in the energetic balance of the laser-plasma interaction, but in principle can be controlled and used for charged particle acceleration. The exact 1D analytical solution of the cold fluid equations for the electrons with fixed ions, for arbitrary intense circularly polarized EM radiation is given in Ref. 56. It corresponds to a single-hump, non drifting RES in an overdense plasma. The relationship between the peak vector potential amplitude a_0 and the radiation frequency ω of the soliton is $a_0 = 2\omega_{pe}(\omega_{pe}^2 - \omega^2)^{1/2}\omega^{-2}$. Its characteristic width is $\approx c/(\omega_{pe}^2 - \omega_s^2)^{1/2}$ and the minimum electron density n_0 at the centre of the soliton is related to ω by the equation $n_0 = n_\infty[1-4(1-\omega^2)^2/\omega^4]$, where n_∞ is the unperturbed electron density. The maximum field amplitude $a_m = 3^{1/2}$ corresponds to the complete expulsion of the electrons, that is, $\omega = (2/3)^{1/2}$, and $n_0 = 0$. This class of solutions of the cold fluid equations undergoes wavebreaking for amplitudes larger than a_m. Indeed, 1D PIC simulations demonstrate a long-term robustness of the analytical solution for $a_0 = 1.5 < a_m$. On the contrary, a soliton-like structure with an initial amplitude $a_0 = 3$ breaks and its energy is converted into the kinetic energy of fast electrons up to several m_ec^2. The analysis of Ref. 56 has been extended to the case of a soliton in a uniform magnetised plasma [57]. If the radiation is circularly polarized, due to the inverse Faraday effect [58-60] a magnetic field normal to the polarization plane is produced and it may affect the soliton properties. The 1D cold fluid equations for the magnetised plasma electrons, coupled with the Maxwell equations, have been reduced to a single nonlinear second order ordinary differential equation for the normalized vector potential amplitude a. We wish to notice that *i*) the generalised momentum conservation writes $(1 - \Omega/\omega\ \gamma)p = a$, where $\Omega = eB/m_ec$ is the electron cyclotron frequency, and $p = p_e/m_ec$, is the normalised electron momentum; *ii*) the relativistic factor $\gamma = (1 - v^2/c^2)^{-1/2}$, once written in terms of the vector potential, satisfies the algebraic equation $(\gamma^2 - 1)(\gamma - \Omega/\omega)^2 - \gamma^2 a^2 = 0$; for $B = \Omega = 0$ we recover the familiar $\gamma = (1 + a^2)^{1/2}$. Again there is a maximum allowed amplitude corresponding to the zero electron density. However, in the magnetised case this value can be controlled by the magnetic field intensity and by its sign (through the ratio Ω/ω). For negative values of Ω/ω, corresponding to radiation frequencies lower than those allowed for $B = 0$, solitons with amplitudes appreciably larger than in the unmagnetised case exist, if $|\Omega/\omega| \sim O(1)$.

The previous investigations [14,15] on multi-humped, moving one-dimensional solitons in a plasma with movable electrons and ions have been reconsidered and new results on the consequences of the ion motion on the soliton structure have been found [61,62]. By numerically integrating the system of two second-order ordinary differential equations for the vector potential amplitude and the electrostatic potential, an extensive study of the RES and of the corresponding electron and ion density spatial distributions has been performed. The study of the spectra of the multi-humped solitons ($p \geq 1$), that is their existence condition in the frequency-velocity $\{\omega^2(1-V^2/c^2)/\omega_{pe}^2, V/c\}$ plane, shows that i) for velocities smaller than a critical value V_{bif} (where a bifurcation occurs), no solution can be found; an important consequence is that the results found in Ref. 56 for non drifting ($V=0$) solitons are not continuously connected with those of the case $V \neq 0$. ii) For $V_{bif} < V < V_{br}$ two solitary waves are found: the upper solution is that of the fixed-ion case, slightly modified by the ion dynamics, while the lower branch appears only if the ion dynamics is taken into account; iii) for $V > V_{br}$, high frequency solution is found only. Here V_{br} is the velocity of soliton breaking: for values of $V \rightarrow V_{br}^+$ on the low frequency branch, field and density spatial distributions peak and a singularity appears at V_{br} (for $p=1$, $V_{br} \approx 0.32c$, for $p=2$, $V_{br} \approx 0.46$). The ions tend to pile up at the centre of the soliton, while the electrons accumulate at its edges (the electron "walls" of Refs.14,15). A potential hump, moving at $V \approx V_{br}$ is thus built up whose maximum peak value is $e\phi_{br} = [1-(1-V_{br}^2/c^2)^{1/2}]m_i c^2$. When the soliton breaks, the associated electromagnetic energy is expected to be released in the form of accelerated ions whose energy can be estimated as $E_{ion} \approx e\phi_{br}/(1-V_{br}^2/c^2)$, corresponding to several tens of MeV.

The models retaining the effects of a finite plasma temperature are likely to smooth out or change several critical features of the cold fluid descriptions, as for example negative density values or the onset of wavebreaking. Moreover, if we interpret the moving sub-cycle soliton as an EM wave-packet, whose front excites a relativistic plasma wave, which is subsequently absorbed by its rear, the transition from solitons travelling with a group velocity of the order of the speed of light, towards slowly moving EM structures should necessarily deal with the problem of wave excitation in a finite temperature plasma. Finally, since the temporal evolution of post-solitons results in the particle acceleration and the plasma heating [45], it is worth developing hot plasma models. This task has been pursued in two recent papers devoted to the search for RES in a multi-component plasma with arbitrary temperatures [63,64]. In the first paper [63] the set of relativistic hydrodynamic equations for a hot plasma coupled with the Maxwell's equations have been derived from first principles, starting from the conservation laws of the particle number and of the energy-momentum tensor (the approach of Ref. 65 has been extended to a conducting fluid). The *adiabatic* closure of the fluid equations leads to the entropy conservation for each species. Similar equations have been derived in Ref. 66, by calculating the equations for the evolution of the moments of the relativistic Vlasov equations. There, the closure was based on the use of the relativistic Maxwellian for the calculation of the average particle energy. Subsequently, the model developed in Ref. 66 was used to investigate RES in hot plasmas [26-29-67]. It is natural to define an effective particle mass for each component s, $m_{eff,s} = m_s \gamma_s R_s$, where $\gamma_e = (1+p_e^2+a^2/R_e^2)^{1/2}$ and $\gamma_i = (1+\rho^2 p_i^2+\rho^2 Z^2 a^2/R_i^2)^{1/2}$ are the relativistic factors for electrons and ions, respectively, $p_s = p_{\parallel s}/m_e c^2$ is the normalised parallel particle momentum, $\rho = m_e/m_i$, and the function $R_s = 1+\alpha_s T_s/m_s c^2$ has been introduced, where $\alpha_s = \Gamma_s/(\Gamma_s-1)$, Γ_s is the adiabatic index, and T_s is the temperature. Notice that the temperature is a dependent variable of the problem, then it is not possible to get γ_s in explicit form as functions of the electrostatic and the vector potentials, only. The conservation law for the transverse component of the generalised momentum for each species writes $R_s p_{s\perp} + q_s A_\perp/c = const$. The system has been reduced for studying 1D, circularly-polarized, non-drifting RES in an overdense electron-positron ($s=e,p$) plasma of arbitrary temperature. Since the two species have equal masses and absolute values of the charge, for $T_e = T_p$, the plasma does not develop any charge separation. The existence conditions for RES have been studied for different radiation frequencies and unperturbed values of the temperatures, $T_{\infty,s}$. In a strongly overdense plasma ($\omega^2 << \omega_{pe}^2$), the ultrarelativistic kinetic pressure (the scaling $T_{\infty,s} \propto \omega^{-2}$ holds) and the ponderomotive pressure balance each other, giving rise to extremely large amplitude solitons. In the limit of vanishingly small temperatures ($T_{\infty,s} << m_e c^2$), even a modest trapped field amplitude ($a << 1$) is able to dig a deep plasma hole up to the full plasma expulsion at the centre of the soliton. A similar analysis performed in the slowly varying approximation, in underdense electron-positron plasmas has been developed in Ref. 67.

An alternative model (*isothermal*) where the plasma temperature is spatially uniform has been developed by giving the particle distribution function $f_s(W_s, P_{s\perp}) = [N_s/2m_s K_1(\beta_s^{-1})] \delta(P_{s\perp}) exp(-W_s/T_s)$, which is an exact solution of the relativistic Vlasov equation, for circularly-polarized EM radiation in a 1D plasma [64]. Here, $W_s(x,t) = m_s c^2 \gamma_s + q_s \phi(x,t)$ is the total particle energy, $P_s(x,t) = p_s + q_s A(x,t)/c$ is the particle generalised momentum, and $\gamma_s = (1 + p_s^2/m_s^2 c^2)^{1/2}$ is the relativistic factor. K_1 is the modified Bessel function of first order and $\beta_s = T_s/m_s c^2$. It represents a highly anisotropic particle distribution function with a finite parallel temperature, T_s, and a transverse beam-like distribution in the momentum space, with a zero perpendicular energy spread. The above distribution function is used to calculate the sources of the fields in the Maxwell equations, that is, the charge and the current densities, and to obtain the closed set of two ordinary differential equations for the electrostatic potential and for the scalar amplitude of the vector potential. The case of RES in an isothermal electron-positron plasma has been investigated. Equilibria with extremely high field intensities in strongly overdense plasmas have been obtained. In contrast to the adiabatic case [63], no lower limit occurs on the temperature in order to have solitons. Moreover, the isothermal model [64] predicts the possibility of the full plasma cavitation in an extended spatial region of width $\sim 1/\omega$, at sufficiently small plasma temperature. This model is presently being worked out to investigate RES in quasineutral electron-ion plasmas [68].

CONCLUSIVE REMARKS

In the present review we have described the main lines of the recent theoretical, analytical and numerical, researches on the relativistic electromagnetic solitons, which have had new impulse with the fast development of ultra-high intensity laser pulses and with the consequent possibility of investigating exotic regimes of the strong radiation-matter interaction, previously outside the experimental capabilities of terrestrial laboratories. Very recently [69] the first experimental evidences of the formation of long-living macroscopic bubble-like structures have become possible by means of a newly developed diagnostic, the *proton imaging* [70]. It is based on the deflection of a proton beam due to the quasistatic electric field produced by the charge separation associated with the localised EM wave-packet. The dynamics of such structures is in good agreement with the multi-dimensional PIC numerical simulations of clusters of RES. These multi-dimensional kinetic codes have revealed themselves excellent tools to investigate such peculiar regimes of laser-plasma interaction.

From the point of view of the analytical studies, it would be worth to develop multi-dimensional descriptions of RES for arbitrary amplitude. Indeed, it is a formidable task, due to the strongly nonlinear characters of the relevant equations. However, the clear evidences of the quasi-neutral character of RES over long times, can be a physically well-grounded simplification of the problem. An other aspect of the analytical approaches which needs theoretical support is the study of the stability of the exact solutions. Since there are no general stability criteria for the complicated systems of equations of Refs. 14,57,61,63,64, probably a good way to test the stability of 1D solitons is to use PIC codes with the analytical solutions as initial conditions, as it was done in Ref. 56.

Finally, once the occurrence of RES will have been definitely established, the next step will be to learn how to control such extremely energetic objects and how to use them for applicative purposes.

REFERENCES

[1] – M.D. Perry, G. Mourou, *Science* **264**, 917 (1994).

[2] – G.A. Mourou, C.P.J. Barty, M.D. Perry, *Phys. Today* **51**, 22 (1998).

[3] – T. Tajima, G. Mourou, "Superstrong Field Science" in Proc. of the 2nd *Intern. Conf. On Superstrong Fields in Plasmas*, edited by M. Lontano, G. Mourou, O. Svelto, T. Tajima (AIP, New York, 2002), AIP Conference Proceedings **611**, p. 423.

[4] – S.V. Bulanov, F. Califano, G.I. Dudnikova, *et al.*, "Relativistic Interaction of Laser Pulses with Plasmas" in *Reviews of Plasma Physics* edited by V.D. Shafranov (Kluwer Acad./Plenum Publ., New York, 2001), vol. **22**, p. 227.

[5] – S.V. Bulanov, T.Zh. Esirkepov, D. Farina, *et al.*, "Relativistic Interaction of Ultra-intense Laser Pulses with Plasmas" in Proc. of the 2nd *Intern. Conf. On Superstrong Fields in Plasmas*, edited by M. Lontano, G. Mourou, O. Svelto, T. Tajima (AIP, New York, 2002), AIP Conference Proceedings **611**, p. 104.

[6] – T. Tajima and T. Taniuti, *Phys. Rev. A* **42**, 3587 (1990).

[7] – A.I. Akhiezer, R.V. Polovin, *Sov. Phys. JETP* **3**, 696 (1956).

[8] – P. Kaw, J. Dawson, *Phys. Fluids* **13**, 472 (1970).

[9] – C. Max, F. Perkins, *Phys. Rev. Lett.* **27**, 1342 (1971).

[10] – P. Kaw, G. Schmidt, T. Wilcox, *Phys. Fluids* **16**, 1522 (1973).

[11] – E.J. Valeo, K.G. Estabrook, *Phys. Rev. Lett.* **34**, 1008 (1975).

[12] – J.H. Marburger, R.F. Tooper, *Phys. Rev. Lett.* **35**, 1001 (1975).

[13] – N.L. Tsintsadze, D.D. Tskhakaya, *Sov. Phys. JETP* **45**, 252 (1977).

[14] – V.A. Kozlov, A.G. Litvak, E.V. Suvorov, *Sov. Phys. JETP* **49**, 76 (1979).

[15] – P.K. Kaw, A. Sen, T. Katsouleas, *Phys. Rev. Lett.* **68**, 3172 (1992).

[16] – H.H. Kuhel, C.Y.Zhang, *Phys. Rev. E* **48**, 1316 (1993).

[17] – K. Mima, T. Ohsuga, H. Takabe, *et al.*, *Phys. Rev. Lett.* **57**, 1421 (1986).

[18] – T. Tajima, J.M. Dawson, *Phys. Rev. Lett.* **43**, 267 (1979).

[19] – F. Cattani, A. Kim, D. Anderson, M. Lisak, *Phys. Rev. E* **62**, 1234 (2000).

[20] – A. Kim, F. Cattani, D. Anderson, M. Lisak, *JETP Lett.* **72**, 241 (2000).

[21] – F. Cattani, A. Kim, D. Anderson, M. Lisak, *Phys. Rev. E* **64**, 016412 (2001).

[22] – M.D. Feit, A.M. Komashko, S.L. Musher, *et al.*, *Phys. Rev. E* **57**, 7122 (1998).

[23] – L.N. Tsintsadze, V.I. Berezhiani, *Sov. Plasma Phys. Rep.* **19**, 258 (1993).

[24] – V.I. Berezhiani, V. Skarka, S.M. Mahajan, *Phys. Rev. E* **48**, R3252 (1993).

[25] – V.I. Berezhiani, S.M. Mahajan, *Phys. Rev. Lett.* **73**, 1110 (1994).

[26] – V.I. Berezhiani, S.M. Mahajan, *Phys. Rev. E* **52**, 1968 (1995).

[27] – L.N. Tsintsadze, *Physica Scripta* **50**, 413 (1994).

[28] – S. Kartal, L.N. Tsintsadze, V.I. Berezhiani, *Phys. Rev. E* **53**, 4225 (1996).

[29] – T. Tatsuno, V.I. Berezhiani, S.M. Mahajan, *Phys. Rev. E* **63**, 046403 (2001).

[30] – P.K. Shukla, N.L. Tsintsadze, L.N. Tsintsadze, *Phys. Fluids B* **5**, 233 (1992).

[31] – L.N. Tsintsadze, *Phys. Plasmas* **2**, 4462 (1995).

[32] – L.N. Tsintsadze, K. Nishikawa, *Phys. Plasmas* **4**, 841 (1997).

[33] – S.V. Bulanov, V.I. Kirsanov, A.S. Sakharov, *Sov. J. Plasma Phys.* **16**, 543 (1990).

[34] – S.V. Bulanov, I.N. Inovenkov, V.I. Kirsanov, et al., *Phys. Fluids B* **4**, 1935 (1992).

[35] – S.V. Bulanov, T.Zh. Esirkepov, F.F. Kamenets, N.M. Naumova, *Plasma Phys. Rep.* **21**, 550 (1995).

[36] – K. Mima, M.S. Jovanovic, Y. Sentoku, et al., *Phys. Plasmas* **8**, 2349 (2001).

[37] – S.V. Bulanov, T.Zh. Esirkepov, N.N. Naumova, *et al.*, *Phys. Rev. Lett.* **82**, 3440 (1999).

[38] – S.V. Bulanov, F. Califano, T.Zh. Esirkepov, *et al.*, *J. Plasma Fus. Res.* **75**, 506 (1999).

[39] – Y. Sentoku, T.Zh. Esirkepov, K. Mima, *et al.*, *Phys. Rev. Lett.* **83**, 3434 (1999).

[40] – S.V. Bulanov, M. Lontano, T.Zh. Esirkepov, et al., *Phys. Rev. Lett.* **76**, 3562 (1996).

[41] – D. Farina, M. Lontano, I.G. Murusidze, S.V. Mikeladze, *Phys. Rev. E* **63**, 056409 (2001).

[42] – S.V. Bulanov, I.V. Inovenkov, V.I. Kirsanov, *et al.*, *Physica Scripta* **47**, 209 (1993).

[43] – S.V. Bulanov, F. Califano, T.Zh. Esirkepov, *et al.*, *Physica D* **152-153**, 682 (2001).

[44] – C.A Coverdale, C.B. Darrow, C.D. Decker, *et al.*, *Plasma Phys. Rep.* **22**, 617 (1996).

[45] – N.M. Naumova, S.V. Bulanov, T.Zh. Esirkepov, *et al.*, *Phys. Rev. Lett.* **87**, 185004 (2001).

[46] – N.M. Naumova, S.V. Bulanov, T.Zh. Esirkepov, et al., "Transformation of Laser Radiation into Post-solitons with Ion Acceleration" in Proc. of the 2nd *Intern. Conf. On Superstrong Fields in Plasmas*, edited by M. Lontano, G. Mourou, O. Svelto, T. Tajima (AIP, New York, 2002), AIP Conference Proceedings **611**, p. 170.

[47] – G. Mourou, Z. Chang, A. Maksimchuk, *et al.*, *Plasma Phys. Rep.* **28**, 12 (2002).

[48] – S.V. Bulanov, T.Zh. Esirkepov, K. Nishihara, F. Pegoraro, "Three-dimensional Electromagnetic Solitary Waves in an Underdense Plasma in PIC Simulations" in Proc. of the 2nd *Intern. Conf. On Superstrong Fields in Plasmas*, edited by M. Lontano, G. Mourou, O. Svelto, T. Tajima (AIP, New York, 2002), AIP Conference Proceedings **611**, p. 191.

[49] – T. Esirkepov, K. Nishihara, S. Bulanov, *et al.*, "A Three-dimensional Sub-cycle

Relativistic Electromagnetic Soliton in a Collisionless Plasma", *this Conference.*

[50] – T.Zh. Esirkepov, S.V. Bulanov, K. Nishihara, F. Pegoraro, "Three-dimensional Relativistic Electromagnetic Sub-cycle Solitons", submitted to *Phys. Rev. Lett.* (2002).

[51] – G.S. Sarkisov, V.Yu. Bychenkov, V.N. Novikov, *et al.*, *Phys. Rev. E* **59**, 7042 (1999).

[52] – T.Zh. Esirkepov, Y. Sentoku, K. Mima, *et al.*, *JETP Lett.* **70**, 82 (1999).

[53] – S.V. Bulanov , T.Zh. Esirkepov, F. Califano, *et al.*, *JETP Lett.* **71**, 407 (2000).

[54] – Y. Sentoku, T.V. Liseikina, T.Zh. Esirkepov, *Phys. Rev. E* **62**, 7271 (2000).

[55] – Ya.B. Zel'dovich, Yu.P. Raizer, *Physics of Shock Waves and High-Temperature Hydrodynamic Phenomena* (Academic, New York, 1967).

[56] – T.Zh. Esirkepov, F.F. Kamenets, S.V. Bulanov, N.M. Naumova, *JETP Lett.* **68**, 36 (1998).

[57] – D. Farina, M. Lontano, S.V. Bulanov, *Phys. Rev. E* **62**, 4146 (2000).

[58] – A.Sh. Abdullaev, *Sov J. Plasma Phys.* **14**, 214 (1988).

[59] – I.V. Sokolov, *Sov. Phys. Usp.* **34**, 925 (1991).

[60] – L.M. Gorbunov, R.R. Ramazashvili, *JETP* **87**, 461 (1998).

[61] – D. Farina, S.V. Bulanov, *Phys. Rev. Lett.* **86**, 5289 (2001).

[62] – D. Farina, S.V. Bulanov, *Plasma Phys. Rep.* **27**, 641 (2001).

[63] – M. Lontano, S. Bulanov, J. Koga, *Phys. Plasmas* **8**, 5113 (2001).

[64] – M. Lontano, S. Bulanov, J. Koga, *et al.*, *Phys. Plasmas* **9**, *in press* (2002).

[65] – S. Weinberg, *Gravitation and Cosmology* (Wiley, New York, 1972).

[66] – D.I. Dzhavakhishvili and N.L. Tsintsadze, *Sov. Phys. JETP* **37**, 666 (1973).

[67] – V.I. Berezhiani, S.M. Mahajan, Z. Yoshida, M. Ohhashi, "Self-trapping of strong electromagnetic beams in relativistic plasmas ", arXiv: physics/0102077 v1 23 Feb 2001.

[68] – M. Passoni, M. Lontano, S. Bulanov, "RES in a hot quasi-neutral plasma", to be presented at the *IQEC/LAT 2002,* Moscow, 22-28 June 2002, poster n.1554.

[69] – M. Borghesi, S. Bulanov, D.H. Campbell, *et al.*, *Phys. Rev. Lett.* **88**, 135002 (2002).

[70] – M. Borghesi, A. Schiavi, D.H. Campbell, *et al.*, *Plasma Phys. Contr. Fus.* **43**, A267, (2001).

Electromagnetic localized structures in a relativistic laser-plasma interaction

Lj. Hadžievski*, M. M. Škorić*, M. S. Jovanović† and Lj. Nikolić**

*Vinča Institute of Nuclear Sciences, POB 522, 11001 Belgrade, Yugoslavia
†Faculty of Natural Sciences, University of Niš, 18000 Niš, Yugoslavia
**Graduate University for Advanced Studies, National Institute for Fusion Science, Toki, Japan

Abstract. Localized electromagnetic structures with trapped laser light inside are often found in simulations of relativistic laser-plasmas. First, we examine existence and stability of one-dimensional electromagnetic solitons formed in a relativistic laser interaction with an underdense cold plasma. In a weakly relativistic model, the original equation of the nonlinear Schrödinger type, with local and non-local cubic nonlinearities is derived. Standing electromagnetic solitons are analytically shown to be stable in agreement with the model simulation. Next, we discuss a novel example of a stimulated backscattering of a laser light from the slow trapped electron-acoustic wave, involving a standing (localized) Stokes light sideband. Conditions are examined when above underdense instability dominates over standard stimulated Raman backscattering.

INTRODUCTION

Slow intense electromagnetic structures with trapped laser light inside are frequently observed in simulations of relativistic laser-plasmas. A complex interplay between relativistic electronic instabilities which produces depletion and frequency down-shift of a laser pulse is often followed by a phenomenon of light localization. Such, a laser pulse gets partially transformed into close to zero velocity light structures, such is e.g. a train of ultra-short relativistic solitons. We first discuss formation and stability of electromagnetic solitons in a relativistic laser propagation through a long underdense plasma. In conditions dominated by stimulated Raman scattering (SRS) and relativistic modulational instability standing light solitons are analytically found. Regions of stability for weak and strong relativistic case are discussed and compared with simulations. Next we turn to a new phenomenon of stimulated electron-acoustic wave scattering (SEAS) proposed recently to explain anomalous scattering data. In first particle simulations of SEAS, in moderately underdense plasma $(> n_{cr}/4)$, not accessible to SRS, a strong backscatter at the electron plasma frequency is observed. A mechanism of resonant coupling between laser, standing backscattered light and trapped low-frequency electron-acoustic mode is proposed to explain an onset of SEAS in hot relativistic plasmas. Conditions for SEAS and implication on relativistic laser-plasma scattering and electron heating are discussed.

CP634, *Science of Superstrong Field Interactions*, edited by K. Nakajima and M. Deguchi
© 2002 American Institute of Physics 0-7354-0089-X/02/$19.00

I ONE- DIMENSIONAL ELECTROMAGNETIC SOLITONS

Relativistic electromagnetic (EM) solitons in laser driven plasmas were analytically predicted and found by particle simulations [1-7]. Relativistic solitons are localized structures with EM energy self-trapped by locally modified plasma refractive index via two effects: the relativistic electron mass increase and the electron density drop by the ponderomotive action of intense laser light [1,3]. A large effort was put into studies of one-dimensional (1D) relativistic EM solitons in an ultraintense laser interaction with underdense and overdense plasmas [8-13]. For laser pulses longer than the electron plasma wave wavelength, spatially modulated depleted pulse, due to stimulated Raman scattering, readily in nonlinear stage breaks-up into a slow train of ultra-short 1D relativistic EM solitons [5,8-9]. It was estimated, for ultra-short laser pulses [9], that 30 to 40% of the laser energy can be trapped inside these low-frequency electromagnetic solitons creating a significant channel for laser beam energy conversion. In this section, the problem of existence, stability and dynamics of linearly polarized electromagnetic solitons is studied by 1D analytical model for a weak relativistic nonlinearity.

Nonlinear Schrödinger model

The fully nonlinear relativistic one-dimensional wave equation, the continuity equation and the cold electron momentum equation, in the Coulomb gauge for the plasma with fixed ions, read

$$\left(\frac{\partial^2}{\partial t^2} - c^2 \frac{\partial^2}{\partial x^2}\right) a = -\frac{\omega_p^2}{n_0} \frac{n}{\gamma} a, \tag{1}$$

$$\frac{\partial n}{\partial t} + \frac{\partial}{\partial x}\left(\frac{np}{m\gamma}\right) = 0, \tag{2}$$

$$\frac{\partial p}{\partial t} = -eE_{\parallel} - mc^2 \frac{\partial \gamma}{\partial x}, \tag{3}$$

where $a = eA/mc^2$ is the normalized vector potential in y direction, n is the electron density, p is the electron momentum in x direction, $\gamma = (1 + a^2 + p^2/m^2c^2)^{1/2}$, $E_{\parallel}$ is the longitudinal electric field, n_0 is the unperturbed electron density and $\omega_p = (4\pi e^2 n_0/m)^{1/2}$ is the background electron plasma frequency. In distinction to circular polarization [9,13], linearly polarized waves have odd harmonics of the vector potential a and even harmonics of the electron density δn [1,12].

In a weakly relativistic limit for $|a| << 1$ and $|\delta n| << 1$ we expand the right hand side of the (1) and introduce the normalized perturbed electron density $\delta n = (n - n_0)/n_0$ and dimensionless variables $x \to (c\omega_p^{-1})x$ and $t \to (\omega_p^{-1})t$ to obtain the wave equation for the vector potential envelope A

$$i\frac{\partial A}{\partial t} + \frac{1}{2}A_{xx} + \frac{3}{16}|A|^2 A - \frac{1}{8}(|A|^2)_{xx}A + \frac{1}{24}(A^2)_{xx}A^* = 0. \tag{4}$$

The eq. (4) has a form of the generalized nonlinear Schrödinger (NLS) equation [16] with two extra nonlocal (derivative) nonlinear terms. We calculate the conserved quantities: photon number P

$$P = \int |A|^2 dx, \tag{5}$$

and Hamiltonian H

$$H = \frac{1}{2} \int \left\{ |A_{xx}|^2 - \frac{3}{16}|A|^4 - \frac{1}{8}[(|A|^2)_x]^2 - \frac{1}{6}|A|^2|A_x|^2 \right\} dx. \tag{6}$$

We look for a stationary, localized solution of (4), found in an implicit form

$$\pm \lambda x = \frac{1}{2\sqrt{2}} \ln \frac{\sqrt{1 - \frac{3}{32}\frac{\alpha^2}{\lambda^2}} + \sqrt{1 - \frac{\alpha^2}{3}}}{\left| \sqrt{1 - \frac{3}{32}\frac{\alpha^2}{\lambda^2}} - \sqrt{1 - \frac{\alpha^2}{3}} \right|} - \frac{4}{3}\lambda \ln \frac{\frac{1}{3}\sqrt{32\lambda^2 - 3\alpha^2} + \sqrt{1 - \frac{\alpha^2}{3}}}{\sqrt{\left| 1 - \frac{32}{9}\lambda^2 \right|}}, \tag{7}$$

with a soliton amplitude $a_0 = \frac{4\sqrt{2}}{\sqrt{3}}\lambda$ of a vector potential a oscillating at $\omega = 1 - \lambda^2$, slightly below the plasma frequency. For the soliton strength λ above the critical value $\lambda \geq \lambda_c = \frac{3}{4\sqrt{2}}$ ($a_0 \geq \sqrt{3}$) the solution has a form of a "cusp" soliton [16]; the centrally highly pointed waveform. For small amplitudes $\lambda << \lambda_c$ one neglects the non-local terms and the solution becomes the well-known secant hyperbolic (NLS) soliton.

Soliton Stability and Model Simulations

To check the soliton (7) stability we use the Vakhitov-Kolokolov criterion [16,17]

$$\frac{dP_0}{d\lambda^2} > 0, \tag{8}$$

where P_0 is the soliton photon number defined by (5). The function $P_0(\lambda)$ for the soliton solution (7) is calculated analytically as

$$P_0(\lambda) = \frac{16\sqrt{2}}{3}\lambda + 2\left(1 - \frac{32}{9}\lambda^2\right)\ln\frac{1 + \frac{4\sqrt{2}}{3}\lambda}{\left|1 - \frac{4\sqrt{2}}{3}\lambda\right|}. \tag{9}$$

The curve $P_0(\lambda)$ represents the stationary solution which corresponds to the minimum of the Hamiltonian H for the fixed photon number P. According to the condition (8) the soliton (7) turns out to be stable in the region $\lambda < \lambda_s \approx 0.44$ ($a_0 < a_s \approx 1.44$) indicating that cusp solitons are also unstable ($\lambda_s < \lambda_c$). More generally, we can now conclude that small amplitude linearly polarized solitons ($a_0 < 1$) within the weakly relativistic model are stable. The shape of the curve $P_0(\lambda)$ predicts the soliton instability region $\lambda > \lambda_s \approx 0.44$. The Vakhitov-Kolokolov criterion just solves a soliton linear stability problem therefore, giving no prediction about the nonlinear evolution of unstable solitons or about stability of the arbitrary shape localized structures. According to nonlinear analysis [18] for the generalized NLS equation with a similar shape of $P_0(\lambda)$, besides the stationary solution there exist two other regimes : (a) soliton collapse and (b) long-lived relaxation oscillations around the stable soliton amplitude. In our case, due to the local cubic nonlinear (NLS) term in (4), one can plausibly expect such dynamical regimes. However, the presence of two extra nonlocal nonlinear terms in (4) indicates the possibility of other dynamical states.

To check analytical results and predicted soliton regimes we perform direct simulation of the model equation (4) using an algorithm based on the split-step Fourier method [19], originally developed for NLS equation. Results prove that the initially launched solitons (7) with the parameters inside the stability region $\lambda < \lambda_s \approx 0.44$ remain stable. Solitons with parameters outside the stability region evolve toward the corresponding stable soliton with long-lived relaxation oscillations. The evolution of the initially launched soliton with amplitude $a_0 \approx 1.6$ ($\lambda \approx 0.2$) outside the stability region is shown on Fig. 1a. A similar behavior exhibit initially perturbed solitons with photon number $P < P_{\max} = P_0(\lambda_s) \approx 4.79$ inside the stability region $\lambda < \lambda_s$ or different localized structures with small deviation from the stable equilibrium. As an example of such dynamics, the time evolution of Gaussian structure with the initial amplitude $A_0 = 0.536$ and photon number $P = 2.875$ is shown in Fig. 1b. The evolution is long-lived relaxation oscillations around the stable soliton amplitude (7) with $\lambda \approx 0.2$ ($a_0 \approx 0.653$) which corresponds to the exact value of the photon number $P = 2.875.$. These dynamical regimes exist also for NLS equation and they are analytically predicted and numerically confirmed in [18]. However, when the initial perturbation increases, the period grows with oscillations becoming strongly nonlinear to exhibit new types of long-lived localized dynamical structures (Fig. 1c). Further deviation from stable equilibrium leads to an aperiodic growth and soliton collapse (Fig. 1d). The understanding of different regimes is important for an insight into the low-frequency process of formation of stable relativistic solitons behind the laser pulse front inside the photon condensate [8]. More detailed determination of these regions in the parameter plane (P, λ) and separatrix curves, requires additional analytical and numerical studies [20].

II STIMULATED ELECTRON-ACOUSTIC-WAVE BACKSCATTERING

Intense laser-plasma interaction can be a source of various electronic instabilities [21-25]. Recently, stimulated backscattering from a trapped low-frequency electron-acoustic wave (SEAS) [D. S. Montgomery *et al.*, Phys. Rev. Lett. **87**, 155001 (2001)] was proposed to reinterpret reflection spectra earlier attributed to stimulated Raman scattering (SRS) from unrealistically low densites. Namely, in the linear theory, the so-called electron-acoustic wave (EAW) exists, i.e. a strongly damped linearized Vlasov-Maxwell (VM) mode whose phase velocity is between an electron plasma wave and an ion-sound wave; often neglected in studies of wave-plasma instabilities [26-28]. However, analytical studies of 1D non-linear VM solutions have found that strong electron trapping

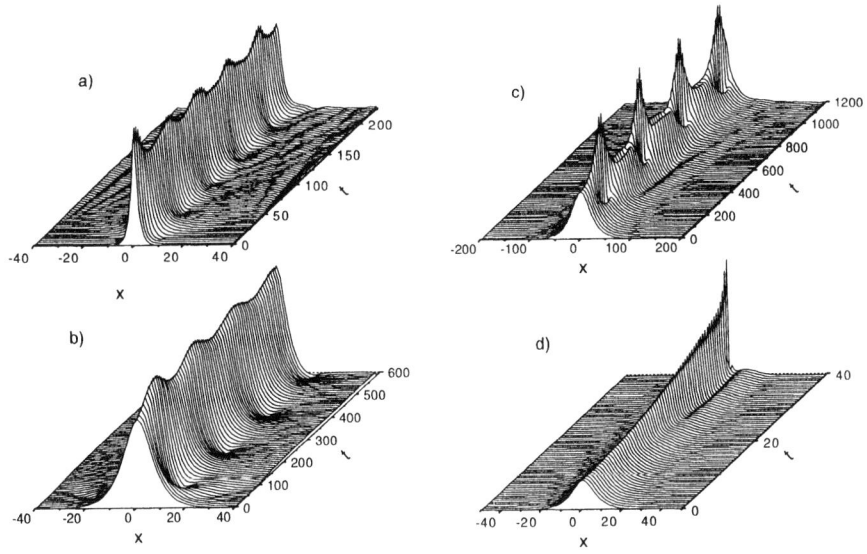

FIGURE 1. Spatio-temporal evolution of different initial structures a) Soliton in the instability region with amplitude $A_0 = 1.6$; b) Gaussian structure with amplitude $A_0 = 0.536$ and photon number $P = 2.875$; c) An example of oscillating dynamical structures and d) An example of collapse dynamics.

can occur even for small amplitude electrostatic wave, resulting in undamped nonlinear traveling waves (BGK-like) [27,28] or, with an inclusion of small dissipation, in weakly damped traveling solutions [29]. The first observation has encouraged further investigation of domains and conditions for SEAS. In this section, excitation of SEAS and its interconnection with SRS instability is addressed by EM relativistic 1d3v particle simulation of linearly polarized laser interacting with a plasma layer placed in vacuum.

From our simulations, SEAS is identified as a resonant 3-wave parametric interaction [30] involving the laser pump (ω_0, k_0), the backscattered lightwave (ω_s, k_s) and the trapped electron-acoustic wave (EAW) (ω_a, k_a). In the linear instability stage, resonant conditions $\omega_0 = \omega_s + \omega_a$ and $k_0 = -k_s + k_a$ are well satisfied, while electromagnetic waves (pump and Stokes wave) satisfy standard dispersion equation $\omega_{0,s}^2 = \omega_p^2 + c^2 k_{0,s}^2$. The backscattered wave is always found to be driven near critical, i.e. $\omega_s \approx \omega_p$ which implies $k_s \approx 0$ and $V_s \approx 0$ (ω_p is the plasma frequency, and $V_s = c^2 k_s / \omega_s$ is the light group velocity). Therefore, the Stokes sideband is a slow, almost standing (localized) electromagnetic wave. The above scheme is observed for wide range of laser intensities, plasma densities and temperatures. It is known that the high temperatures can significantly change the growth rates and sometimes suppress parametric instabilities [31]. However, according to [29], efficient excitation of trapped EAW, is expected in the range $v_{ph}/v_t = 1 - 2$ (v_{ph} and $v_t = (T/m)^{1/2}$ are the phase and electron thermal velocities). Thus, for SEAS excitation at the threshold, high thermal velocities which closely match the EAW phase velocity are important.

Fig. 2a shows the descrete spectrum in an early phase of SEAS instability. The density and the plasma length are $n = 0.6n_{cr}$ ($n_{cr} = n(\omega_0/\omega_p)^{1/2}$)and $L = 40c/\omega_0$, respectively, the longitudinal thermal velocity is $v_t/c = 0.28$ and the laser strength is $\beta = 0.3$. The backscattered EM wave grows at the electron plasma frequency $\omega_p \approx 0.72\omega_0$ (the laser wave at $\omega/\omega_0 = 1$ is not shown), while corresponding EAW is at $\omega_0 - \omega_p \approx 0.28\omega_0$. Note that apart from ES noise around a natural plasma mode ($\omega_p \approx 0.72\omega_0$), ponderomotively driven non-resonant modes are also present (not shown in Fig. 2) at 2-nd, $\omega = 2\omega_0$ and $k = 2k_0$ ($v_{ph}/c \approx 1.3$), as well as at zero-harmonic [32]. From obtained data, it follows that the phase velocity of the EAW is $v_{ph}/c = (\omega_0 - \omega_p)/k_0 \approx 0.45$.

We model SEAS as a resonant parametric 3 -wave coupling $a_i(x,t)\exp[i(k_ix - \omega_i t)]$, which in weakly-varying envelope approximation [32-35], reads

$$\frac{\partial a_0}{\partial t} + V_g^0\frac{\partial a_0}{\partial x} = -M_0 a_s a_a, \tag{10}$$

$$\frac{\partial a_s}{\partial t} - V_g^s\frac{\partial a_s}{\partial x} = M_s a_0^* a_a, \tag{11}$$

$$\frac{\partial a_a}{\partial t} + V_g^a\frac{\partial a_a}{\partial x} + \Gamma_a a_a = M_a a_0^* a_s, \tag{12}$$

where $V_i > 0$ are the group velocities, Γ_a is damping rate for EAW ($\Gamma_0 = \Gamma_a = 0$ for light waves is used), $M_i > 0$ are the coupling coefficients and a_i are the wave amplitudes, where $i = 0, s, a$, stand for the pump, backscattered wave and EAW, respectively. Our model is a short plasma and for high reflectivity, instability has to be Absolute. With boundary conditions $a_0(0) = E_0$, $a_s(L) = a_a(0) = 0$, the backscatter is absolute for

$$L/L_0 > \pi/2, \tag{13}$$

[33-34], where $L_0 = (V_sV_a)^{1/2}/\gamma_0$ is the interaction length and $\gamma_0 = E_0(M_sM_a)^{1/2}$ is the uniform growth rate. Since observed $V_s \approx 0$ for the backscatter, the condition (13) is readily satisfied ($L_0 \approx 0$). However, for EAW (12) a linear dispersion relation in explicit form does not exists [36-39]. Since damping rate $\Gamma_a \neq 0$, the EAW is characterized by the longitudinal absorption length $L_a = V_a/\Gamma_a$, SEAS-backscatter instability becomes absolute under an extra condition [34],

$$L_0/L_a < 2. \tag{14}$$

In linear theory EAW is a highly damped mode, so the absorption length L_a is taking small values. The key factor for an onset and growth of SAES is nearly critical "standing" backward Stokes wave ($L_0 \approx 0$), so that $V_s \approx 0$ also minimizes the threshold E_0 for the instability [34], $\gamma_0 > 0.5\Gamma_a(V_s/V_a)^{1/2}$. Initial discrete EM spectrum, later changes to modulated, broadened blue-shifted profiles, consistent with 3- wave complexity induced by a nonlinear phase, as proposed by some of these authors [25, 35].

The temperature effect is seen near the threshold for SEAS (laser strength $\beta = (eE_0)/(mc\omega_0) \sim 0.3$, where E_0 is the amplitude of the electric field). There is an optimum temperature for perfect matching with an EAW which gives a maximum SEAS reflectivity. For $v_t/c = 0.2$ observed reflectivity is very high - nearly 140% of the incident

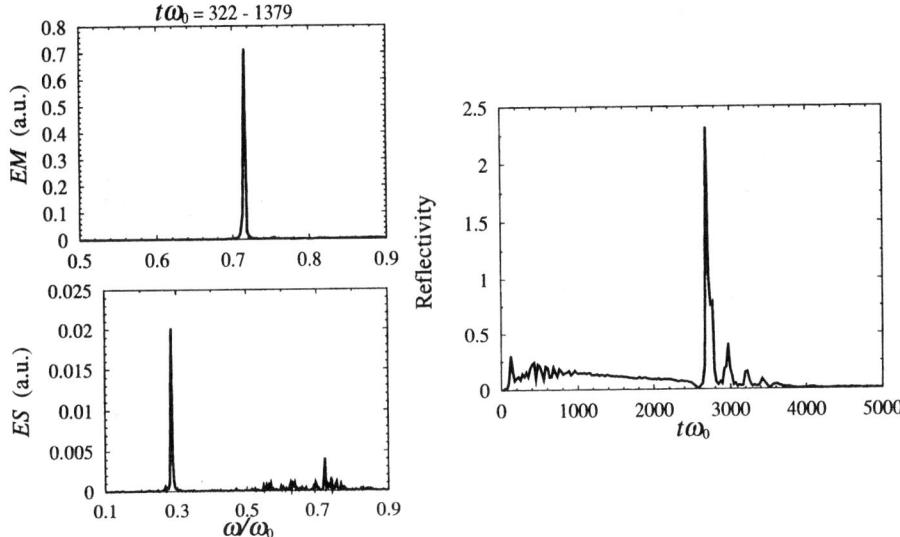

FIGURE 2. a) Spectrum of electromagnetic (top) and electrostatic (bottom) waves in the plasma layer ($n = 0.6n_{cr}$, $l = 40c/\omega_0$) for time interval $t\omega_0 = 322 - 2438$. The initial electron thermal velocity is $v_t/c = 0.28$. b) Reflectivity in time from two connected plasma layers. Initial bursts from ordinary SRS in L_1 are followed at late times by a huge SEAS pulse generated in L_2, after it was heated by SRS-produced hot electrons in L_1.

laser light. One calculates $v_{ph}/v_t = 2.71$, 2.58, 1.84 and 1.72 for $v_t/c = 0.19, 0.20, 0.28$ and 0.30, respectively. For temperatures $\leq v_t/c = 0.18$ and $\beta = 0.3$ the instability was not observed during time period of $t\omega_0 = 5000$. However, for laser intensities well above the threshold there appears no need for high electron temperatures to excite SEAS. For example, already at $T = 500$eV, with a strong relativistic pump $\beta = 0.6$, $n = 0.6n_{cr}$ and $l = 40c/\omega_0$ instability develops fast and quickly saturates within $t\omega_0 = 400$.

Finally, we discuss a coexistence and interrelation between SRS and SEAS. The simulated system consists of two connected underdense plasma layers L_1 and L_2, of the length $l_1 = 20c/\omega_0$ and $l_2 = 80c/\omega_0$ with corresponding densities $n_1 = 0.2n_{cr}$ and $n_2 = 0.6n_{cr}$, respectively. Initial temperature is taken at 500eV. Our choice of densities makes L_1 strongly active for Raman instability, while L_2 (overdense for SRS) is practically in a role of a heat sink. Simulations show common picture, strong SRS with intermittent reflectivity pulsations [25, 35]. The instability eventually gets suppressed by strong electron heating. Hot electrons quickly escape the Raman region to enter and heat the sink (L_2). A striking feature emerges at late times, with an a second intense reflected pulse much larger than the original Raman signal (see Fig.3). This is readily identified as SEAS which comes from the large "sink", once its temperature increased to resonate with EAW and excite SEAS. Therefore, SEAS mediated by SRS becomes a dominant process, as an example of a complex interplay possibly relevant to our understanding of future NIF experiments.

In summary, in first particle simulations in a plasma not accessible to SRS, strong isolated SEAS reflection was observed near the electron plasma frequency. While in reported experiments [26] SEAS to SRS signal ratio was smaller than 10^{-3}, conditions in which SEAS dominates over SRS were found.

ACKNOWLEDGMENTS

This work was supported by Ministry of Science and Technology, Serbia, Project 1964.

REFERENCES

1. A. I. Akhiezer,R. V. Polovin, Soviet Phys. JETP **3**, 696-706 (1956).
2. V. A. Kozlov, A. G. Litvak and E. V. Suvorov, Sov. Phys. JETP **49**, 75 (1979)
3. K. V. Kotetishvili, P. Kaw and N. L. Tsintzadze, Sov. J. Plasma Phys. **9**, 468 (1983)
4. M. Y. Yu, P. K. Shukla and N. L. Tsintzadze, Phys. Fluids **25**, 1049 (1984)
5. S.V. Bulanov, et al. , Phys. Fluids B **4**,1935 (1992)
6. H.H. Kuehl, C.Y. Zhang and T. Katsouleas, Phys. Rev. E **47**, 1249 (1993)
7. P. K. Kaw, A. Sen, T. Katsouleas, Phys. Rev. Lett. **68**, 3172-3175 (1992).
8. K. Mima, M. S. Jovanović, Y. Sentoku, Z. -M. Sheng, M. M. Škorić, T. Sato, Phys. Plasmas, **8**, 2349-2356 (2001).
9. S. V. Bulanov, et al. , Physica D, **152-153**, 682-693 (2001)   (and references therein).
10. D. Farina and S. V. Bulanov, Phys. Rev. Lett. **86**, 5289 (2001).
11. N. M. Naumova, et al. , Phys. Rev. Lett. **87**, 185004-1 (2001).
12. C. D. Decker, W. B. Mori, K. -C. Tzeng, T. Katsouleas, Phys. Plasmas **3**, 2047-2056 (1996).
13. T. Zh. Esirkepov, F. F. Kamenets, S. V. Bulanov, N. M. Naumova, JETP Lett. **68**, 36-41 (1998).
14. Y. Sentoku, et al. , Phys. Rev. Lett. **83**, 3434-3437 (1999).
15. S. V. Bulanov, et al. , Phys. Rev. Lett. **82**, 3440-3443 (1999).
16. A. G. Litvak, A.M. Sergeev, JETP Lett. **27**, 517-520 (1978).
17. N. G. Vakhitov, A. A. Kolokolov, Izv. Vyssh. Uchebn. Zaved. Radiofizika **16**, 1020-1028 (1973).
18. D. E. Pelinovsky, V. V. Afanasijev and Y. S. Kivshar, Phys. Rev. E **53**, 1940-1952 (1996).
19. T. R. Taha, M. J. Ablowitz, J. Comput. Phys. **55**, 203 (1984).
20. Lj. Hadžievski, M. M. Škorić, Phys. of Fluids B, **3** 2452 (1991).
21. W. L. Kruer *et al.*, Phys. Scripta T**75**, 7 (1998).
22. D. S. Montgomery *et al.*, Phys. Plasmas **5**, 1973 (1998).
23. R. L. Berger *et al.*, Phys. Plasmas **5**, 4337 (1998).
24. J. C. Fernandéz, J. A. Cobble, D. S. Montgomery, M. D. Wilke and B. B. Afeyan, Phys. Plasmas **7**, 3473 (2000).
25. H. X. Vu *et al.*, Phys. Rev. Lett. **86**, 4306 (2001).
26. D. S. Montgomery *et al.*, Phys. Rev. Lett. **87** (2001).
27. H. Shamel, Phys. Plasmas **7**, 4831 (2000).
28. J. P. Holloway and J. J. Doring, Phys. Rev. A **44**, 3856 (1991).
29. H. A. Rose and D. A. Russell, Phys. Plasmas **8**, 4784 (2001).
30. K. Mima and K. Nishikawa, Basic Plasma Physics II, eds. A. A. Galeev and R. N. Sudan, (North-Holland, Amsterdam, 1984), p. 451
31. Z.-M. Sheng *et al.*, Phys. Rev. E **61**, 4361 (2000).
32. F. W. Sluijter and D. Montgomery, Phys. Fluids **8**, 551 (1965).
33. R. W. Harvey and G. Schmidt, Phys. Fluids **18**, 1395 (1975).
34. D. W. Forslund *et al.*, Phys. Fluids **18**, 1002 (1975).
35. S. Miyamoto *et al.*, J. Phys. Soc. Jpn. **67**, 1281 (1998); M.M. Škorić *et al.*, Phys. Rev. E **53**, 4056 (1996).

A Three Dimensional Simulation of Solitary Waves in the Laser Wake

T. Esirkepov[1,2], K. Nishihara[2], S. Bulanov[3], F. Pegoraro[4], F. Kamenets[1], N. Knyazev[1]

[1]Moscow Institute of Physics and Technology, Institutskij per. 9, Dolgoprudny 141700, Russia
[2]Institute of Laser Engineering, Osaka University, 2-6 Yamada-oka, Suita, Osaka 565-0871, Japan
[3]General Physics Institute of RAS, Vavilov str. 38, Moscow 119991, Russia
[4]University of Pisa and INFM, pz.Torricelli 2, Pisa 56100, Italy

Abstract. Three dimensional relativistic electromagnetic sub-cycle solitons were observed in Particle-in-Cell simulations of an intense short laser pulse propagation in an underdense plasma. Their structure resembles that of an oscillating electric dipole with a poloidal electric field and a toroidal magnetic field that oscillate in-phase with the electron density with frequency below the Langmuir frequency. On the ion time scale the soliton undergoes a Coulomb explosion of its core, resulting in ion acceleration, and then evolves into a slowly expanding quasi-neutral cavity.

INTRODUCTION

Particle-in-Cell (PIC) simulation show different types of coherent structures in the wake of an intense short laser pulse propagating in underdense plasmas [1]. These are electrostatic wakefields, quasistatic magnetic fields associated with electron fluid vortices, bubbles in the plasma density associated with electromagnetic solitary waves and soliton traines. In this paper we discuss the formation and the evolution of electromagnetic solitary waves ('solitons' for short) in the wake of an intense laser pulse in the three dimensional (3D) case.

The nonlinear wave evolution in the three dimensional regimes differs drastically from their evolution in the one- and two-dimensional approximations, as exemplified by the problem of wave collapse [2] and by the problem of the transverse stability of solitons [3]. Relativistic electromagnetic solitons are now routinely observed in two dimensional (2D) PIC simulations [4-7] in the wake of an intense short laser pulse propagating in an underdense plasma. Solitons attract a great attention because they are of fundamental importance for nonlinear science [8] and are considered to be a basic component of turbulence in plasmas [9]. Thus the numerical identification of solitons, among the different kinds of coherent structures formed by an intense laser pulse in a plasma, stimulated a renewed interest in developing an analytical model [10,11] and in envisaging ways of detecting solitons experimentally [12]. In principle, knowing the properties of various types of two-dimensional solitons, one could guess the topological structure of a three-dimensional soliton. However, the main problem

CP634, *Science of Superstrong Field Interactions*, edited by K. Nakajima and M. Deguchi
© 2002 American Institute of Physics 0-7354-0089-X/02/$19.00

that arises here stems from the difficulty of confining the vector electromagnetic fields inside a finite 3D domain.

As was stressed in Ref. [13], the complexity of the strong electromagnetic wave interaction with plasmas, due to the high dimensionality of the problem, to the lack of symmetry and to the importance of nonlinear and kinetic effects, makes analytical methods unable to provide a detailed description. On the other hand, powerful methods for investigating the laser-plasma interaction have become available through the advent of modern supercomputers and the developments of applied mathematics. In the case of ultra-short relativistically strong laser pulses, simulations with 3D Particle-in-Cell codes provide a unique opportunity for describing the nonlinear dynamics of laser plasmas adequately, including the generation of coherent nonlinear structures such as relativistic solitons.

This has motivated us to perform 3D simulations of the formation and evolution of laser induced sub-cycle relativistic solitons.

In the following section we briefly summarize the recent developments in the investigation of relativistic solitons in plasmas. After that, we describe the simulation code. Then we discuss a 3D relativistic electromagnetic sub-cycle soliton seen in PIC simulation. In the last section we present a conclusion.

RELATIVISTIC ELECTROMAGNETIC SUB-CYCLE SOLITON

As observed in PIC simulations, the solitons generated in a laser-plasma interaction are very important for the theory of laser plasma dynamics (more generally - for the theory of nonlinear waves) and for laser plasma applications because they acquire a significant part of the laser pulse energy; they exist for many electron plasma wave periods; their formation and evolution are associated with the change of the laser pulse frequency; they cause the so called Coulomb explosion of plasma in their locations producing resulting in the heating of ions.

The soliton seen in PIC simulations is an isolated cavity in the plasma containing an oscillating electromagnetic field. The size of the cavity is of the order of the plasma wavelength and the oscillation frequency is below the Langmuir frequency of the surrounding unperturbed plasma. Exactly one half of a spatial electromagnetic wave is trapped inside the cavity, therefore we use the term 'sub-cycle'. The dimensionless amplitude of the electromagnetic field in the soliton is of the order of $a=eE/(m_e\omega c)\approx1$ or higher, here E is the magnitude of the electric field inside the soliton, ω is the soliton frequency, e and m_e are electric charge and mass of electron, c is the speed of light. The mass of an electron in such a field $\sim m_e(1+a^2)^{1/2}$ is significantly greater than its rest mass, therefore we use the term 'relativistic'. The soliton under consideration moves with a non-relativistic velocity or stands still at the place where it is formed, and so we sometimes use the term 'slow'.

We use the term 'soliton' instead of 'solitary wave' for brevity, even if the evolution of these structures can violate their strict definition as solitons in the sense of the inverse-scattering method in the theory of solitons.

The theory of intense electromagnetic solitons in plasmas has been developed in many papers, see Refs. [10,11,14], and also references in Ref. [1]. Their best known application is the use of electromagnetic 'fast' solitons with group velocity close to the speed of light for charged particle and photon acceleration. Many analytical models are based on the so called 'envelope' approximation and predict solitons in the form of a 'fast' wave packet containing many spatial cycles. Here we primarily discuss the sub-cycle slow solitons that are observed in Particle-in-Cell simulations. We emphasize that we have not yet observed 'fast' multi-cycles solitary waves in the PIC simulations.

One-dimensional (1D) sub-cycle relativistic electro-magnetic solitons in underdense plasma were observed for the first time in PIC simulation in Ref. [15]. The solitons are the cavities in the electron density with trapped electro-magnetic field oscillating with frequency below the unperturbed Langmuir frequency. The mechanism of soliton formation and the structure of circularly polarized solitons were investigated in Ref. [16]. An intense laser pulse propagating in an underdense plasma undergoes Raman scattering. The laser pulse loses its energy rapidly while the number of photons in the pulse is conserved. Thus the local frequency of the rear part of the pulse decreases down to the local Langmuir frequency. As a result, the low-frequency part of the electromagnetic radiation of the laser pulse is trapped inside the cavity in the electron density, and a sub-cycle soliton is formed. Two "pure" types of one-dimensional solitons were found: the linearly polarized L-solitons and circularly polarized C-solitons.

The exact analytical solution of the electron hydrodynamic – Maxwell equations representing one-dimensional sub-cycle circularly-polarized relativistic electromagnetic soliton was obtained in Ref. [10] in perfect agreement with the results of 1D PIC simulation.

Two-dimensional relativistic electromagnetic solitons were discovered in the PIC simulations presented in Ref. [4]. Two "pure" types of two-dimensional solitons were found: S-soliton with transverse electric field and azimuthal magnetic field, and P-soliton with the opposite structure — transverse magnetic field and azimuthal electric field. In contrast to the electron vortices, which move across a density gradient, solitons move along the density gradient towards the lowest density. At some critical density the soliton emits its energy in the form of a low-frequency short electomagnetic burst [5]. One can imagine "a solitonic gun" when the laser pulse propagates in the direction perpendicular to the plasma density gradient, and solitons are emitted in the direction opposite to the density gradient.

The interaction of two 2D S-solitons leads to their merging. The resulting soliton acquires the total energy of the merged solitons [17]. The interaction of two counter-propagating 1D solitons is completely different: they preserve their shape and velocity but lose their phase, similarly to a well-known "phase shift" in the theory of KdV solitons.

In an electron-ion plasma 1D and 2D solitons evolve into post-solitons in an ion time-scale due to the ion acceleration by the time-averaged electrostatic field inside the soliton. This effect leads to the formation of slowly expanding bubbles in plasma density [7].

Presumably sub-cycle soliton cavities were detected recently in the brilliant experiment of Dr. M. Borghesi et al., [12]. In this experiment a new powerful diagnostic tool - proton imaging - was applied. The intense laser pulse is divided into two beams, one interacts with the target under investigation and produces a plasma, the other irradiates a thin metal foil producing fast protons with a quasi-thermal energy spectrum corresponding to the effective temperature 3-8 MeV. This proton beam passes through the target, and is then captured by a detector, which consists of many thin films. Each film of the detector captures the protons with the corresponding energy because the proton slowing down in the matter depends on its energy. So, we obtain a "proton image" in each film. The initial transverse size of the proton beam is tens of microns, the size of the image formed is up to several millimeters. Inside the soliton the ponderomotive pressure of the trapped electromagnetic radiation acts on the electrons and results in the formation of the quasistatic electric field. The electric fields in the target under study deflect the protons and therefore the density of protons in different parts of the image is changed accordingly. This allows us to reconstruct the structure of the strong quasistatic electric fields in the target.

3D PARTICLE-IN-CELL CODE «REMP»

We use three-dimensional massively parallel and fully vectorized Relativistic Electro-Magnetic Particle-mesh code (REMP), written by one of the authors (T.E.). The code is based on Particle-in-Cell method [18], and exploits 'Density Decomposition' — a new scheme of the electric current density assignment, described in Ref. [19]. The scheme guarantees the exact conservation of the local electric charge, and is valid for an arbitrary form-factor (shape) of a quasi-particle. Currently we use the second-order polynomial form-factor (parabolic spline). This technique allows us to significantly reduce unphysical numerical effects of the PIC method such us numerical heating, sampling noise, numerical collisions, etc. This advantage is very important for identifying hydrodynamic-like long-living coherent structures such as solitons and electron vortices.

The code is written in Fortran-90. It is parallelized via Message Passing Interface (MPI) using domain decomposition method in the framework of the symmetric SIMD (single image – multiple data) model. The code is fully vectorized. The electric current density (or electric charge density) assignment is vectorized using the modified pre-sorting method and advantages of *Density Decomposition* scheme. Interprocessor communications are optimized using vector algorithm of binary sortkey sorting.

We developed the system of automatic visualization of data. This system is optimized for processing large three-dimensional data. The system uses various techniques to visualize three-dimensional data such as vector and scalar fields, distribution functions, phase spaces, etc. Currently it is based on Interactive Data Language package (IDL, RSI inc.), which works on a PC or graphics workstation. We routinely use animations, which are very desirable to reveal the features and mechanisms of a quickly varying physical process.

The code is tested and used on CRAY-T3E 1200 (CINECA, Bologna, Italy), SGI Origin-2000 (SNS, Pisa, Italy), Hitachi SR-8000, NEC SX-5 (ILE, Osaka University, Japan), and Intel-based PC or clusters under Windows or Linux.

3D SOLITON SEEN IN PIC SIMULATION

In this section we present the results of three-dimensional Particle-in-Cell simulation of laser induced 3D sub-cycle relativistic electromagnetic solitons. Simulations were performed with the code REMP, described above. In the simulation the laser pulse is gaussian and linearly polarized along the z-axis, its dimensionless amplitude is $a=eE/(m_e\omega c)=1$ corresponding to the peak intensity $I=1.38\times10^{18}W/cm^2\times m^2/\lambda^2$, its FWHM size is $8\lambda\times5\lambda\times5\lambda$. The laser pulse propagates along the x-axis. The focal plane of the laser pulse is placed in front of the plasma slab at the distance of 3λ. The density of plasma is $n=0.36n_{cr}$, where $n_{cr}=m_e\omega^2/(4\pi e^2)$ — critical density. Ions and electrons have the same absolute charge, the mass ratio is $m_i/m_e=1836$. The simulation grid is $660\times400\times400$, the mesh size is 0.05λ. The plasma slab length is 13λ. The total number of quasiparticles is $425\cdot10^6$. The boundary conditions along the y- and z-axes are periodic and absorbing for electromagnetic radiation and for the particles along the x-axis. The simulation were performed on 16 processors of the NEC SX-5 vector supercomputer at CMC, Osaka University. In the figures the time and space units are the period $2\pi/\omega$ and the wavelength λ of the incident radiation, respectively. The structure of 3D soliton is clearly seen in the animations produced from the data (420 stills with the period 0.05).

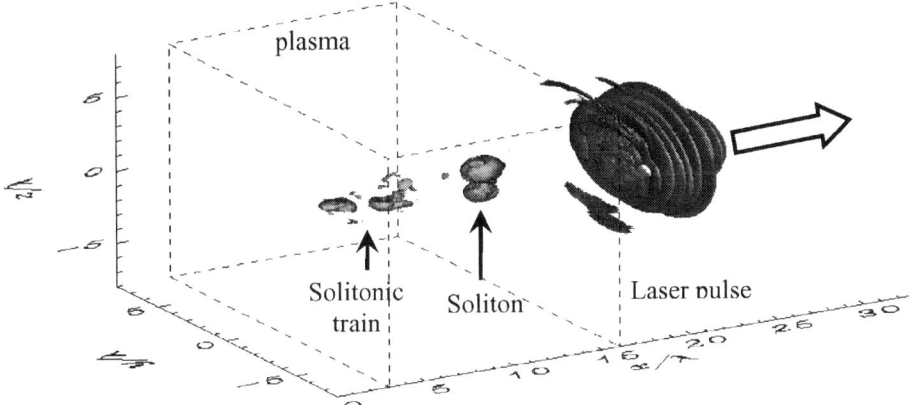

FIGURE 1. Iso-surface of the electromagnetic energy density corresponding to the dimensionless value of $(E^2+B^2)/8\pi=0.09/8\pi$ at t $= 33.75\cdot2\pi/\omega$.

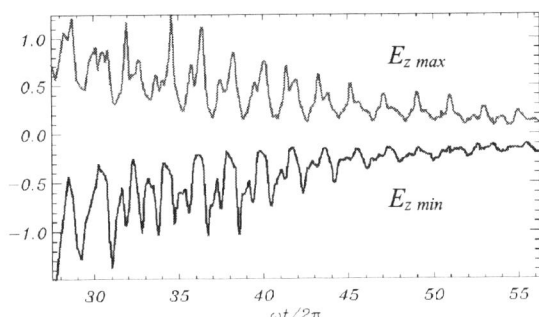

FIGURE 2. Electric field component E_z minimal and maximal values in the soliton versus time.

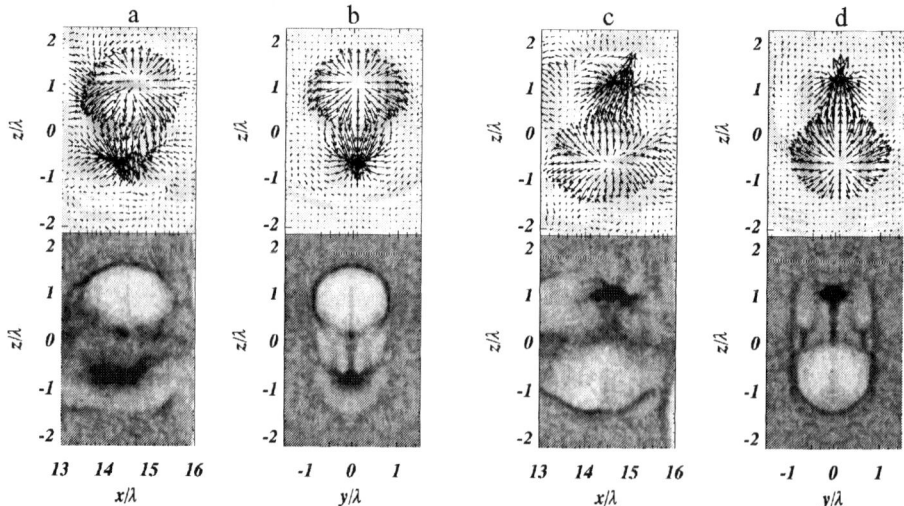

FIGURE 3. The soliton structure. Upper raw: arrows represent planar electric field, background is transverse magnetic field. Bottom: electron density. Columns (a),(c) are cross-sections at $y=0$; (b),(d) – at $x=14.5$. (a),(b) correspond to $t=39.3$; (c),(d) – to $t=40.2$.

The laser pulse undergoes a weak self-focusing in the plasma slab. Behind the laser pulse we see a soliton train and one isolated soliton in Fig. 1. These structures have oscillating electric field polarized in the same direction as the laser pulse, Fig. 2. The frequency of the oscillations is below the Langmuir frequency of the surrounding unperturbed plasma $\omega_s \approx 0.93\,\omega_{pe}$. Therefore the soliton oscillation is not-resonant with the plasma waves and radiation loss due to the wave transformation is not significant.

Figures 3 and 4 show the structure of the isolated soliton. The soliton resembles an oscillating electric dipole. Although we see a strong electrostatic component, the soliton is electromagnetic as one can see directly from the oscillations of the electric and magnetic fields in the soliton shown in Fig. 4. The electric field in the soliton is poloidal and the magnetic field is toroidal. This structure of the electromagnetic field can be considered as that of the lowest eigenmode of a cavity resonator with a deformable wall, and thus we call this structure a Transverse Magnetic (TM) soliton. We note that since the cavity crosses the equatorial plane twice within one oscillation period, the field at the equatorial plane oscillates with frequency $2\omega_s$. The 3D soliton can be described by the 4-component electromagnetic potential (φ, A) and by the 4-component electric current density (ρ, j), which are both confined inside a finite domain. In the equatorial plane the structure of the three dimensional soliton is similar to that of a two dimensional s-soliton, while that in the perpendicular planes is similar to a two dimensional p-soliton. Considering the soliton as a wave packet, it has only half of one cycle in space, so we use the term "sub-cycle soliton". We also note that the soliton is azimuthally symmetric but in the x-z plane the symmetry is perturbed, as seen in Fig. 3. This may be due to the fact that the laser pulse has propagated through the plasma in the x-direction. Namely the soliton dynamics is slightly affected by the quasistatic magnetic field and by the remnants of the density filament [20] induced by the laser pulse.

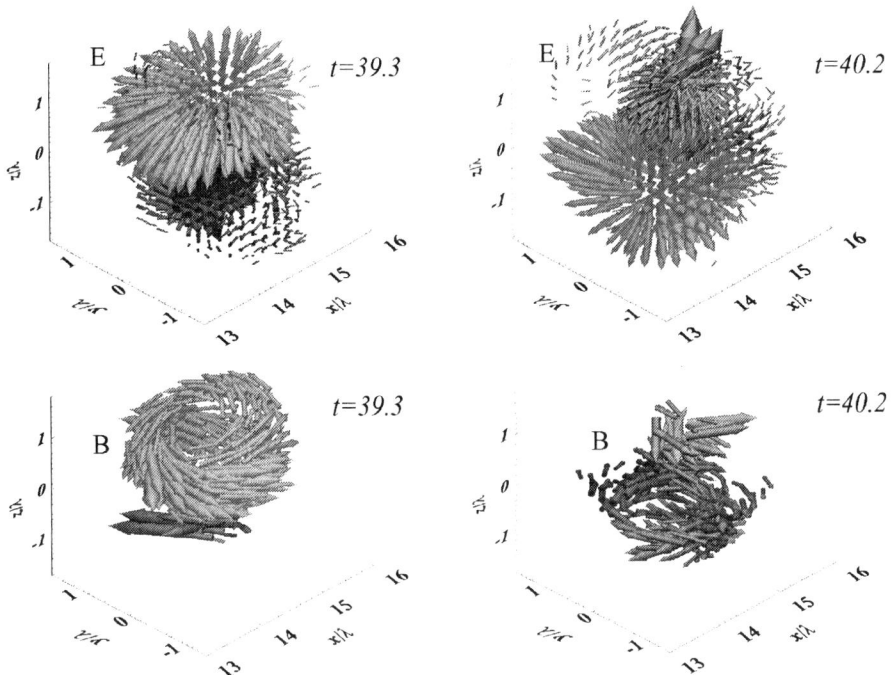

FIGURE 4. Three dimensional structure of the electric and magnetic fields in the soliton during the soliton period. Electric field is poloidal, magnetic field is toroidal.

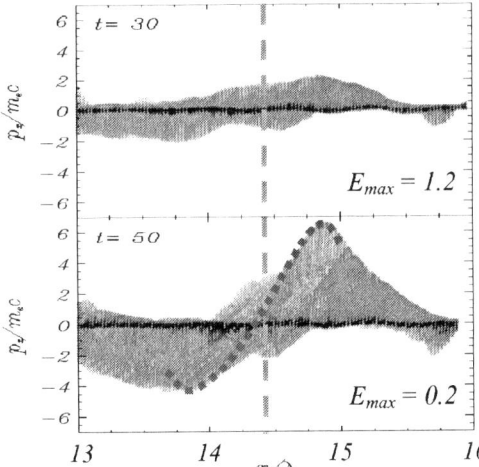

FIGURE 5. Ion acceleration inside the soliton during 10 soliton periods, ion momentum is expressed in $m_e c$.

On the ion time-scale the soliton evolves into the post-soliton. The magnitude of the soliton decreases, as shown in Fig. 2, and ions acquire momentum, see Fig. 5. As the last stage we see a slowly expanding ($v \approx 3 \cdot 10^{-3}c$) cavity in plasma density. In the soliton train almost isolated structures considered as solitons tend to merge and form a foam of bubbles with relatively high-density ($2\text{-}5n_{cr}$) walls.

CONCLUSION

In conclusion we have provided the demonstration of the existence of 3D sub-cycle relativistic electromagnetic solitons in a collisionless cold plasma. The solitons are generated as a consequence of the frequency downshift of an intense laser pulse propagating in an underdense plasma. A substantial part of the pulse energy is transformed into solitons, approximately *25-30%* of the incident laser pulse. The core of the soliton is positively charged on average in time and the soliton undergoes a Coulomb explosion in an ion time-scale. This process results in heating of the plasma ions. Then the soliton evolves into a post-soliton, which is a slowly expanding quasi-neutral cavity in the plasma. Merging solitons can form a slowly expanding foam.

ACKNOWLEDGMENTS

We appreciate the help of ILE computer group and Cyber Media Center of Osaka University (Japan). This work was supported by Japan Society for the Promotion of Sciense, Istituto Nazionale di Fisica della Materia (Italy), Russian Ministry of science and technology, Russian Fund of Basic Research.

REFERENCES

1. Bulanov, S. V., et al., in *Reviews of Plasma Physics,* New York: Kluwer Academic / Plenum Publishers, 2001, edited by V. D. Shafranov, Vol. 22, p. 227.
2. Zakharov, V. E., *Sov. Phys. JETP* **35**, 908 (1972).
3. Kadomtsev, B. B., and Petviashvili, V. I., *Doklady AN SSSR* **192**, 753 (1970); Zakharov, V. E., and Rubenchik, A. M., *Sov. Phys. JETP* **38**, 494 (1974).
4. Bulanov, S. V., et al., *Phys. Rev. Lett.* **82**, 3440 (1999).
5. Sentoku, Y., et al., *Phys. Rev. Lett.* **83**, 3434 (1999).
6. Naumova, N. M., et al., *Phys. Plasmas* **8**, 4149 (2001).
7. Naumova, N. M., et al., *Phys. Rev. Lett.* **87**, 185004 (2001).
8. Novikov, S. et al., *Theory of Solitons: the Inverse Scattering Method,* New York: Consultants Bureau, 1984; Taniuti, T., and Nishihara, K., *Nonlinear Waves,* Boston: Pitman Advanced Publishing Program, 1983.
9. Mima, K., et al., *Phys. Plasmas* **8**, 2349 (2001).
10. Esirkepov, T. Zh., et al., *JETP Lett.* **68**, 36 (1998).
11. Farina, D., et al., *Phys. Rev.* **E 62**, 4146 (2000); Farina, D., and Bulanov, S. V., *Phys. Rev. Lett.* **86**, 5289 (2001); *Plasma Phys. Rep.* **27**, 680 (2001); Poornakala, S., et al., *Phys. Plasmas* **9**, 1820 (2002).
12. Borghesi, M., et al., *Phys. Rev. Lett.* **88**, 135002 (2002).
13. Dawson, J. M., and Lin, A. T., in: *Basic Plasma Physics,* edited by M. N. Rosenbluth and R. Z. Sagdeev, North-Holland, Amsterdam, 1984, Vol. 2, p. 555; Dawson, J. M., *Phys. Plasmas* **6**, 4436 (1999).
14. Gersten, J. I., and Tzoar, N., *Phys. Rev. Lett.* **35**, 934 (1975); Kozlov, V. A., et al., *Sov. Phys. JETP* **49**, 75 (1979); Kaw, P. K., et al., *Phys. Rev. Lett.* **68**, 3172 (1992).
15. Bulanov, S.V., et al., *Phys. Fluids* **B 4**, 1935 (1992).
16. Bulanov, S.V., et al., *Plasma Phys. Rep.* **21**, 600 (1995).
17. Bulanov, S.V., et al., *Physica* **D 152–153**, 682–693 (2001).
18. Hockney, R. W., and Eastwood, J. W., *Computer Simulation Using Particles,* McGraw-Hill Inc., 1981.
19. Esirkepov, T.Zh., *Comput. Phys. Comm.* **135**, 144 (2001).
20. Kuznetsov, A.V., et. al., *Plasma Phys. Rep.* **27**, 3 (2001).

Low energy nuclear transitions initiated by femtosecond laser plasma

A.B. Savel'ev, A.V.Andreev, V.M.Gordienko, and P.M.Mikheev

International Laser Centre & Physics Faculty, M.V.Lomonosov Moscow State University, Vorobyevy gory, Moscow, 119899, Russia, Phone:7(095)9395318; Fax:7(095)9393113; e-mail:savelev@femto.phys.msu.su

Abstract: We discuss origins of low energy nuclear transitions taking place at irradiation of solids with moderate intensity femtosecond laser pulses. Possible applications of these phenomena to population inversion on nuclear transitions are also presented.

INTRODUCTION

During past few years nuclear physics with ultra short laser pulses became one of the most "hot" and intensively investigated area of super strong laser-matter interaction[1]. A wide range of nuclear processes was observed experimentally: from nuclear fusion and fission to photo-excitation and positron production. Among them, there exist nuclear processes initiated by electrons, x-rays or ions of less than few tenths of keV energy. Such processes we would further refer as *low energy nuclear processes*. One does not need relativistic intensities to enter this regime of femtosecond laser-matter interaction with solids: even at intensities of 10^{16}-10^{17} W/cm^2 hot electron temperature ranges from 3 up to 10 keV. Accelerated fast ions gain nearly the same energy per nucleon from these electrons. Moreover it is someway "optimal" to use moderate intensities for low energy nuclear processes initiation in plasma, and relativistic intensity regime seems even useless. Thus, under irradiation of solids by femtosecond laser pulses at intensity around 10^{16} W/cm^2 X-ray photo-excitation of 6.238 keV level of ^{181}Ta nuclei took place[2]. At the same conditions neutrons emerging from DD fusion in structured D-enriched Ti target[3] were detected.

In this paper, we discuss experimental and theoretical aspects while initiating low energy nuclear transitions in solids using moderate intensity laser pulses. We start by presenting our data for thorough characterization of hot electrons production under moderate intensity femtosecond laser-plasma interaction. We then move to discussion of low energy nuclear transition excitation by plasma x-rays. Finally we discuss two-, three- and four level schemes for population inversion on nuclear transitions of 1-30 keV energy.

CP634, *Science of Superstrong Field Interactions*, edited by K. Nakajima and M. Deguchi
© 2002 American Institute of Physics 0-7354-0089-X/02/$19.00

EXPERIMENTAL CHARACTERIZATION OF HOT ELECTRONIC COMPONENT

In our experiments we characterize hot electrons production under femtosecond laser-plasma interaction by three different methods: by direct detection of hot electrons escaping from plasma, with the help of double-channel x-ray scheme [4], and from time-of-flight ionic measurements.

Figure 1 sketches our experimental setup. P-polarized radiation (200 fs, 616 nm) was tightly focused onto flat target at 45° angle of incidence with maximum intensity up to $3 \cdot 10^{16}$ W/cm^2. Energy contrast ratio of the pulses was better than 1000 that yield in better than 5 orders for intensity contrast. Target was placed inside vacuum chamber with residual pressure less than 10^{-5} Torr. Different diagnostics were placed around the target.

Fig.1 Experimental set up for hot electron component study.

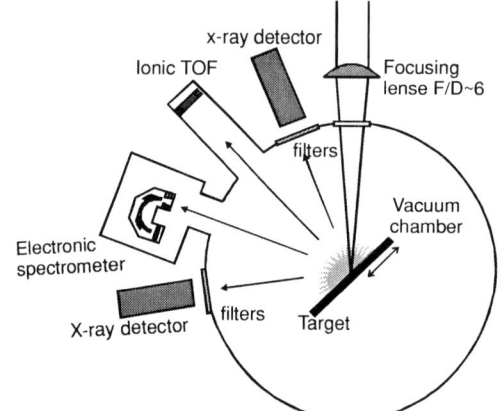

Two hard x-ray detectors equipped with different filter sets were mounted out of the vacuum chamber looking at the target through beryllium windows 100 μm thick. Double channel x-ray detection scheme and algorithms providing for hot electron temperature assessment in a single laser shot were described elsewhere [4,5]. Simple ionic time-of-flight (TOF) spectrometer consists of chevron-type double plate MCP detector VEU-7 was placed 22 cm apart from the target along the target normal. This enables measurement of ion velocities up to 5×10^8 cm/s [6].

Direct detection of electronic plasma currents was realized with special spectrometer providing for both time-of-flight and energy resolution capabilities. It consists of cylindrical capacitor-like half-circle energy analyzer with energy resolution of 8% and detection range of 0.1-40 keV. This analyzer having itself 25 cm electron path was placed 50 cm apart from the target. Its entrance diaphragm was right up on the target normal. Electron detection was fulfilled with chevron-type double-plate MCP detector VEU-7.

Fig.2 presents results from double-channel x-ray detection scheme in the case of Si target. Characteristic slope of the curve $E_2 \sim I^{0.8}$ corresponds well to the known data and theoretical predictions if laser energy absorbed in anomalous skin-effect regime [7]. Here we are using quantity "mean energy" instead of "electron temperature" because the latter could be defined provided hot electrons follow maxwellian type distribution, that is natural only for a system in equilibrium. At the same time, quantity "mean energy" is applicable for any electron ensemble regardless of exact form of the distribution function. Moreover, within frames of our algorithms special filter choice makes assessment of E_2 nearly independent on exact distribution form [5].

Fig. 2 Dependence of mean energy of hot electrons E_2 on laser intensity I estimated by double channel hard X-rays yield (o) and ionic TOF measurements (•).

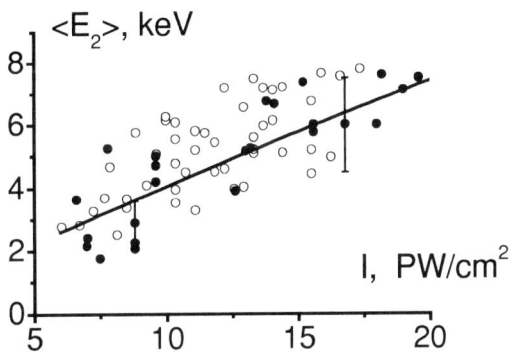

Ionic TOF measurements allow within frames of simple model of plasma expansion estimating both energy E_2 and also the ratio of thermal to hot electron concentration [5]. Figure 2 displays E_2 estimation from ionic TOF measurements. The coincidence of the this estimate with the one obtained by double-channel x-ray detection is excellent.

Finally we made direct detection of hot electron spectrum by electron analyzer. Figure 3 shows such a spectrum at laser pulse intensity of 2×10^{16} W/cm^2. Fit of experimental data was done by the bi-Maxwellian function. This gives values of thermal electrons mean energy $<1.5T_1> \sim 300 \pm 100$ eV, hot electrons mean energy $<E_2> \sim 9 \pm 4$ keV. These values reasonably fits to the data plotted in the Figure 2.

LOW ENERGY NUCLEAR EXCITATION

Among other low energy nuclear processes, the isomeric level excitation seems the most interesting one because of both fundamental physics involved and a wide set of possible applications. First, even for low energy levels of stable isotopes their characteristics could be still shady or yet unknown (for example only calculations exists for both lifetime and conversion coefficient of ^{201}Hg 1.56 keV nuclear level). Usually, still fewer data are known for metastable isotopes. These mean that excitation in hot dense laser plasma could provide for new methods of nuclear spectroscopy of low energy nuclear levels.

Fig. 3. Electronic spectrum of laser plasma at I~2x10^{16} W/cm^2, λ=616 nm. Solid line – bi-maxwellian fit: T_1=350 eV, T_2=6 keV.

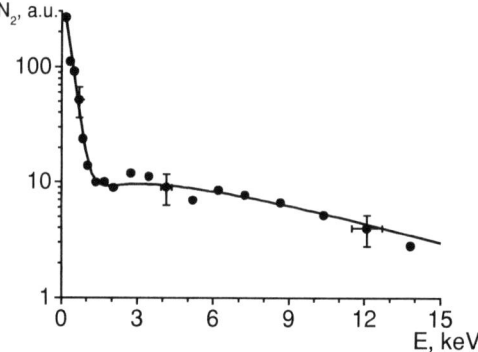

Electrons and X-rays emitted under a femtosecond laser -plasma interaction can not only initiate nuclear reactions but also excite nuclear levels. These processes are of interest for a number of promising applications such as the isotope separation[8] and the creation of population inversion[9]. In particular, X-rays emitted by the USP plasma can be efficiently used for excitation of metastable isotopes to the closely spaced level accompanied by its decay to the ground state. This was realized in paper[10], where the metastable level of ^{178}Hf with the energy of 2.446 MeV and the lifetime of 31 year was excited with a cw X-ray tube.

To our knowledge, no experiments have been performed in this field so far at relativistic intensities. The hot-electron temperature reaches 10 keV already at a moderate intensities, which is sufficient for direct excitation of low-lying nuclear levels of both stable and metastable isotopes (we call the nuclear levels with energy Eg < 20 keV the low-lying nuclear levels). The standard methods of nuclear spectroscopy of such levels are based on the indirect population via the states with energy ε_γ > 100 keV using electron and ion accelerators [11] or on the direct photoexcitation using synchrotron radiation sources[12, 13]. Note that the parameters of low-lying nuclear levels are often unknown (even when the ground state of a nucleus is stable) (see Table 1).

For metastable isotopes, the situation when a number of parameters are unknown is encountered even more often. An important positive factor is that the low-lying nuclear levels in a plasma are excited `instantly' because the relation τ_g>τ_p is satisfied (τ_g is the total lifetime of the excited state and τ_p is the lifetime of a hot dense plasma). The possibility of excitation of low-lying nuclear levels in the laser plasma has been discussed over twenty years[14, 15, 16]. Experimental attempts to use the low- density laser plasma were scarce[17, 18] and unsuccessful[19]. The main problems are related to the low efficiency of excitation by nanosecond laser pulses because of a low electron temperature of the plasma and its low density. The plasma arisen from femtosecond laser plasma interaction is free from these disadvantages and it has at the same time a sufficiently high electron temperature and the nuclear density that is close to that of a solid.

Table 1. Low lying nuclear level data for stable isotopes: ε_γ -excitation energy, M –moment and parity, τ_γ- full half-lifetime, β - internal conversion coefficient, Π - transition multiplicity.

Isotope	ε_γ, keV	M	τ_γ, ns	Π	β
$^{201}_{80}Hg$	1.556	1/2-	1-10	M1+E2	$(2\div5)\cdot10^4$
$^{181}_{73}Ta$	6.238	9/2-	6050	E1	70.5
$^{169}_{69}Tm$	8.4103	3/2+	4.08	M1+E2	285
$^{83}_{36}Kr$	9.396	7/2+	147	M1+E2	17.09
$^{187}_{76}Os$	9.746	3/2-	2.38	M1(+E2)	264
$^{73}_{32}Ge$	13.275	5/2+	2950	E2	1120
$^{57}_{26}Fe$	14.4129	3/2-	98.3	M1+E2	8.56
$^{151}_{63}Eu$	21.541	7/2+	9.6	M1+E2	28

The low-lying nuclear levels in the laser plasma can be excited via different channels, both nuclear-electronic and nuclear-photonic, but the excitation by X-rays emitted by the plasma dominates in the hot dense femtosecond plasma.

POPULATION INVERSION SCHEMES FOR NUCLEAR TRANSITIONS

The study of low-lying nuclear levels of metastable isotopes substantially expands the field of investigations. One of the problems can be a search for candidates suitable for the production of population inversion at nuclear transitions. One can see from Table 1 that the lifetime of low-lying nuclear levels greatly exceeds that of the plasma, so that the population inversion can be produced during the plasma lifetime. At the same time, to obtain the population inversion for the nuclear level with energy less than 10 keV, when the nuclear density is close to the density of a solid, the X-ray intensity within the nuclear transition band should be $>10^{16}$ W cm^{-2}, which corresponds to the laser radiation intensity $I > 10^{21}$ W cm^{-2}. For such intensities, the electron temperature exceeds 10 MeV and it strongly differs from the optimal temperature $T_h \approx \varepsilon_\gamma$ for the energy range of transitions under study. From the point of view of observation of stimulated emission at nuclear transitions, the isotopes presented in Table 1 can be divided into two groups. In the case of nuclei with the high conversion coefficient at the transition to the ground state (^{201}Hg,

^{181}Ta, ^{169}Tm, ^{187}Os, ^{73}Ge), one can hope to obtain generation only in a plasma layer containing ions with a high degree of ionization, because the IC process can be completely or strongly suppressed in this case. In the case of isotopes with a relatively low conversion coefficient (^{83}Kr, ^{57}Fe, ^{161}Dy), it is preferable to attempt to produce inversion in a solid sample. Such an inversion can be achieved, for example, using a multicomponent target consisting of a metal foil emitting X-rays and an active body, which has no thermal contact with the foil. This substantially decreases the Doppler broadening and allows, in principle, the development of schemes of the Mossbauer gamma laser.

When ultrashort relativistic laser pulses are used, the hot-electron temperature achieves several hundreds of keV, providing the possibility for excitation of higher-lying nuclear levels. Consider nuclei of isotopes ^{73}Ge, ^{83}Kr (Table 2), and ^{107}Ag, ^{109}Ag, ^{119}Sn. These nuclei have a number of common features. First, the radiative decay of the first two excited levels is cascade, i.e., the gamma transition from the second excited level occurs to the first level, whereas the transition to the ground state is forbidden. Second, all the isotopes have a long-lived isomeric state with the lifetime of the order of 1 s and longer. Third, these transitions have a rather high electron con- version coefficient. In the scheme for producing the population inversion under study, the temperature T_h should provide cascade excitation of the second nuclear level by X-rays emitted from the plasma or its direct excitation due to inelastic electron scattering. The expanding plasma containing highly ionized ions is deposited on a surface where the ions are completely neutralized. During the lifetime of isotopes ^{73}Ge, ^{83}Kr, and ^{119}Sn in the second excited state, a rather large number of nuclei can be produced in the isomeric state. The ions formed due to the conversion decay at the $2\rightarrow1$ transition are pulled out by a focusing electric field and are deposited on a substrate, where the active body of a gamma laser is formed.

The population inversion can also be achieved for the four-level lasing schemes (^{153}Eu, Table 2). In this case, nuclei at the level 2 are selected due to the conversion decay at the $3 \rightarrow 2$ transition, while the lasing should occur at the $2 \rightarrow 1$ or $2 \rightarrow 0$ transition. A substantial difference of the four-level scheme from the three- level scheme is, for example, the fact that the third excited level of isotopes ^{153}Eu, ^{155}Gd, and ^{189}Os is coupled with the ground state of the nucleus by a radiative transition having a low conversion coefficient. Therefore, the nucleus can be excited to the third level in a multi-component target by filtered X-rays emitted by the laser plasma, which excludes the necessity to neutralize plasma ions. Another advantage of the four-level scheme with the $2 \rightarrow 1$ lasing transition is that this scheme is insensitive to the presence of nuclei in the ground state in the active region of the gamma laser. This substantially decreases technological requirements imposed on the process of preparation of the active region.

Among stable isotopes with low-lying nuclear levels with energies less than several hundreds kiloelectronvolts there are isotopes that permit the use of three- and four-level lasing schemes, which are typical for the visible range. The first five excited levels in the ^{161}Dy isotope are coupled by radiative transitions with the ground state and, therefore, they can be excited by laser plasma X-rays (see Figure 4). It seems that the most

120

Table 2. Low lying nuclear levels of isotopes: ε'_γ and I_γ - γ transition energy and intensity.

Isotope	ε_γ, keV	M	τ_γ, ns	γ transition			β
				Π	I_γ	ε'_γ	
$^{73}_{32}Ge$	0	9/2+					
	13.275	5/2+	2950	E2	100	13.275	1120
	66.716	1/2-	0.499 s	M2	100	53.440	8.67
	68.752	(7/2)+	1.74			55.42	
	68.752	(7/2)+	1.74	M1+E2	100	68.752	0.227
$^{83}_{36}Kr$	0	9/2+					
	9.396	7/2+	147	M1+E2	100	9.396	17.09
	41.543	1/2-	1.83 h	E3	100	32.1473	2035
$^{153}_{63}Eu$	0	5/2+					
	83.36720	7/2+	0.793	M1+E2	100	83.36717	3.82
	97.43103	5/2-	0.198	E1	0.12	14.06383	11.2
	97.43103	5/2-	0.198	E1	100	97.43100	0.307
	103.18016	3/2+	3.85	E2		19.81296	3290
	103.18016	3/2+	3.85	M1+E2	100	103.18012	1.72

promising are the transitions terminating on the second excited level because the lifetime of this level (0.83 ns) is 35 times shorter than that of the first excited state. The ^{171}Yb isotope is of great interest for the development of lasing schemes for nuclear transitions because its third excited level with energy of 95.272 keV has a lifetime of 5.25 ms. This level is coupled by the radiative transition only with the second excited level with energy of 75.878 keV. However, the third level can be excited due to inelastic electron -ion and ion - ion collisions or due to the radiative transition via the fourth excited level with energy of 122.418 keV. The latter level is coupled with the third level by an intense radiative transition with the multiplicity E1.

CONCLUSIONS

Thus hot electrons produced under interaction of femtosecond laser pulses with dense plasma gain enough energy to excite nuclear transitions in wide range of stable and metastable isotopes and isomers. At laser intensities above 10 PW/cm^2 mean energy of hot electrons reaches 6-8 keV providing for generation of hard x-rays up to 10-20 keV.

X-ray photoexcitation of low energy nuclear levels could be treated as new tool for nuclear spectroscopy of such levels. Besides, different schemes for population inversion on gamma-transitions can be proposed, while more consecutive treatment of these ideas suffers from luck of nuclear data on low lying nuclear levels.

ACKNOWLEDGEMENTS

This work was supported by Russian Foundation for Basic Research (grants ## 00-02-17302, 02-02-17138a, and 02-02-06104) and in part by EOARD.

Fig.4. Photoexcitation and decay of ^{161}Dy isotope.

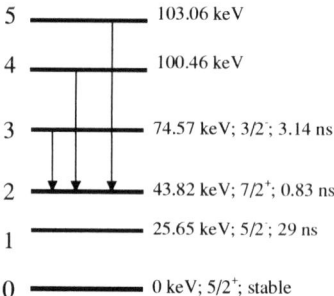

5	103.06 keV
4	100.46 keV
3	74.57 keV; 3/2⁻; 3.14 ns
2	43.82 keV; 7/2⁺; 0.83 ns
1	25.65 keV; 5/2⁻; 29 ns
0	0 keV; 5/2⁺; stable

REFERENCES

1 Andreev A.V., Gordienko V.M., Savel'ev A.B., Quantum electronics, **31**, 941-956 (2001).

2 Andreev A.V., Volkov R.V., Gordienko V.M., Mikheev P.M., Savel'ev A.B., Tkalya E.V., Chutko O.V., Shashkov A.A., Dykhne A.M., JETP Lett., **69**, 371-376 (1999).

3 Volkov R.V., Golishnikov D.M., Gordienko V.M., Mikheev P.M., Savel'ev A.B., Sevastyanov V.D., Chernysh V.S., Chutko O.V., JETP Lett., **72**, 401-404 (2000).

4 Volkov R.V., Gordienko V.M., Mikheev P.M., and Savel'ev A.B., Quantum electronics **30**, 896-900 (2000).

5 Gordienko V.M., Lachko I.M., Mikheev P.M., Savel'ev A.B., Volkov R.V. and Uryupina D.S., to be published.

6 Gavrilov S.A., Golishnikov D.M., Gordienko V.M., Mikheev P.M., Savel'ev A.B., Serov A.A., and Volkov R.V., Quantum Electronics **31**, 24-246 (2001).

7 Andreev A.A., Gamaly E.G., Novikov V.N., Semakin A.N., and Tikhonchuk V.T., JETP **74**, 963 (1992).

8 Andreev A.V., Gordienko V.M., Savel'ev A.B., Laser Physics 10 557 (2000).

9 Andreev A.V., Vestn. Mosk. Univ. Ser. R: Fiz. Astron. 35 (3) 28 (1994).

10 Collins C B, Davanloo F, et al. Phys. Rev. C 61 4305 (2000).

11 Attallah F., Aiche M., Chemin G. F., et al. Phys. Rev. Lett. 75, 1715 (1995).

12 Chumakov A. I., Baron A. Q. R., Arthur J., et al. Phys. Rev. Lett. 75 549 (1995).

13 Seto M., Yoda Y., Kikuta S., et al. Phys. Rev. Lett. 74 3828 (1995).

14 Morita M., Prog. Theor. Phys. 49 1574 (1973)

15 Letokhov V.S., Sov. J.Quantum Electron. 3 360 (1973).

16 Tkalya E.V., Zh. Eksp. Teor. Fiz. 102 379 (1992).

17 Izawa Y., Yamanaka C., Phys. Lett. B 88 59 (1979).

18 Arutyunyan R. V., Bol'shov L. A., Vikharev V. D., et al. Yadernaya. Fizika. 53 36 (1991).

19 Tkalya E. V., Pis'ma Zh. Eksp. Teor. Fiz. 53 441 (1991).

Laser Induced Nuclear Physics

K.W.D.Ledingham

Dept of Physics and Astronomy, University of Glasgow, Glasgow G12 8QQ, Scotland
and
AWE plc, Aldermaston, Reading, Berkshire RG7 4PR, UK

Abstract. *The interaction of ultra-intense focused laser beams with solid targets is a new field of research resulting in the production of exotic plasma conditions similar to the conditions which exist in the interior of some stellar objects. The lasers generate very high energy electrons and ions which can subsequently produce γ-rays, positrons, neutrons and pions. The paper will show that results obtained from these studies have major implications to fundamental plasma physics and high energy accelerator physics as well as important technological potential for the production of compact sources of protons, neutrons, positrons and isotopes. One of the applications considered at some length will be the production of protons and the possibility of designing a table top laser to produce radioactive sources for positron emission tomography (PET) as well as proton oncology. The exciting new physics which can be carried out at laser intensities of $10^{22-23} Wcm^{-2}$ will also be briefly discussed.*

PREAMBLE

This story started nearly fourteen years ago when two of us KWDL and Ravi Singhal (Glasgow) were discussing with Joe Magill (Karlsruhe) the seminal theoretical paper written by Boyer, Luc and Rhodes [1] which discussed the possibility of using a high intensity focused laser beam (10^{21} Wcm^{-2} and 248 nm) to induce fission in ^{238}U.

Many probes had been used to induce fission, particularly neutrons both fast and slow. Other nuclear probes [2], had also been used e.g. protons, deuterons, α particles and heavy ions as well as non nuclear beams e.g. γ-rays, electrons and muons. It is perhaps not surprising for a sufficiently intense light source to induce fission especially since a number of lasers had pulse intensities of a terawatt (10^{12} W) or even a petawatt (10^{15}W).

This initiated, for the research team at Glasgow working in collaboration with Imperial College and the plasma physics group at the Rutherford Appleton Laboratory and now many other groups around the world, an exciting new area of physics that has been called laser induced nuclear physics.

CP634, *Science of Superstrong Field Interactions*, edited by K. Nakajima and M. Deguchi
© 2002 American Institute of Physics 0-7354-0089-X/02/$19.00

INTRODUCTION

The mechanism of the interaction of charged particles with intense electromagnetic fields has been considered for more than fifty years. This was one of the first explanations put forward by the early workers to explain the origin and energies of cosmic rays, e.g. [3]. Simply the idea is as follows: a charged particle in an intense electromagnetic field is accelerated initially along the direction of the electric field. The vxB force causes the particle s path to be bent into the direction of travel of the wave. In large fields the particle s velocity rapidly approaches the velocity of light and tends to travel with the EM wave gaining energy from it. In astrophysical situations the solar corona was thought to be one of the sources of the electromagnetic waves. Charged particles which are known to exist throughout space become entrained in a long wavelength wave and can be accelerated to energies as high as 10^{21}eV. These astrophysical phenomena are the counterparts of the machines built on the earth to accelerate particles to high energies.

In 1971 the possibility of accelerating electrons in focused laser fields was first proposed by Feldman and Chiao [4]. They showed theoretically that an electron could gain energies as high as 30 MeV after a single pass through the focus of a diffraction limited laser beam of power 10^{12} W and wavelength 1μ. Chan [5] similarly calculated that an intense laser beam could be used as an energy booster for relativistic charged particles showing that a 10 MeV electron can absorb 40 MeV from a laser beam of 1μ wavelength and an electric field of 3×10^{10} Vcm^{-1} in a distance of 1.3 mm. The problem with these schemes is that it is still not possible to maintain the required intensities over the necessary distances in vacuum, even with the highest energy laser systems which exist today.

In the late seventies, this problem was overcome by the seminal work of Tajima and Dawson [6] who realised that by focusing laser light into a plasma medium, much higher accelerated energies could be generated than by focusing into the vacuum alone. They proposed the construction of a laser-electron accelerator which could be created when an intense laser pulse produced a wake of plasma oscillations (volumes of low and high densities of electrons). Similar to a boat creating a bow wave or wake as it moves through water, a bunch of high velocity electrons creates a wake of plasma waves as it passes through a plasma. They demonstrated with computer simulations that existing glass lasers of 10^{18} Wcm^{-2} directed on plasmas of densities 10^{18}cm^{-3} could yield electrons of 100 GeV energy per m of acceleration. It is known that conventional accelerators are limited by electrical breakdown at fields of about 20 MV/m; at these fields the electrons are torn from the nuclei in the accelerator s support structure. Thus these plasma particle accelerators promise fields more than 1000 times stronger that those of the most powerful conventional accelerators, Dawson (1989).

The reason for the current and burgeoning interest in high intensity laser-plasma interactions is not only their relevance to advanced concepts of plasma high energy particle accelerators but also to a number of other fields e.g. laser induced nuclear photo-physics, astrophysics and inertial confinement fusion (ICF) e.g. Takabe [7] and Tabak et al [8]. Specifically the question of carrying out nuclear physics using a laser source has been addressed by a number of authors using high repetition rate, fs lasers e.g., Shkolnikov *et al* [9] Umstadter [10] Schwoerer *et al* [11] and by using single shot ultra-high intensity lasers like NOVA, LULI and VULCAN e.g. Ledingham and Norreys [12], Ledingham *et al* [13], Cowan *et al* [14] and Gauthier [15].

BACKGROUND INFORMATION

Photo-nuclear Physics

In a nucleus, nucleons (protons and neutrons) are bound in shells similar to the electrons in Bohr s picture of the atom. The most weakly bound nucleon in a typical medium mass nucleus is about 10 MeV. If a nucleon absorbs energy from a beam of incident γ-rays of energy >10 MeV, it can escape from the nucleus. Neutrons can escape much more readily than protons from a nucleus since protons in addition must burrow through a positive Coulomb barrier before being liberated. Thus photon induced nuclear physics generates large numbers of neutrons via (γ,n) processes.

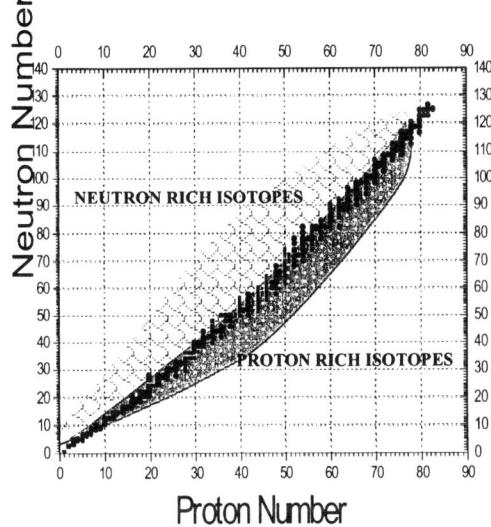

Figure 1. A plot of the stable isotopes in order of increasing neutron and proton number.

When all the stable nuclei are arranged in order of increasing neutron and proton number in the Segre chart as shown in **Figure 1** a line of stability exists. It can be seen that for light stable nuclei, the number of neutrons and protons are equal but as the mass increases however, many more neutrons must be added to reduce the unstable effect of the repulsive protons. Above the line of stability lies a sea of isotopes which are neutron rich. These isotopes are unstable and have the ability to transform themselves back to stability by β⁻ radioactivity in which a neutron decays to a proton plus an electron (with an accompanying neutrino).

Below the line of stability we have a sea of proton rich isotopes which reach stability when one of the protons transforms into a neutron plus a positron (β^+ radioactivity). When γ-rays of tens of MeV energy interact with solid targets, beams of neutrons are produced with the formation of proton rich isotopes which decay by β^+ emission.

A nucleus which decays by emitting positrons can be detected in an unambiguous way. A positron combines with an electron to produce two simultaneous 511 keV γ-rays emitted at 180°. These can be readily detected in coincidence by two radiation counters like NaI scintillation detectors. A typical coincidence system is shown in **Figure 2.**

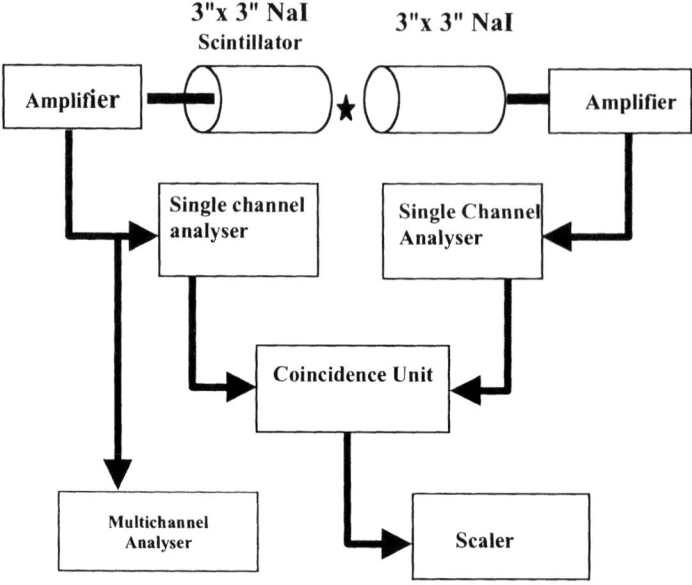

Figure 2. A 3 x3 NaI coincidence system for measuring positrons. A positron annihilates with an electron to form two 511 keV γ rays in time coincidence emitted at 180° to each other. The star indicates the position of the radioactive source

When an intense laser beam interacts with a high Z target, the high electric fields associated with the laser, accelerates electrons from the preformed plasma on the front face of the target into the bulk which then decelerate emitting high energy bremsstrahlung photons. These are the photons which knock neutrons from secondary targets resulting in proton rich isotopes decaying by positron emission.

EXPERIMENTAL RESULTS

Proton Production

The procedure used the Chirped Pulse Amplification (CPA) beam line of the VULCAN Nd:Glass laser [16] incident on a thin aluminium or CH foil target, at 45° incidence with p-polarised light within an evacuated target chamber. This system delivered pulses of energies up to 120 J, duration 0.9 —1.2 ps with a wavelength of 1.053 μm. When focused to a spot size of radius 6μm on the target using a f/4 off-axis parabolic mirror at 45°, the laser yielded intensities of up to 10^{20} Wcm^{-2}. The contrast ratio between the peak and ASE energies was measured by a third order autocorrelator and found to be 1: 10^{-6}. This is sufficient to generate a pre-plasma on the target surface, a few picoseconds before the arrival of the main pulse. The main pulse then interacts with this pre-plasma, causing electrons, and subsequently protons, from hydrocarbon impurities on the target surface, to be accelerated.

To characterise the proton beam, and to produce radioisotopes, different activation samples were placed in the target chamber, behind the target, in the straight through direction at a distance 10 mm from the target, and in front of the target, directed along the target normal, at a distance 50 mm from the target, in the blow-off direction. These samples were then transferred to a nuclear laboratory for analysis. A diagram of the Vulcan laser and the target chamber inside which is the high Z target is shown in **Figure 3.**

The energy spectrum, and spatial distribution of the accelerated protons were determined by placing activation stacks in both the blow-off and straight through directions. These consisted of 9 pieces of copper foil, 100 μm thick, plus a 1mm piece.

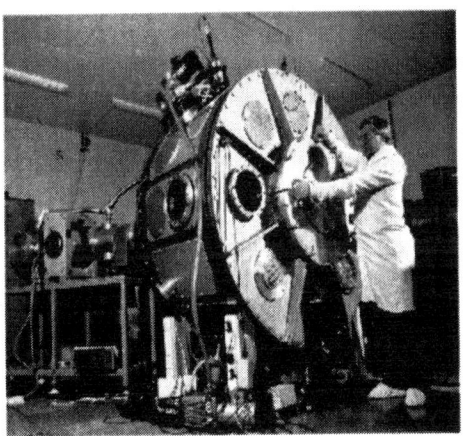

Figure 3. The target chamber and some of the Nd glass disc amplifiers in the present VULCAN laser

Copper undergoes the nuclear reaction $^{63}Cu(p,n)^{63}Zn$ when bombarded with protons of energies greater than 4.15 MeV. The ^{63}Zn isotope produced is radioactive, and decays via β^+ emission with a half life of 38.1 mins. The efficiency of the coincidence system was measured using a calibrated ^{22}Na source, and hence the absolute activity, i.e. the number of (p,n) reactions in the copper pieces in the stack could be determined. Knowing the cross section, the number of protons incident on each piece of copper in the stack can be found, and hence a proton energy spectrum can be built up. Each piece of copper in the stack represents an energy bin , because protons are slowed down in copper. **Figure 4** shows typical energy spectra of the emitted protons from the front and rear of the target. It is clear that more energetic protons are observed in the straight through direction, having energies up to a diagnostic limit of 37 MeV. The first copper foils in each stack are shown as insets in Figure 4 The white circles observed are the result of aluminium debris from the target, and correspond to the dimensions of the proton beam, and hence provide spatial information about the proton beam.

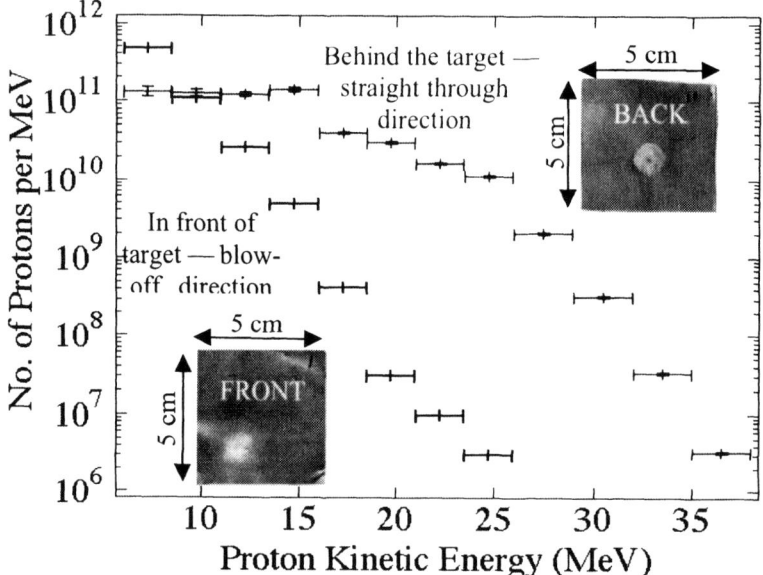

Figure 4 Proton energy spectra obtained from copper activation stacks in front of and behind the target. Insets show the first copper pieces in each stack, containing spatial information

The protons were used to produce radio-active sources of ^{11}C by (p,n) reactions of activity $>10^5$Bq per pulse. For commercial viability, PET source activities some 1000 times greater than this are required. Potentially this could be carried out using table-top OPCPA lasers with high repetition rates [17]. These lasers could also provide proton energies up to 150 MeV that could be used for proton oncology.

REVIEW OF EXCITING NEW NUCLEAR AND PARTICLE PHYSICS AT LASER INTENSITIES OF 10^{22-23} Wcm^{-2}

Ultrahigh magnetic field generation

Tatarakis *et al* [18] have recently demonstrated the production of huge magnetic fields using the VULCAN laser at intensities of up to 9×10^{19} Wcm^{-2} onto thin solid targets. Such fields are more than an order of magnitude larger than any previously observed in the laboratory. At 10^{22} Wcm^{-2} simulations have suggested that magnetic fields greater than 10^{9} gauss may be possible. Magnetic fields generated in this way could soon approach those needed for testing astrophysical models of neutron stars and white dwarfs [19].

Direct interaction with the nucleus

A number of experiments have been carried out at different laboratories to produce laser induced high energy gamma rays and protons. These beams have been used to produce radioactive isotopes via (γ, n) and (p,n) reactions. However as the laser intensity is increased above 10^{22} Wcm^{-2}, the oscillating electric field can make the protons in the nucleus quiver in exactly the same way as the electrons in the plasma. At 10^{22} Wcm^{-2} the direct interaction of the laser electric field and a high Z nucleus causes energy shifts of the order of 250 eV. At 10^{24} Wcm^{-2} this has increased to 2.5 keV and at 10^{28} Wcm^{-2} shifts of the order of 250 keV occur. Such shifts could radically alter nuclear energy levels and decay half-lives.

Fusion by direct laser acceleration of ions

When deuterium (D) and tritium (T) ions collide with sufficient energy they fuse. If these ions are oscillating in the field of a laser beam they can gain sufficient energy from the field to cause fusion to take place. At 10^{21} Wcm^{-2} the collision energy is on average about 8 keV with a corresponding fusion cross section of $\sim 10^{-4}$ barn. However while the ion energies remain non relativistic the collision energies scales directly with the laser intensity and hence at 10^{22} Wcm^{-2} the collision energy is about 80 keV. The peak of the DT fusion cross section (5 barns) occurs at about 100 keV. Most fusion reactions take place in very large scale facilities associated with the inertial confinement programme but now it will be possible to study fusion reactions in this novel way for the first time for a range of low Z nuclei with small OPCPA laser systems.

Particle Physics

As the intensity of the laser increases, the energy of the accelerated electrons also increases. At 10^{20} Wcm^{-2}, electron energies up to 100 MeV have been measured. This is very much greater than the expected ponderomotive energy of about 10 MeV and hence there are accelerating mechanisms like plasma wave acceleration that can result in greatly increased electron energies. At 10^{22} Wcm^{-2} the radiation pressure on electrons in a thin target can reach values greater than 1 Tbar and for very thin targets, resulting in low plasma densities, electron energies in excess of 100 GeV are possible. It is even possible that electron energies as high as 100 TeV can be generated by lasers at intensities of $10^{26}\,\mathrm{Wcm^{-2}}$ and even 10 PeV at intensities of $10^{28}\,\mathrm{Wcm^{-2}}$.

EPILOGUE

Since we embarked on this exciting journey of laser induced nuclear physics and its applications, we have never ceased to be amazed at the world-wide interest it has generated. At this time, we feel that among the most important applications for this technology is the generation of proton and neutron beams which I have not managed to cover in this paper. However as with many scientific endeavours, the most important results that come from the study of a new technology are likely to be totally unexpected..

ACKNOWLEDGMENTS

We would like to acknowledge the hard work and dedication of all the members of our team without whom none of this would have been possible and the support of the engineering and target area staff of the Central Laser Facility. In addition we would like to thank Prof. Steve Rose for help with the calculations which has lead to the potential experiments described in the latter part of the article dealing with super intense laser beams.

REFERENCES

1. Boyer, K, et al, Phys.Rev.Lett **60**, 557, (1988)
2. Lynn, J.E., Nature **333**, 116, (1988)
3 Fermi, E., Phys.Rev., **75**, 1169, (1949
4 Feldman, M.J., and Chiao, R.Y., Phys.Rev A **4**, 352 (1971)
5 Chan, Y.W., Phys.Lett., **35A**, 305, (1971)
6 Tajima, T., and Dawson, J.M., Phys.Rev.Lett, **43**, 267 (1979)
7 Takabe, H., Progress of Theor.Phys. Supplement, **143**, 202, (2001)
8 Tabak, M., et al, Phys.Plasmas, **1**, 1626 (1994)
9 Shkolnikov, P.L., et al, Appl.Phys.Lett **71**, 3471 (1997)
10 Umstadter, D., Physics of Plasmas, **8**, 1774, (2001)
11 Schwoerer, H., et al, Phys.Rev.Lett., **86**, 2317, (2001)
12 Ledingham, K.W.D. and Norreys, P. A., Contemporary Physics, **40**, 367, (1999)
13 Ledingham, K.W.D., et al, Phys.Rev.Lett. **84**, 899, (2000)
14 Cowan, T.E., Phys.Rev.Lett., **844**, 903, (2000)
15 Gauthier, J.C., Laser and Plasma Beams **17**, 195, (1999)
16 Danson , C.N., et al J of Modern Optics **45**, 1653, (1998)
17 Ross, I.N., et al, Optics Communications **144**, 125, (1997)
18 Tatarakis, M., et al, Nature, **415**, 280, (2002)
19 Lai, D., and Salpeter, E.E., Astrophysics J., **491**, 270, (1997)

Particle Acceleration and Radiation with Super-Strong Field Interactions

Igor V. Smetanin[1] and Kazuhisa Nakajima[1,2]

[1] *Advanced Photon Research Center, JAERI, 8-1 Umemidai, Kizu, Kyoto 619-0215 Japan*
[2] *High-Energy Accelerator Research Organization (KEK), 1-1 Oho, Tsukuba, Ibaraki 305-0801 Japan*

Abstract. New concepts of particle acceleration and coherent x-ray generation in the super – strong laser - beam interactions are investigated, the quantum effects are treated using the Klein – Gordon theory. Particles are accelerated up to TeV energy range in the scattering by the leading edge of the laser pulse propagating at the group velocity which is less than the speed of light. The quantum regime in the x-ray Compton FEL is investigated, in which the FEL becomes the two-level quantum oscillator with a completely inverted active medium.

INTRODUCTION

The physics of interaction of high-power laser pulses with relativistic particle beams is of the great importance for the progress both in the accelerator technology and in the development of novel radiation sources. The study of these processes have been initiated by P. L. Kapitza and P. A. M. Dirac [1]. Recent advances in high-intensity laser systems [2] have renewed interest in the next generation of high-gradient laser-driven accelerators [3,4]. The laser synchrotron source concept, utilizing the Compton scattering of laser photons by relativistic electrons, opens a prospective route for the development of intense high-brightness sources of x-ray and γ radiation [5,6].

In this paper, the quantum effects in laser – electron beam interactions at relativistic intensities are investigated with the use of Klein-Gordon field theory. In the second section, we discuss a new concept of TeV-range laser ponderomotive acceleration [7], in which particles are accelerated in the point-like scattering by the leading front of the laser pulse, propagating at the group velocity which is less than the vacuum speed of light. The acceleration is treated as the scattering of de Broglie waves by the laser ponderomotive potential, in analogy with the scattering of electromagnetic wave by the overdense plasma. The peak laser intensity determines the threshold and the quantum probability of acceleration, and have no influence on the energy gain, which is determined by the group velocity and the initial energy of a particle.

The quantum operation regime in the x-ray Compton free-electron laser (FEL) is analyzed in the third section. This regime emerges when the quantum emitted exceeds the energy spread of the electron beam and FEL homogeneous linewidth, determined by the optical undulator length [8]. Such a FEL becomes a two-level quantum oscillator with a completely inverted active medium. In the nonlinear interaction regime, inversion can be completely removed in one pass, in analogy with π-pulse formation in coherently amplifying medium. Estimates show the quantum FEL is a promising compact high-power high-brightness source of coherent x-ray radiation.

HIGH-ENERGY LASER PONDEROMOTIVE ACCELERATION IN A PLASMA

In the conventional theory of laser ponderomotive acceleration (LPA) in vacuum [9], the laser pulse group velocity is assumed to be equal to the vacuum speed of light c, $\beta^* = v_{gr} / c = 1$. The acceleration thus originates from the *transverse gradient* of the ponderomotive potential. Expelled particle inevitably has a transverse velocity. At sufficiently high intensities, the quiver amplitude of motion in laser wave becomes comparable to the beam waist, and electron is effectively expelled out from the focal region due to the transverse gradient of the laser field. The laser-particle interaction is thus terminated within a wavelength, and the gain in energy of the accelerated particle increases monotonously with the laser peak intensity.

In a plasma, the laser pulse group velocity is less than c, $\beta^* < 1$. With an increase in laser intensity, the particle velocity inevitably reaches v_{gr}. The expelled particle has begun to move in front of the laser pulse, and qualitatively a new LPA regime has to occur, in which the particle is accelerated as a result of the scattering by the leading front of the pulse [10].

Kinematics Of The High-Energy LPA

Let us assume a high-power ultrashort laser pulse of the normalized vector potential amplitude a_0 propagating in a plasma at a definite group velocity, $\beta^* < 1$. In a new reference frame, which is moving with the pulse group velocity in the same (z) direction, the laser pulse ponderomotive potential becomes static, so the interaction can be considered now as just an elastic potential scattering. Let the initial energy and velocity of the particle be γ_0 and β_0, respectively. The initial collision angle is α_0, $\alpha_0 = 0$ corresponds to a co-propagating laser pulse and accelerated particle. The energy of the particle in the moving frame is

$$\gamma' = (1 - \beta^* \beta_0 \cos\alpha_0)\gamma_0\gamma^* \qquad (1)$$

where $\gamma^* = (1 - \beta^{*2})^{-1/2}$ is the characteristic γ-factor of the laser pulse. According to the classical scattering theory, if this energy is less than the peak ponderomotive potential,

$$\gamma' \le \left(1 + a_0^2 / 2\right)^{1/2} \qquad (2)$$

the particle has to be reflected, and the final energy γ of the particle does not change after scattering (at infinity). In the laboratory frame, we have

$$\gamma(1 - \beta^* \beta \cos\alpha) = \gamma'/\gamma^* \qquad (3)$$

In this implicit relation, β is the particle absolute velocity, and α is the final scattering angle, where $\alpha=0$ corresponds to the co-propagation case. For the particle initially at rest, the maximum energy gain is at $\alpha = 0$, i.e., $\gamma_{max} = (1 + \beta^{*2})\gamma^{*2} \approx 2\gamma^{*2}$, and the particle velocity exceeds the laser pulse group velocity $\beta_{max} = 2\beta^*/(1 + \beta^{*2}) > \beta^*$. Thus, the accelerated particle is in front of the laser pulse during all the interaction process, which is qualitatively different than that for conventional vacuum LPA [9].

The energy gain for accelerated particle does not depend on the laser pulse intensity, but on the initial energy of the particles and the pulse group velocity only. From Eqs. (1) and (3) we obtain for $\alpha = 0$

$$\gamma_{\pm} = [(\beta^* \pm \beta_0)^2 + \gamma_0^{-2}]\gamma_0\gamma^{*2} \qquad (4)$$

where the signs correspond to an initially counter- ("+") and co-propagating ("-") laser pulse and accelerated particle. This result coincides with the one-dimensional approach [10].

The value of the laser peak intensity determines the *possibility* of the acceleration according to the classical scattering threshold condition (2). In a more correct quantum picture, the value of ponderomotive potential determines the probability of scattering, which, in fact, gives the number of accelerated particles.

Scattering Of De Broglie Waves By Ponderomotive Potential

To calculate the quantum probability of scattering by the ponderomotive potential, i.e., the probability of high-energy acceleration, we seek a solution of the Klein-Gordon equation in the moving frame in the form of de Broglie waves $R(z)\exp[-i(\varepsilon/\hbar)t]$, where $\varepsilon = m\gamma'c^2$ is the particle energy and γ' is given by Eq.(1). Let us consider the most interesting case of zero scattering angle $\alpha=0$. We assume a stationary flow of particles scattered by the potential, and seek the reflected and transmitted fractions with the following asymptotic behavior of the wave function $R_{z \to \pm\infty} \sim \exp(\pm ik_0 z)$, as asymptotic wavenumbers $k_0 = \sqrt{\varepsilon^2 - m^2c^4}/\hbar c$ should be the same for all three (i.e., incident, transmitted and reflected) waves due to the energy conservation. The spatial part of the wave function is then governed by the well-known equation

$$\frac{d^2R}{dz^2} + \left(k_0^2 - \frac{e^2}{\hbar^2c^2}A^2(z)\right)R(z) = 0 \qquad (5)$$

which is analogous to the case of scattering of the electromagnetic wave by the plasma boundary. The second term in the brackets can be regarded as the square of "photon plasma frequency", which is proportional to the photon density, i.e., the laser intensity.

Under the condition (2), when the energy of particle is less than the ponderomotive potential, i. e., $k_0 < eA_0 / \hbar c\sqrt{2}$, where A_0 is the peak amplitude of the laser vector potential, we find for the reflection probability in the WKB-approximation [11]

$$R \approx 1 - \exp\left[-2\int_{z_1}^{z_2}\left(\frac{e^2 A^2(z)}{2\hbar^2 c^2} - k_0^2\right)^{1/2} dz\right] \qquad (6)$$

Here, the integration is over a classically prohibited zone, between the turning points z_1 and z_2, where the integrand becomes zero. In this case, most of particles, except an exponentially small fraction, are reflected, and consequently, accelerated. In fact, the physical meaning of the integrand is the characteristic inverse "skin depth" of de Broglie waves in the "photon plasma".

In the opposite case, when the energy is well above the ponderomotive barrier, $k_0 \gg eA_0 / \hbar c\sqrt{2}$, the result of perturbation theory is [11]

$$R \approx \left|\frac{1}{k_0}\int_{-\infty}^{+\infty}\frac{e^2 A^2(z)}{\hbar^2 c^2}\exp(2ik_0 z)dz\right|^2 \sim \left(\frac{eA_0}{\hbar k_0 c}\right)^4 \ll 1 \qquad (7)$$

Expression in the brackets is, in fact, the ratio of the quiver momentum to the initial electron momentum in the moving frame. The transition between these two limiting cases is shown in the Fig.1, in which the dependence of the reflection coefficient R on the laser peak amplitude is calculated for a specific intensity profile of the laser pulse $A^2(z) = A_0^2 \cosh^{-2}(z/l^*)$, where $l^* = \gamma * \beta * \tau_{pc}$ is the characteristic pulse length in the moving frame.

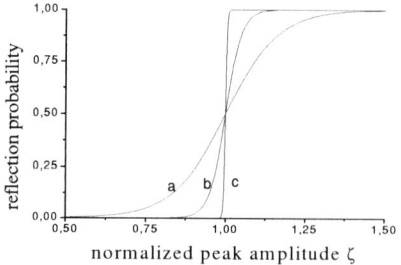

FIGURE 1. The dependence of the scattering probability (12)-(13) on the normalized peak amplitude $\varsigma = (1 + a_0^2 / 2)^{1/2} / \gamma'$ for a) $k_0 l^* = 3$, b) $k_0 l^* = 10$, and c) $k_0 l^* = 20$.

TeV Laser Ponderomotive Acceleration Of Electrons In A Plasma

The LPA scheme under discussion makes attainable high energy gains using current laser technologies which makes this scheme a prospect for the next generation of particle accelerators.

Let us assume an electron beam with the initial energy 150 MeV ($\gamma_0 = 300$) injected into the interaction region in the direction of the laser pulse propagation. Using Eq. (4), one can easily estimate the laser beam and plasma parameters which are necessary to get the GeV-range energy gains. The laser pulse group velocity determines the required plasma density, $n_e[cm^{-3}] \approx 1.115 \times 10^{21} \gamma*^{-2} \lambda^{-2}[\mu m]$. The normalized laser field amplitude is determined by the threshold condition (2), and the corresponding laser peak intensity is $I[W/cm^2] = 1.37 \times 10^{18} a_0^2 \lambda^{-2}[\mu m]$. The dependencies of the threshold normalized amplitude and laser peak intensity on the final energy of electrons are shown in Fig. 2a ($\lambda \sim 1\mu m$), and the required plasma density and Lorentz factor $\gamma*$, corresponding to the pulse group velocity, are shown in

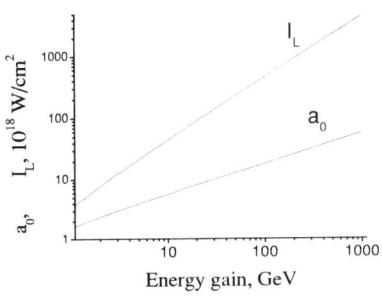
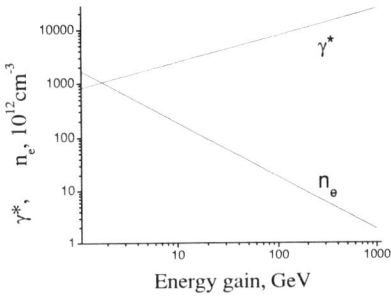

FIGURE 2. The dependence of the threshold laser peak intensity and normalized amplitude (a), and of the plasma density and the corresponding Lorentz factor $\gamma*$ (b) on the energy gain. Electrons are injected in the direction of the pulse propagation at 150 MeV the initial energy.

The regime proposed makes possible acceleration of electrons up to the TeV energy range using modern laser technology. For the injection energy 275MeV ($\gamma_0 \sim 550$) and plasma density $n_e \sim 1.115 \times 10^{12} cm^{-3}$ ($\gamma* \sim 3.2 \times 10^4$) the energy gain is ~1TeV ($\gamma \sim 2 \times 10^6$). The threshold normalized laser field amplitude required is then $a_0 \sim 42.3$ which corresponds to a laser peak intensity of $\sim 2.47 \times 10^{21} W/cm^2$. The plasma density (and consequently the plasma frequency) is sufficiently low, and no plasma instability and relativistic self-focusing can affect the pulse propagation at a reasonable Rayleigh lengths.

X-RAY COMPTON FEL IN THE QUANTUM REGIME

In the x-ray Compton FEL, the high-intensity laser pulse is scattered by counter propagating relativistic electron beam [12], and the high-frequency component is generated as a result of resonant interaction of electrons with the slow ponderomotive wave of frequency $\omega = \omega_s - \omega_i$ and wavenumber $k = k_s + k_i$, $\omega_{s,i}$ and $k_{s,i}$ are frequencies and wave numbers of pumping (i) and signal (s) waves respectively. The dependencies of the emission and absorption probabilities for high-frequency photons in FEL on the beam energy, i.e., laser absorption and emission lines, are thus of resonant nature and situated at the energies $\varepsilon_{a,e} = \varepsilon_0 \pm \hbar\omega/2$, where the energy $\varepsilon_0 = mc^2(1 - \omega/kc)^{-2}$ corresponds to the synchronism between the electron beam and ponderomotive wave.

The quantum properties of an FEL are determined by the ratio of the separation $\hbar\omega$ between the absorption and emission lines and their effective width,

$$\eta = \frac{\hbar\omega}{\Delta\varepsilon}, \tag{8}$$

which in X-ray Compton FEL is mainly determined by the beam energy spread $\Delta\varepsilon = mc^2\Delta\gamma$. In the classical FEL operation regime $\eta \ll 1$, the emission and absorption profiles practically coincide which lead to the classical gain coefficien formula [12,13].

The quantum operation regime emerges under the condition $\eta \gg 1$, when the emission and absorption lines are completely separated, so the state of electron changes dramatically within the scale of $\hbar\omega$. The FEL dynamics reduces to the quantum transitions between these two energy levels, ε_e and ε_a. Adequate description of lasing in this regime thus requires the quantum theory and can not be done within the classical electrodynamics [8].

To describe correctly two-level x-ray FEL dynamics in the quantum regime, we will seek the electron wave function as a sequence of de Broglie waves with the momentum $p_n = p_0 + nk$ and energy $\varepsilon(p_n) = \left(p_n^2 c^2 + m^2 c^4\right)^{1/2}$ correspondent to the quasienergy states. The mass of an electron in the field of relativistically intenselaser pulse is $m = m_0(1 + a_i^2/2)^{1/2}$, a_i is the normalised vector potential. The Klein Gordon equation leads to the following set of equation for amplitudes of these de Broglie waves [13]

$$i\dot{a}_n = g_n a_{n-1} \exp(-i\Delta_n t) + g_{n+1} a_{n+1} \exp(i\Delta_{n+1} t)$$
$$g_n = (e^2/\hbar) A_i A_s (\varepsilon_{n-1}\varepsilon_n)^{-1/2} \tag{9}$$

Here, $A_{i,s}$ are the vector potential amplitudes of the pumping (i) and x-ray signal (s) coherent waves. Eqs. (9) describes an anharmonic oscillator, which is characterized by the detuning from the resonance, increasing with the number of level

136

$$\Delta_n = (\varepsilon_n - \varepsilon_{n-1})/\hbar = 4\frac{\varepsilon(p_0) - \varepsilon_e + n\hbar\omega}{\varepsilon_0} c/\lambda_i \qquad (10)$$

Under the condition $\eta \gg 1$, all the exponents in (9) with $n \neq 0$ oscillates sufficiently fast during the interaction time, so the excitation of correspondent states is negligible. As a result, when electrons are injected at the optimum energy $\varepsilon \approx \varepsilon_a$, the Compton FEL becomes a two-level ($n = 0,-1$) quantum oscillator with a completely inverted active medium [8].

Material equations (9) at $n = 0,-1$ joint with the wave equation for the x-ray signal amplitude form the Maxwell - Bloch system of equations, solution to which is usually seeking in the form of coherent oscillations $a_0 = \cos(\chi(z)/2)$, $a_0 = \cos(\chi(z)/2)$, with the "pulse area" of the signal

$$\chi(z) = \frac{\mu}{c\hbar}\int_0^z E_s(z)dz, \qquad \mu = e^2 E_i/m\omega_i^{1/2}\omega_s^{3/2} \qquad (11)$$

here μ is an effective dipole moment of the two-level system of electrons. The pulse area is guided then by the wave equation which becomes in our case the pendulum equation [8]

$$\frac{\partial^2 \chi}{\partial z^2} = \alpha^2 \sin\chi \qquad \alpha = \left(\frac{2\pi m_e e^4 E_i^2}{\hbar\omega_s^2 \omega_i m^2 c^2}\right)^{1/2} \qquad (12)$$

Note, the linear gain coefficient α contains the Plank constant in explicit form. It means that the regime under consideration has essentially quantum nature and qualitatively differs from all the conventional FEL operation regimes.

Generation Of Coherent X-ray π-Pulses In The Compton FEL

The coherent nonlinear oscillations of electron and signal wave amplitudes lead to the effect of π-pulse generation, which is well-known in the conventional nonlinear optics. When the total square of the pulse reaches the value $\chi(L) = \pi$, the system becomes in the lower energy state, and the complete removal of inversion emerges.

Using the solution of Eq. (12), one can determine the nonlinear evolution of the signal intensity over the interaction region

$$I_s(z) = dn^{-2}\left(\frac{\alpha z}{\sqrt{\kappa}},\kappa\right)I_s(0) \qquad (13)$$

where the parameter for the Jacobi function is $\kappa = 1/(1 + I_s(0)/I_b)$, and $I_b = n_e\hbar\omega_s c$. The solution (13) is a periodical function of z, the maximum value of which is $I_{s\max} = I_s(0) + I_b$. That is, I_b is the maximum x-ray intensity which can be emitted by the bunch of electrons, when each "two-level" electron emits one x-ray quanta. The optimum interaction length, corresponding to the maximum of intensity, is determined by the half-period of the Jakobi dn-function

$$L_{opt} = K(\kappa)\sqrt{\kappa}\alpha^{-1} \tag{14}$$

where $K(x)$ is the complete elliptic integral.

The coherent generation of x-ray photons in the quantum FEL can be realized at conditions close to the recent experiment on nonlinear laser-beam Thomson scattreing at BNL & KEK [14]. In this experiment, the ~6 keV spontaneous photons were produced in the scattering of 600 MW 200 ps CO_2 laser pulses ($\lambda_i = 10.6\mu$m) by the 1nC 1ps bunches of 60 MeV electrons. Taking for estimate the electron density in the bunch $\sim 10^{17} cm^{-3}$, one can easily find that the coherent interaction effects dominates $\alpha L_{sp} > 1$ (here $L_{sp}^{-1} = (4\pi/3)(e^2/\hbar c)\lambda_i^{-1}a_i^2(1 + a_i^2/2)^{-1}$ is the characteristic length of spontaneous scattering [8]) in the domains of normalized laser amplitudes $a_i < 0.5$ and $a_i > 2.4$. At $a_i = 0.5$ ($I_i = 3 \times 10^{15}$ W/cm^2) the linear gain coefficien is $\alpha \approx 7.6\ cm^{-1}$, so that the characteristic interaction length can be choosen close to the optimum length L_{opt}. In the x-ray π-pulse each of $\sim 6 \times 10^9$ bunch electrons will emit x-ray quantum coherently, which corresponds to the signal intensity $I_{s,max}$ $\sim 2.9 \cdot 10^{12}$ W/cm^2. Due to coherence, the divergency of x-ray pulse is determined by the diffraction limit <<20μrad that lead to the high brightness of the quantum x-ray FEL. Really, the estimate for the photon flux in the spherical angle determined by divergency is $\sim 6 \times 10^{27}$ mm^{-2} mrad^{-2} sec^{-1}.

REFERENCES

1. Kapitza P. L. and Dirac P. A. M., *Proc. Cambridge Philos. Soc.* **29**, 297 (1933).
2. Perry M. D. and Mourou G., *Science* **264**, 917 (1994).
3. Tajima T. and Dawson J. M., *Phys. Rev. Lett.* **43**, 267 (1979).
4. Esarey E., Sprangle P., Krall J., Ting A., *IEEE Trans. Plasma Sci.* **24**, 252 (1996).
5. P. Sprangle, A. Ting, E. Esarey, A. Fisher., *J. Appl. Phys.,* **72**, 503 (1992).
6. I. V. Pogorelsky, I. Ben-Zvi, X. J. Wang, T. Hirose, *NIMA*, **455**, 176 (2000)
7. I. V. Smetanin, C. Barnes, K. Nakajima, Proc. HEACC'2001, Tsukuba, 2001
8. I. V. Smetanin, Laser Physics, **7**, 318 (1997); E.M.Belenov, S.V.Grigor'ev, A.V.Nazarkin, and I.V.Smetanin. Quantum Electronics **20**, 733 (1990); JETP **78**, 431 (1994).
9. Hartemann F. V., Fochs S. N., Le Sage G. P. et al., *Phys. Rev. E* **51**, 4833 (1995); *Phys. Rev. E* **58** 5001 (1998).
10. McInstrie C. J. and Startsev E. A., *Phys. Rev. E* **54**, R1070 (1996); *Phys. Rev. E* **56**, 2130 (1997).
11. Migdal A. B., *Qualitative Methods in Quantum Theory*, W.A. Benjamin, Inc., Reading, 1977.
12. P.Dobiasch, P.Meystre, and M.O.Scully. IEEE J.Quant.Electron. **19** (1983) 1812; J.Gea-Banacloche, G.T.Moore, R.R.Schlichter et al. IEEE J.Quant.Electron. **23** (1987) 1558.
13. M.V.Fedorov. Progr. Quant. Electron. **7** 73 (1981); M.V.Fedorov and J.McIver. JETP **49**, 1012 (1979).
14. T. Kimura, Y. Kamiya, T. Hirose. Proc 21st ICFA Beam Dynamics Workshop on laser-beam interactions, Stony Brook, USA 11-15 June 2001.

Atomic Processes and High-power Radiation

Richard M. More
National Institute for Fusion Science
Oroshi, Toki, Gifu, Japan 509-5292
and
Hitoki Yoneda
University of Electrocommunications
Chofugaoka, Chofu, Tokyo, Japan 182-8585

ABSTRACT

What scientific research can be done with strong electric fields from new pulsed laser and x-ray sources?   In this paper we examine three possibilities: first, nuclear fusion directly driven by high-intensity laser radiation in a DT gas target; second, nonlinear (multiphoton) x-ray processes in high-charge ions irradiated by high intensity x-ray sources; third, the interaction of ultra-short pulse lasers with solid targets.   In the third case, some experiments have been performed and while theory agrees with experiment for simple metal targets, unexpected behavior occurs for targets having more complex chemical structure.   These laser absorption experiments illustrate the new science that is already emerging from experiments with high-power lasers.

1.) Strong Electric Fields

Nature makes strong electric fields.   Near a uranium nucleus,

$$E_{peak} \sim 2 \; 10^{19} \text{ Volt/cm} \qquad (1)$$

When an electron passes near the uranium nucleus, the nuclear field causes an acceleration of $\sim 10^{31}$ times the Earth's gravity g, but doesn't create black holes or cause other general-relativity effects.   The nuclear electric field exceeds 10^{16} Volt/cm over a distance of about one electron Compton length $\hbar / mc$ and yet it is not strong enough to pull electrons and positrons from the vacuum.   The nuclear field cannot even pull protons out of the nucleus.   Since at present the strongest laser fields are millions of times weaker, lasers probably cannot do these exciting things either.

The nuclear field can be used for accurate scientific experiments.   For

CP634, *Science of Superstrong Field Interactions,* edited by K. Nakajima and M. Deguchi
© 2002 American Institute of Physics 0-7354-0089-X/02/$19.00

example, fine-structure splitting of electron energy levels is caused by the spin-orbit interaction,

$$H_{so} = - \frac{e}{2m^2c^2} \vec{s} \cdot \vec{E}(\vec{r}) \times \vec{p} \qquad (2)$$

Eq. (2) includes Thomas precession as well as the electron spin interaction with the **vxE** magnetic field experienced by an electron moving in the nuclear electric field. These effects are well-known corrections to the non-relativistic Schroedinger equation, and Eq. (2) shows that fine-structure splitting directly measures the electric field. The Dirac equation and quantum electrodynamics also include smaller effects such as the Lamb shift. Spin-orbit splitting has been carefully measured in hydrogen-like uranium ions. The average electric field near a 2p electron is less than Eq. (1), but still large compared to laser electric fields. Recent experiments find no observable difference from QED theory to a precision of about a part per thousand.[1]

Comparing to this strong natural electric field we are skeptical about the possibility of easy discoveries with lasers producing fields of 10^{10} - 10^{13} Volts/cm. The targets for laser research must be chosen carefully.

The effect of an electric field depends on the space and time scales over which the field exists, and the scientific application depends on having access to perform experiments. From this viewpoint lasers are interesting because laser fields extend over a larger distance (about a micron) and there is good access to the high-field region.

Atoms are easily affected by an applied electric field. The Stark splitting of spectral lines is well-known. Already in the 1930's the Stark effect was measured for hydrogen emission lines at fields up to 10^6 Volt/cm. [2] For hydrogen-like (one-electron) ions the energy levels are independent of angular momentum and the classical electron orbits are ellipses fixed in space, so an orbit pointing along the field has a different energy from an orbit perpendicular to the field. The result is a large linear Stark splitting. For many-electron ions, the orbits are precessing ellipses, and the precession causes the first-order effects of the applied field to cancel, leading to a (smaller) quadratic Stark effect.

When the applied electric field is large it can pull electrons from the atom, a process called Stark ionization. For time-dependent fields, the mechanism of ionization depends on the frequency of the field, and one talks of tunneling ionization, "over-the-barrier" (OTB) ionization and multiphoton ionization.[3]

Other atomic processes are affected by electric fields. For example, in dielectronic recombination, free electrons attach to an ion in large orbits of low angular momentum. An externally applied electric field can transfer these electrons to large angular momentum where they do not autoionize, causing an enhancement of the recombination. Recent experiments on electron-ion recombination in storage rings find that radiative recombination also depends on environmental electric and magnetic fields.[4]

Recent research on atomic processes in laser fields has explored above-threshold ionization (ATI) and the generation of high harmonics. [3]

2.) Fusion in strong laser fields

In search of something more exotic, we have considered fusion reactions induced by strong electric fields in DT gas irradiated by high-power lasers.[5]   We imgine laser intensities $I_{peak} \sim 10^{22}$ W/cm^2 with pulse lengths of $\tau \sim 1$ psec, to be created by future lasers in the multi-PWatt power range.

If such a laser is focused in DT gas, the electrons immediately reach relativistic quiver energies and essentially decouple from the ions.   The ions also have a quiver motion.   All deuterium ions will oscillate in phase, but because of the mass difference between D and T ions, the unlike ions have a relative velocity which can be estimated by

$$V_D - V_T = \left(\frac{1}{M_D} - \frac{1}{M_T} \right) \frac{eE}{\omega} \cos(\omega t) \qquad (3)$$

If the laser intensity ($I = c\, E^2/8\pi$) is large, this velocity difference can be large enough to initiate fusion reactions.   At present this is a theoretical idea which we call **ion quiver-velocity (IQV) fusion**. [5]

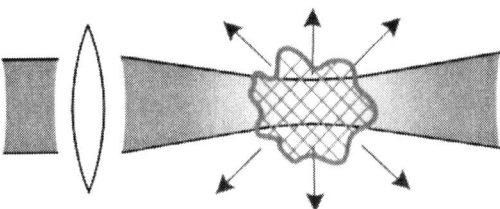

Figure 1.) IQV fusion: the laser directly shakes D, T nuclei to fusion velocities.

IQV fusion would occur against a strong background of fusion reactions from ion acceleration by electrostatic fields generated by the target interaction.   Such electrostatic (beam-target) fusion has already been observed, so it is likely these reactions will be numerous at the high intensities considered here.   Special features of IQV fusion may make it visible despite this strong background.   IQV fusion reactions can only occur for nuclei with unequal mass.   Since IQV fusion can occur only in the high-intensity focal-spot, the IQV reactions are strongly localized in space and time, and this localization might help distinguish IQV fusion from background.   If IQV fusion can be measured, they may provide a way to directly verify the existence of high intensities in the laser focal-spot.

To calculate IQV fusion, we must find the focal-spot diffraction pattern, calculate fusion for oscillating Maxwell distributions, and form the space-time integral. It is assumed that D, T ions have random spatial locations (i.e., are uncorrelated) consistent with the high energy-density in the focal spot. We neglect reactions which occur outside the focal-spot. At present we also neglect reactions due to fusion alpha knock-ons, although such reactions might increase the fusion yield because the alpha energy-loss to relativistic electrons is so small. We also neglect possible enhancement of the fusion cross-section by the laser electric field. The following examples illustrate the possibilities. The gas targets are taken to have a density of 80% of the critical density.

Case 1: **100 TW laser**   F/2 lens
D-T gas target   0.24   reactions

Case 2: **1 PW laser**   F/2 lens
D-T gas target $8.5 \ 10^3$ reactions

Case 3: **10 PW laser**   F/6 lens
D-T gas target $9 \ 10^5$ reactions

Case 4: **10 PW laser**   F/6 lens
DT droplet target (0.1 liquid density) $5 \ 10^8$ reactions

From these calculations it appears that IQV fusion should be observable. It has scientific interest as new kind of fusion at exotic conditions. The predictions deserve an experimental test, but with the yields calculated here, this would be scientific research rather than energy engineering.

3.) X-Ray Nonlinear Optics

We next consider atomic effects of high-intensity X-rays. New x-ray sources (x-ray lasers, high harmonic sources and undulators proposed for electron linear accelerators) may offer the possibility of experiments with short-wavelength radiation at very high intensities in the coming decade.

We consider a 50 fsec pulse of kilovolt X-rays focused to 10^{17} W/cm^2. To reach this intensity it would be necessary to focus 50 microjoules of X-rays into a $1\mu^2$ target area. This may not be easy, but is in the range of conditions being discussed by scientists developing new x-ray sources. Existing x-ray laser sources have photon energies ~ 100 eV.

Another possible approach for this research would be visible laser irradiation of ions in a high-energy ion storage ring. We consider ions trapped in a storage ring, having very high velocity so their motion is relativistic with a gamma-parameter of about 50.   Such a storage ring does not exist today but may be possible in the future. If a visible laser shines on such ions from the correct direction, the light is Doppler-shifted to 100 eV or so (and the intensity is raised) in the ion rest-frame.   The moving ion would be irradiated by high-intensity x-rays whose parameters could be controlled very precisely.

If high-intensity x-rays interact with an ion, can they produce nonlinear optical phenomena like those caused by visible lasers?   We should not expect an exact repetition of visible-light nonlinear optics, because the parameter $I\lambda^2$ is so much smaller. For this reason we cannot expect to observe high harmonics or above-threshold ionization resulting from x-ray irradiation.

Let us ask about the simplest nonlinear effects.   Will there be two-photon absorption?   Harmonic production?   How do the cross-sections for these processes depend on photon frequency hv and target atomic number Z?

To study non-linear processes caused by high-intensity x-rays, we have solved the time-dependent Schroedinger equation for a hydrogen-like ion irradiated by a Gaussian pulse of 1 keV x-rays.   The target ion is taken to be hydrogen-like Silicon in its 1s groundstate.   The computer code includes up to 60 basis states.   The X-rays are described by time-dependent classical fields E(t), B(t).   The x-ray pulse length is taken to be 50 fsec (FWHM).   For peak intensities in the range 10^{15} - 10^{19} W/cm^2, we find strong two-photon absorption 1s --> 2s, a nearly resonant interaction for H-like Silicon ions.

In these exploratory calculations, the population of the 2s state oscillates rapidly during the x-ray pulse, a nonlinear Rabi oscillation, and the probability to remain in the 2s state at the end of the laser pulse oscillates rapidly with the peak x-ray intensity. The average excitation probability is a few percent.

A second type of exploratory calculation uses the Thomas-Fermi theory of a many-electron ion in strong electric field.   If the x-ray frequency is low compared to the atomic excitation frequencies, it is at least qualitatively reasonable to treat the x-ray electric field as a slowly-varying static field.   We performed 2-dimensional Thomas-Fermi self-consistent field calculations of ions in strong static electric fields.

The induced electric dipole moment depends on the electric field:

$$\mathbf{p} \; = \; \alpha E \; + \; \gamma E^3 \; + ... \qquad\qquad (4)$$

where the functions $\alpha = \alpha(Z, Q)$, $\gamma = \gamma(Z, Q)$ are linear and nonlinear polarizabilities. Using Thomas-Fermi theory, we calculated α and γ for many ions and obtained general interpolation formulas.

For an oscillating electric field, the cubic polarization oscillates at frequency 3ω and generates third-harmonic radiation.   We divide the third-harmonic emission by the incident intensity to obtain a cross-section for third-harmonic conversion; this cross-section is proportional to the x-ray intensity squared.   The formulas are

$$P = \frac{2}{3c^3}(\ddot{p})^2 \qquad (5)$$

$$\sigma_{3\omega} = 3^3 2^7 (2\pi)^7 \left(\frac{\gamma(Z,Q)\ I_X}{\lambda^2 c}\right)^2 \qquad (6)$$

Eq. (6) predicts large cross-sections for heavy ions in low charge states. However these ions are easily ionized by three-photon processes, and the approximation of using the nonlinear static polarizability for x-ray response is least appropriate.   For very high charge ions, the cross-section is smaller, comparable to the Compton cross-section.   Further analysis is underway.

4.) Short-Pulse Laser Interaction with Solid Targets

Short-pulse lasers can be used to heat any solid material.   If we measure the absorption (the fraction of laser light absorbed) we learn the heat energy deposited in the target and obtain an indirect determination of the high-temperature electrical properties of the target material.   The temperature can be controlled by adjusting the laser heating energy.   The target remains near its initial solid density if the pulse is sufficiently short, especially at the lower intensities.

Such experiments have been performed in several laboratories.   We can model such experiments by solving the Maxwell equations for laser light interacting with a heated target material.   It is necessary to make assumptions about the optical properties (dielectric function) of the hot material and also about the density profile resulting from target hydrodynamic expansion.   Typically this expansion cannot be neglected even for 100 fsec laser pulses unless the intensity is low.

In our calculations, we assume a local, linear relation

$$\mathbf{j} = \sigma\,\mathbf{E} \qquad (7)$$

At the laser frequency the (AC) conductivity is a complex number.   We solve Maxwell equations with a complex dielectric function $\varepsilon(x)$ that depends on the depth x from the target surface,

$$\varepsilon(x) = 1 + \frac{4\pi i \sigma(x)}{\omega} \qquad (8)$$

The computer code solves the Maxwell equations for s- and p-polarized light at any angle of incidence.   If there is sufficient experimental data, including dependence of the absorption upon the angle of incidence, polarization and laser intensity, we are able to deduce consistent values for the hydrodynamic scale length and optical coefficients of the high-temperature target material.

Even if we determine the dielectric function $\varepsilon = \varepsilon_r + i\,\varepsilon_i$ by analysis of experimental data, we must ask: Is the result unique? Can we determine the microscopic mechanisms which control the dielectric function? These are difficult questions. However, if we can determine the dielectric function over a range of frequencies, $\varepsilon(\omega)$, these questions will have better answers. Thus experiments with a broad-band probe pulse would be very valuable.

Accurate and detailed modeling should include the spatial averaging over the range of intensities in the focal spot, time-dependence of the laser and probe pulses, and electron thermal conduction in the target material. Simpler models are appropriate for conditions of low intensity or low target temperature.

Experiments underway in the University of Electro-communications use a 300 fsec pump pulse of .248 micron radiation (at normal incidence) and a 120 fsec probe pulse at 0.75 micron wavelength (0.65 degrees) to measure reflectivity of heated targets.[8] The laser system has a 0.01~0.1 kHz repetition rate so it can take data rapidly. An example of time-dependent reflectivity (taken with varying delays of the probe pulse) for an aluminum layer (on glass) is shown in Figure 2.

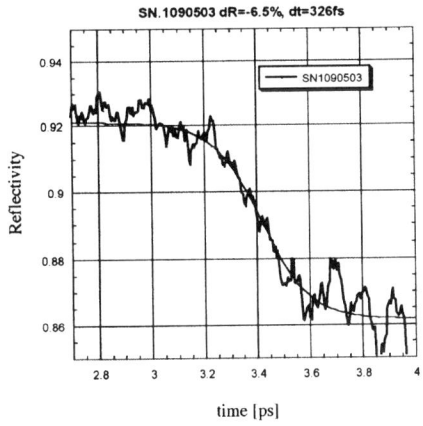

Figure 2. Time-dependent reflectivity for aluminum targets measured with the laser system described in the text.

In these experiments, the electric fields and currents induced in the targets are not small. For example, at a typical moderate laser intensity $I = 3 \cdot 10^{14}$ W/cm^2, for a laser wavelength of $\lambda = 0.25\,\mu$ the calculations give an electric field $\mathbf{E}_o \sim 10^8$ Volt/cm and a current density $\mathbf{j} \sim 7 \cdot 10^{10}$ Amp/cm^2 in the target material. There is an interesting question whether Eq. (7) remains valid at such high currents.

These experiments expose solid targets to pulsed high electric fields in a well-controlled manner.   At high intensities, most targets have the same behavior and agree with theory.[6]   These conditions correspond to target temperatures of perhaps 50 eV or higher.   At lower intensities, we observe interesting new properties of warm condensed matter.   These properties include anomalous high absorption by ionic solids including glass, quartz and sodium chloride, and anomalous absorption features in the response of transition metals. [See REF. 6]

The high absorption by ionic materials is especially surprising because it differs strongly from the behavior of these materials in the usual solid state.

In summary, we see that real-world experiments can be performed today with enough precision to yield new scientific information about the optical properties of matter at temperatures higher than previous accurate experiments.   The challenge is to assemble and interpret enough of this data to teach us something about the nature of the transition from the usual solid state to hot dense plasma.

REFERENCES

[1.] J. P. Briand, P. Chevallier, P. Indelicato, K. P. Ziock and D. D. Dietrich, Phys. Rev. Letters **65**, 2761 (1990).

[2.] H. Bethe and E. Salpeter, **Quantum Mechanics of One- and Two-electron Atoms**, Academic Press, Inc., New York, 1957.

[3.] See article by C. Joachain in **Atoms, Solids and Plasmas in Super-Intense Laser Fields**, Ed. by D. Batani, C. Joachain, S. Martellucci and A. Chester, Kluwer Academic/Plenum Publishers, 2001; and in **Laser Interaction with Atoms, Solids and Plasmas**, Ed. by R. More, Plenum Publishers, 1992

[4.] See article by G. Dunn in **Recombination of Atomic Ions**, Ed. by W. Graham, W. Fritsch, Y. Hahn and J. Tanis, Plenum Publishers, 1992.

[5.] T. Kenmotsu, T. Nishikawa and R. More, in **Inertial Fusion Science and Applications**, Kyoto, September, 2001 (to be published).

[6.] D. F. Price, R. M. More, R. S. Walling, G. Guethlein, R. L. Shepherd, R. E. Stewart and W. E. White, Physical Review Letters **75**, 252 (1995).

[7.] A. Forsman, A. Ng, G. Chiu and R. More, Physical Review **E 58**, R1248 (1998).

[8] H. Yoneda, S. Ohta, K. Ueda and R. More, Proc. of LAMP 2002, Special Session 3, Osaka, May 2002 (to be published in SPIE).

Cluster Dynamics in Strong Fields and Its Application to Fusion Science

Y. Kishimoto[1*†], T. Masaki[2*] and T. Tajima[†]

Naka Fusion Research Establishment, Japan Atomic Energy Research Institute, Naka, Ibaraki 311-0193, Japan
†*Advenced Photon Research Center, Japan Atomic Energy Research Institute, Kizu, Kyoto 619-0215, Japan*

Abstract. Key physical processes of laser-cluster interaction, i.e. the absorption of laser energy and the highly efficient energy conversion to cluster ions and resulting rapid expansion, are theoretically and computationaly investigated. Depending on the interrelationship between cluster size a and the electron excursion length ξ_e (which is related to the laser field amplitude), the expansion characteristics is categorized into the *Coulomb explosion* ($a \ll \xi_e$) and alternative hydrodynamic *ambi-polar expansion* ($a \gg \xi_e$), revealing different feature in the ion energy distribution. The cluster state is *small particle system* where the level of fluctuation is high compared with that of usual gas and/or plasma. Due to this fact, the interacion is affected by the microscopic cluster distributions, so that the observed quantity has to be determined from the statistical average over possible disordered probabilistic distributions. An application of the laser-cluster interaction to nuclear fusion for short pulse high flux neutron generation is proposed and the feasibility to obtain $10^{12} - 10^{13}/cm^3$ per laser shot is discussed.

INTRODUCTION

A cluster medium is known as a *small particle system* from the statiatical view point where the level of fluctuation is high comparted with other preparation of the same materials such as gas and/or plasma. A number of salient phenomena have been observed in the laser irradiation of clusters. These include : the Coulomb explosion of clusters [1,2], enhanced emission of X-rays [3,4,5], generation of energetic electrons [6], and energetic ions [7]. There are indications that some strong nonlinear interaction of laser and clusters do occur upon the intense short pulse irradiation [8,9] . It has been pointed out [10] that the polarization induced on the cluster surface by the laser plays an essential role in the interaction. It has been further pointed out [11] that such polarization of clusters gives rise to the emergence of a new branch of propagating electromagnetic (EM) waves even in the frequency that is otherwise cutoff. Prior to the contemporary interest, there has been a long standing research on the properties of cluster materials since Doremus[12] and Kubo et. al.[13]. Perhaps the most spectacular demonstration of the enhanced laser-cluster interaction among the series of latest experiments is the

---

[1] kishimoy@fusion.naka.jaeri.go.jp
[2] also Division of Advanced Plasma Simulation, Research Organization for Information Science & Technology (RIST), Tokai, Naka 319-1109, Japan

report of copious fusion neutron generation upon the laser irradiation of deuterium clusters[14,15]. The theoretical and computational studies for the cluster based fusion reaction and a possible reactor design were performed by the present authors [16,20]. The production of energetic ions by the irradeation of intense short-pulse laser on clusters is through the efficient creation of high energy electrons and their efficient conversion of energy into high energy (fusionable) ions in the clustered preparation.

In this paper, based on our theory and 2- and 3- dimensional numerical simulation, we investigate the fundamental optical properties of laser-cluster interaction, i.e. the absorption process of laser energy, and energy conversion to cluster electrons and ions. Specifically, due to the significant level of fluctuation originating from the small particle system, we found that the interaction depends on the microscopic initial cluster distributions, so that the physical observation quantity is determined from the statistical average.

We also discuss a feasibility to employ such a laser-cluster interaction for generating short pulse high flux neutrons in the range of $N \geq 10^{12}/\text{cm}^2$ for single laser shot. This can be comparable to an intense neutron source required for studying the first wall neutron damage for fusion reactor. In order to realize such a high flux neutron source, we need to optimize the cluster medium and laser system, so that cluster ions are efficiently accelerated and fusion reaction is optimally enhanced.

FUNDAMENTAL PROCESS OF LASER-CLUSTER INTERACTION

A difference of the cluster medium from the conventional plasma is the variety of controllable parameters which characterize the interaction. A cluster medium is characterized by : 1. its material (the charge state Z, and density $n_e = Zn_{cl}$, n_{cl} is the cluster ion density), 2. spatial size (the radius a), 3. surface and internal structures (for example, multi-layer coatings using different materials), 4. packing fraction (ratio that clusters occupy to the system volume $f \equiv 4\pi a^3 N_{cl}/3V$, N_{cl} is the number of cluster in the volume V) 5. spatial configuration or distribution of clusters (ordered or disordered/random distribution). In relation to the interaction with laser field, following parameters such as : 6. cluster size to laser wavelength (a/λ_l), 7. electron skin depth to cluster size (δ_e/a, $\delta_e \equiv \omega_p/c$), 8. electron excursion length to cluster size (ξ_e/a), 9. inter-cluster distance to electron excursion length [ξ_e/R, $R = (3V/4\pi N_{cl})^{1/3}$], 10. laser cut-off density to local cluster density (Zn_{av}/n_c), etc, regulate the interaction. Such a large number of parameters widens applications utilizing the laser-cluster interaction. We demonstrate to utilize such laser-cluster interaction for efficient neutron production by employing the deuterium-deuterium(DD) nuclear fusion.

In the laser-cluster interaction, there exist two possible regimes depending on the above relations (specifically 7. and 8.), i.e. *Coulomb explosion* [3,7-9] triggered for $\xi_e >> \delta_e \geq a$ and alternatively *ambi-polar expansion* [17] triggered for $a >> \delta_e \sim \xi_e$. In the Coulomb explosion, since the electron skin depth is greater than the cluster size ($a < \delta_e$) and the laser intensity is high enough to fulfill the relation $a < \xi_e$, the laser field penetrates to the interior of the cluster and a large fraction of cluster electrons can leave the host cluster within its optical cycle. When this happens, the cluster becomes highly non-neutral and ions explode due to its Coulomb repulsive force. On the other hand, when the skin depth is such as $a > \delta_e$, the laser field does not fully penetrate to the interior of the cluster, but interacts only with a limited peripheral region. Then, the coronal plasma with a hot electron temperature is produced at the ion front, and the ion front is accelerated by strong ambi-polar field and expanded into the vacuum. Depending on the regime, the feature of the expansion and resulting physical processes, such as absorption of laser energy, conversion to electron and ion energy, and the energy distribution functions are different.

NUMERICAL SIMULATION OF LASER-CLUSTER INTERACTION

In order to analyze the laser-cluster interaction and to explore an operating regime optimal for each application, we carry out 2- and 3-dimensional particle in cell (PIC) simulation, where the laser pulse propagates in the x-direction, while the polarization of electric field is in the y-direction[18]. The packing fraction is given by $f_{2D}(\equiv n_{av}/n_{cl}) = a^2/R^2$ with $R = (S/\pi N_{cl})^{1/2}$ for the 2D case and $f_{3D} = a^3/R^3$ with $R = (3V/4\pi N_{cl})^{1/3}$ for the 3D, where N_{cl} represents the number of cluster in the area $S(= L_x L_y)$ for the 2D case and in the volume $V = L_x L_y L_z$ for the 3D one, respectively.

Analysis of the numerical simulation should be careful since the laser-cluster interacion and resultant expansion process are qualitatively different between 2D and full 3D cases. The average potential energy per ion is given in the 2D and 3D case as follows[9]:

$$w_i^{(2D)} = \left(\frac{\pi Z^2 e^2}{4}\right) n_{cl} a^2 \left(f_{2D} - \ln f_{2D} - 1\right), \tag{1}$$

$$w_i^{(3D)} = \left(\frac{2\pi Z^2 e^2}{5}\right) n_{cl} a^2 \left(2 - 3 f_{3D}^{1/3} + 3 f_{3D}\right). \tag{2}$$

Namely, in the 2D case, due to the rod structure of cluster that infinitely extends in the z-direction, the potential energy diverges with a logarithmic dependence as $R \to \infty$ (or $f_{2D} \to 0$, i.e. infinite free space). However, as discussed in Ref.[9], although the potential energy and corresponding ion kinetic energy in the 2D case overestimates compared with that in the 3D case (about 3-6 times), the ratio of potential energy between 2D and 3D cases, i.e. $w_i^{(3D)}/w_i^{(2D)}$, weakly depends on the packing fraction f over a wide range of our interest. This suggests that 2D simulation could be used to grasp fundamental features of laser-cluster interaction.

Expansion dynamics of cluster medium

At first, we present typical 3D numerical simulations for the case of two different laser field amplitudes, i.e. $a_0 = 0.25(I_L \simeq 1.2 \times 10^{17} \text{W/cm}^2, \xi \simeq 32\text{nm})$ in Fig.1 and $a_0 = 0.6(I_L \simeq 6.8 \times 10^{17} \text{W/cm}^2, \xi \simeq 71\text{nm})$ in Fig.2, where different four snap shots

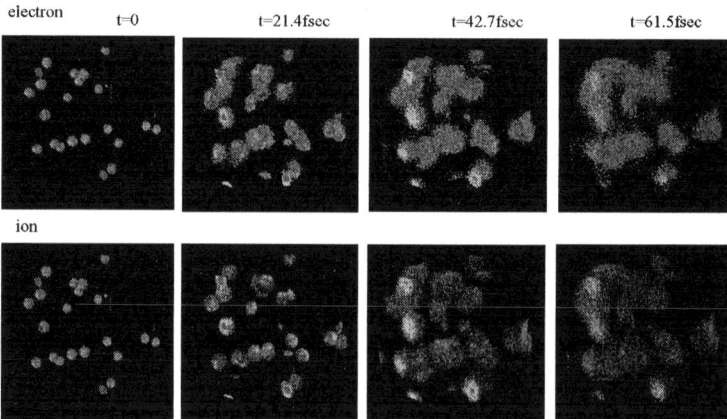

FIGURE 1. The snap shots of electron and ion distribution in the case of $a_0 = 0.25$ (ambi-polar expansion regime). Clusters with the radius $a = 32$nm are randomly distributed with the packing fraction of $f \simeq 4.8 \times 10^{-3}$.

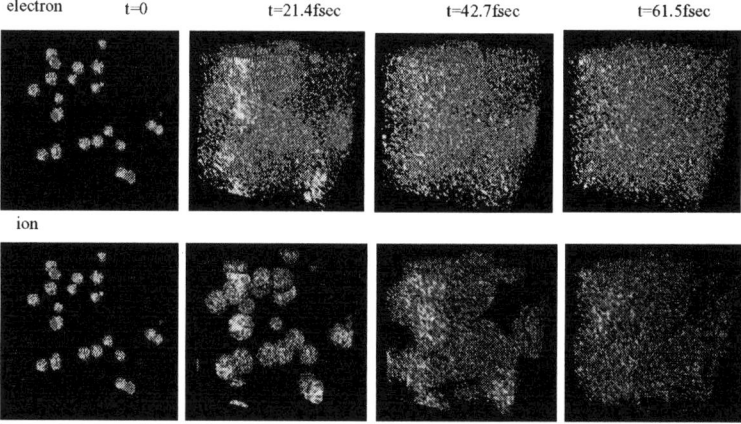

FIGURE 2. The snap shots of electron and ion distribution in the case of $a_0 = 0.6$ (Coulomb explosion regime). The initial cluster distribution is same os Fig.1.

of electron and ion distributions are shown. We employ a solid deuterium cluster with the radius of $a = 32$nm, and the density $n_{cl} \simeq 4.56 \times 10^{22}$cm^{-3}, where the ratio with respect to the cut-off density for 820nm laser wavelength is $n_{cl}/n_c \simeq 27.5$ ($n_c = 1.66 \times 10^{21}$cm^{-3}) and the skin depth is $\delta_e = 25$nm. The packing fraction and average density are given as $f \simeq 4.78 \times 10^{-3}$ and $n_{av}/n_c = 0.13$ ($n_{av} = 2.19 \times 10^{20}$cm^{-3}). The relative relation among the cluster radius, electron excursion length, and skin depth is given as $\delta_e < a \sim \xi_e$ in Fig.1 and $\delta_e < a < \xi_e$ in Fig.2, and therefore, the result of Fig.1 and Fig.2 may correspond to the ambi-polar expansion and Coulomb explosion regime,

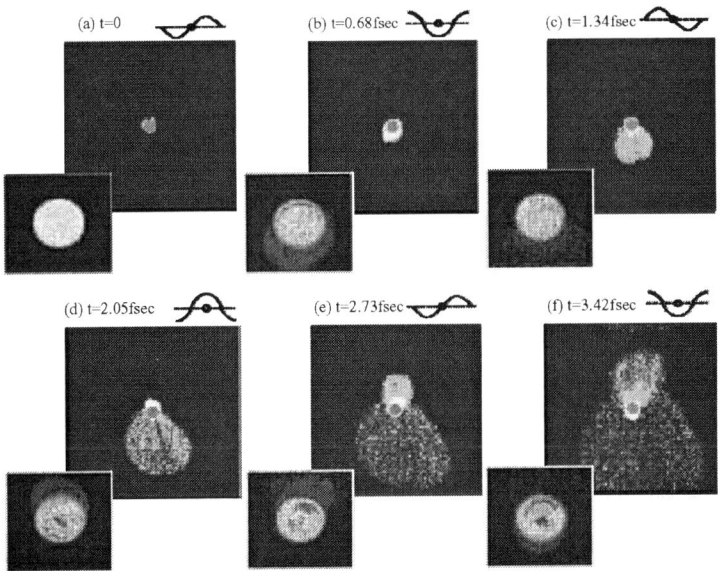

FIGURE 3. The snap shot of elctron distribution of the cluster during initial wave period.

respectively.

In the case of weak laser field amplitude $a_0 = 0.25$ (Fig.1), a part of cluster electron leaves their host cluster due to the laser field (see t=21.4fsec) and ions start to expand. In this power regime, the ion expansion follows that of electron, suggesting that the expansion dynamics is characterized by the ambi-polar expansion as we discussed in the above. Since the clusters are probabilistically distributed at t=0, heated electrons ejected from one particular cluster interacts with the adjacent clusters that locate in the neighborhood. As a result, a colony-like structure in which several and/or more clusters are contained is formed (t=42.7fsec). Such a structure bocomes more prominant as the expansion is developed as seen at t=61.5fsec. Some heated electrons bocomes meandering among clusters. On the other hand, in the case of higher laser field $a = 0.6$ (Fig.2), electrons are instantaneously expelled from each host cluster and distributed over the whole system as seen at t=21.7fsec in Fig.2. Ions are subsequently exploded in faster time scale than that observed in Fig.1, suggesting that the expansion dynamics is dominant by Coulomb explosion.

In order to investigate the details of the interaction, we perform the simulation of single cluster-laser interaction in 2D configuration. Cluster parameters are same as that in Fig.1, and the system size L is chosen to the wavelength $L = \lambda_l$. Figure 3 shows snap shots of the electron distribution during an initial wave period. It is found that the electrons deviates from the cluster surface as seen in (b) and (c). However, when the laser field becomes zero at the cluster center due to the node of the laser field, i.e. (c), some electrons are found not to return to the original location as seen in (d) and (e). This suggests that besides the laser field, complicated local cluster fields may affect on the interaction even if classical inverse Blemsstrung and/or collisional absorption is absent.

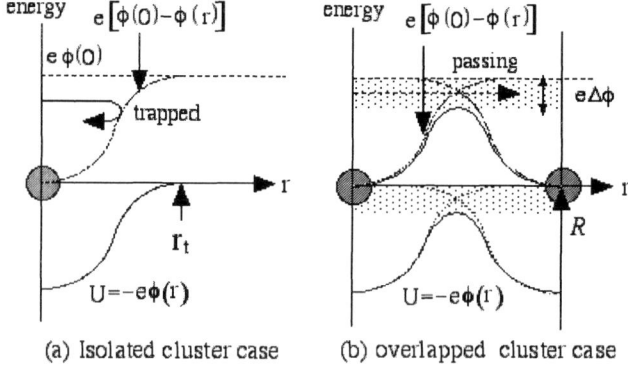

(a) Isolated cluster case   (b) overlapped cluster case

FIGURE 4. The schematic picture of electrostatic potential formed by isolated single cluster located $r = 0$ (a) and by two clusters located at $r = 0$ and $r = R$ (b).

It is also found that the electron distribution inside the cluster is also changed during one wave period so that the plasma oscillation excited in the cluster contributes to the dynamics.

Potential structure around the cluster

In order to understand the fundamental aspect of laser-cluster interaction, here we describe the expansion dynamics by introducing a simple isotropic electrostatic potential localized at one particular cluster, $U(r) = -e\phi(r)$, as schematically illustrated in Fig.4. In early interaction phase, some cluster electrons are expelled from the host cluster. Then, the potential energy at one particular cluster drops to a value $U(0) = U_0 < 0$ as illustrated in Fig.4(a). Here, we define the coordinate of the electron front $r = r_t$ determined from the electron with maximum energy. Assuming that the electron front does not overlap at this time with that of adjacent clusters, i.e. $r_t < R/2$ (R:inter-cluster distance), we set $U(r = r_t) = 0$. Here, the potential energy U_0 is related to the maximum electron energy as

$$\varepsilon_{e,max} = e[\phi(0) - \phi(r_t)] = -U_0, \qquad (3)$$

which increases with the maximum electron energy. However, since the electron front also expands with time and overlaps with that of adjacent clusters, the potential at the electron front is lowered from $U(r_t) = 0$ to some value $U(r \simeq R/2) = -e\Delta\phi_0 < 0$, as illustrated in Fig.4(b). Then, some high energy electrons in the range of $e\phi(0) \geq \varepsilon_e \geq e[\phi(0) - \Delta\phi_0]$ meander among clusters without being trapped by particular cluster being similar to that of conduction band in solid state lattice system[19]. This may correspnd to the electron distribution for exsample at t=21.4fsec in Fig.2. The level of the potential lowering, $\Delta\phi_0$, may depend on the inter-cluster distance R and also on the details of the heating process. As the ion expansion evolves, the potential energy is exhausted to

152

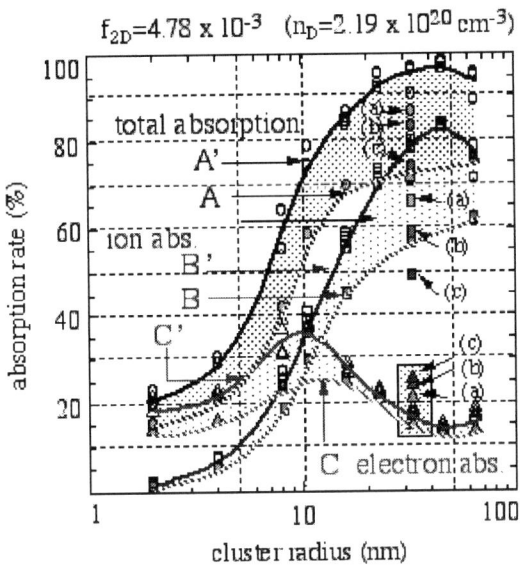

$f_{2D}=4.78 \times 10^{-3}$ $(n_D=2.19 \times 10^{20} \text{cm}^{-3})$

FIGURE 5. The ion, electron and total absorption rate of laser energy for different cluster size in the case $a_0 = 0.25$. A, B and C is the result of ordered cluster distribution, and A', B' and C' is the result of disordered (random) one. (a), (b) and (c) correspond to the random distribution discussed in Fig.7.

accelerate ions and therefore U_0 decreases. However, note that $\Delta\phi_0$ may increase since the overlapping is enhanced. Resultantly, the number of untrapped electrons increases.

Effect of probabilistic cluster distribution on interaction

The detail of the parametric dependence of laser-cluster interaction were discussed in Refs.[9,20]. One of important characteristics of the laser-cluster interaction is a high fluctuation level originated from a small number of cluster particles from statistical view point. Namely, the effective number of particle, i.e. the number of cluster, becomes much smaller than that of usual gas and/or plasma for a given average density and this fact leads to a high fluctuation in the interaction. Therefore, physical quantities resulting from the interaction depend on the micro-scopic initial distribution. Figure 5 shows the result of 2D simulation of absorption rate of laser energy to electrons and ions (and also the total absorption rate) as a function of initial cluster radius for the laser amplitude fixed to $a_0 = 0.25$. In the figure, the dotted line A represents the total absorption rate in the case of the ordered (and/or lattice like) initial cluster distribution. The corresponding electron and ion absorption rates are also shown by dotted line, B and C, respectively. Those results, i.e. A, B and C, are same as that discussed in Ref.9 (see Fig.10 in Ref.9). Here, we investigate the absorption rate in the disordered distribution case where the initial cluster distribution is probabilistically determined by

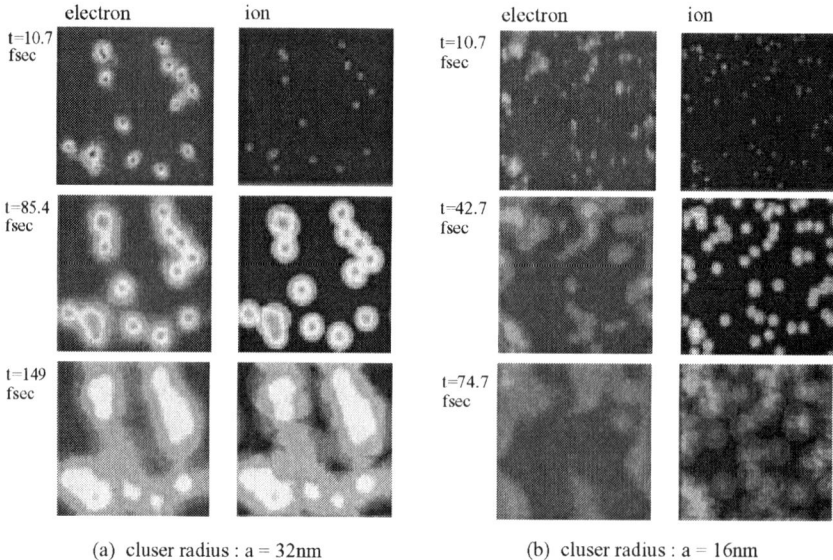

	electron	ion		electron	ion
t=10.7 fsec			t=10.7 fsec		
t=85.4 fsec			t=42.7 fsec		
t=149 fsec			t=74.7 fsec		

(a) cluser radius : a = 32nm   (b) cluser radius : a = 16nm

FIGURE 6. Electron and ion density distribution for disordered distribution case for given packing fraction $f = 4.78 \times 10^{-3}$.

the random number generation. One of those many random distributions is shown in Figs.6(a) and (b), where the cluster radius is chosen to $a = 32$nm (a) and $a = 16$nm (b) for the given packing fraction $f_{2D} = 4.78 \times 10^{-3}$. Note that the number of cluster in Fig.6(b) is 4 times larger than that in Fig.6(a). The absorption rate for 4 cases of disordered distribution are also plotted in Fig.5 by fitted sold lines A', B', and C'. As found from Fig.5, although there is a scattering in the absorption rate resulting from the fluctuation, the total absorption rate shows higher values (about 15-20% in ambi-polar regime of $a > 15$nm and about 10% in Coulomb explosion one) than that in the ordered distribution case. Note that the ordered distribution provides smaller and/or smallest absorption rate. Specifically, the ion absorption rate is found to drastically increase in the ambi-polar regime, showing that the Coulomb energy is effectively released though the high absorption rate due to the enhanced stochasticity of electron motion. On the other hand, only the electron absorption increases in the Coulomb explosion regime, due to the fact that the whole ion energy initially contained in the cluster is released without depending on the initial distribution.

Figure 7 shows the time history of normalized laser field energy and ion energy for three different disordered distribution cases, i.e. (a), (b) and (c), which are plotted also in Fig.6 ($a = 32$nm). Here, the ordered distribution case that corresponds to A, B and C in Fig.5 is also shown. The employed distributions for this simulation are shown in Fig.7. Namely, the same number of cluster is more closely packed in the small area, i.e. (b) 25% and (c) 14.1% of total area that is used in the case (a). This means that the cases, (b) and (c), have more free space as seen in Fig.7. As the clusters are packed in smaller space, even if the distribution is random, the interaction become weak. For example,

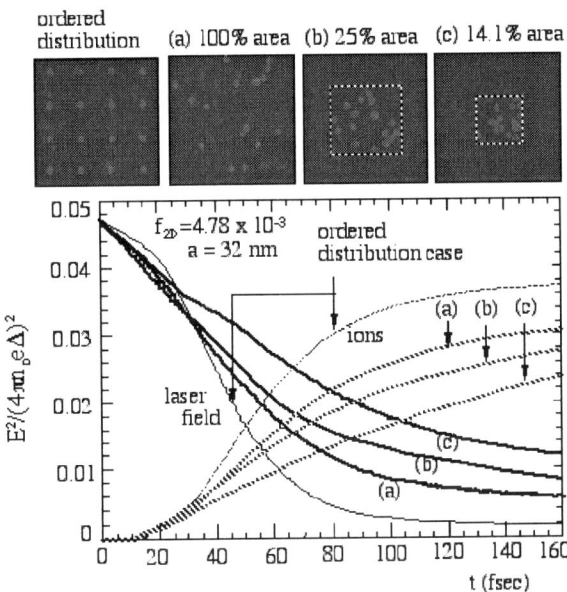

FIGURE 7. Time history of normalized laser energy and ion kinetic energy for four different cluster distribution, i.e. ordered distribution case and three disordered one, (a), (b) and (c). The cluster radius is $a = 32$nm, corresponding to the absorption rate (a), (b) and (c) in Fig.5.

in the case (c), a long interaction time is necessary to reach a saturation. Those results support the idea that the interacton sensitively depends on the microscopic initial cluster distrubution.

Here, we define an initial cluster distribution $\mathbf{R}_n^{(j)}$ where $\mathbf{R}_n^{(j)} \equiv (\mathbf{R}_1, \mathbf{R}_2, \cdots, \mathbf{R}_N)_j$ represents a set of coordinate of N cluster from j-th sample out of many possible set of distribution. A physical quantity A characterized by the laser-cluster interaction (such as absorption of laser energy, conversion rate to plasma particles, etc.) is a function of the initial cluster distribution as $A(\mathbf{R}_n^{(j)})$. The observed quantity from the experiment $< A >$ may be described by an ansamble average over many micro-scopic distributions as

$$< A >= \sum_j A(\mathbf{R}_n^{(j)})P\{\mathbf{R}_n^{(j)}\} \tag{4}$$

where $P\{\mathbf{R}_n^{(j)}\}$ is a probability that the clusters takes the microscopic distribution $\{\mathbf{R}_n^{(j)}\}$. The distribution chosen in Fig.6 is concidered to be one of possible distributions which has a probability $P\{\mathbf{R}_n^{(j)}\}$. For exsample, it is considered that the probability to have the distribution such as (a), (b) and (c) may depend on the detailed physical process when cluster medium is established from the gas jet ejected into the vacuum.

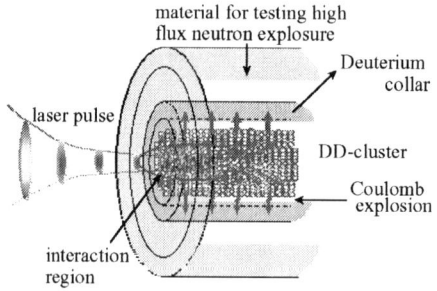

FIGURE 8. Schematic picture of the set-up of laser-cluster interaction for producing fusion neutrons. Some material for testing high flux neutron exposure is placed in the outside region.

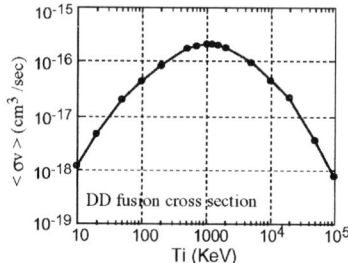

FIGURE 9. DD fusion cross-section $\langle\sigma v\rangle$ averaged over Maxwellian velocity distribusion as a function of deuterium ion temperature.

ESTIMATE OF FUSION NEUTRON BY LASER-CLUSTER INTERACTION

Here, we apply the laser-cluster interaction to achive nuclear fusion to obtain short pulse high flux neutron source. An experimental set-up we are proposing is schematically illustrated in Fig.8. In this picture, high intensity short pulse laser is irradiated to the surface of the cluster medium, and the small spot with high temperature and density is produced through direct laser-cluster interaction. The hot spot expands triggering further cluster explosion to the surrounding region so that cluster ions are accelerated to high energies to induce nuclear fusion. Here, we restrict our discussion to deuterium-deuterium (DD) fusion by employing deuterium clusters. However, the idea developed here is straightforwardly applied to deuterium-tritium(DT) fusion where the cross-section is higher than that of the DD fusion. The deuterium fusion cross-section σ is a smooth function of the deuteron energy E_i and the average of the cross section over the Maxwellian velocity distribution, i.e. $\langle\sigma v\rangle^{(M)}$, is illustrated in Fig.9, as a function of the deuterium temperature. The peak of $\langle\sigma v\rangle^{(M)}$ apears at around $T_i = 1\text{MeV}$, and the cross-section around there is approximately $\langle\sigma v\rangle \simeq 3 \times 10^{-16}\text{cm}^3/\text{sec}$. As seen in Fig.9, a wide range of deuterium temperature around $0.1\text{MeV} \leq T_i \leq 10\text{MeV}$ is found to contribute for producing neutron yield. The total neutron yield is roughly estimated by

$$N^{(cluster)} \sim \int dt \int \langle \sigma v \rangle n_{av}^2 d\mathbf{X}$$

$$\simeq \frac{\tau_d}{2} \langle \sigma v \rangle f^2 n_{cl}^2 V , \qquad (5)$$

where $\langle \sigma v \rangle$ is the average over the arbitrary deuterium energy distribution function, V and τ_d are the effective interaction volume and confinement and/or disassembly time. Here, we assume that $\langle \sigma v \rangle$ and n_{av} are spatially uniform and the average deuterium density n_{av} is given by $n_{av} = f n_{cl}$. The packing fraction f is described as $f = a^3/R^3$. The disassembly time τ_d is approximately estimated by $\tau_d \sim V^{1/3}/C_s$ where $C_s = \sqrt{T_e/M}$ is the deuterium ion sound speed with the eletron temperature T_e. Note that since the deuterium energy distribution resulting from the laser-cluster interaction is not generally described by the Maxwellian distribution, the direct velocity space integral has to be performed to evaluate the cross section. From Eq.(5), in order to increase the neutron yield, besides $\langle \sigma v \rangle$, we need to increase the packing fraction f, the effective intetaction volume V, and also the disassembly time τ_d. Note that the interaction volume is not restricted to the region where the laser directly interacts with clusters, but is extended to the surrounding region that is secondly heated by accelerated high energy particles.

The number of deuterium ions which contribute to nuclear fusion is quite small compared with the total number of heated deuterons (i.e. $n_{av}V$). Therefore, in order to efficiently utilize heated ions, it is desirable to introduce a solid deuterium (and/or deuterium doped) collar that surrounds the cluster medium. This is also schematically illustrated in Fig.8. In this case, the neutron yield produced by the collar is estimated as

$$N^{(collar)} \sim \langle \sigma l_m \rangle n_D n_{av} V , \qquad (6)$$

where $n_D (= n_{cl})$ is the solid deuterium collar density and l_m represents the velocity dependent mean free path of an accelerated deuterium ion in the solid collar. The mean free path of the energetic deuterium ion is roughly evaluated from the stopping power in the plasma given by $dE_i/dt = -Z^2 e^4 n_e \ln \Lambda / 4\pi \varepsilon_0^2 m v_i$ as

$$l_m \simeq \int_0^{E_i} \frac{dE_i}{dE_i/dx} \sim -\frac{4\pi \varepsilon_0^2}{Z^2 e^2 n_e \ln \Lambda} \left(\frac{m}{M} \right) E_i^2 , \qquad (7)$$

assuming that ions are primarily stopped by electrons. Here, the MKS unit is used, and $E_i \equiv M v_i^2/2$ and $\ln \Lambda$ represent the ion kinetic energy and Coulomb logarithm, respectively. When the surrounding collar is introduced, the total neutron yield is a sum of Eq.(5) and Eq.(6), i.e. $N^{(cluster)} + N^{(collar)}$. The ratio of the neutron yield resulting from the cluster region and collar one is

$$\frac{N^{(cluster)}}{N^{(collar)}} \sim \frac{\langle \sigma v \rangle \tau_d}{\langle \sigma l_m \rangle} \frac{n_{av}}{n_D} \sim \left(\frac{v_i \tau_d}{l_m} \right) f \sim \left(\frac{V^{1/3}}{l_m} \right) f , \qquad (8)$$

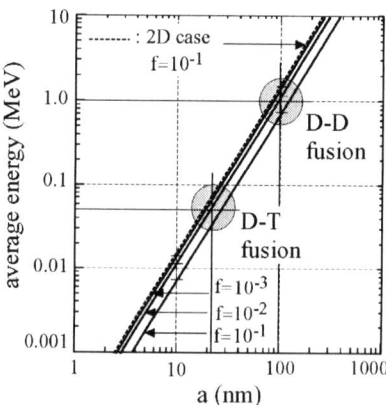

FIGURE 10. Maximum average deuterium kinetic energy obtained by the Coulomb explosion as a function of cluster radius for different packing fraction. Circles in the figure denote the region where DD and DT fusion cross-sections are maximized.

assuming that the cross-section is almost the same both regions. Since the mean free path is propotional to E_i^2, the contribution from the collar fusion is expected to become large for higher laser power regime.

Figure 10 shows the average accelerated ion energy from the Coulomb explosion estimated from Eq.2. For example, the cluster size of $a \simeq 100$nm is required to obtain the kinetic energy around $E_i \sim 1$MeV. Note that such a cluster size is larger by an order of magnitude than that achieved so far in experiments using gas jet technique[14,15]. This suggest that the neutron yield can be drastically increased by enlarging the cluster size and utilizing strong laser field to expell electrons from the cluster.

In order to perform a rough estimate of the neutron yield, we assume a plasma filament of the ion and electron temperature given by $T_i \sim T_e \sim 0.5$MeV and the average density of $n_{av} = 4.56 \times 10^{21}/\text{cm}^3 (f = n_{av}/n_D = 0.1)$, which diameter and axial length are given by $d \sim 20\,\mu$m and $l \sim 40\,\mu$m (the interaction volume : $V = 1.26 \times 10^{-8}\text{cm}^{-3}$). The stored energy in such plasma filament is around $W_p \simeq 10$J and then the disassembly time is estimated as $\tau_d \sim V^{1/3}/C_s \sim 5$psec. Assuming the cross section $\langle \sigma v \rangle \sim 10^{-16}$ at $T_i \sim 0.5$MeV, the neutron yield is estimated as $N^{(cluster)} \sim 1.3 \times 10^8$ per laser shot, which corresponds to 1.3×10^7 neutrons per joule. When we introduce the solid deuterium collar as illustrated in Fig.8, the stopping range is about $l_m \sim 20\mu$m from Eq.(7) and the neutron yield is estimated as $N^{(cluster)} \sim 5.2 \times 10^8$, which is the same order as that obtained from the inside cluster fusion.

If we surround the interaction region including the collar by some material at the radius $r_m = 50\mu$m to which we like to expose neutrons as illustrated in Fig.8, the average neutron flux on the material surface is $N/\pi r_m^2 l \sim 5.2 \times 10^{12}/\text{cm}^2$ per laser shot. To obtain high energy ions around 1MeV, we need to employ a large size cluster around $a \sim 100$nm and choose the packing fraction $f \simeq 10^{-2} - 10^{-1}$. The laser field will be determined to expell electrons from the cluster to trigger the Coulomb explosion.

As found in our numerical simulation, ions are efficiently accelerated in a low packing fraction case where the effect of inter-cluster interaction is rather small. In this case, the laser pulse can deeply propagate along the axial direction since the average density is low. On the other hand, when the packing fraction is high, although the interaction region as well as the ion accerelation are suppressed, electrons are efficiently heated to high temperature, so that such high energy electrons penetrate to surrounding clusters and secondarily heat and explode the clusters. This is an electron-triggered explosion of clusters, as opposed to the direct laser-triggered explosion. As a result, we can expect wider interaction region. The propagation of such a heated spot to surrounding clusters is a complicated problem where we need to take into account the collision effect and this remains a feature works. In the above example, the stored energy of the plasma filament is $W_p \sim 10J$ and this corresponds to 500TW laser with $\tau_l \simeq 20fsec$. In order to explode the large cluster as $a \simeq 100nm$, the excursion length of heated electrons has to satisfy the relation $\xi_e = (c/\omega)\arcsin(a_0/\sqrt{1+a_0^2}) \geq a(\simeq 100nm)$. Once the cluster which size is around $a \simeq 100nm$ is exploded in the Coulomb explosion regime, the average temperature becomes $T_e \sim T_i \sim 1MeV$. Assuming that high power laser around $I_0 \simeq 10^{21}W/cm^2(a_0 \simeq 20)$ is required to explode the cluster with $a = 100nm$ by considering the relation $a > \delta_e$, the interaction area and the radius are estimated as $S_l(= \pi r_l^2) \simeq W_P/I_0\tau_l \simeq 5 \times 10^{-7}cm^2$, and then $r_l \simeq 4\mu m$. Then, as we discussed, we expect that this small spot expands to the above plasma filament of the volume V. Further numerical simulation is nessessary to precisely predict the plasma filament and the ion energy distribution that is discussed in Ref.20.

CONCLUDING REMARKS

We investigated the laser-cluster interaction to clarify key physical issues essential to understand and optimize the cluster fusion. The laser-cluster interaction is found to essentially different from the conventional prepatation of laser-plasma and/or laser-solid interaction. One of the crucial advantage to use the cluster material is that there exists the freedom to control the interaction, or more specifically to control electron and ion energy distribution. By properly choosing the cluster size, packing fraction and laser field amplitude, we can efficiently accelerate cluster ions to the high energy where the fusion cross-section is maximized.

The cluster medium shows characteristics as small particle system, where the level of fluctuation is significantly higher than that of usual gas and/or plasmas. Therefore, the interaction depends on the microscopic cluster distribution. Such a microscopic distribution may be probabilistically determined through the detailed physical process by which the clustered medium is formed from the gas state. It is found that the absorption of the laser energy for a random distribution case bacome generally higher than that for an ordered lattice-like distribution case. This is due to the fact that the inter-cluster interaction becomes important and the stochasticity of the electron motion is enhanced by complicated electrostatic and electromagnetic field strunture in a random cluster distribution. As a result, a physical quantity resulting from laser-cluster interaction is determined as a expectation value averaged over possible distributions.

We also investigated the feasibility to use laser-cluster interaction for generating high flux neutrons. From the simple (and also rough) estimate, we evaluate about 10^7 neutrons per joule. It is expected that this value may be further enhanced by properly adjusting the paramerers and by incorporating the surrounding clusters and collar effects. For example, a laser with 10J energy with repetition rate 100Hz irradeating DD clusters surrounded by a collar produces the intensity of neutron 10^{14}/cm^2/sec at a distance of 50μm. If we adopt DT clusters, the intensity of nertrons is $\sim 10^{15}$/cm^2/sec. It is worth while to try to increase the confinement and/or disassembly time, for example, by introducing strong confinement magnetic field. These, however, remain as future works.

ACKNOWLEDGEMENTS

One of authors, Y. Kishimoto, acknowledges Dr. A. Kitsunezaki and Dr. M. Kikuchi of Naka Fusion Research Establishment, and also Y. Kato of Advanced Photon Research Center of JAERI for supporting our works. This work is supported by NEXT(Numerical Experimental Tokamak) research project of Naka fusion program. We used SGI Origin3800 massively parallel computer of JAERI to perform present simulations.

REFERENCES

1. N. G. Gotts, D. G. Lethbridge, and A. J. Stase. J. Chem. Phys. **96**, 408 (1962).
2. J. Purnell, E. M. Snyder, S. Wei, and A. W. Caslemand, Chem. Phys. Lett. **229**, 33e (1994).
3. T. Ditmire, T. Donnelley, R. W. Falcone, and M. D. Perry, Phys. Rev. Lett. **75**, 3122 (1995).
4. S. Dobosz, M. Lezius, M. Schmidt, P. Meynadier, M. Perdrix, and D. Normand, Phys. Rev. A **56**, R2526 (1997).
5. C. Deutsch, Plasma Phys. Control. Fusion **41**, A195 (1999).
6. Y. L. Shao et. al. Phys. Rev. Lett. **77**, 3343 (1996).
7. T. Ditmire, J.W.G. Tisch, E. Springate, M.B. Mason, N. Hay, R.A. Smith, J. Marangos, and M.H.R. Hutchinson, Nature **386**, 54 (1996).
8. Y. Kishimoto and T. Tajima: in *High Field Science*, eds. T. Tajima, K. Mima, and H. Baldis (Kluwer, NY, 2000) p.83.
9. Y. Kishimoto, T. Masaki and T. Tajima, Phys. Plasmas, Vol.9, No.2, 589 (2002).
10. T. Ditmire, T. Donnelley, A. M. Rubenchuk, R. W. Falcone, and M. D. Perry, Phys. Rev. A **53**, 3379 (1996).
11. T. Tajima, Y. Kishimoto, and M.C. Downer, Phys. Plasma **6**, 3759 (1999).
12. R. H. Doremus, J. Chem. Phys. **40**, 2389 (1964).
13. A. Kawabata and R. Kubo, J. Phys. **40**, 1765 (1966).
14. T. Ditmire, J. Zweibeck, V.D. Yanovsky, T. E. Cowan, G. Hays, and K. B. Wharton, Nature **398**, 489 (1999).
15. J. Zweibeck, R.A. Smith, T. E. Cowan, G. Hays, K. B. Wharton, V.P. Yanovsky, and T. Ditmire, Phys. Rev. Lett. **84**, 2634 (2000).
16. T. Tajima, Y. Kishimoto, and T. Masaki, Physica Scripta. **T89**, 45 (2001).
17. Y. Kishimoto, K. Mima, T. Watanabe, and K. Nis hikawa, Phys. Fluids **26**, 2308 (1983).
18. Y. Kishimoto (private communication): 2-dimensional fully relativistic electromagnetic particle-in-cell code, EM2D-EB, where the Maxwell equations are directly solved with respect to the electric and magnetic fields ($\mathbf{E}, \mathbf{B}$).
19. C.Kittel : Introduction to solid state physics (John Wiley & Sons, 1971).
20. Y. Kishimoto, T. Masaki and T. Tajima, AIP Conference Proceedings 611 titled "Second Interaction Confernce on Superstrong Field in Plasmas".

Particle Acceleration and Cosmic-Ray Origin in the Galaxy

Toru Tanimori

Dept. of Physics, Graduate School of Science, Kyoto University, Kitashirakawa-Oiwakechou Sakyo-ku, Kyoto, 606-8502, Japan

Abstract.
In 1990's Very High Energy Gamma-ray Astrophysics has dramatically advanced due to the Imaging Air Čerenkov Telescopes(IACTs). After the first detection of TeV gamma-ray emission from the Crab nebula in 1989, several type of TeV gamma-ray sources, Active Galactic Nuclei(AGN), young pulsar, and SuperNova Remnant(SNR), have been detected. In those discoveries, recent detections of both synchrotron X-rays and TeV gamma-ray emissions from several SNRs are very significant. SNR has been widely believed to be an unique candidate of galactic cosmic-ray origin since the beginning of cosmic-ray physics, whereas little observational evidences have been reported so far. Those are expected to be a clue of not only the galactic cosmic-ray origin but also the understanding of the particle acceleration due to a diffusive shock.

Here I present the recent results obtained by our group, CANGAROO, about the evidences of electron and proton acceleration in SNRs.

INTRODUCTION

As is well known, X-ray astronomy is an established field in astronomy, where several hundred thousands of X-rays sources have been found so far. On the other hands, the growth of gamma-ray astronomy had been relatively slow until the launch of the advanced gamma-ray satellite, Compton Gamma-Ray Observatory(CGRO), in 1991[1]. EGRET, a detector observing GeV gamma rays on-board CGRO, found more than 200 gamma-ray sources, whereas only 20 sources had been known before then. Emissions of high energy gamma rays were observed not only from galactic sources such as pulsar/nebulae and supernova remnants but also the more than hundred extragalactic sources. Figure 1 shows the GeV gamma-ray source catalog detected by EGRET in galactic coordinates. You see that hundreds of AGN were located out of the galactic plane, and moreover along the galactic plane there concentrated lots of unidentified sources.

Also several TeV gamma-ray sources have been recently established: Crab pulsar/nebula[2], Mrk421[3], Mrk501[4], PSR1706-44[5]. Celestial objects emitting very high energy gamma rays of energies greater than TeV had been expected as a natural consequence of the existence of cosmic rays, and they have been searched for since the middle of 20th century using the ground based detectors such as scintillation-counter arrays and air Čerenkov telescopes. Since huge background of hadron showers overwhelms the tiny signal of celestial gamma rays, no persistent TeV gamma-ray source had been found until the discovery of TeV gamma-ray emission from the Crab

CP634, *Science of Superstrong Field Interactions*, edited by K. Nakajima and M. Deguchi

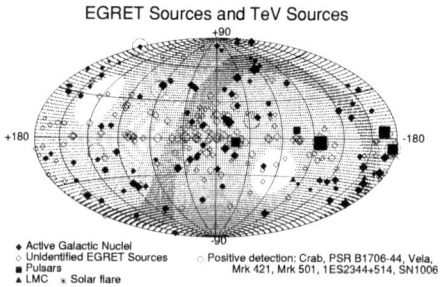

EGRET Sources and TeV Sources

- ◆ Active Galactic Nuclei
- ◇ Unidentified EGRET Sources
- ■ Pulsars
- ▲ LMC   ✶ Solar flare
- ○ Positive detection: Crab, PSR B1706-44, Vela, Mrk 421, Mrk 501, 1ES2344+514, SN1006

FIGURE 1. GeV gamma-ray source catalog detected by EGRET in galactic coordinates, where six TeV gamma-ray sources are also drawn by a circle.

by the Whipple group in 1989[2]. The Whipple group developed imaging Čerenkov technique[6], and hence the rejection power of hadron showers was greatly improved. In 1990's several type of TeV gamma-ray sources have been detected with high statistics by adopting this imaging technique. In particular, successive discoveries of AGN emitting TeV gamma rays were astonishing[3, 4].

Our group, CANGAROO, the collaboration of Japanese and Australian institutes, has observed TeV gamma-ray sources in the southern hemisphere since 1992 in South Australia[7]. The southern hemisphere provides us a good chance to observe lots of galactic objects such as pulsar/nebulae, supernova remnants (SNRs), black holes, the galactic center, and so on. In fact we have found several galactic TeV gamma ray sources as listed in the review article of Weekes[8]. In particular recent noteworthy discoveries are the several reports on the detection of TeV gamma-ray emissions from shell-type SNRs in both southern and northern skies[9, 11, 12], to which people have been eagerly looking forward ever since the beginning of cosmic-ray physics.

COSMIC-RAY ORIGIN AND SHOCK ACCELERATION

High energy particles are known to fill all over the galaxy and maybe over the surrounding halo, and play non-negligible roles on almost all phenomena in the universe. Moreover, high energy particles coming to the earth, "cosmic rays", are surely affected on the circumstance of the earth. However, nobody know how and where such high energy particles are generated in the universe. These questions are unresolved yet, and always important issues in astrophysics in spite of the long history of the study of cosmic rays.

Radio and X-ray observations have revealed lots of high energy phenomena in the universe, where huge amount of energy is consumed to accelerate particles up to more than GeV energies, since high energy electrons in the magnetic field emit synchrotron radiation of radio to X-ray. Also high energy ions (mainly proton) must surely exist in the galaxy since almost cosmic rays are ions.

For a long time, supernova remnants (SNRs) have been believed to be a favored site for accelerating cosmic rays up to 10^{15}eV, because only they can satisfy the required energy input rate to the galaxy among several galactic objects[13]. In addition, a shock

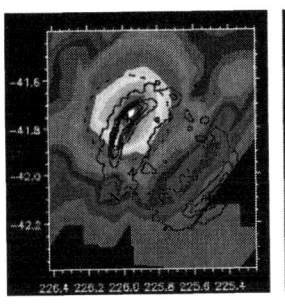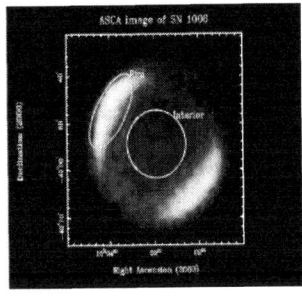

FIGURE 2. Map of TeV gamma-ray (right) and ASCA X-ray (left) emissions of SN1006.

acceleration theory was established around 1980s, in which particles are accelerated with the collisions between particles and plasma gas moving at a supersonic velocity in the space[14]. SNRs are a just extended and heated gas system accompanied by very strong shocks. Shocks are very common phenomena in the universe, and hence the shock acceleration has been widely applied for high energy phenomena in the universe. Thus shock-acceleration mechanism has been a standard theory for particle acceleration in astrophysics. Although this theory looks very simple and reliable, an observational evidence is still very sparse.

In order to investigate the shock-acceleration mechanism, SNR is an unique and ideal laboratory because it is quite simple and a well-understood astronomical object. The evolution of SNR can be fairly explained with several observables such as explosion energy of the SNR, total mass of the ejecta, density of the inter stellar medium (ISM) around the SNR and the age after the explosion[15]. In addition, the resoluble size of a SNR enables us to directly observe the geometrical structure of the shock front accelerating particles, which provides lots of significant physical parameters quantitatively (absolute value of the magnetic field, index of power law, maximum energy of particle acceleration, and so on).

SN1006:first evidence of electron acceleration up to $\geq$ TeV

The first evidence for the very high-energy particle acceleration in a SNR was presented by the observation of the strong synchrotron emission of SN1006 by the Japanese X-ray satellite ASCA in 1995[16]. By assuming the magnetic field of several μ gauss, observed synchrotron X-ray emission strongly supported the existence of high energy electrons of tens or hundreds of TeV. Those high energy electrons must emit not only synchrotron radiation but also high-energy gamma rays due to Inverse Compton (IC) process by the hard collision with 2.7K Cosmic Microwave Background (CMB). Scattered photons acquire the energy of about one tenth of primary electrons, and hence their energies reach near 10 TeV in SN1006. They could be detected by the CANGA-ROO telescope in the TeV region [17, 18, 19, 20]. In 1996 and 1997, CANGAROO succeeded in detecting the TeV gamma-ray emission from the north rim of SN1006[9]

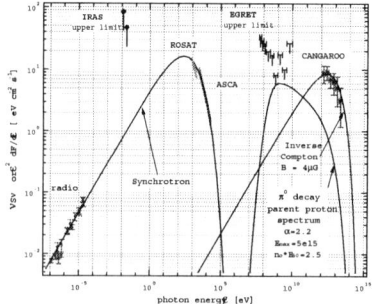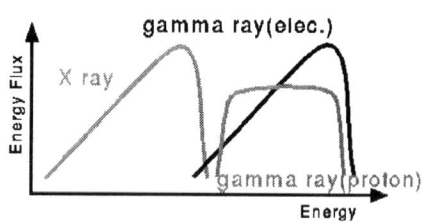

FIGURE 3. Energy spectrum at the north rim of SN1006 from radio to TeV with the fitting results based on the shock model (left), and the expected energy spectra from both IC process and π° decay with synchrotron spectrum (right).

as shown in Fig.2.

There exists a simple but useful formula connecting among relativistic electrons, high energy photons scattered by IC process, synchrotron photons and soft seed photons. The emission powers of synchrotron radiation and IC scattering, P_{sync} and P_{IC} are respectively expressed as follows,

$$P_{sync} = \frac{4}{3}\sigma_T c \gamma^2 \beta^2 U_B, \quad P_{IC} = \frac{4}{3}\sigma_T c \gamma^2 \beta^2 U_{soft}, \tag{1}$$

where σ_T is the Thomson cross section, $\sigma_T = 6.7 \times 10^{-25} cm^2$, and $c\beta$ and γ are the velocity and Lorentz factor of the electron. U_B and U_{soft} are energy densities of the magnetic field and the soft seed photons, respectively. Since the energy density of soft seed photons (CMB) is well-known, an observed gamma-ray flux provides a good estimation of the magnetic field strength at the acceleration site in SN1006. Figure 3(left) shows the wide band energy spectrum at the north rim of SN1006 from radio to TeV, and also the fitting result based on the IC model[21]. All data are fitted very well, and several significant parameters, magnetic field(B), power index(a), maximum energy (E_{max}) were determined independently: $B = 4.3\mu$ gauss, a = 2.2, and $E_{max} = 60$ TeV.

Here we assumed that the detected TeV gamma-ray emission is mainly due to IC process with very high energy electrons considering the tenuous shell of ($\leq\sim 0.4\,cm^{-3}$)[10]. In fact the expected spectrum from π° decay generated by high energy proton conflicts with the upper limit in the GeV (Fig.3left).

In order to verify the origin of cosmic rays, we have to obtain a clear evidence of proton acceleration. The identification of the parent particles of TeV gamma-rays (electron or proton) will be possible by observing the wide spectrum from sub to multi TeV region as shown in Fig.3(right). Gamma-ray spectrum flatter than $E^{-2.0}$ in this region is surely due to IC process, while that due to π° decay generated by collision between intersteller matter (ISM) and high energy proton is expected to be steeper than $E^{--2.0}$.

RX J1713.7-3946: FIRST EVIDENCE OF PROTON ACCELERATION UP TO THE TEV REGION

RXJ1713.7-3946 was observed as the strongest synchrotron X-ray emitter among SNRs by ASCA in 1997[22], and subsequently TeV emission was detected from the maximum X-ray emission point[11]. Although this SNR emits an intense synchrotron X-ray similar to SN1006, those two SNRs looks different. Its morphology is obviously more complex than that of SN1006, of which north parts might interact with the molecular cloud observed by the radio telescope[23]. Therefore this TeV emission might be ascribed to the π° decay generated by the collision of accelerated protons with the molecular cloud.

To clarify the nature of accelerated particles, we have observed this point again in 2000 and 2001 using the new CANGAROO 10m telescope[24], and the result has been recently published[25]. Here summary is presented.

The differential fluxes of TeV gamma-rays from RX J1713.7-3946 are plotted in Fig.4(left) with previous data[11], and the best fit is

$$dF/dE = (1.63 \pm 0.15 \pm 0.32) \times 10^{-11} (E/1TeV)^{-2.84 \pm 0.5 \pm 0.20} \text{ TeV}^{-1}\text{cm}^{-2}\text{s}^{-1}, \quad (2)$$

where the first errors are statistical and the second are systematic. You note that RX J1713.7-3946 is one of the brightest galactic TeV gamma-ray sources discovered so far, and the power-law spectrum increases monotonically in a single power law as energy decreases. This feature is in contrast to the Tev gamma-ray spectrum of SN 1006, which flattens below 1 TeV[26], and is well consistent with synchrotron/Inverse Compton (IC) models[17, 18, 19, 20, 27]. While both SNRs emit intense X-rays via the synchrotron process, different TeV spectra suggest that different emission mechanisms may act respectively.

Morphology of the gamma-ray emitting region is shown by the thick- solid contours in Fig.4(right), together with the synchrotron X-ray ($\geq$ 2keV) contours by ASCA[28] and infrared ones from IRAS 100μm results[29] which possibly indicates the density distribution of the inter stellar matter. The observed TeV gamma-ray intensity peak coincides with the maximum point NW- rim observed in X-ray, but the TeV gamma ray emission extends over the ASCA contours. A possible extension towards the CO cloud in the north-east[30] can also be seen.

The broad band energy spectrum is plotted in Fig.5(left) with theoretical predictions (described below). Also in this figure, other data has been shown using data from the ATCA (Australia Telescope Compact Array)[31], ASCA[22, 28],and EGRET[32].

In order to explain the broad-band spectrum, three mechanisms, the synchrotron/IC process, bremsstrahlung, and π° decay produced by proton-nucleon collisions are considered, where the momentum spectra of incident particles (electrons and protons) are assumed to be

$$\frac{dN}{dp} = N_0 \left(\frac{p}{mc}\right)^{-\alpha} \exp\left(-\frac{p}{pmax}\right) [cm^{-2}eV^{-1}c] \quad . \quad (3)$$

based on the shock acceleration model. The effects of acceleration limits from the age and size of the SNR and energy losses of particles are included in the exponential term.

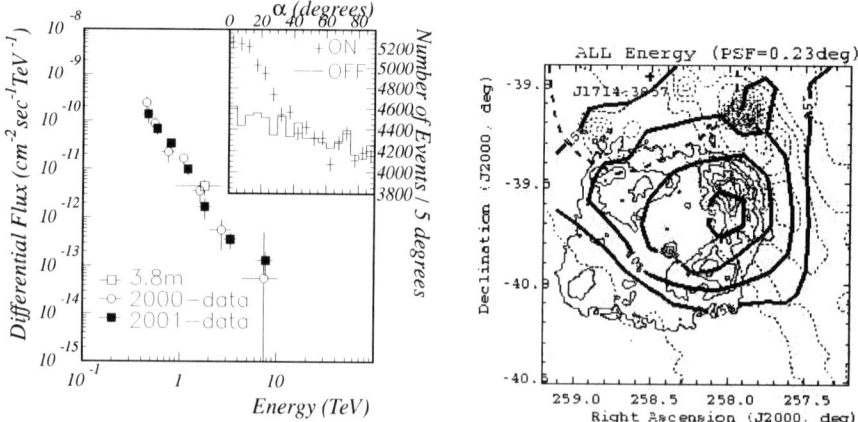

FIGURE 4. Differential fluxes obtained by this experiment together with that of CANGAROO-I, where inserted graph is the excess events determined from the plots of image orientation angle (left). Also profile of emission of TeV gamma rays around the NW-rim of RX J1713.7-3946 (solid thick contours) are shown (right).

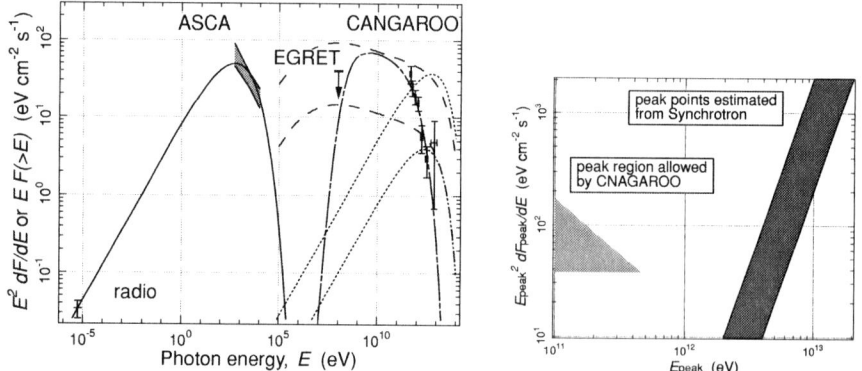

FIGURE 5. Multi-band emission with models is presented (left). TeV gamma-ray points are from this work. Lines show model calculations: synchrotron emission (solid line), Inverse Compton emission (dotted lines), bremsstrahlung (dashed lines) and emission from π° decay (dot-dashed line). Also allowed regions in parameter space (peak flux versus peak energy) is drawn (right); the peak energies and fluxes for IC processes are allowed in the above parameter space, and also the allowed region from experimental flux corresponds to the lower-right corner of the shaded area.

Best-fit values of free parameters in (2) for the radio and X-ray fluxes due to synchrotron radiation from electrons are obtained 2.08[34] for α, 126 for $(p_{max}/TeV)(B/\mu G)^{0.5}$, where c is the speed of light, and 2.00 for $N_\circ(V/4\pi d^2)(B/\mu G)^{(\alpha+1)/2}$, where V is the volume of the radiation region and d is the distance from the Earth.

Thus the resultant best fit is plotted in Fig.5(left) (the solid line). We initially assumed the 2.7 K CMB as the seed photons for IC scattering. Calculated inverse Compton Spectra are plotted with dotted lines in Figure 2 for two typical magnetic field strengths, 3 and $20\mu G$. Note that these models are far from consistent with the observed sub-TeV spectrum. In general synchrotron/IC process gives a clear correlation between the peak fluxes and its energies of synchrotron and IC emissions as a function of the strength of the magnetic field. We investigated the allowed region of peak flux of IC emission taking into account the uncertainties of IR emission for the IC seed photons. The calculation was carried out both with and without this IR maximum energy emission in the IC process, assuming various incident-electron spectra. The hatched area in Fig.5(right) is the resulting theoretically allowed region of the peak flux, and also the experimentally allowed region are plotted by the shaded area. The predictions of synchrotron/IC models are obviously inconsistent with our experimental data by an order of magnitude. The bremsstrahlung spectrum was calculated assuming that it occurs in the same region as the synchrotron radiation. A material density of $\sim 300 protons/cm^3$ was assumed. The dashed lines in Fig.5(left), for magnetic fields of 3 and $20\mu G$, are both inconsistent with our observation.

Thus, electron-based models fail to explain the observational results and so we examined π° decay models. The π°s are produced in collisions of accelerated protons with interstellar matter. A model[33] adopting Δ-resonance and scaling was used. We adopted parameters for Equation (2) of $\alpha = 2.08$ and $p_{max} = 10$ TeV, considering the plausible parameter regions of typical shock acceleration theory. The result is shown by the dot-dashed curve in Fig.5(left). The best fit parameters for the total energy of accelerated protons E_0 and matter density n_0 must satisfy $(E_0/10^{50}[ergs]) \cdot (n_0[protons/cm^3]) \cdot (d/6[kpc])^{-2} = 300$, where d is the distance to RX J1713.7-3946. A value of $E_0 \sim 10^{50}[ergs]$ gives n_0 of the order of 10 or $100[protons/cm^3]$ for distances of 1 or 6 kpc, respectively. Both cases are consistent with the molecular column density estimated from Fig. 7 of ref.6. Thus the π° decay model alone readily explains our results, which provide the first observational evidence that protons are accelerated in SNR to at least TeV energies.

We can guess more about the galactic cosmic-ray origin. Using observed TeV gamma-ray flux, total energy of protons in the SNR was estimated to be $E_0 \sim 10^{50}erg$, assuming the distance of 6 kpc and target molecular cloud density of several times of $10^2 protons/cm^3$. The energy input rate for proton acceleration estimated from this observation can represent the energetics of cosmic rays in the Galaxy. In other words, if such SNR are born every 100 years and accelerate protons, the SNR can provide all the energy of cosmic rays inside the Galaxy, where we adopt $1 \times 10^{40}erg/sec$ for the luminosity of galactic cosmic-rays.

SUMMARY

Recent detections of TeV gamma rays from SNRs will obviously advance the studies of the galactic cosmic-ray origin and the shock acceleration mechanism by stimulating multi-wave length observations for SNRs. In particular combination of the morphological studies in both X-ray and gamma-rays will be a key for these studies; advanced X-ray telescope satellites, Chandra and Newton, are now providing excellent images and spectroscopy, also the stereo observation by the 10m-class IACTs will soon provide good quality TeV gamma-ray images with the angular resolution of ≤ 0.1 degree. In southern hemisphere, CANGAROO and H.E.S.S groups will begin the stereo observations using 10m-class IACTs in 2002.

REFERENCES

1. R.C. Hartmann, et al., *Astrophys. J. Suppl.*,123,79, (1999)
2. G. Vacanti,et al., *Astrophys. J.*,377,467, (1991)
3. M. Punch,et al., *Nature*,358, 477, (1992)
4. J. Quinn ,et al.,*Astrophys. J.*,456, L85, (1996)
5. T. Kifune ,et al.,*Astrophys. J.*,438, L91, (1995)
6. A.M, Hillas, *Proc. 19th Int. Cosmic-Ray Con.*(La Jolla), 3, 445, (1985)
7. T. Hara,et al., *Nucl. Instrum. Methods Phys. Rev.*, A332, 300, (1993)
8. T.C.Weekes, *Proc. of GeV-TeV Gamma Ray Astrophysics Workshop* (Snowbird),ed B. Dingus, 3
9. T. Tanimori ,et al., *Astrophys. J.*,497, L25, (1998)
10. R. Willingale et al., *Mon. Not. R. Astron. Soc.*,278, 749, (1996)
11. H. Muraishi ,et al., *Astron. Astrpophys.*,354, L57, (2000)
12. F. Aharomian et al., *Astron. Astrpophys.*,370, L12, (2001)
13. L.O'C. Drury, W.J. Markiewicz and W.J. Volk, *Astron. Astrpophys.*,225, 179, (1989)
14. R.D. Blandford and D. Eichler, *Phys. Rep.*,154, 1, (1987)
15. L.I. Sedov, *LaTeX: Similarity and Dimensional Methods in Mechanics* (Academic Press, 1959).
16. K. Koyama,et al., *Nature*,376, 255, (1995)
17. M. Pohl, *Astron. Astrpophys.*,307, L57, (1996)
18. A. Mastichiadis, *Astron. Astrpophys.*,305, L53, (1996)
19. A. Mastichiadis and O.C. Jager, *Astron. Astrpophys.*,311, L5, (1996)
20. T.Yoshida and S.Yanagita, *Proc. 2nd INTEGRAL Workshop*, ESA SP382, 85, (1997)
21. T. Naito,et al., *Aston. Nachr.*,320, 205, (1999)
22. K. Koyama et al, *Publ. Astron. Soc. Japan*, 49, L7, (1997)
23. P. Slane et al., *Astrophys. J.*,525, 357, (1999)
24. Tanimori, T. *Prog. of Theoretical Phys. Suppl.*, 143, 78, (2001)
25. Enomoto, R., et al., *Nature*, 416, 823, (2002)
26. Tanimori, T., et al., *Proc. on 27th ICRC2001*, 6, 2465, (2001)
27. Naito, T., Yoshida, T., Mori, M., and Tanimori, T., *Astron. Nachr.*, 320, 205 (1999)
28. Tomida, H., Synchrotron Emission from the Shell-like Supernova Remnants and the Cosmic-Ray Origin, Ph. D. thesis, Kyoto Univ. (1999).
29. Skyview, NASA, at <http://skyview.gsfc.nasa.gov>.
30. Butt, Y. M., Torres, D. F., Combi, J. A., Dame, T., and Romero, G., *Astrophys. J.*, 562, L167, (2001)
31. Ellison, D. C., Slane, P., and Gaensler, B. M., Preprint astro-ph/0106257 (2001)
32. Hartman, R.C. et al.,Astrophys. J. Suppl., 123, 79, (1999)
33. Naito, T., and Takahara, F., *J. Phys.* G, 20, 477, (1994)
34. Rybichi, G. B. and Lightman, A. P., Radiation Processes in Astrophysics, John Wiley and Sons (1979), and references therein.

Direct Particle Acceleration in Astroplasmas

M. Hoshino

Department of Earth and Planetary Science, University of Tokyo, 7-3-1 Hongo, Bunkyo, Tokyo 113-0033 Japan

Abstract. The high energy particle acceleration mechanisms are discussed by focusing on the direct acceleration in the astrophysical context. We specifically argue that the relativistic magnetic reconnection and the shock surfing/surfatron processes can efficiently accelerate charged particles to a relativistic energy, and that those mechanisms may produce a non-thermal, power-law energy spectrum.

INTRODUCTION

The particle acceleration is known to occur in a diverse variety of astrophysical sites: the nearest-at-hand examples are the terrestrial aurorae whose typical energies of $\sim$ keV, the terrestrial bow shock and magnetosphere of energies of ~ 1 MeV, and the radiation belt of energy as high as 100 MeV. In the pulsar magnetosphere and its surrounding synchrotron nebula, TeV $\gamma-$rays are observed. Cosmic-ray particles with energies in excess of 10^{20} eV have been detected [1].

Charged particles can be accelerated by utilizing several kinds of free energies. The most common free energy source is bulk kinetic energy in the form of jets and winds etc. Shock waves play an significant role on the energy conversion from the bulk kinetic to the thermal/non-thermal energies, and the diffusive/Fermi shock acceleration has been discussed as one of the key processes producing the high-energy particles [2, 3, 4]. It predicts a power-law energy spectrum with the power-law index of 2 that depends weakly on the shock Mach number, and the index suggested by the theory is believed to be very close to the observational spectra. The above supports the diffusive/Fermi shock acceleration model as an origin of cosmic-ray, but the disadvantage is that the acceleration process is slow, and that it is hard to keep up with energy losses against radiation processes.

Many other different acceleration processes are also known to exist: magnetic reconnection, electrostatic double layer, unipolar induction, shock surfing/surfatron, plasma wakefield, and self-focusing etc. In these mechanisms, charged particles may be accelerated directly to high energy by continuously traveling toward an electric field direction, and this kind of acceleration process is called as the "direct" acceleration, contrary to the "stochastic" acceleration such as the diffusive shock acceleration. The direct acceleration has the advantage of being fast, but it was usually not obvious how to get an efficient non-thermal particle acceleration with a power-law energy spectrum. In this paper, we argue recent progress on the

CP634, *Science of Superstrong Field Interactions*, edited by K. Nakajima and M. Deguchi
© 2002 American Institute of Physics 0-7354-0089-X/02/$19.00

direct acceleration that may produce a non-thermal, power-law energy spectrum. Specifically, we discuss two mechanisms in the family of the direct acceleration, magnetic reconnection and the shock surfing acceleration.

MAGNETIC RECONNECTION

Let us discuss magnetic reconnection as the first example of the direct acceleration. Many plasma states in universe have magnetic fields whose structure contains the neutral sheet where the magnetic field polarity changes its direction. Magnetic reconnection is believed to play a very important role by converting the magnetic field energy into plasma kinetic energy. The solar corona and the terrestrial plasma sheet are the best-studied examples of this process. In addition, for astrophysical objects, the relativistic reconnection, whose Alfvén speed becomes almost the speed of light, has been recently paid the attention. For example, the relativistic reconnection in the pulsar wind region of neutron stars has been studied [5, 6]. We discuss the non-thermal particle acceleration in the relativistic reconnection.

Nonthermal Particles in Relativistic Reconnection

Figure 1 shows a snapshot of the non-linear evolution of the relativistic magnetic reconnection obtained by a two-dimensional, particle-in-cell code in the (X, Z) plane [7]. The plasma is assumed to consist of positrons and electrons, and the modified Harris equilibrium model is used as the initial condition [7, 8]. The thickness of the initial plasma sheet is set to be the positron/electron gyro-radius. The magnetic field lines and the density are shown in Figure 1. An X-type neutral line is formed at the center of the simulation box, and plasmas are ejected from the X-type region toward both $\pm X$ directions. The behavior of the reconnecting

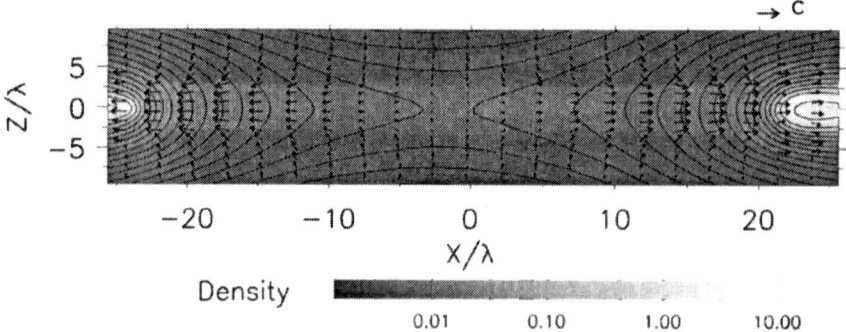

FIGURE 1. Snapshot of relativistic magnetic reconnection. The solid lines and the vectors show magnetic field lines and plasma flow vectors, respectively. The density contour represents the plasma density, which is normalized by the initial density in the plasma sheet.

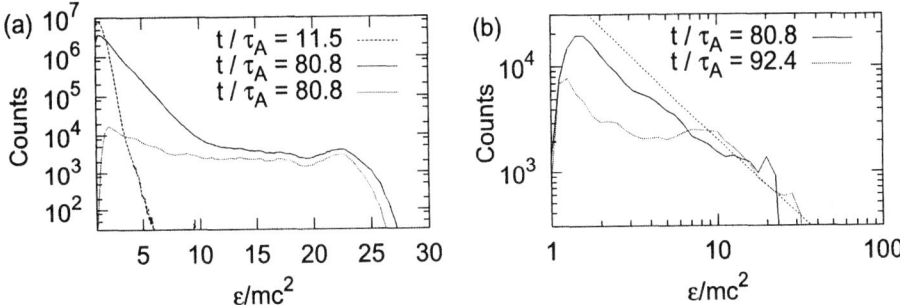

FIGURE 2. Energy spectra in the reconnection region. (a) Dashed and solid lines indicate energy spectra over the whole simulation box at $t/\tau_A = 11.5$ and 80.6, respectively. Dotted line shows the partial energy of particles around the X-type region. (b) Energy spectra in the double logarithmic scales. Note that the spectra are approximated by the power-law spectrum with its index of 1.

magnetic field is basically same as those studied in non-relativistic MHD and particle simulations. Note that the outflow speed in this relativistic reconnection case reaches up to 0.91c.

Figure 2 shows the energy spectra integrated over pitch angle in the reconnection region. The dashed line in Figure 2a shows the spectrum at $t/\tau_A = 11.5$ in the early stage of reconnection, which is well described by a Maxwellian distribution, while the solid line represents the spectrum at $t/\tau_A = 80.8$ in the non-linear stage. τ_A is the Alfvén transit time of the plasma sheet. One can find not only strong plasma heating but also non-thermal high energy tail. The dotted line represents a partial energy spectrum obtained from the particles around the X-type region, and we find that most of the high energy particles are produced around the X-type region. Figure 2b shows the partial energy spectra around the X-type region in the double log scales at $t/\tau_A = 80.8$ and 92.4. Both spectra are well approximated by a power-law spectrum $N(\varepsilon) \propto \varepsilon^{-s}$ with $s \simeq 1$. We also observed that the maximum energy grows in time, and the energy is described by $\varepsilon_{max} \sim e E_0 c (t - t_0)$, where e, E_0, c and t_0 are the electric charge, the electric field around the X-type region, the speed of light, and the onset time of reconnection, respectively.

Acceleration in X-type Region

Let us discuss how the non-thermal power-law energy spectrum can be produced near the X-type region. First we assume that the particle can be accelerated by the reconnection electric field E_y, and gain quickly a relativistic energy. The acceleration rate of the relativistic particle with the energy ε can be expressed by,

$$\frac{d\varepsilon}{dt} \sim e E_y c. \tag{1}$$

Secondly we evaluate the life time of particle around the X-type region. Since the particle motion around the X-type region can be described by a Speiser orbit, we may assume that the life time of particle can be expressed by the gyro-period of the relativistic particle. Then the loss rate with which the accelerated particles escape from the X-type acceleration region can be given by

$$\frac{1}{N(\varepsilon)}\frac{dN(\varepsilon)}{dt} \sim -(\frac{eB_z}{mc})(\frac{mc^2}{\varepsilon}),\tag{2}$$

where N is the particle number and B_z is the reconnecting magnetic field component.

From the above two equations, we can obtain the energy spectrum as

$$N(\varepsilon) \propto \varepsilon^{-s}, \qquad s \sim B_z/E_y \sim 1.\tag{3}$$

Since we discuss the particle acceleration around the X-type region where $B_z \sim E_y$, we get the power-law index $s \sim 1$. The particle trajectory near the X-type region is very sensitive to the reconnecting magnetic field structure, and it would be required for further discussion to determine the energy spectrum in details.

SHOCK SURFING ACCELERATION

The next example of the direct acceleration is the shock surfing acceleration. It is believed that the wave-particle interaction inside the shock front layer where strong plasma wave/turbulence is expected can provide the non-thermal particle acceleration as well as the plasma heating/thermalization. As to the shock front acceleration processes, the shock drift acceleration and the shock surfing acceleration are often discussed. Both mechanisms utilize a motional electric field for accelerating a charged particle. In the case of the shock surfing, the particle can be effectively trapped in the shock front region, and its energy would be expected to reach up to a relativistic energy.

Nonthermal Electrons in High Mach Number Shock

Figure 3 shows a snapshot of the nonlinear stage of a magnetosonic shock front region, obtained by the one-dimensional, particle-in-cell simulation where both ions and electrons are treated as particle [9, 10]. In the simulation, a low-entropy, high-speed plasma consisting of electrons and ions is injected from the left boundary region which travels towards positive x. The downstream right boundary condition is a wall where particles and waves are reflected. Under the interaction between the plasma traveling towards positive x and the reflected particles and waves from the right-hand boundary, the shock wave is produced, and it propagates backward in the $-x$ direction. The left-hand panel in Figure 3 shows the shock front region from $X/(c/\omega_{pe}) = 150$ to 300, while the right-hand panel is its enlarged picture.

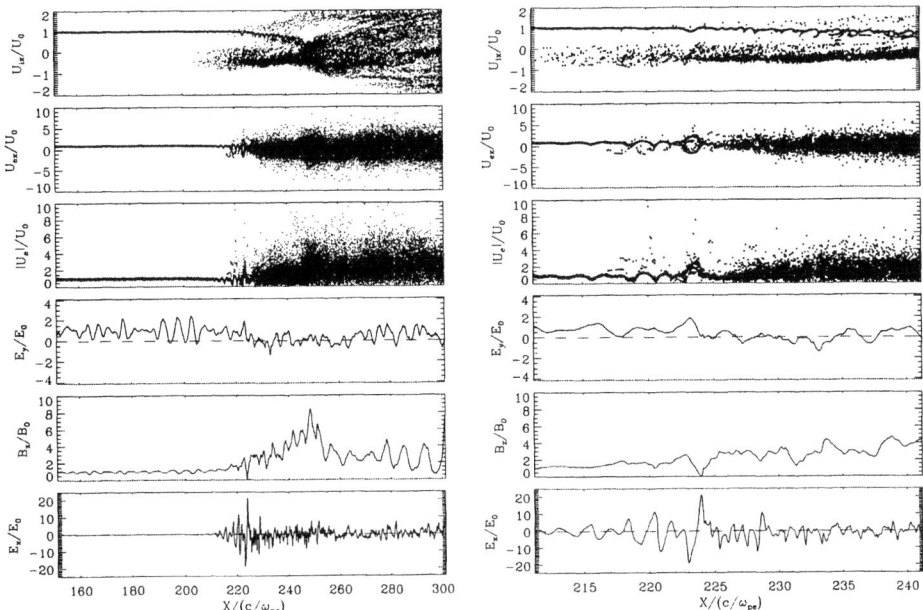

FIGURE 3. High Mach number, magnetosonic shock structure for $M_A = 32$: an overall shock front region (left), and an enlarged picture in the shock transition region (right).

The leftward and the rightward regions are respectively the shock upstream and downstream. From the top, the ion phase space diagram in (X, U_{ix}), the electron phase space diagrams in (X, U_{ex}), and in $(X, |U_e|)$ where $|U_e|$ is the magnitude of electron velocity. The bottom three panels are the transverse electric field E_y, the magnetic field B_z, and the electric field E_x. The plasma four velocity U is normalized by the upstream flow velocity $U_0 = v_0/\sqrt{1 - (v_0/c)^2}$. The magnetic field, and the electric field are normalized by the upstream magnetic field B_0 and the upstream motional electric field $E_0 = v_0 B_0/c$, respectively. The spatial scale is normalized by the electron inertia length c/ω_{pe} in upstream.

In the shock front region in Figure 3, we can find two ion components in the ion phase space diagram; one is the cold ion flowing into the shock downstream, the other is the reflected ion going away from the shock front. A part of the incoming ions is reflected due to both the polarization electrostatic field and the compressed magnetic field in shock downstream [11, 12]. The magnetic overshoot structure can be found around $X/(c/\omega_{pe}) \sim 248$. In addition to such ion dynamics around this shock transition region, we find the electron hole in the electron phase space diagram in (X, U_{ex}) around $X/(c/\omega_{pe}) \sim 223$, and its corresponding electrostatic solitary wave (ESW) in the electric field E_x. It is understood that the ESW can be formed by a nonlinear Buneman instability between the reflected ions and the incoming electrons for a super-critical shock of $M_A > 3$ [9].

Looking at the electron hole structure in the shock transition region in details,

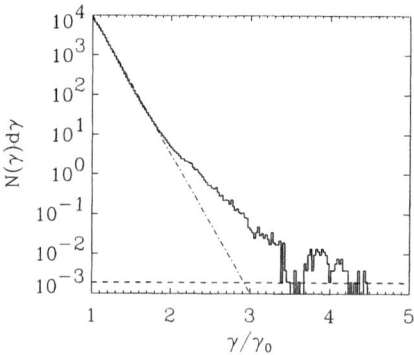

FIGURE 4. Shock downstream energy spectrum for electrons. Dotted lines indicate Maxwellian distribution as reference.

we find that some electrons are accelerated and gain a large amount of energy. After the strong acceleration in the shock transition region, the electrons are transported downstream. Figure 4 shows the downstream electron energy spectrum. The energy $\gamma = \sqrt{1+U_e^2}$ is normalized by the incident electron bulk energy $\gamma_0 = 1/\sqrt{1-(v_0/c)^2}$. The dot-dashed line is the Maxwellian fit for the spectrum as reference, and the bottom dotted line is the so-called one-count level. We can clearly find the enhancement of the nonthermal population above the Maxwellian level over $\gamma/\gamma_0 > 1.8$.

Electron "Shock Surfing Acceleration"

The above electron acceleration can be understood by the shock surfing/surfatron acceleration. Let us quickly review the idea of the shock surfing [13, 14]. The shock surfing mechanism has been extensively studied for the ion acceleration [15, 16, 17, 18, 19, 20]. Due to the inertia difference between ions and electrons flowing into the shock, the polarization electric field normal to the shock front is formed. An ion having a small velocity normal to the shock front, i.e., $1/2Mv_x^2 < e\phi_s$, can be reflected from the shock front to the shock upstream, where ϕ_s is the electrostatic shock front potential induced by the inertia difference between ions and electrons. During the gyro-motion in upstream, ions gain their energy from the motional electric field parallel to the shock front. If the width of the shock front potential ϕ_s is thin, the electric force can overcome the Lorentz force, and then the multiple reflection can occurs.

What we discuss here is the "electron" shock surfing acceleration [10, 21, 22]. The above "standard" shock surfing acceleration cannot apply for the electron acceleration, because the electron cannot be reflected from the shock front by

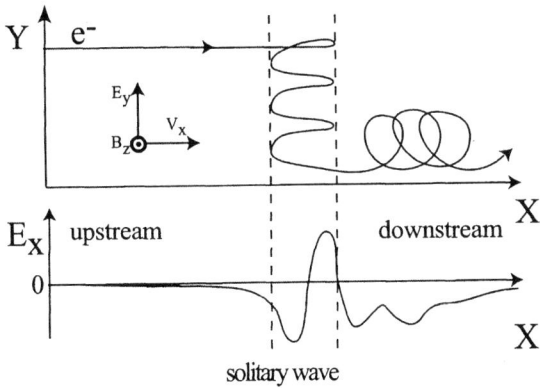

FIGURE 5. An illustration of electron shock surfing mechanism under an action of electrostatic solitary wave.

the shock front potential ϕ_s. We propose a new scheme of "electron" shock surfing acceleration under the action of ESW. Since the electron hole is a positively charged structure, and an electron can be trapped if $m_e v_x^2/2 < e\phi_{esw}$, where ϕ_{esw} is the scalar potential for the electrostatic solitary wave (ESW). Furthermore, the propagation velocity of ESW differs from the plasma bulk velocity, and ESW together with the trapped electrons can stay longer time in the shock transition region. Namely, in the frame moving with ESW, the convection electric field is not zero. Therefore, we think that the "shock surfing" mechanism is occurring for electrons.

Figure 5 summarizes our idea of the electron shock surfing mechanism. Top panel shows an electron's trajectory in the $x - y$ plane. The magnetic field is polarized perpendicular to the $x - y$ plane, and the plasmas are convecting towards positive x. The shock upstream is the left-hand side, while the downstream is the right-hand side. Bottom panel shows the electric field E_x along the x axis, and ESW in association with its electron hole in phase space is depicted in the center. Due to the nature of the electron hole, the electron charge density is slightly lower than the ion one, and ESW has a bipolar signature with diverging electric field. If an electron convecting towards the ESW structure is reflected by both the Lorentz force and the electric field E_x and is trapped inside the ESW structure, it is successively accelerated towards the negative E_y direction. As increasing the electron's velocity v_y by the shock surfing/surfatron acceleration, it can be de-trapped from ESW when the Lorentz force $ev_y B_z/c$ becomes larger than the electric force eE_x, and then it is convecting towards downstream and becomes quickly an isotropic, gyrotropic distribution.

To illustrate the efficiency of the above electron shock surfing mechanism, let us discuss the x component of the equation of motion,

$$dP_x/dt = e(E_x + v_y B_z/c), \qquad (4)$$

where $P_x = m_e U_{ex}$ is the electron momentum. In our interest, E_x on the right-hand side can be replaced by the electrostatic solitary wave (ESW) produced in the shock transition region. If the electric force of eE_{esw} is larger than the Lorentz force of $(e/c)v_y B_0$, the electrons are trapped and gain their energy. During the non-adiabatic acceleration phase, the velocity v_y increases. If the electron satisfies the condition of $eE_{esw} < (e/c)v_y B_0$, it can escape from ESW and will be convected downstream.

The amplitude of ESW may be estimated by equating the wave energy density to the drift energy density of the incoming electron [23], and we obtain, $E_{esw}^2/8\pi \sim m_e n V_d^2 \alpha/2$, where $V_d \sim 2v_0$ is the relative velocity between the reflected ions and the incoming electrons. The energy conversion factor α is of order of $O(10^{-1})$, and we assume $\alpha = 1/4 \sim (m_e/m_i)^{1/3}$[23, 24]. The electric force $F_E = eE_x$ can be given by,

$$F_E \sim eE_{esw} \sim 2\left(\frac{v_0}{V_A}\right)eB_0\sqrt{\alpha\frac{m_e}{m_i}}, \tag{5}$$

and the Lorentz force F_B in the shock transition region is,

$$F_B \sim ev_y B_0/c. \tag{6}$$

By equating F_E to F_B, we obtain the maximum velocity for $v_{y,max}$ as,

$$v_{y,max} \sim \begin{cases} 2cM_A\sqrt{\alpha\frac{m_e}{m_i}} & \text{for } 2M_A\sqrt{\alpha(m_e/m_i)} < 1 \\ c & \text{for } 2M_A\sqrt{\alpha(m_e/m_i)} \geq 1 \end{cases} \tag{7}$$

If $M_A > (1/2)\sqrt{m_i/(m_e\alpha)}$, then $F_E > F_B$ is always satisfied, and the maximum speed becomes the speed of light. In this case, the electrons cannot escape from the ESW region, and the electrons continue to accelerate until they reach the edge of a global shock structure. The condition of the "unlimited electron acceleration" for the real mass ratio of $m_i/m_e = 1830$ is given by $M_A > 43 \sim 75$.

DISCUSSIONS

The relativistic particle acceleration mechanism discussed in this paper may be applied for more general situations. For simplicity, we discussed the relativistic reconnection for the electron-positron plasmas, but the similar acceleration would be possible for the ion-electron plasmas. If $V_A \sim c$, then the ion gyro-radius on the reconnecting magnetic field becomes comparable to the size of the acceleration region with $E > B$, and we can expect a strong non-thermal ion acceleration as well.

In the shock surfing acceleration, we discussed the electron acceleration through the trapping of electron by the electrostatic solitary wave. This process can occur only for the ion-electron shocks. For a shock wave in the electron-positron plasmas that can be expected in the high energy wind region such as the Crab nebula, the electrostatic solitary wave cannot be excited in the shock transition layer due to the symmetry of the electric charge [25, 26]. However, it is discussed that the Alfvénic structure with a current sheet can be formed in the shock front region, and that the current sheet plays a key role as a counter part of the electrostatic solitary wave in the ion-electron plasmas [21]. The formation of a power-law energy spectrum in the shock surfing is still an open question, but it has been pointed out that shock surfing can produce a power-law energy spectrum for the ion shock surfing case [18]. It would be occurring the similar process for the other shock surfing cases.

The non-thermal particle enegization in the direct acceleration process is still controversial, and we have only begun to investigate the potentially rich consequences of the astroplasma particle acceleration mechanism. It obviously will be important to understand the relationship between the stochastic acceleration and the direct acceleration.

REFERENCES

1. Hillas, A. M., ARA & A, 22, 425–444 (1984).
2. Axford, W. I., Leer, E., and Skadorn, G., *Proc. 15th Internat. Cosmic Ray Conf.*, 11, 132 (1977).
3. Bell, A. R., Mon. Not. R. Astro. Soc., 182, 147–156 (1978).
4. Blandford, R. D., and Ostriker, J. P., Astrophys. J., 221, L29–L32 (1978).
5. Coroniti, F. V., Astrophys. J., 349, 538–548 (1990).
6. Lyubarsky, Y. and Kirk, J. G., Astrophys. J., 547, 437–448 (2001).
7. Zenitani, S. and Hoshino, M., Astrophys. J. Lett, 562, L63–L66 (2001).
8. Harris, E. G., Nuovo Cimento, 23, 115–121 (1962).
9. Shimada, N., and Hoshino, M., Astrophys. J. Lett, 543, L67–L71 (2000).
10. Hoshino, M. and Shimada, N., Astrophys. J., 572, in press (2002).
11. Leroy, M., Winske, D., Goodrih, C. C., Wu, C. S., and Papadopoulos, K., Geophys. Res. Lett., 87, 5081–5091 (1982).
12. Sckopke, N., Paschmann, G., Bame, S. J., Gosling, J. T., and Russell, C. T., J. Geophys. Res., 88, 6121–6136 (1983).
13. Sagdeev, R. Z. *Revies of Plasma Physics*, vol.4, edited by M. A. Leontovich, Consultants Bur., New York, 1966, p. 23.
14. Sugihara, R., and Mizuno, Y., J. Phys. Soc. Jpn., 47, 1290–1295 (1979).
15. Katsouleas T., and Daswon, J. M., Phys. Rev. Lett., 51, 392–395 (1983).
16. Lembège B., and Dawson, J. M., Phys. Rev. Lett., 53, 1053–1056 (1984).
17. Ohsawa, Y., Phys. Fluids, 28, 2130–2136 (1985).
18. Zank, G. P., Pauls, H. L., Cairns, I. H., and Webb, G. M., J. Geophys. Res., 101, 457–478 (1996).
19. Lee, M. A., Shapiro, V. D., and Sgdeev, R. Z., J. Geophys. Res., 101, 4777–4790 (1996).
20. Ucer, D., and Shapiro V. D., Phys. Rev. Lett., 87, 075001 (2001).
21. Hoshino, M., Prog. Theoretical Physics Suppl., 143, 149–181 (2001).
22. McClements, K. G., M. E. Dieckmann, Ynnerman, A., Chapman, S. C., and Dendy, R. O., Phys. Rev. Lett., 87, 255002 (2001).
23. Ishihara, O., Hirose, A., and Langdon, A. B., Phys. Fluids, 24, 452–464 (1981).
24. Dieckmann, M. E., Ljung, P., Ynnerman, A., and McClements, K. G., Physics of Plasmas, 7, 5171–5181 (2000).
25. Gallant, Y. A., Hoshino, M., Langdon, A. B., Arons, J., and Max, C. E., Astrophys. J., 391, 73–101 (1992).
26. Hoshino, M., Arons, J., Gallant, Y. A., and Langdon, A. B., Astrophys. J., 390, 454–479 (1992).

COSMIC PLASMA WAKEFIELD ACCELERATION [1]

Pisin Chen*, Toshiki Tajima† and Yoshiyuki Takahashi**

*Stanford Linear Accelerator Center
Stanford University, Stanford, CA 94309
†Advanced Photon Research Center
Japan Atomic Energy Research Institute, Kyoto, 619-0215 Japan
**Department of Physics, University of Alabama, Huntsville, AL 35899

Abstract. A cosmic acceleration mechanism is introduced which is based on the wakefields excited by the Alfven shocks in a relativistically flowing plasma. We show that there exists a threshold condition for transparency below which the accelerating particle is collision-free and suffers little energy loss in the plasma medium. The stochastic encounters of the random accelerating-decelerating phases results in a power-law energy spectrum: $f(\varepsilon) \propto 1/\varepsilon^2$. As an example, we discuss the possible production of super-GZK ultra high energy cosmic rays (UHECR) in the atmosphere of gamma ray bursts. The estimated event rate in our model agrees with that from UHECR observations.

Ultra high energy cosmic ray (UHECR) events exceeding the Greisen-Zatsepin-Kuzmin (GZK) cutoff[1] (5×10^{19}eV for protons originated from a distance larger than ~ 50 Mps) have been found in recent years[2, 3, 4, 5]. Observations also indicate a change of the power-law index in the UHECR spectrum (events/energy/area/time $\propto \varepsilon^{-\alpha}$) from $\alpha \sim 3$ to a smaller value, at energy around $10^{18} - 10^{19}$eV. These present an acute theoretical challenge regarding their composition as well as their origin[6].

So far the theories that attempt to explain the UHECR can be largely categorized into the "top-down" and the "bottom-up" scenarios. In addition to relying on exotic particle physics beyond the standard model, the main challenges of top-down scenarios are their difficulty in compliance with the observed event rates and the energy spectrum[6], and the fine-tuning of particle lifetimes. The main challenges of the bottom-up scenarios, on the other hand, are the GZK cutoff, as well as the lack of an efficient acceleration mechanism[6]. To circumvent the GZK limit, several authors proposethe "Z-burst" scenario[7] where neutrinos, instead of protons, are the actual messenger across the cosmos. For such a scenario to work, it requires that the original particle, say protons, be several orders of magnitude more energetic than the one eventually reaches the Earth.

Even if the GZK-limit can be circumvented through the Z-burst, the challenge for a viable acceleration mechanism remains, or becomes even more acute. This is mainly because the existing paradigm for cosmic acceleration, namely the Fermi mechanism[8], as well as its variants, such as the diffusive shock acceleration[9], are not effective in reaching ultra high energies[10]. These acceleration mechanisms rely on the random

---

[1] Work supported by Department of Energy contracts DE–AC03–76SF00515 and DE-AC03-78SF00098.

CP634, *Science of Superstrong Field Interactions*, edited by K. Nakajima and M. Deguchi
© 2002 American Institute of Physics 0-7354-0089-X/02/$19.00

collisions of the high energy particle against magnetic field domains or the shock media, which necessarily induce increasingly more severe energy losses at higher particle energies.

From the experience of terrestrial particle accelerators, we learn that it takes several qualifications for an accelerator to operate effectively. First, the particle should gain energy through the interaction with the longitudinal electric field of a subluminous ($v \le c$) electromagnetic (EM) wave. In such a setting the accelerated particle can gain energy from the field over a macroscopic distance, much like how a surfer gains momentum from an ocean wave. It is important to note that such a longitudinal field is Lorentz invariant, meaning that the acceleration gradient is independent of the instantaneous energy of the accelerating particle. Second, such a particle-field interaction should be a non-collisional process. This would help to avoid severe energy loss through inelastic scatterings. Third, to avoid excessive synchrotron radiation loss, which scales as particle energy squared, the accelerating particle should avoid any drastic bending beyond certain energy regime. We believe that these qualifications for terrestrial accelerators are also applicable to celestial ones.

Although they are still in the experimental stage, the "plasma wakefield accelerator" concepts[11, 12], promise to provide all the conditions stated above. Plasmas are capable of supporting large amplitude electro-static waves with phase velocities near the speed of light. Such collective waves, or "wakefields", can be excited by highly concentrated, relativistic EM energies such as lasers[11] and particle beams[12]. A trailing particle can then gain energy by riding on this wakefield. Although hard scatterings between the accelerating particle and the plasma medium is inevitable, under appropriate conditions, as we will demonstrate below, the particle can be collision-free.

In this Letter we demonstrate that magneto-shocks (Alfven shocks) in a relativistic plasma flow can also excite large amplitude plasma wakefields, which in turn can be highly efficient in accelerating ultra high energy particles.

It is well-known that an ordinary Alfven wave propagating in a stationary magnetized plasma has a velocity $v_A = eB_0/(4\pi m_i n_p)^{1/2}$, which is typically much less than the speed of light. Here B_0 is the longitudinal magnetic field and n_p is the density of the magnetized plasma. The relative strength between the transverse E and B fields of the Alfven wave is $E/B = v_A/c$. Although the two components are not equal, being mutually perpendicular to the direction of propagation they jointly generate a non-vanishing ponderomotive force that can excite a wakefield in the plasma, which is slow: $v_{ph} = v_A \ll c$. For the purpose of ultra high energy acceleration, such a wakefield would not be too useful, for the accelerating particle can become quickly out of phase with the accelerating field.

In the case where the bulk flow of the plasma approaches the speed of light, however, the Alfven waves acquire a phase velocity close to c and enhances the ratio of E/B to $\sim V_p/c \le 1$, and it becomes indistinguishable from a bona fide EM wave. Preliminary results from simulations indicate that such relativistic Alfven waves can indeed excite plasma wakefields[13]. In this relativistic flow the excited wakefields are all in one direction, which contributes to the unidirectional acceleration. With our applications to astrophysical problems in mind, the Alfven-wave-plasma interaction relevant to us is in the nonlinear regime.

The plasma wakefield in the nonlinear regime has been well-studied[14]. The non-linearity is determined by the driving EM wave's *ponderomotive* potential, which is governed by its normalized vector potential $a_0 = eE/mc\omega$. When this parameter exceeds unity, nonlinearity is strong[11] so that additional important physics incurs. For a stationary plasma, the maximum field amplitude that the plasma wakefield can support is

$$E_{\max} \approx E_{\mathrm{wb}} a_0 = \frac{m_e c \omega_p}{e} a_0 ,\tag{1}$$

which is enhanced by a factor a_0 from the cold wavebreaking limit (the naively assumed maximum field), $E_{\mathrm{wb}} = m_e c \omega_p / e$, of the linear regime. In a relativistic plasma flow with a Lorentz factor Γ_p, the cold wavebreaking field is reduced by a factor $\Gamma_p^{1/2}$ due to Lorentz contraction. The maximum "acceleration gradient" G experienced by a singly-charge particle riding on this plasma wakefield is then

$$G = eE'_{\max} \approx a_0 m_e c^2 \sqrt{\frac{4\pi r_e n_p}{\Gamma_p}} .\tag{2}$$

The plasma wavelength, in the mean time, is stretched also by a factor a_0 from that in the linear regime. So in a plasma flow the wavelength is

$$\lambda_{pN} = \frac{2}{\pi} a_0 \lambda'_p \approx a_0 \sqrt{\frac{\pi \Gamma_p}{r_e n_p}} ,\tag{3}$$

where $r_e = e^2/m_e c^2 = 2.8 \times 10^{-13}$cm is the classical electron radius.

To determine the maximum possible energy gain, we need to know how far can a test particle be accelerated. At ultra high energies once the test particle encounters a hard scattering or bending, the hard-earned kinetic energy would most likely be lost. The scattering of an ultra high energy proton with the background plasma is dominated by the proton-proton collision. Existing laboratory measurements of the total pp cross section scales roughly as $\sigma_{pp} = \sigma_0 \cdot \{1 + 6.30 \times 10^{-3}[\log(s)]^{2.1}\}$, where $\sigma_0 \approx 32$mb and the center-of-mass energy-squared, s, is given in $(\mathrm{GeV})^2$. In our system, even though the UHE protons are in the ZeV regime, the center-of-mass energy of such a proton colliding with a comoving background plasma proton is in the TeV range, so it is safe to ignore the logarithmic dependence and assume a constant total cross section, $\sigma_{pp} \sim \sigma_0 \sim 30$ mb in the ZeV energy regime. Since in astrophysical settings an out-bursting relativistic plasma dilutes as it expands radially, its density scales as $n_p(r) = n_{p0}(R_0/r)^2$, where n_{p0} is the plasma density at a reference radius R_0 . The proton mean-free-path can be determined by integrating the collision probability up to unity,

$$1 = \int_{R_0}^{R_0 + L_{\mathrm{mfp}}} \frac{\sigma_{pp} n_p(r)}{\Gamma_p} dr = \int_{R_0}^{R_0 + L_{\mathrm{mfp}}} \frac{\sigma_{pp} n_{p0}}{\Gamma_p} \frac{R_0^2}{r^2} dr .\tag{4}$$

We find

$$1 = \frac{\sigma_{pp} n_{p0} R_0}{\Gamma_p} \left[1 - \frac{R_0}{R_0 + L_{\mathrm{mfp}}}\right] .\tag{5}$$

Since L_{mfp} is positive definite, $0 < [1 - R_0/(R_0 + L_{mfp})] < 1$. Therefore the solution to L_{mfp} does not exist unless the coefficient, $\sigma_{pp} n_{p0} R_0/\Gamma_p > 1$. That is there exists a threshold condition below which the system is collision-free:

$$\frac{\sigma_{pp} n_{p0} R_0}{\Gamma_p} = 1 . \tag{6}$$

When a system is below this threshold, a test particle can in principle be accelerated unbound. In practice, of course, other secondary physical effects would eventually intervene.

In a terrestrial accelerator, the wakefields are coherently excited by the driving beam, and the accelerating particle would ride on the same wave crest over a macroscopic distance. There the aim is to produce near-monoenergetic final energies (and tight phase-space) for high energy physics and other applications. In astrophysical settings, however, the drivers, such as the Alfven shocks, will not be so organized. A test particle would then face random encounters of accelerating and decelerating phases of the plasma wakefields excited by Alfven shocks.

The stochastic process of the random acceleration-deceleration can be described by the distribution function $f(\varepsilon, t)$ governed by the Chapman-Kolmogorov equation[15, 16]

$$\frac{\partial}{\partial t} f = \int_{-\infty}^{+\infty} d(\Delta\varepsilon) W(\varepsilon - \Delta\varepsilon, \Delta\varepsilon) f(\varepsilon - \Delta\varepsilon, t)$$
$$- \int_{-\infty}^{+\infty} d(\Delta\varepsilon) W(\varepsilon, \Delta\varepsilon) f(\varepsilon, t) - \nu(\varepsilon) f(\varepsilon, t) . \tag{7}$$

The first term governs the probability per unit time of a particle "sinking" into energy ε from an initial energy $\varepsilon - \Delta\varepsilon$ while the second term that "leaking" out from ε. The last term governs the dissipation due to collision or radiation, or both. As we will demonstrate later, the astrophysical environment that we invoke for the production of UHECR is below the collision threshold condition, and so accelerating particles are essentially collision-free.

The radiation loss in our system is also negligible. As discussed earlier, in a relativistic flow the transverse E and B fields associated with the Alfven shock are near equal in magnitude. Analogous to that in an ordinary EM wave, an ultra relativistic particle (with a Lorentz factor γ) co-moving with such a wave will experience a much suppressed bending field, by a factor $1/\gamma^2$. Furthermore, the plasma wakefield acceleration takes place in the region that trails behind the shock (and not in the bulk of the shock) where the accelerating particle in effect sees only the longitudinal *electrostatic* field colinear to the particle motion[14]. We are therefore safe to ignore the radiation loss entirely as well. We can thus ignore the dissipation term in the Chapman-Komogorov equation and focus only on the purely random plasma wakefield acceleration-deceleration.

Assuming that the energy gain per phase encounter is much less than the final energy, i.e., $\Delta\varepsilon \ll \varepsilon$, we Taylor-expand $W(\varepsilon - \Delta\varepsilon, \Delta\varepsilon) f(\varepsilon - \Delta\varepsilon)$ around $W(\varepsilon, \Delta\varepsilon) f(\varepsilon)$ in the sink term and reduce Eq.(9) to the Fokker-Planck equation

$$\frac{\partial}{\partial t} f = \frac{\partial}{\partial \varepsilon} \int_{-\infty}^{+\infty} d(\Delta\varepsilon) \Delta\varepsilon W(\varepsilon, \Delta\varepsilon) f(\varepsilon, t) + \frac{\partial^2}{\partial \varepsilon^2} \int_{-\infty}^{+\infty} d(\Delta\varepsilon) \frac{\Delta\varepsilon^2}{2} W(\varepsilon, \Delta\varepsilon) f(\varepsilon, t) . \tag{8}$$

We now assume the following properties of the transition rate $W(\varepsilon, \Delta\varepsilon)$ for a purely stochastic process:

a) W is an even function;
b) W is independent of ε;
c) W is independent of $\Delta\varepsilon$.

Property a) follows from the fact that in a plasma wave there is an equal probability of gaining and losing energy. In addition, since the wakefield amplitude is Lorentz invariant, the chance of gaining a given amount of energy, $\Delta\varepsilon$, is independent of the particle energy ε. Finally, under a purely stochastic white noise, the chance of gaining or losing any amount of energy is the same. Based on these arguments we deduce that

$$W(\varepsilon, \Delta\varepsilon) = \frac{1}{2c\tau^2 G} , \tag{9}$$

where τ is the typical time of interaction between the test particle and the random waves and G is the maximum acceleration gradient (cf. Eq.(4)). We note that there is a stark departure of the functional dependence of W in our theory from that in Fermi's mechanism, in which the energy gain $\Delta\varepsilon$ per encounter scales linearly and quadratically in ε for the first-order and second-order Fermi mechanism, respectively.

To look for a stationary distribution, we put $\partial f / \partial t = 0$. Since W is an even function, the first term on the RHS in Eq.(10) vanishes. To ensure the positivity of particle energies before and after each encounter, the integration limits are reduced from $(-\infty, +\infty)$ to $[-\varepsilon, +\varepsilon]$, and we have

$$\frac{\partial^2}{\partial\varepsilon^2} \int_{-\varepsilon}^{+\varepsilon} d(\Delta\varepsilon) \frac{\Delta\varepsilon^2}{2} W(\varepsilon, \Delta\varepsilon) f(\varepsilon) = 0 . \tag{10}$$

Inserting W from Eq.(11), we arrive at the energy distribution function that follows power-law scaling,

$$f(\varepsilon) = \frac{\varepsilon_0}{\varepsilon^2} , \tag{11}$$

where the normalization factor ε_0 is taken to be the mean energy of the background plasma proton, $\varepsilon_0 \sim \Gamma_p m_p c^2$. The actually observed UHECR spectrum is expected to be degraded somewhat from the above idealized, theoretical power-law index, $\alpha = 2$, not only due to possible departure of the reality from the idealized model, but also due to additional intermediate cascade processes that transcend the original UHE protons to the observed UHECRs.

We note that a power-law energy spectrum is generic to all purely stochastic, collisionless acceleration processes. This is why both the first and the second order Fermi mechanisms also predict power-law spectrum, if the energy losses, e.g., through inelastic scattering and radiation (which are severe at ultra high energies), are ignored. The difference is that in the Fermi mechanism the stochasticity is due to random collisions of the test particle against magnetic walls or the shock medium, which necessarily induce *reorientation* of the momentum vector of the test particle after every diffusive encounter, and therefore should trigger inevitable radiation loss at high energies. The stochasticity in our mechanism is due instead to the random encounters of the test particle with different accelerating-decelerating *phases*. As we mentioned earlier, the phase vector of the

182

wakefields created by the Alfven shocks in the relativistic flow is nearly unidirectional. The particle's momentum vector, therefore, never changes its direction but only magnitude, and is therefore radiation free in the energy regime that we consider for proton acceleration.

We now apply our acceleration mechanism to the problem of UHECR. GRBs are by far the most violent release of energy in the universe, second only to the big bang itself. Within seconds (for short bursts) about $\varepsilon_{GRB} \sim 10^{52}$erg of energy is released through gamma rays with a spectrum that peaks around several hundred keV. Existing models for GRB, such as the relativistic fireball model[17], typically assume neutron-star-neutron-star (NS-NS) coalescence as the progenitor. Neutron stars are known to be compact ($R_{NS} \sim O(10)$km) and carrying intense surface magnetic fields ($B_{NS} \sim 10^{12}$G). Several generic properties are assumed when such compact objects collide. First, the collision creates sequence of strong magneto-shocks (Alfven shocks). Second, the tremendous release of energy creates a highly relativistic out-bursting fireball, most likely in the form of a plasma.

The fact that the GRB prompt (photon) signals arrive within a brief time-window implies that there must exists a threshold condition in the GRB atmosphere where the plasma becomes optically transparent beyond some radius R_0 from the NS-NS epicenter. Applying Eq.(8) to the case of out-bursting GRB photons, this condition means

$$\frac{\sigma_c n_{p0} R_0}{\Gamma_p} = 1 \, , \tag{12}$$

where $\sigma_c = (\pi r_e^2)(m_e/\omega_{GRB})[\log(2\omega_{GRB}/m_e) + 1/2] \approx 2 \times 10^{-25}$cm^2 is the Compton scattering cross section. Since $\sigma_{pp} < \sigma_c$, the UHECRs are also collision-free in the same environment. There is clearly a large parameter space where this condition is satisfied. To narrow down our further discussion, it is not unreasonable to assume that $R_0 \sim O(10^4)$km. A set of self-consistent parameters can then be chosen: $n_{p0} \sim 10^{20}$cm^{-3}, $\Gamma_p \sim 10^4$, and $\varepsilon_0 \sim 10^{13}$eV $\equiv \varepsilon_{13}$.

To estimate the plasma wakefield acceleration gradient, we first derive the value for the a_0 parameter. We believe that the megneto-shocks constitute a substantial fraction, say $\eta_a \sim 10^{-2}$, of the total energy released from the GRB progenitor. The energy Alfven shocks carry is therefore $\varepsilon_A \sim 10^{50}$erg. Due to the pressure gradient along the radial direction, the magnetic fields in Alfven shocks that propagate outward from the epicenter will develop sharp discontinuities and be compactified[18]. The estimated shock thickness is $\sim O(1)$m at $R_0 \sim O(10^4)$km. From this and ε_A one can deduce the magnetic field strength in the Alfven shocks at R_0, which gives $B_A \sim 10^{10}$G. This leads to $a_0 = eE_A/mc\omega_A \sim 10^9$. Under these assumptions, the acceleration gradient G (cf. Eq.(4)) is as large as

$$G \sim a_0 mc^2 \sqrt{\frac{4\pi r_e}{\sigma_c R_0}} \sim 10^{16} \left(\frac{a_0}{10^9}\right) \left(\frac{10^9 \text{cm}}{R_0}\right)^{1/2} \text{eV/cm} \, . \tag{13}$$

Although the UHE protons can in principle be accelerated unbound in our system, the ultimate maximum reachable energy is determined by the conservation of energy and our assumption on the population of UHE protons. Since it is known that the coupling between the ponderomotive potential of the EM wave and the plasma wakefield is efficient, we assume that the Alfven shock energy is entirely loaded to the plasma wakefields after propagating through the plasma. Furthermore, we assume that the energy in the plasma wakefield is entirely reloaded to the UHE protons through the stochastic process. Thus the highest possible UHE proton energy can be determined by energy conservation

$$\eta_a \varepsilon_{\mathrm{GRB}} \sim \varepsilon_{\mathrm{A}} \sim \varepsilon_{\mathrm{UHE}} \sim N_{\mathrm{UHE}} \int_{\varepsilon_{13}}^{\varepsilon_m} \varepsilon f(\varepsilon) d\varepsilon . \tag{14}$$

which gives

$$\varepsilon_m = \varepsilon_{13} \exp(\eta_a \varepsilon_{\mathrm{GRB}} / N_{\mathrm{UHE}} \varepsilon_{13}) . \tag{15}$$

This provides a relationship between the maximum possible energy, ε_m, and the UHE proton population, N_{UHE}. We assume that $\eta_b \sim 10^{-2}$ of the GRB energy is consumed to create the bulk plasma flow, i.e., $\eta_b \varepsilon_{\mathrm{GRB}} \sim N_p \Gamma_p m_p c^2 \sim N_p \varepsilon_{13}$, where N_p is the total number of plasma protons. We further assume that $\eta_c \sim 10^{-2}$ of the plasma protons are trapped and accelerated to UHE, i.e., $N_{\mathrm{UHE}} \sim \eta_c N_p$. Then we find $\varepsilon_m \sim \varepsilon_{13} \exp(\eta_a / \eta_b \eta_c)$. We note that this estimate of ε_m is exponentially sensitive to the ratio of several efficiencies, and therefore should be handled with caution. If the values are indeed as we have assumed, $\eta_a / \eta_b \eta_c \sim O(10^2)$, then ε_m is effectively unbound until additional limiting physics enters. Whereas if the ratio is $\sim O(10)$ instead, the UHE cannot even reach the ZeV regime. The validity of our assumed GRB efficiencies then relies on the consistency check against observations.

In addition to the energy production issue, equally important to a viable UHECR model is the theoretical estimate of the UHECR event rates. The NS-NS coalescence rate is believed to be about 10 events per day in the entire Universe[19, 20]. This frequency is consistent with the observed GRB events, which is on the order of $f_{\mathrm{GRB}} \sim 10^{3.5}$ per year.

In the Z-burst scenario an initial neutrino energy above 10^{21}eV[7] or 10^{23}eV[21] is required (depending on the assumption of the neutrino mass) to reach the Z-boson threshold. For the sake of discussion, we shall take the necessary neutrino energy as $\varepsilon_\nu > 10^{22}$eV. Such ultra high energy neutrinos can in principle be produced through the collisions of UHE protons with the GRB background protons: $pp \to \pi + X \to \mu + \nu + X$. All UHE protons with energy $\varepsilon_{>22} \geq 10^{22}$eV should be able to produce such neutrinos. The mean energy (by integrating over the distribution function $f(\varepsilon)$) of these protons is $\langle \varepsilon_{>22} \rangle \sim O(100) \varepsilon_{22}$. Therefore the multiplicity of neutrinos per UHE proton is around $\mu_{(p \to \nu)} \sim O(10) - O(100)$. At the opposite end of the cosmic process, we also expect multiple hadrons produced in a Z-burst. The average number of protons that Z-boson produces is ~ 2.7[22]. Finally, the population of UHE protons above 10^{22}eV is related to the total UHE population by $N_{>22} \sim (\varepsilon_{13} / \varepsilon_{22}) N_{\mathrm{UHE}} \sim \eta_b \eta_c \varepsilon_{\mathrm{GRB}} / \varepsilon_{22}$.

Putting the above arguments together, we arrive at our theoretical estimate of the expected UHECR event rate on earth,

$$N_{\mathrm{UHECR}}(> 10^{20}\mathrm{eV}) = f_{\mathrm{GRB}} \mu_{(p \to \nu)} \mu_{(Z \to p)} N_{>22} \frac{1}{4\pi R_{\mathrm{GRB}}^2}$$

184

$$\sim \quad f_{\text{GRB}}\mu_{(p\to\nu)}\mu_{(Z\to p)}\eta_b\eta_c\frac{\varepsilon_{\text{GRB}}}{\varepsilon_{22}}\frac{1}{4\pi R_{\text{GRB}}^2} \ . \qquad (16)$$

The typical observed GRB events is at a redshift $z \sim O(1)$, or a distance $R_{\text{GRB}} \sim 10^{22}$km. Our estimate of observable UHECR event rate is therefore

$$N_{\text{UHECR}}(> 10^{20}\text{eV}) = O(1)/100\text{km}^2/\text{yr}/\text{sr} \ , \qquad (17)$$

which is consistent with observations, or in turn this observed event rate can serve as a constraint on the various assumptions of our specific GRB model.

We have demonstrated that plasma wakefields excited by Alfven shocks in a relativistic plasma flow can be a very efficient mechanism for cosmic acceleration, with a power-law energy spectrum. When invoking GRBs as the sites for UHECR production with a set of reasonable assumptions, we show that our estimated UHECR event rate is consistent with observations. This cosmic acceleration mechanism is generic, and can in principle be applied to other astrophysical phenomena, such as *blazars*[23]. It is generally believed that the AGN jets are relativistic plasmas. The observed "lumps", or density concentrations, in the jet may well serve as the driver to excite plasma wakefields. These wakefields can accelerate electrons as well as protons to multi-TeV energies. Bent by the confining helical magnetic fields in the jet, these high energy electrons can radiate hard photons in the TeV range, while the protons can cascade into high energy neutrinos. We will present a more detailed discussion on blazars in a separate paper.

We appreciate helpful discussions with J. Arons, R. Blandford, P. Meszaros.

REFERENCES

1. K. Greisen, Phys. Rev. Lett. **16**, 748 (1966); G. T. Zatsepin and V. A. Kuzmin, Pis'ma Zh. Eksp. Teor. Fiz. **4**, 114 (1966) [JETP Lett. **4**, 78 (1966)].
2. D. J. Bird et al., Phys. Rev. Lett. **71**, 3401 (1993); Astrophys. J. **424**, 491 (1994); **441**, 144 (1995).
3. M. Takeda et al., Phys. Rev. Lett. **81**, 1163 (1998); Astrophys. J. **522**, 225 (1999).
4. T. Abu-Zayyad et al., Int. Cosmic Ray Conf. **3**, 264 (1999).
5. M. A. Lawrence, R. J. O. Reid, and A. A. Watson, J. Phys. G Nucl. Part. Phys. **17**, 773 (1991).
6. A. V. Olinto, Phys. Rep. **333-334**, 329 (2000).
7. T. Weiler, Astropart. Phys. **11**, 303 (1999); D. Fargion, B. Mele, and A. Salis, Astrophys. J. **517**, 725 (1999).
8. E. Fermi, Phys. Rev. **75**, 1169 (1949); Astrophys. J. **119**, 1 (1954).
9. W. I. Axford, E. Leer, and G. Skadron, in Proc. 15th Int. Cosmic Ray Conf. (Plovdic) **11**, 132 (1977); G. F. Krymsky, Dokl. Acad. Nauk. SSR **234**, 1306 (1977); A. R. Bell, Mon. Not. R. Astro. Soc. **182**, 147 (1978); R. D. Blandford and J. F. Ostriker, Astrophys. J. Lett. **221**, L29 (1978).
10. A. Achterberg, in Highly Energetic Physical Processes and Mechanisms for Emission from Astrophysical Plasmas, IAU Symposium, vol. 195, P. C. H. Martens and S. Tsuruta, eds. (1999).
11. T. Tajima and J. M. Dawson, Phys. Rev. Lett. **43**, 267 (1979).
12. P. Chen, J. M. Dawson, R. Huff, and T. Katsouleas, Phys. Rev. Lett. **54**, 693 (1985).
13. P. Romenesko, P. Chen, T. Tajima, in preparation (2002).
14. E. Esarey, P. Sprangle, J. Krall, and A. Ting, IEEE Trans. Plasma Sci. **24**, 252 (1996).
15. K. Mima, W. Horton, T. Tajima, and A. Hasegawa, in Proc. Nonlinear Dynamics and Particle Acceleration, eds. Y.H.Ichikawa and T.Tajima (AIP, New York, 1991) p.27.
16. Y. Takahashi, L. Hillman, T. Tajima, in High Field Science, eds. T. Tajima, K. Mima, H. Baldis, (Kluwer Academic/Plenum 2000) p.171.

17. M. J. Rees and P. Meszaros, Mon. Not. R. Astro. Soc. **158**, P41 (1992); P. Mezsaros and M. J. Rees, Astrophys. J. **405**, 278 (1993).
18. A. Jeffrey and T. Taniuti, Nonlinear Wave Propagation, (Academic Press, NY, 1964).
19. T. Piran, ApJ Letters 389, L45 (1992); A. Shemi and T. Piran, Atrophys. J. **365**, L55 (1990).
20. LISA collaboration, http://www.lisa.uni-hannover.de/
21. G. Gelmini and G. Varieschi, UCLA/02/TEP/4 (2002), unpublished.
22. Particle Data Group, Phys. Rev. D54, 187 (1996).
23. M. Punch et al., Nature 358, 477 (1992); J. Quinn et al., ApJ Letters 456, L63 (1996); C. M. Urry, Advances in Space Research **21**, 89 (1998).

Poster Presentations

Coherent Phonons At A Semiconductor Surface

Kazuya Watanabe, Dimitre T. Dimitrov, Noriaki Takagi, and
Yoshiyasu Matsumoto

Department of Photoscience, School of Advanced Science, The Graduate University for Advanced Studies (Sokendai), Hayama, Kanagawa, 240-0193, Japan

Abstract. Coherent surface phonon at a GaAs(100)-c(8×2)-Ga reconstructed surface has been investigated by time resolved second harmonic generation (TRSHG). The phonon mode is impulsively excited by an ultrashort laser pulse and subsequent coherent nuclear motion is monitored through the intensity modulation of the second harmonics of a probe pulse. Oscillatory traces are clearly observed in TRSHG signals and their Fourier transformation show two components, bulk LO-phonon at 8.8 THz and the surface phonon at 6.0-8.6 THz. The frequency of the surface phonon shows red shift as the pumping power increases. The shifts are indicative of a marked electron phonon interaction or anharmonicity of the surface phonon modes.

INTRODUCTION

Laser pulses with the duration sufficiently shorter than the period of a vibrational mode allow us to excite the mode with a high degree of spatial and temporal coherence and to follow the nuclear motion directly in the time domain. In contrast to a number of works on coherent bulk phonon excitation reported [1], studies on the excitation of coherent phonons or vibrations at surfaces are scarce. Recently, Chang et al. have shown that time resolved second harmonic generation (TRSHG) is a powerful method to probe surface coherent phonons [2-4]. In this work, we focus on TRSHG on a GaAs(100)-c(8×2)-Ga surface, and describe new features in TRSHG traces of this surface that has not been reported in the previous works [2-4].

EXPERIMENT

The experiment was performed with a UHV chamber evacuated to a base pressure better than 3.0×10^{-10} Torr by cascaded turbo-molecular pumps. It is equipped with a retractable LEED and a retractable cylindrical mirror analyzer for Auger electron spectroscopy (AES). A GaAs(100) sample (*n*-type, Si doped, 1-5 × 10^{18} cm^{-3}) was held by a Ta foil welded to Ta wires. It could be heated up to 1000 K by resistive heating of the Ta wires, and could be cooled down to 110 K by liquid nitrogen. Sample was cleaned by cycles of Ar$^+$ sputtering (1μA, 500 V) and annealing. The surface condition was checked by LEED and AES. A GaAs(100)-c(8×2)-Ga

CP634, *Science of Superstrong Field Interactions*, edited by K. Nakajima and M. Deguchi
© 2002 American Institute of Physics 0-7354-0089-X/02/$19.00

reconstructed surface was obtained by annealing the sputtered surface to higher than 900 K.

We used a chirped-mirror-compensated femtosecond Ti:sapphire laser (Femtosource, Femto-pro, 800 nm, 12 fs, 75 MHz, 800 mW) as a light source. The output laser beam was passed through a prism pair for compensating further chirping by optical components mounted between the laser and the sample. A part (30 %) of the output beam was used as a probe beam, and the rest was passed through a variable neutral density filter and used as a pump beam. The time delay t between the pump and the probe pulses was controlled by a computer-controlled mechanical mirror stage. The pump and the probe beams were introduced in the chamber through a window of a 1mm-thick quartz plane plate. Then both of the beams were focused onto the sample surface by a quartz lens of the 100-mm focal length held in the chamber.

The incidence angles of the two beams onto the sample surface were ~70°. The electric vectors of the beams were in the incidence plane (p-polarization) parallel with the [0-11] direction. The p-polarized component of the second harmonics (SH) of the probe beam generated on the sample surface was passed through a band pass filter to reject the fundamental light and detected by a photomultiplier thermoelectric-cooled down to –30° C. The output signal of the photomultiplier was amplified by a current pre-amplifier and fed into a lock-in amplifier synchronized with an optical chopper that was inserted in the optical path of the pump beam. The sample temperature was kept at 180 K during the measurements.

RESULTS

FIGURE 1(a) shows a typical TRSHG trace obtained on the GaAs(100)-c(8×2)-Ga surface. A clear oscillatory component is superimposed on the background component with a rapid rise and a slow decay. The derivative of the curve in FIGURE 1(a) is plotted in FIGURE 1(b) after subtracting low frequency (<3THz) components. While a large amplitude oscillation component appears immediately after the excitation pulse almost decays within $t < 1$ ps, a small oscillation component is persistent in $t > 1$ ps.

Power spectra are obtained by Fourier transformation of the derivative curve in FIGURE 1(b). Here we plot the power in two different time domains: one in the range from 70 fs to 5.8 ps in FIGURE 2(a) and the other from 1.0 ps to 5.8 ps in FIGURE 2(b). Two peaks at 8.9 THz and 8.4 THz are prominent and a sharp dip at 8.7 THz appears in FIGURE 2(a). On the other hand, the power spectrum of a longer delay times after the rapid damping of the initial large oscillation (FIGURE 2(b)) shows only one peak at 8.8 THz. The frequency of a bulk LO-phonon was reported to be 8.8 THz by time-resolved linear reflectivity measurements of GaAs [5]. Chang et al. have also observed the bulk LO-phonon peak at 8.8 THz in the TRSHG measurements, in addition to the four surface phonon modes in the region of 6.2-8.2 THz [4].

We simulated the observed oscillatory traces, $D(t)$, with a linear combination of two underdamped modes, as in eq (1).

$$D(t) = A_1\sin(\omega_1 t + \phi_1)\exp(-t/\tau_1) + A_2\sin(\omega_2 t + \phi_2)\exp(-t/\tau_2) \qquad (1)$$

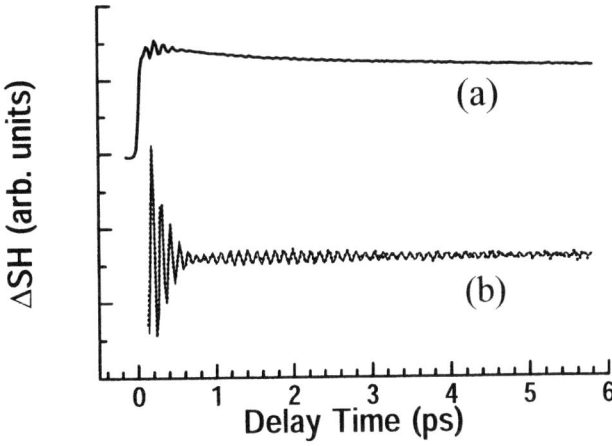

FIGURE 1. (a) A typical TRSHG trace taken from a GaAs (100)-c(8×2) surface. The pump power was 140 mW, and the SH signal with p-polarization was detected. The incidence plane was parallel to [0-11] direction. (b) Dotted curve: A derivative curve of the trace (a). Solid curve: a simulation curve with two components α and β in eq. (2). See text for the parameters employed in the simulation.

A simulated curve (solid curve) is superimposed on the measured data in FIGURE 1(b). The employed parameters are as follows: $\omega_1 = 8.67$ THz, $\tau_1 = 3$ ps, $\phi_1 = -12°$, $\omega_2 = 8.52$ THz, $\tau_2 = 0.22$ ps, $\phi_2 = -128°$, and $A_1:A_2 = 1:23$. The power spectrum of the simulated curve is also plotted in FIGURE 2(a). It is obvious that the essential features of the oscillatory curve measured by TRSHG are well represented by a linear combination of the two underdamped oscillators. Hereafter we denote the first one as the α component that has a decay time of about 3 ps and a center frequency at 8.7 THz, and the second one as the β component that has a decay time of 220 fs and a center frequency at about 8.5 THz.

Note that the initial phase of the α components (ϕ_1) is shifted relatively from that of the β component (ϕ_2) by about 120°. The relative phase shift is crucial to obtain the satisfactory simulation of the waveform. The asymmetric spectral shape and the slight frequency shift of the peak at 8.9 THz and a dip at 8.7 THz in FIGURE 2(a) are due to the interference between the α and the β components. The β component has a center frequency lower than that of the α component, and possesses much shorter decay times (200-400 fs). The β component is reduced by surface sputtering and recovers its intensity by annealing.

FIGURE 3 shows the excitation power dependence of the power spectra. At the lowest pump power of 70 mW, two peaks appear at 8.6 THz and 8.9 THz, and as the pump power is increased, the lower frequency peak increases in its relative intensity and are shifted to red. All the waveforms under the three different pump power conditions could be reproduced qualitatively by a linear combination of two underdamped oscillators, eq. (1). The power spectra of these simulated curves are also shown in FIGURE 3. The parameters used for the simulations are listed in TABLE I.

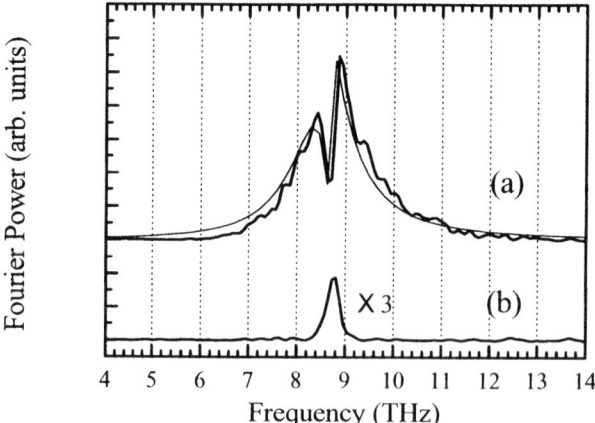

FIGURE 2. The power spectra obtained from the Fourier transformations of the TRSHG traces in FIGURE 1(b). (a) The transformation is performed for the data in the range from 70 fs to 5.8 ps of the measured (thick curve) and the simulated (thin curve) traces. (b) The power spectrum obtained from the experimental data in FIGURE 1(b) in the range from 1.0 ps to 5.8 ps.

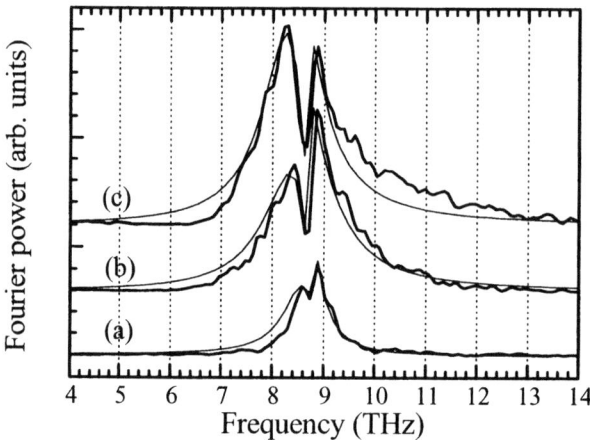

FIGURE 3. The pump power dependence of the Fourier power spectra of TRSHG traces. Thick curves represent the results taken from the surface with 3 hours annealing after the Ar^+ sputtering. The pump power for each curve is (a) 70 mW, (b) 140 mW, and (c) 210 mW, respectively. Thin curves represent the results simlulated by the two components analysis with eq. (1). The used parameters for the components α and β are tabulated in TABLE I.

TABLE 1. Parameters Used For The Simulation In FIGURE 3.

Pump power (mW)	α component		β component	
	Center Frequency (THz)	Decay Time (ps)	Center Frequency (THz)	Decay Time (ps)
70	8.80	2.8	8.65	0.40
140	8.67	3.0	8.52	0.22
210	8.66	2.3	8.34	0.24

In the simulated curves, the α component is always with a decay time of 2- 3 ps and with a center frequency of 8.7-8.8 THz. Differences of the initial phases of the α and β components are 69°, 116°, and 103° for pump power of 70, 140, and 210 mW, respectively, and these phase differences give a dip at 8.7 THz in each spectrum.

DISCUSSION

Origins Of The Oscillatory Components

The observed features of the oscillatory signals apparent in the TRSHG traces are summarized as follows. The oscillatory parts are decomposed into two underdamped modes: the components α and β. The α component has a frequency of 8.7-8.8 THz and a decay time of 2-3 ps, and appears irrespective of the surface sputtering and annealing, and the pump power. Thus, this component is assigned to the bulk LO-phonon mode. On the other hand, the β component is very sensitive to long-range regularity of the surface. In addition, the β component changes its frequency and the width (decay time) as a function of the pump power.

There are some candidates for the origin of the β component: surface phonon, and the bulk plasmon–bulk phonon coupled mode [2-4,6-8]. Since the intensity of the β component is strongly reduced by surface sputtering, the bulk mode cannot be the origin of the signal. Thus, this is due to surface phonon modes localized in a few atomic layers near the surface.

Here we discuss the initial phases of the observed modes. We observed the surface phonon components exhibit different initial phases from that of the bulk LO-phonon component. The initial phase of a coherent phonon oscillation depends on its excitation mechanism [9]. Mechanisms for coherent optical phonon excitation are categorized as either a Raman or a non-Raman process. In the Raman process, the initial phase of coherent nuclear motion depends on the detuning of laser photon energy from the resonant energy for an electronic transition. It is claimed that the driving force is purely displacive for $|\omega_R-\omega_0| \ll \delta_R$, but impulsive for $|\omega_R-\omega_0| \gg \delta_R$, where ω_R is the resonance frequency of the electronic transition, ω_0 is the excitation laser frequency, and δ_R is the width of the resonance [9]. Thus, in the former case, coherent nuclear motion is described as $\cos(\omega t)$, while in the latter case, it is described as $\sin(\omega t)$.

On the other hand, some non-Raman processes have been proposed to explain coherent phonon excitation in opaque solids [1,10]. One of the most prominent models

193

is denoted as Displacive Excitation of Coherent Phonons (DECP), which is based on interband excitation from bonding to antibonding states. Since the lattice equilibrium position changes abruptly by the excitation, the coherent amplitude is modulated as $\cos(\omega t)$. As for the space charge layer in semiconductors, another non-Raman mechanism for coherent bulk LO-phonon generation is well accepted, which is based on the ultrafast screening of a depletion field by photocarriers [11]. The initial phase of the nuclear motion induced by this mechanism is similar to that of DECP.

Following the discussion given in ref. 11, the ultrafast screening of the depletion field at the GaAs(100)-c(8×2) surface is expected to be a dominant mechanism for driving the coherent bulk LO-phonon. On the other hand, the phase difference between the α and the β components clearly indicates that the excitation mechanism of the surface component is different from that of the bulk phonon. In fact, Chang et al. have claimed that the surface phonon components are generated via the Raman process [2], since the relative contributions of the surface phonon modes depend on the pump polarization and the crystal azimuth angle in the light incidence plane.

Excitation Power Dependence

In the above discussion, the β component is ascribed to the surface phonons confined in a few atomic layers. Here, we focus on the frequency shift of the β component as a function of the pump power. There are two possible origins for the frequency shift: (1) the softening of the bonds at the surface owing to excitation of the surface states, and (2) the anharmonicity of the bonds at the surface.

If electrons (holes) are generated in antibonding (bonding) surface states by photoexcitation, the surface phonon modes are softened. A similar red shift of coherent bulk phonon frequency of Tellurium as a function of pumping power has been reported [12]. The excitation of about 2% of valence electrons of bulk Tellurium causes the weakening of the crystal lattice and results in the low frequency shift from 3.6 THz to 3.0 THz. In the present case, although the bulk carrier density is too low to modify the bulk phonon frequency, the 2D confined surface mode might be perturbed by the excess carrier density in the unoccupied surface states.

Next we consider a possible role of the anharmonicity of surface bonds. As has been discussed earlier, the excitation mechanism of the coherent surface phonon on the GaAs surface is considered to be the impulsive Raman scattering. In the classical damped oscillator picture, the amplitude of the induced oscillation would be proportional to the fluence of the excitation laser. The coherent phonon amplitude becomes larger with increase of the excitation power, and finally the red shift of the phonon frequency is expected by the anharmonicity of the potential at the surface. Therefore, another possibility for the red shift is due to the anharmonicity of surface bonds.

Generally, the anharmonicity of surface phonon modes is considered to be greater than that of bulk phonon modes. Baddorf et al. have revealed that the anharmonicity for the motion normal to the surface on a Cu(110) surface, is 4-5 time greater than that in bulk copper [13].

Chang et al. have suggested that the coherent phonon amplitudes observed at the surface and in the bulk are the same order of magnitudes on the GaAs [4]. Since the excitation conditions in this study are close to those in ref. 4, the situation would be similar. Since the α component (bulk LO-phonon) does not show significant power dependence of its frequency, the employed pump power is not so intense as to realize the anharmonic effect in the bulk. As in FIGURE 3, the β component frequency shifts from 8.6 THz to 8.3 THz with the pump power changing from 70 mW to 210 mW. This shift is larger than that of the bulk mode observed between at 10 K and 300 K, in which occupation at the phonon mode of 300 cm^{-1} increases from ~0 to 0.3 [14]. It is intriguing that such a large shift is induced with the moderate pump power conditions. This could be a manifestation of larger anharmonicity at the surface than that in the bulk.

CONCLUSION

We performed TRSHG measurements on a GaAs(100)-c(8×2) surface, and identified the surface phonon mode in the time domain. The surface phonon signals show dramatic dependence on the surface annealing, which indicates that the observed mode is sensitive to the long-range regularity of the surface. Clear interference dips were observed in the power spectra, which are indicative of the initial phase difference between the bulk and the surface phonon modes. The surface phonon frequency showed a red shift as the pump power increases. This might be ascribed to the change of the interatomic force constant by the photocarriers in the antibonding unoccupied surface state, or to the anharmonicity of the potential energy curve for the surface lattice motion.

ACKNOWLEDGMENTS

This work was supported in part by the Grants-in-aid for Scientific Research by Japan Society for the Promotion of Science (JSPS) (11304041, 12874086, and 13874068). KW is grateful for the financial support from The Kao Foundation For Arts And Sciences. DTD is also grateful to JSPS for the postdoctoral fellowship.

REFERENCES

1. Dekorsy, T., Cho, G. C., and Kurz, H., in *Light Scattering in Solids VIII*, edited by M. Cardona and G. Güntherodt , Springer-Verlag, Berlin, 2000, pp. 169-209.
2. Chang, Y. M., Xu, L., and Tom, H. W. K., *Phys. Rev. Letters* **78**, 4649 (1997).
3. Chang, Y. M., Xu, L., and Tom, H. W. K., *Phys. Rev.* B **59**, 12220 (1999).
4. Chang, Y. M., Xu, L., and Tom, H. W. K., *Chem. Phys.* **251**, 283 (2000).
5. Cho, G. C., Kütt, W., and Kurz, H., *Phys. Rev. Letters* **65**, 764 (1990).
6. Cho, G. C., Dekorsy, T., Bakker, H. J., Hövel, R., and Kurz, H., *Phys. Rev. Letters* **77**, 4062 (1996).
7. Hase, M., Mizogushi, K., Harima, H., Miyamaru, F., Nakashima, S., Furusawa, R., Tani, M., and Sakai, K., *J. Luminescence* **76-77**, 68 (1998).

8. Wan, K. and Young, J. F., *Phys. Rev.* B **41**, 10772 (1990).

9. Merlin, R., *Solid State Comm.* **102**, 207 (1997).

10. Zeiger, H. J., Vidal, J., Cheng, T. K., Ippen, E. P., Dresselhaus, G., and Dresselhaus, M., *Phys. Rev.* B **45**, 768 (1992).

11. Pfeifer, T., Dekorsy, T., Kütt, W., and Kurz, H., *Appl. Phys.* A **55**, 482 (1992).

12. Hunsche, S., Wienecke, K., Dekorsy, T., and Kurz, H., *Phys. Rev. Letters* **75**, 1815 (1995).

13. Baddorf, A. P., and Plummer, E. W., *Phys. Rev. Letters* **66**, 2770 (1991).

14. Verma, P., Abbi, S. C., and Jain, K. P., *Phys. Rev.* B **51**, 16660 (1995).

Coincidence Imaging of Coulomb Explosion of CS$_2$ in Intense Laser Fields

Hirokazu Hasegawa, Kyoko Doi, Akiyoshi Hishikawa and
Kaoru Yamanouchi

Department of Chemistry, School of Science, The University of Tokyo,
7-3-1 Hongo, Bunkyo-ku, Tokyo 113-0033, Japan

Abstract. The coincidence imaging technique is applied to direct determination of the momentum vectors of all the fragment ions produced through every event of the Coulomb explosion of a single molecular ion, CS$_2$$^{z+}$ (z = 2 - 4) formed in intense laser fields (0.36 PW/cm^2). The molecular structure of CS$_2$$^{3+}$ just before the three-body Coulomb explosion is reconstructed from the measured momentum vectors of the atomic fragment ions. It was confirmed that the skeletal deformation of CS$_2$ is induced to a large extent in intense laser fields

INTRODUCTION

Recent studies on molecules in intense laser fields ($\sim 10^{12}$ - 10^{15} W/cm^2) revealed that the geometrical structure of polyatomic molecules is deformed to a large extent within the duration of the ultrashort laser pulses [1]. Such a structural deformation process could be studied in more detail if the momentum vectors of respective fragment ions ejected from a single parent ion are measured simultaneously by coincidence-type measurements.

In the present study, we report the double and triple ion coincidence imaging of the fragment ions produced from CS$_2$ in intense laser fields using a position-sensitive detector (PSD) with delay-line readout anodes, and demonstrate that the coincidence imaging using a two-dimensional detector is a direct and efficient method to investigate the dynamics of molecules in intense laser fields [2,3].

EXPERIMENT

Intense laser fields (0.36 PW/cm^2) are generated by focusing linearly polarized femtosecond laser pulses (800 nm, 60 fs, 0.18 mJ/pulse, 1 or 3 kHz) into an ultra-high-vacuum (UHV) main chamber. The CS$_2$ vapor is introduced effusively into the UHV chamber through a skimmer. The ions produced through the interaction with the laser field are guided to the PSD through a three equally spaced parallel-plate electrodes in the velocity mapping configuration. The number of events per laser shot is kept less than unity in order to detect securely all the fragment ions ejected from a single parent ion. The signals are converted to the (x, y) positions on the PSD and the time of flight,

CP634, *Science of Superstrong Field Interactions*, edited by K. Nakajima and M. Deguchi
© 2002 American Institute of Physics 0-7354-0089-X/02/$19.00

T, of the respective ions, from which momentum vectors of the fragment ions are determined.

RESULTS AND DISCUSSION

The fragment ions such as CS^+, S^+, S^{2+} and C^+ are produced from CS_2^{z+} through a three-body (p, q, r) Coulomb explosion pathway, $CS_2^{z+} \rightarrow S^{p+} + C^{q+} + S^{r+}$ ($z = p + q + r$), and/or through a two-body (m, n) Coulomb explosion pathway, $CS_2^{z+} \rightarrow CS^{m+} + S^{n+}$ ($z = m + n$). When an imaging map of fragment ions is measured by a non-coincidence-type method, the contributions from a number of different pathways overlap. By adopting the coincidence imaging method, it becomes possible to identify a specific single Coulomb explosion event. Therefore, it is straightforward to determine separately the respective momentum vector distributions of fragment ions produced through different explosion pathways.

It becomes also possible to derive geometrical structure of parent molecules just before the Coulomb explosion. For a given set of initial structural parameters (two C-S bond lengths, r_1 and r_2, and the $\angle$S-C-S bond angle, γ), the resultant momentum vectors of the ejected fragment ions are calculated by solving numerically the classical equation of motion. After iterative calculation, optimized structural parameters are determined as those reproduce the observed momentum vectors.

For example, the averaged structural parameters of CS_2^{3+} just before the (1, 1, 1) Coulomb explosion are determined to be $<r_1> = <r_2> \sim 2.5$ Å and $<\gamma> = 145°$. This means that CS_2^{3+} is deformed largely from neutral CS_2 in the $\tilde{X}^1\Sigma_g^+$ state whose averaged structural parameters are $<r> = 1.555$ Å and $<\gamma>_0 = 174.0°$.

It should be noted that the correlation among the variations of the structural parameters can also be determined from the coincidence imaging measurements. It is found that the deformation of CS_2^{3+} proceeds so that γ decreases while r_1 and r_2 increases. This dynamical behavior is considered to reflect the temporal variation of the shape of the light-dressed potential energy surfaces formed in intense laser fields.

ACKNOWLEDGMENTS

The present work was supported by the CREST (Core Research for Evolutionary Science and Technology) fund from the Japan Science and Technology Corporation and by the research grant from Graduate School of the University of Tokyo.

REFERENCES

1. Yamanouchi, K., *Science* **295**, 1659-1660(2002).
2. Hasegawa, H., Hishikawa, A., and Yamanouchi, K., *Chem. Phys. Lett.* **349**, 57-63 (2001).
3. Hishikawa, A., Hasegawa, H., and Yamanouchi, K., *Chem. Phys. Lett.* in press.

Direct Observation of Molecular Alignment in Intense Laser Fields by Pulsed Gas Electron Diffraction

Kennosuke Hoshina, Keiko Kato, Tomoya Okino, Kaoru Yamanouchi

Department of Chemistry, School of Science, The University of Tokyo,
7-3-1 Hongo, Bunkyo-ku, Tokyo 113-0033, JAPAN

Abstract. The anisotropic two-dimensional (2D) electron diffraction pattern of jet-cooled CS_2 in an intense nanosecond Nd:YAG laser field (1064 nm, ~0.64 TW/cm^2, 10 ns) was measured by a short-pulsed 25 keV electron beam packet (~7 ns) generated by irradiating a tantalum photocathode with the 4th harmonics of another nanosecond pulsed YAG laser light. It was confirmed from the numerical simulation of the observed anisotropic diffraction image that the alignment of the S-C-S molecular axis proceeds in the intense laser field along the laser polarization direction.

INTRODUCTION

Recent studies of molecules in intense laser fields ($>10^{11}$ W/cm^2) generated by short pulsed high-power laser radiation have revealed that molecules undergo characteristic processes such as alignment in the direction of laser polarization, structural deformation, multiple ionization, and Coulomb explosion forming singly or multiply charged atomic cations [1]. In the previous studies on the alignment [2] and geometric deformation [3] of molecules in intense laser fields, momentum vector distributions of fragment ions were measured. In most cases geometric structures of molecules just before the Coulomb explosion were estimated by solving the classical equation of motion of charged atomic fragments under the assumption that the force imposed among them was Coulombic. Therefore, a method for direct observation of the geometric structure of molecules during the laser-molecule interaction period has been awaited. We demonstrate that the alignment process of CS_2 molecules along the polarization direction of the intense nanonsecond Nd:YAG laser field (~1 TW/cm^2) can be recorded as an anisotropic electron diffraction pattern [4,5].

EXPERIMENTAL

A pulsed electron beam was generated by the photoelectric effect [6-8] by irradiation of a tantalum photocathode surface of a Pierce type electron gun with the 4th harmonics (266 nm, ~50 µJ/pulse, 7 ns) of a Nd:YAG laser (GCR130 Spectra-Physics). The accelerated electron beam packets (25 keV) passed through an aperture (0.1 mm ϕ) and collided with a jet cooled CS_2 molecules in intense laser fields (~

CP634, *Science of Superstrong Field Interactions*, edited by K. Nakajima and M. Deguchi
© 2002 American Institute of Physics 0-7354-0089-X/02/$19.00

10^{12} W/cm^2) produced by focusing a fundamental output (1064 nm, $\sim$500 mJ/pulse, 10 ns) of another Nd:YAG laser. The temporal width of the electron beam packets was FWHM $\sim$7 ns.

The 2D pulsed electron diffraction was projected onto a phosphor (P-31) screen with a diameter of 96 mmϕ located L = 31.6 cm downstream from the electron-sample crossing point. The diffraction image appearing as an illumination pattern on the screen was projected through a camera lens onto an ICCD detector.

RESULTS AND DISCUSSIONS

In the diffraction image obtained when CS$_2$ molecules were irradiated with a linearly polarized nanosecond YAG laser with an average intensity of $\bar{F} \sim 0.64$ TW/cm^2, a small but significant anisotropy was identified; the central region (s = 1.5 $\sim$ 2 Å^{-1}) of the diffraction image is extended along the direction perpendicular to the laser polarization direction, where s = $(4\pi/\lambda)\sin(\theta/2)$ with a de Broglie wavelength of electrons λ and a scattering angle θ. When the 2D diffraction image of CS$_2$ was measured without the laser field, the recorded halo pattern was concentric as expected for free molecules whose molecular axes are randomly oriented in space. Therefore, the anisotropic pattern observed in the intense laser fields indicates that the anisotropy was induced in the spatial distribution of the S-C-S molecular axis by the linearly polarized laser field.

In order to represent the aligned spatial distribution of the molecular axis, a distribution function, $P(\chi)=1+\beta P_2(\cos\chi)$, was adopted, where χ is an angle between the molecular S-C-S axis and the laser polarization direction. Assuming $r(S\cdots S)$ = $2r(C\text{-}S)$, two parameters, $r(C\text{-}S)$ and β, were treated as independent parameters in the trial-and-error simulation of the observed anisotropic diffraction pattern, and the optimized parameters, $r(C\text{-}S)=1.56(3)$ Å and $\beta = 0.17(7)$, were obtained

The determined $r(C\text{-}S)$ value agreed well with that of neutral CS$_2$ in the ground state, r_g = 1.5592(22) Å, indicating that the geometric structure of CS$_2$ does not change largely in an intense laser field ($\sim$1 TW/cm^2). The $<\cos^2\chi>$ =0.36(1) value converted from the alignment parameter $\beta= 0.17(7)$ was well reproduced when the temporal and spatial variations of both the laser pulses and the electron beam packets were taken into account.

REFERENCES

1. Yamanouchi K., *Science* **295**,1659-1660(2002)
2. Iwasaki A., Hishikawa A, and Yamanouchi K., *Chem. Phys. Lett.* **346**,379-386(2001)
3. Hishikawa A., Iwamae A., and Yamanouchi K., *Phys. Rev. Lett.* **83**,1127-1130 (1999)
4. Hoshina K., Yamanouchi K., Ohshima T., Ose Y., and Todokoro H., *Chem. Phys. Lett.* **353**,33-39(2002)
5. Hoshina K., Yamanouchi K., Ohshima T., Ose Y., and Todokoro H., *Chem. Phys. Lett.* **353**,27-32(2002).
6. Mourou G., and Williamson S., *Appl. Phys. Lett.* **41**, 44-45(1982).
7. Ewbank J.D., Faust W.L., Luo J.Y., English J.T., Monts D.L. Paul D.W., Dou Q., and Schäfer L., *Rev. Sci. Instrum.* **63**,3352-3358(1992).
8. Williamson J.C., Dantus M., Kim S.B., and Zewail A.H., *Chem. Phys. Lett.*, **28**, 529-534(1992).

Reaction dynamics of aniline-ammonia cluster ions in intense laser fields by tandem mass spectroscopy

Ryuji Itakura, Jun Watanabe, Taiki Asano, and Kaoru Yamanouchi

Department of Chemistry, School of Science, The University of Tokyo, 7-3-1 Hongo, Bunkyo-ku, Tokyo 113-0033, Japan

Abstract. Mass-selected aniline cations and $[\text{aniline-}(NH_3)_n]^+$ (n = 1 and 2) cluster cations are exposed to intense femtosecond laser fields ($\lambda \sim 395$ nm, $I \sim 4 \times 10^{15}$ W/cm^2) using a tandem-type time-of-flight spectrometer. It is found that a bare aniline cation decomposes into $C_5H_6{}^+$, while the decomposition is suppressed when one or two ammonia molecules are attached to the aniline cation. This suppression is interpreted in terms of an intermolecular energy flow within the cluster through the hydrogen bonding.

INTRODUCTION

In the last decade, characteristic molecular responses to intense laser fields such as alignment along the laser polarization direction, and multiple ionization leading to the Coulomb explosion, have stimulated researchers to investigate dynamical behavior of molecules in intense laser fields [1]. Along with the expansion of this research field, it has been revealed that the ionization and fragmentation mechanism of organic compounds in intense laser fields can be influenced significantly by a resonance coupling between electronic states of a parent molecule with a specific charge number [2, 3].

In the present study, we focus our attention on an effect of hydrogen bonding on the response of aniline-$(NH_3)_n$ cluster cations to intense laser fields.

EXPERIMENT

Aniline cations and aniline-$(NH_3)_n$ (n = 1, 2) cations generated by irradiating with a fourth harmonics of Nd:YAG laser output are mass-selected in the first stage of a tandem-type mass spectrometer, and are exposed to intense laser fields ($\Delta t \sim 50$fs, $\lambda \sim 395$ nm, $I \sim 10^{15}$W/cm^2). The resultant fragment ions and multiply charged parent ions are mass-separated in the second stage of the tandem-type time-of-flight mass spectrometer and detected by a MCP detector.

CP634, *Science of Superstrong Field Interactions*, edited by K. Nakajima and M. Deguchi
© 2002 American Institute of Physics 0-7354-0089-X/02/$19.00

RESULTS AND DISCUSSION

It was found that a small portion of aniline cations in intense laser fields ($\lambda \sim 395$ nm) undergoes the decomposition into $C_5H_6^+$ and HNC. This fragmentation is considered to occur through a resonant coupling between the electronically excited $^2B_1$ and ground $^2B_1$ states. When hydrogen bonded [aniline-$(NH_3)_n$]$^+$ ($n = 1, 2$) cluster ions are exposed to the intense laser fields, the decomposition of the aniline cation moiety is significantly suppressed, and instead the evaporation of NH_3 proceeds.

As shown in table 1, the yields of $C_5H_6^+$ from the [aniline-$(NH_3)_n$]$^+$ ($n = 1, 2$) cluster ions with respect to the corresponding parent ions are only $0.56(1) \times 10^{-3}$ and $2.2(2) \times 10^{-3}$, respectively, which are 11 and 44 % of the corresponding ratio, $5.2(3) \times 10^{-3}$, for the bare aniline cations.

The formation of $C_5H_6^+$ from the planar aniline cation is expected to proceed through a transition state in which the geometrical structure around the amino group is substantially deformed, since one of the two H atoms in the amino group needs to migrate into the ortho-carbon site by simultaneously carrying a positive charge. The suppression of decomposition of an aniline moiety in the cluster cations indicates that the NH_3 attachment to the amino group hinders largely the migration of H atom in the amino group.

TABLE 1. Relative yields of $C_5H_6^+$ with respect to the signal intensities of parent ions.

Parent ions	Relative yields
aniline cation	$5.2(3) \times 10^{-3}$
[aniline-(NH_3)]$^+$	$0.56(1) \times 10^{-3}$
[aniline-$(NH_3)_2$]$^+$	$2.2(2) \times 10^{-3}$

In view of the intra-cluster energy flow, it is possible that a part of the energy gained from the photon fields is distributed to the NH_3 moiety through the intermolecular hydrogen bonding, and eventually, its small portion is consumed for the breaking of one or two intermolecular hydrogen bonds. Such an energy flow may lower the energy kept in the aniline cation moiety below the threshold energy for its decomposition.

REFERENCES

1. Yamanouchi, K., Science **295**, 1659 (2002).
2. Itakura, R., Watanabe, J., Hishikawa, A., and Yamanouchi, K., *J. Chem. Phys.* **114**, 5598 (2001).
3. Watanabe, J., Itakura, R., Hishikawa, A., and Yamanouchi, K., *J. Chem. Phys.* **116**, 9697 (2002).

High-power femtosecond-laser pulse propagation in silica

Kenichi Ishikawa[1*], Hiroshi Kumagai* and Katsumi Midorikawa*

*Laser Technology Laboratory, RIKEN (The Institute of Physical and Chemical Research), Hirosawa 2-1, Wako-shi, Saitama 351-0198, Japan

Abstract. We present a numerical study of the $(2 + 1)$-dimensional propagation dynamics of femtosecond laser pulses in silica. Pulses whose power is tens to hundreds of times higher than the threshold for self-focusing is split into multiple cones during its propagation. This new structure is formed as a result of the interplay of strong Kerr self-focusing and plasma defocusing. The number of cones increases with incident pulse energy. The uncertainty which may be contained in the evaluation of plasma response and multiphoton band-to-band transition cross section does not affect our results much.

INTRODUCTION

The propagation of light in a medium has been fundamental in pure physics as well as important in applied science for centuries. In early days, available light sources were so weak that its propagation was dominated by linear effects. Since the advent of intense femtosecond laser pulses, complex linear and nonlinear effects originating from their broad spectral bandwidths, short pulsewidths, and high peak power have posed significant challenges. For several years researchers have intensively investigated the propagation of intense fs pulses in gases [1, 2, 3, 4, 5, 6, 7], especially in air, and have found various phenomena such as filamentation and pulse splitting. An important parameter is the critical power $P_{cr} = \lambda_0^2 / 2\pi n_0 n_2$ for self-focusing in the continuous wave limit, where λ_0 denotes the laser wavelength, n_0 the linear refractive index, and n_2 the nonlinear refractive index. The value of P_{cr} is about 3 GW for air at $\lambda_0 = 800$ nm, and work [1, 2, 3, 4, 5, 6, 7] for gas has been performed using a power up to a few tens of times P_{cr}. The propagation in solids [8, 9, 10, 11, 12] is less studied. Zozulya [9] simulated the propagation of fs pulses with a power of $3.9 - 5.4$ MW in the context of multiple splitting, coalescence, and continuum generation. Ranka *et al.* [10] and Tzortzakis *et al.* [11] also studied the propagation of a pulse with a power of several MW and found processes similar to those in gas. The laser power used in these studies, however, is by orders of magnitude lower than those used in the studies for air. On the other hand, an experiment for silica glass using a power up to 1 GW, several hundred

---

[1] Present address: Department of Quantum Engineering and Systems Science, Graduate School of Engineering, University of Tokyo 7-3-1 Hongo, Bunkyo-ku, Tokyo 113-8656, Japan. Email address: ishiken@q.t.u-tokyo.ac.jp

times P_{cr} ($= 2.2$ MW), was recently performed [12]. In the present study we report the results of our numerical simulations of the propagation of cylindrically symmetric intense fs laser pulses in silica in such a high-power regime. In this regime the laser pulse propagation is qualitatively different. After self-focusing, the pulse is multiply split in time as well as in space to form multiple cones. This structure originates from a large variation of the refractive index caused by nonlinear response of silica and plasma formation.

MODEL

Assuming propagation along the z axis, we model the evolution of the complex envelope $E(\mathbf{r},z,t)$ of the electric field $\mathscr{E}(\mathbf{r},z,t) = E(\mathbf{r},z,t)\exp(ik_0 z - i\omega_0 t)$ with the extended nonlinear Schrödinger equation [1, 7, 3, 4] in a reference frame moving at the group velocity,

$$\frac{\partial E}{\partial z} + \frac{i}{2}\beta_2\frac{\partial^2 E}{\partial t^2} - \frac{1}{6}\beta_3\frac{\partial^3 E}{\partial t^3} - \frac{i}{2n_0 k_0}\left(\frac{\partial^2}{\partial r^2} + \frac{1}{r}\frac{\partial}{\partial r}\right)\left(1 - \frac{i}{\omega_0}\frac{\partial}{\partial t}\right)E$$

$$= in_2 k_0\left(1 + \frac{i}{\omega_0}\frac{\partial}{\partial t}\right)(|E|^2 E)$$

$$- \frac{ik_0}{2}\left(1 - \frac{i}{\omega_0}\frac{\partial}{\partial t}\right)\left(\frac{\rho}{\rho_{cr}}E\right) - 3\sigma_6\left(\rho_0 - \rho\right)\left(\frac{|E|^2}{\hbar\omega_0}\right)^5 E. \quad (1)$$

E is normalized in such a way that $|E|^2$ gives an intensity in W/cm² for convenience, and c denotes the vacuum velocity of light. The second through fourth terms on the left-hand side describe the normal and third-order group velocity dispersion, and transverse diffraction, respectively. The terms on the right-hand side describe the nonlinear Kerr response, the contribution of plasma formation to the refractive index, and loss due to multi-photon absorption. The terms containing an operator $\frac{i}{\omega_0}\frac{\partial}{\partial t}$ account for the correction beyond the slowly varying envelope approximation [8, 3, 4], necessary to treat short pulses considered in the present study.

We choose a laser operating wavelength of $\lambda_0 = 800$ nm, which corresponds to the angular frequency $\omega_0 = 2.35$ fs⁻¹ and the vacuum wave number $k_0 = 7.85\,\mu$m⁻¹. The material parameters used in our simulations for silica are as follows: $\beta_1 = 4.894\,\mu$m⁻¹fs, $\beta_2 = 0.034, \mu$m⁻¹fs², $n_0 = 1.4533$, $n_2 = 2.66 \times 10^{-16}$ cm²/W. The critical electron density ρ_{cr} for $\lambda_0 = 800$ nm is 1.74×10^{21} cm⁻³, and the initial electron density ρ_0 in the valence band is 2.23×10^{22} cm⁻³. ρ is the density of electrons produced through six-photon band-to-band transitions (the band gap is 9.0 eV). The cross section σ_6 for this process is evaluated to be 2.6×10^{-180} s⁵cm¹² from the Keldysh theory [13]. Equation (1) is coupled with an equation describing the conduction electron density evolution,

$$\frac{\partial \rho}{\partial t} = \sigma_6\left(\frac{|E|^2}{\hbar\omega_0}\right)^6 (\rho_0 - \rho). \quad (2)$$

204

In principle, a term for avalanche ionization should be added to this equation. The classical models [14, 15] of optical breakdown, however, cannot be applied in the fs-pulse regime [16, 17]. Although several new models [16, 17] have been proposed, their validity is yet to be established. On the other hand, no laser-induced damage was observed in the experiments by Kumagai et $al.$ [12] under the conditions similar to those in our simulations. Moreover, in our results, the maximum intensity achieved during the propagation is much lower than the threshold for the laser-induced breakdown [18] as we will see later in Fig. 1, indicating that multi-photon band-to-band transition dominates plasma formation. We have, therefore, neglected avalanche ionization. The loss of electrons due to diffusion and recombination is also negligible in the fs regime [18]. The temporal intensity profile of the incident pulse is proportional to sech^2 with a full width at half maximum (FWHM) of 130 fs. Its lateral profile is gaussian with a FWHM of 235.5 μm. The geometrical focus is at a propagation distance z of 10.9 mm. The numerical scheme to solve Eq. (1) is based on the split-step Fourier method [19]. The diffraction term is treated with the Peaceman-Rachford method [20]. The nonlinear terms as well as Eq. (2) are integrated using the Runge-Kutta method. We have checked the numerical results by doubling the space and time resolution, which led to no significant changes in the behavior of the results presented here. We have also tried a few other integration schemes for Eq. (1) and confirmed that the results remain virtually the same.

RESULTS AND DISCUSSIONS

Evolution of the pulse profile

Figure 1 illustrates the spatio-temporal laser intensity profile at different propagation distances z for the input energy of 135 μJ. This translates to the input power $P \approx 1$ GW or $4.7 \times 10^2 P_{cr}$. By $z = 3200 \mu$m, the pulse energy is concentrated near the beam axis due to self-focusing, and the peak is shifted toward the trailing edge due to self-steepening [19]. As the self-focusing proceeds and the local intensity increases further, conduction (plasma) electrons are produced through multi-photon absorption. Since the plasma formation has a negative contribution to the refractive index, this leads to defocusing near the trailing edge ($z = 3300 \mu$m) and results in the formation of a cone-like structure ($z = 3400 \mu$m). So far the pulse evolution is similar to those for the input power a few times higher than P_{cr}. What follows, however, is a dramatic new feature that emerges only when P exceeds P_{cr} by orders of magnitude. A second cone has been formed outside the first cone at $z = 3500 \mu$m. With pulse propagation, more and more cones are formed, resulting in the formation of multiple-cone-like structure by $z = 4000 \mu$m. This structure is formed rapidly during a short interval of several hundreds of μm and propagates stably at least for 2 mm. The intensity achieved is at most 1.5×10^{13} W/cm^2, and the maximum value of the fluence is 0.9 J/cm^2. This is much smaller than the experimentally obtained threshold fluence for optical breakdown (30 J/cm^2) for a 150-fs pulse [18].

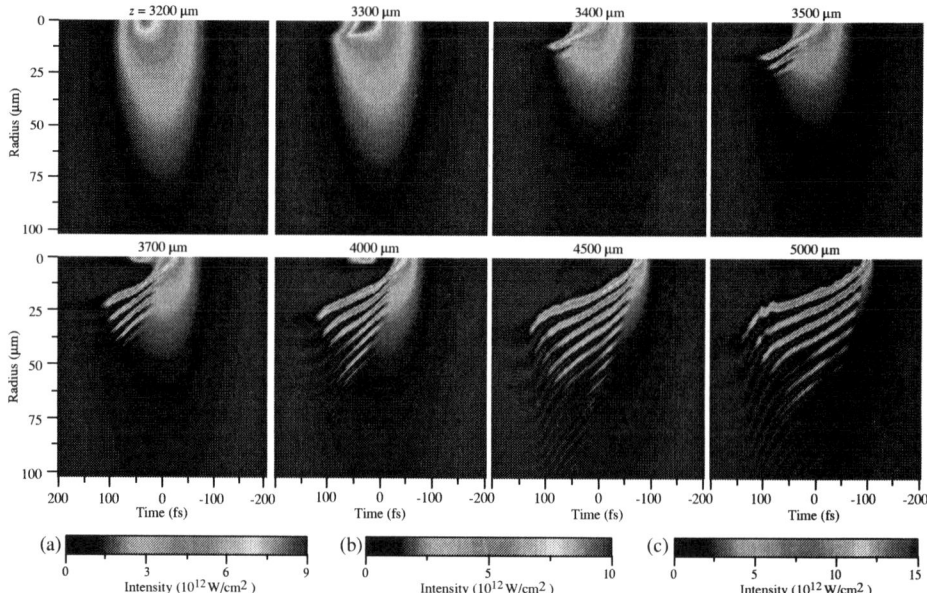

FIGURE 1. Spatio-temporal intensity distribution of an initially gaussian pulse with an energy of $135\mu J$ propagating in silica at 8 different propagation distance z indicated above each image. The scale of vertical and horizontal axes is common for all the images. The gray-scale palette (a) of intensity applies for $z = 3200\,\mu m$, (b) for $z = 3300\,\mu m$, and (c) for the other values of z. Note that the distribution is cylindrically symmetric around the beam axis $r = 0$.

To explore the mechanism of the multiple-cone formation, in Fig. 2 we show the lateral distribution of the laser-induced refractive index change Δn given by,

$$\Delta n = n_2|E|^2 - \frac{1}{2}\frac{\rho}{\rho_{cr}}, \tag{3}$$

at $t = 44\,\mathrm{fs}$ and $z = 3340$ and $3360\ \mu m$, as well as the corresponding intensity distribution. It follows from Eq. (3) that the increase in intensity has a positive contribution, and the increase in conduction electron density, i.e., plasma formation has a negative contribution. In Fig. 2, at $z = 3340\,\mu m$, the maximum in Δn is slightly outside the intensity maximum (i.e., the first cone), and Δn is nearly flat in the range $r = 9 - 12\,\mu m$ while the intensity gradually decreases with increasing r. These are because the electron density is higher for smaller value of r. At $z = 3360\,\mu m$, the self-focusing is so strong that the first intensity peak takes up much energy from its vicinity. As a result, the peak becomes more prominent, and a second local maximum in Δn is formed around $r = 11.3\,\mu m$. Once the second peak is formed, the local self-focusing effect leads to the grow-up of the second cone and to the formation of a third maximum in Δn. The avalanche of this process continues as long as the part of the pulse outside the outermost cone contains sufficient power.

206

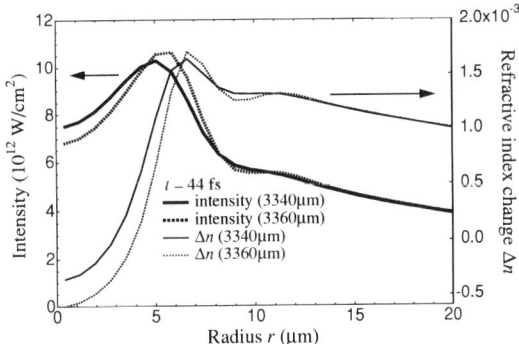

FIGURE 2. Radial distribution of intensity (thick lines, left axis) and refractive index change Δn (thin lines, right axis) at $t = 44$ fs. Solid lines are for the propagation distance $z = 3340\,\mu$m, and dotted lines for $z = 3360\,\mu$m.

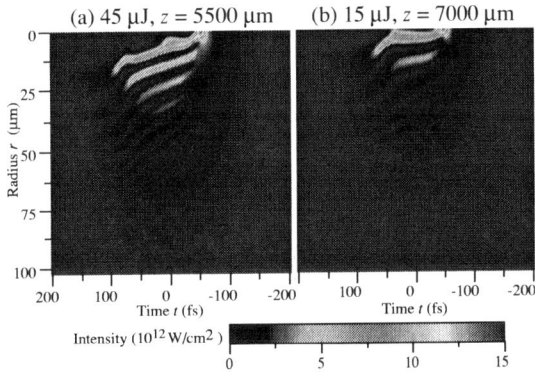

FIGURE 3. Spatio-temporal intensity distribution of an initially gaussian pulse (a) with an energy of $45\,\mu$J at the propagation distance $z = 5500\,\mu$m and (b) with an energy of $15\,\mu$J at $z = 7000\,\mu$m.

Figure 3 shows the spatio-temporal intensity distribution for the pulses with an incident energy of 15 and $45\,\mu$J at the propagation distance at which the multiple-cone formation is nearly saturated. With decreasing pulse energy, the number of cones decreases, and cones are more parallel to the propagation direction. Though not shown in Fig. 3, the multiple-cone formation ceases when we further decrease the incident pulse energy.

Effect of the saturation of the electron drift velocity

In Eq. (1) we have assumed a free-electron-like response of conduction electrons in silica. To our knowledge, no quantitative data are available on their response to intense laser fields. The electron drift velocity in a static electric field is proportional to the field up to 2×10^5 V/cm with a mobility of $21 \, \text{cm}^2/\text{Vs}$ [21], and much larger than the maximum velocity of a free electron oscillating in a laser field with the same strength. On the other hand, when the field is higher than a few MV/cm, the drift velocity is saturated at about 2×10^7 cm/s [21], i.e., the maximum free electron velocity in a laser field with an intensity of $I_{th} = 10^{12}$ W/cm^2. Above this intensity, the response of conduction electrons might be significantly different from that of free electrons. To examine the impact of this correction, we have performed simulations by replacing $\frac{\rho}{\rho_{cr}}$ in Eq. (1) with $\frac{\rho}{\rho_{cr}} f(|E|)$, where,

$$ f(|E|) = \frac{1}{\sqrt{1 + |E|^2/I_{th}}}, \tag{4} $$

models the saturation of electron drift velocity in a simple manner. In Fig. 4 (a), we show the obtained intensity distribution at $z = 4000 \, \mu$m for the input energy of 135μJ. Although the cones are shifted toward the beam axis due to smaller plasma defocusing, the image is similar to the corresponding one in Fig. 1.

Effect of the multi-photon absorption cross section

Among the material parameters used in the present study, the value of σ_6 is the most uncertain. The Keldysh theory may underestimate or overestimate the multiphoton ionization cross section even by two orders of magnitude. We have examined how the error in σ_6 affects our results by simulating the pulse propagation with a hundred times smaller value of σ_6 ($2.6 \times 10^{-182} \, s^5 \text{cm}^{12}$). The result is shown for $z = 3500 \, \mu$m in Fig. 4 (b). The difference from the corresponding image in Fig. 1 is again relatively small. The surprisingly small effect of the error in σ_6 is explained as follows: from the viewpoint of the band-to-band transition rate in Eq. (2) the decrease in σ_6 is recovered by the increase in intensity only by $\sqrt[6]{100} = 2.2$. This does not affect the propagation behavior significantly.

It follows from the preceding discussion that the qualitative features of the results presented in Fig. 1 do not depend on the details of the response and the production rate of conduction electrons. It is expected, therefore, that the essence of our results, especially the multiple-cone formation, probably holds for many other solids and liquids. Now we can divide the fs pulse propagation without breakdown into three different power regimes. In the low-power regime ($P < P_{cr}$) the propagation is predominantly linear. In the intermediate regime($P \approx$ several times P_{cr}), the filamentation and the temporal splitting is observed. In the high-power regime ($P \approx$ several tens to hundreds of times P_{cr}), the multiple-cone structure is formed.

CONCLUSIONS

We have presented numerical simulations of the fs pulse propagation in silica in the high-power regime, where the power of the incident pulse is several hundred times higher than P_{cr}. The pulse is split many times both temporally and spatially, and, as a result,

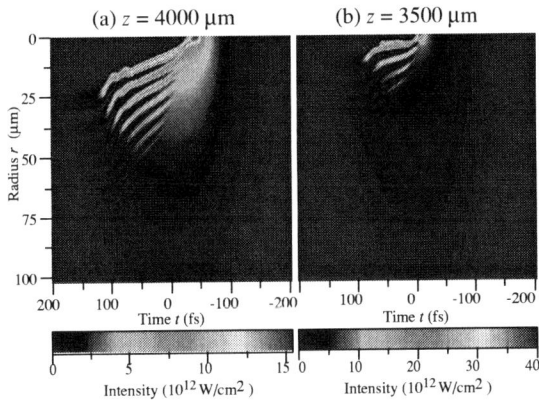

FIGURE 4. Spatio-temporal intensity distribution of an initially gaussian pulse with an energy of $135 \mu J$ (a) at the propagation distance $z = 4000 \mu m$ obtained by taking account of the saturation of conduction electron drift velocity, (b) at $z = 3500 \mu m$ obtained with a value of σ_6 which is 100 times smaller than in Fig. 1.

the intensity distribution contains multiple cones. The cones are formed rapidly within several hundred μm by the interplay of Kerr self-focusing and plasma defocusing. The account of the drift velocity saturation of conduction electrons and the modification of the multiphoton band-to-band transition cross section do not affect our results much; the multiple-cone structure is formed in any case. This observation strongly supports the physical reality of this new feature.

ACKNOWLEDGMENTS

This work was supported by the Special Postdoctoral Researchers Program of RIKEN.

REFERENCES

1. Bergé, L., and Couairon, A., *Phys. Rev. Lett.*, **86**, 1003–1006 (2001).
2. Tzortzakis, S., Bergé, L., Couairon, A., Franco, M., Prade, B., and Mysyrowicz, A., *Phys. Rev. Lett.*, **86**, 5470–5473 (2001).
3. Ranka, J. K., and Gaeta, A. L., *Opt. Lett.*, **23**, 534–536 (1998).
4. Rothenberg, J. E., *Opt. Lett.*, **17**, 1340–1342 (1992).
5. Mlejnek, M., Wright, E. M., and Moloney, J. V., *Phys. Rev. E*, **58**, 4903–4910 (1998).
6. Mlejnek, M., Kolesik, M., Moloney, J. V., and Wright, E. M., *Phys. Rev. Lett.*, **83**, 2938–2941 (1999).
7. Aközbek, N., Scalora, M., Bowden, C. M., and Chin, S. L., *Opt. Comm.*, **191**, 353–362 (2001).
8. Christov, I. P., Kapteyn, H. C., Murname, M. M., Huang, C.-P., and Zhou, J., *Opt. Lett.*, **20**, 309–311 (1995).
9. Zozulya, A. A., Diddams, S. A., van Engen, A. G., and Clement, T. S., *Phys. Rev. Lett.*, **82**, 1430–1433 (1999).
10. Ranka, J. K., Schirmer, R. W., and Gaeta, A. L., *Phys. Rev. Lett.*, **77**, 3783–3786 (1996).

11. Tzortzakis, S., Sudrie, L., Franco, M., Prade, B., and Mysyrowicz, A., *Phys. Rev. Lett.*, **87**, 213902 (2001).
12. Kumagai, H., Cho, S.-H., Ishikawa, K., and Midorikawa, K., *unpublished* (2002).
13. Keldysh, L. V., *Sov. Phys. JETP*, **20**, 1307–1314 (1965).
14. Thornber, K. K., *J. Appl. Phys.*, **52**, 279–290 (1981).
15. Docchio, F., Regondi, P., Capon, M. R. C., and Mellerio, J., *Appl. Opt.*, **27**, 3661–3668 (1988).
16. Noack, J., and Vogel, A., *IEEE J. Quantum Electron.*, **35**, 1156–1167 (1999).
17. Fan, C.-H., and Longtin, J. P., *Appl. Opt.*, **40**, 3124–3131 (2001).
18. Du, D., Liu, X., Korn, G., Squier, J., and Mourou, G., *Appl. Phys. Lett.*, **64**, 3071–3073 (1994).
19. Agrawal, G. P., *Nonlinear Fiber Optics*, Academic, San Diego, 1995, second edn.
20. Koonin, S. E., Davies, K. T. R., Maruhn-Rezwani, V., Feldmeier, H., Krieger, J., and Negele, J. W., *Phys. Rev. C*, **15**, 1359–1375 (1977).
21. Hughes, R. C., *Solid-State Electron.*, **21**, 251–258 (1978).

Rare gas cluster explosion in an intense laser field

Kenichi Ishikawa[1*] and Thomas Blenski[†]

*Laser Technology Laboratory, RIKEN (The Institute of Physical and Chemical Research),
Hirosawa 2-1, Wako-shi, Saitama 351-0198, Japan
†CEA-Saclay, DSM/DRECAM/SPAM, Bât. 522, 91191 Gif-sur-Yvette Cedex, France

Abstract. We study rare gas cluster explosion in an intense, femtosecond laser pulse using Monte Carlo particle dynamics simulations. The obtained dependence of ion energy on ion charge is approximately quadratic for lower charge states and linear for higher ones. This behavior, also observed in experiments, can be entirely explained on the basis of the Coulomb explosion mechanism. Our results also show that electron impact ionization plays only a minor role in ionization.

INTRODUCTION

The intense laser interaction with clusters has been extensively studied for several years [1, 2, 3, 4, 5, 6]. Multiply charged energetic ions [1], high electron temperature [2], and x-ray emission [3] have been observed in experiments. The laser-cluster interaction involves two processes, i.e., ionization and explosion. Several theories have been developed on the ionization mechanism which leads to the production of high charge states. In a coherent electron motion model by McPherson *et al.* [3], multiple ionization arises from impact by coherently moving electrons, behaving like a quasi-particle. Ditmire *et al.* [2, 4] proposed a " nanoplasma" model, in which ions are ionized mainly through impact of hot electrons. Rose-Petruck *et al.* [5] introduced an "ionization ignition (II)" model, where ionization is driven by the combined field of the laser, the other ions, and the electrons. Concerning the explosion dynamics, related to the high ion energy, the nanoplasma model [2, 4] suggests that the cluster expands in a hydrodynamic manner by the pressure of hot electrons confined inside the cluster. Last and Jortner [6], however, showed that electrons relatively quickly leave even large clusters containing over 2000 atoms and that the genuineness of the nanoplasma is questionable.

In the present study, we investigate the explosion mechanism of rare gas clusters (Ar_{55}, Ar_{147}, Xe_{55} and Xe_{147}) in an intense laser pulse using an improved version of the Monte Carlo particle dynamics simulations used previously by Ditmire[4]. We include ion-electron recombination in addition to tunneling and electron impact ionization, and use real (singular) Coulomb interaction between particles. Our results indicate that electron impact ionization plays only a minor role in ionization and that the experimentally observed charge-energy relation [1] can be entirely explained on the basis of the

---

[1] Present address: Department of Quantum Engineering and Systems Science, Graduate School of Engineering, University of Tokyo 7-3-1 Hongo, Bunkyo-ku, Tokyo 113-8656, Japan. Email address: ishiken@q.t.u-tokyo.ac.jp

CP634, *Science of Superstrong Field Interactions*, edited by K. Nakajima and M. Deguchi

Coulomb explosion mechanism.

MODEL

Since the simulation model is detailed in Ref. [7], we summarize it only briefly here.

The nonrelativistic equation of motion of each ion and electron is integrated using the force calculated as the sum of the contributions from all the other particles and the laser electric field. We model the field from an ion of charge state Q as a real Coulomb one from an effective nuclear charge $Q_{eff}(r)$ of the form,

$$Q_{eff}(r) = \begin{cases} Z(1 - r/r_a) & \text{for} \quad r < (Z - Q)r_a/Z, \\ Q & \text{for} \quad r \geq (Z - Q)r_a/Z, \end{cases} \tag{1}$$

with Z being the atomic number and r_a the "atomic radius" (1.3 a.u. for Ar and 2.0 a.u. for Xe). This potential is more suitable for the description of ion-electron interaction than a soft Coulomb potential used by Ditmire[4]. To handle an ion-electron close encounter, we resort to Kustaanheimo-Stiefel regularization[8], widely used in astrophysics.

Free electrons appear through tunnel ionization or elecron impact ionization (EII). We evaluate the tunneling rate via an analytic formula [9] using the total electric field seen by each ion, and calculate the EII cross section using fitting formulas by Lennon *et al.* [10] for Ar and the Lotz formula [11] for Xe. Upon ionization a new electron is placed near the ion in such a way that the total energy is conserved. Free electrons may be recombined with ions. A pair of an ion of charge Q and an electron is replaced by an ion of charge $Q - 1$ if the distance between them is decreasing, if there is no potential barrier between them, and if the total energy of the electron is negative.

The pulse used in our simulations has a field envelope proportional to sine squared with a full width at half maximum of 100 fs and a wavelength of 780 nm. The clusters used in our simulations are Ar_{55}, Ar_{147}, Xe_{55}, and Xe_{147} with a closed-shell icosahedral structure [12] with an atom spacing of 3.76 Å for Ar and 4.32 Å for Xe [13].

IONIZATION

Figure 1 shows the temporal evolution of the mean ion charge state for the case of an Ar_{147} cluster (solid line) and individual Ar atoms (dashed line) irradiated by a laser pulse with a peak intensity of $1.4 \times 10^{15} W/cm^2$. The mean charge state obtained in a cluster gas is considerably higher than that in an atomic gas. Moreover, high charge states up to 8+ were obtained from the cluster gas though it is not explicitly indicated in the figure. In order to examine the importance of EII, we have performed a simulation with EII switched off, whose result is shown as a dotted line in Fig. 1. It can be seen that EII has practically no effect on the mean charge state. Moreover, the highest charge state produced by EII was only 5+. This result is easy to understand from the viewpoint of mean free path. The electron mean free path with respect to EII of Ar^{5+} in an Ar cluster takes the minimum value of 580 Å at an incident energy of 300 eV. This is even larger

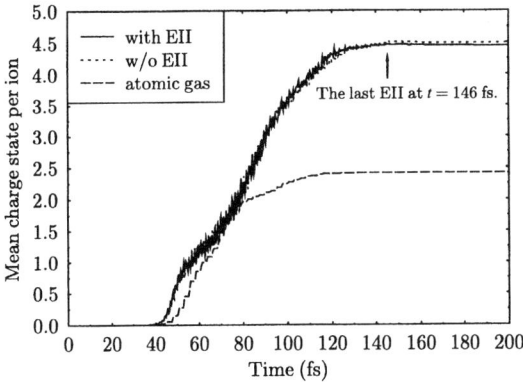

FIGURE 1. Evolution of the mean charge state per ion in Ar_{147} (solid line) and an atomic gas of Ar (dashed line) irradiated by the pulse with a peak intensity of 1.4×10^{15} W/cm². The dotted line is the result for Ar_{147} with electron impact ionization switched off.

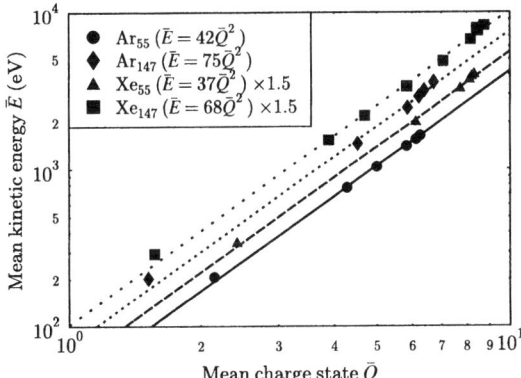

FIGURE 2. Relation between the mean ion energy $\bar{E}$ and the mean charge state $\bar{Q}$ obtained using different intensities (starting from the lowest value of $\bar{Q}$, 0.35, 1.4, 3.2, 5.6, 8.8, 13 × 10¹⁵W/cm² for Ar_{55}, Ar_{147}, and Xe_{55}, and 0.35, 0.79, 1.0, 1.4, 1.9, 3.2, 13, 5.6, 8.8 × 10¹⁵W/cm² for Xe_{147}). The values of $\bar{E}$ for Xe clusters were multiplied by 1.5 for clarity.

than the diameter of Ar_{10^6} (410 Å). Thus, the contribution of EII to the production of high charge states is negligible. A similar argument holds true for Xe clusters. On the other hand, once several atoms are ionized, the total electric field strength seen by each ion can be significantly larger than the laser field alone. This leads to high charge states, driving further tunneling ionization, just as in the II model [5].

CHARGE-ENERGY RELATION

In Fig. 2 we plot the relation between the mean ion energy $\bar{E}$ and the mean charge state $\bar{Q}$. The relation can be modeled with $\bar{E} \approx \alpha \bar{Q}^2$, where α is a constant. This indicates that ions are accelerated mainly through a Coulomb explosion mechanism. In fact, electrons quit the cluster before they exchange significant energy with ions, which excludes a hydrodynamic scenario. A 147-atom cluster consists of a central atom and three closed atomic shells. The outermost and middle shells are composed of three and two subshells, respectively. In Fig. 3 we plot the mean energy $E_s(Q)$ of the ions with a charge of Q in subshell s ($s = 1, \cdots, 6$) of Xe_{147} in the case of a peak intensity $8.8 \times 10^{15} W/cm^2$. We can model the relation with $E_s(Q) \approx \beta_s Q$, where a constant β_s depends on s. In Fig. 3 we have also plotted the average $E(Q)$ of $E_s(Q)$ over the entire cluster as filled circles. Again $E(Q)$ is approximately proportional to Q. We have found that this holds more or less also for Ar_{55}, Ar_{147}, and Xe_{55} and for other values of laser intensity. We simulated the pure Coulomb explosion of Xe_{147}, neglecting the electronic fields in the ionic equations of motion but taking account of the ion charge history obtained for the case of Fig. 3. The resulting energy-charge relation is very similar to that in Fig. 3. This confirms that the approximately linear dependence is due to the mutual Coulomb repulsion between ions.

Lezius et al. [1] found a quadratic charge dependence of ion energy for lower charge states and a linear dependence at higher charge states in their experiments with Xe clusters. If we take a spatial profile of the laser intensity into account, we observe a similar behavior in our simulation results. This is illustrated in Fig. 4, which shows the ion charge-energy relation that we obtained by taking average over the simulation results with different values of intensity as in Fig. 2. It can be seen from Fig. 4 that the relation is quadratic for lower charges and linear for higher charges. This behavior can be entirely explained as a consequence of Coulomb explosion as follows: for lower charges, the contribution to each charge state Q mainly comes from such a spatial region where the

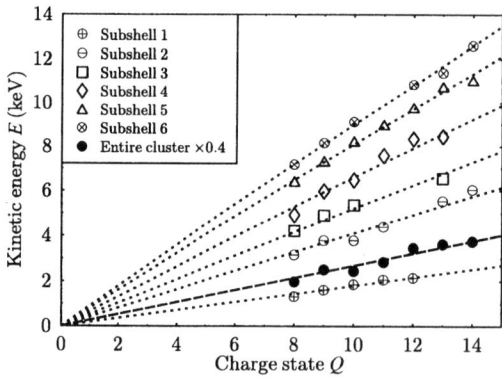

FIGURE 3. Mean ion kinetic energy $E(Q)$ as a function of charge state Q for each subshell of Xe_{147} and for the entire cluster (filled circles, multiplied by 0.4 for clarity.) in the case of a peak intensity $8.8 \times 10^{15} W/cm^2$. The subshells are enumerated outwards starting from the innermost one.

214

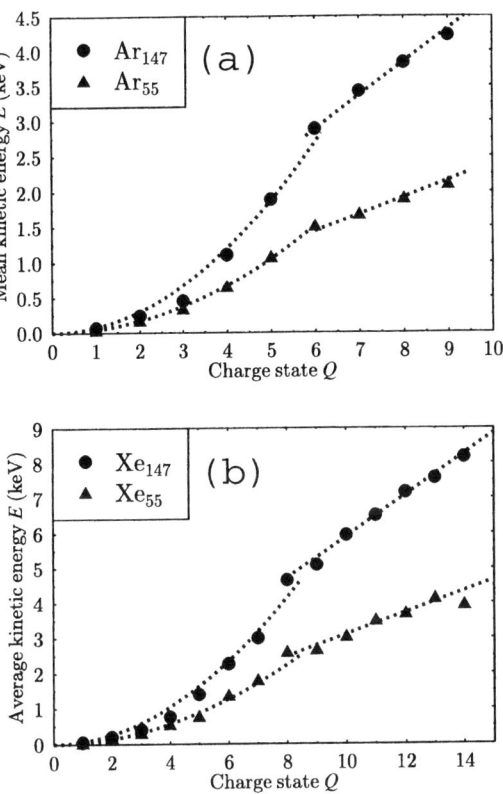

FIGURE 4. Charge dependence of ion energy of (a) Ar_{55}, Ar_{147} and (b) Xe_{55}, Xe_{147} irradiated by the laser pulse with a peak intensity of $1.3 \times 10^{16} W/cm^2$, with account of the spatial intensity profile.

mean charge state equals to Q. As a result, we obtain a quadratic relation, corresponding to the one in Fig. 2. On the other hand, ions of higher charges come from the same region where the intensity is highest. This leads to a linear relation, corresponding to the one in Fig. 3. In Ref. [1] the relation was reported to be quadratic in the entire range of $1 \leq Q \leq 8$ in the case of Ar. This is probably because the authors of Ref. [1] used sufficiently high intensity ($5 \times 10^{17} W/cm^2$) to produce Ar^{8+} while the yield of Ar^{Q+} ($Q \geq 9$), for which a linear relation should be expected, was very low since the ionization potential of Ar^{8+} (422 eV [14]) is much higher than that of Ar^{7+} (143 eV [14]).

The preceding discussion has an important impact on the interpretation of the experimental results by Lezius *et al.* [1]. These authors interpreted the linear relation found for higher charge states as a sign of hydrodynamic expansion. Our results, however, indicate that this is not necessarily correct.

CONCLUSIONS

We have studied the explosion of rare gas clusters containing up to 147 atoms in an intense laser field via Monte Carlo particle dynamics simulations. Electron impact ionization plays only a minor role in ionization. The entire charge dependence of ion kinetic energy, including a linear part at higher charge states, formerly attributed to hydrodynamic expansion, can be understood on the basis of Coulomb explosion.

It is true that our simulation results cannot directly exclude the possibility of hydrodynamic expansion in larger clusters containing over 1000 atoms. However, it is striking that the overall feature of the charge-energy relation in Fig. 4 is similar to the one obtained by Lezius *et al.* [1] Last and Jortner [6] showed that electrons quickly leave even large clusters containing 2097 atoms. Moreover, according to our results, the mean energy of ions is 10-20% higher when they are emitted along the direction of laser polarization than perpendicular to it, which agrees with a recent experimental finding [15]. These agreements strongly indicate that the cluster explosion may be essentially governed by Coulomb explosion also in large clusters.

REFERENCES

1. Lezius, M., Dobosz, S., Normand, D., and Schmidt, M., *Phys. Rev. Lett.*, **80**, 261–264 (1998).
2. Ditmire, T., Donnelly, T., Rubenchik, A. M., Falcone, R. W., and Perry, M. D., *Phys. Rev. A*, **53**, 3379–3402 (1996).
3. Thompson, B. D., McPherson, A., Boyer, K., and Rhodes, C. K., *J. Phys. B*, **27**, 4391–4400 (1994).
4. Ditmire, T., *Phys. Rev. A*, **57**, R4094–R4097 (1998).
5. Rose-Petruck, C., Schafer, K. J., Wilson, K. R., and Barty, C. P. J., *Phys. Rev. A*, **55**, 1182–1190 (1997).
6. Last, I., and Jortner, J., *Phys. Rev. A*, **60**, 2215–2221 (1999).
7. Ishikawa, K., and Blenski, T., *Phys. Rev. A*, **62**, 063204 (2000).
8. Aarseth, S. J., "Direct Methods for *N*-Body Simulations," in *Multiple Time Scales*, edited by J. U. Brackbill and B. I. Cohen, Academic Press, Orlando, 1985, pp. 377–418.
9. Perelomov, A. M., Popov, V. S., and Terent'ev, M. V., *Sov. Phys. JETP*, **23**, 924–934 (1966).
10. Lennon, M. A., Bell, K. L., Gilbody, H. B., Hughes, J. G., Kingston, A. E., Murray, M. J., and Smith, F. J., *J. Phys. Chem. Ref. Data*, **17**, 1285–1363 (1988).
11. Lotz, W., *Z. Phys.*, **216**, 241–247 (1968).
12. Northby, J. A., *J. Chem. Phys.*, **87**, 6166–6177 (1987).
13. Bondi, A., *J. Phys. Chem.*, **68**, 441–451 (1964).
14. Cowan, R. D., *the Theory of Atomic Structure and Spectra*, Univ. of California Press, Berkeley, 1981.
15. Springate, E., Hay, N., Tisch, J. W. G., Mason, M. B., Ditmire, T., Hutchinson, M. H. R., and Marangos, J. P., *Phys. Rev. A*, **61**, 063201 (2000).

Vacuum Ultraviolet Ar$_2$* Excimer Excited by a High Intensity Laser

Masanori Kaku, Takeshi Higashiguchi, Shoichi Kubodera, and Wataru Sasaki

Department of Electrical and Electronic Engineering and Photon Science Center, Miyazaki University, Gakuen Kibanadai Nishi 1-1, Miyazaki, 889-2192 Japan

Abstract. We have observed Ar$_2$* emission at 126 nm using a high intensity laser as an excitation source. The Ar$_2$* production kinetics by this method is same as the three-body association process in the case of electron beam excitation. The increase of Ar$_2$* emission intensity was observed by use of a hollow fiber.

INTRODUCTION

Practical vacuum ultraviolet (VUV) lasers are desired in various fields, such as industrial and scientific fields. VUV lasers would be applicable for photochemistry, biological science, and new materials processing since they produce high energy photons with high photon flux. Currently available practical VUV lasers are the ArF laser at 193 nm and the F$_2$ laser at 157 nm, both of which are operated by a compact discharge device with high repetition.

In addition to these lasers, rare gas excimers are known to be one of the few VUV laser media. Their center wavelengths are 126, 147, and 172 nm for Ar$_2$*, Kr$_2$*, and Xe$_2$*, respectively. Among these, the Ar$_2$* laser produces the highest photon energy of 9.8 eV. The corresponding wavelength is long enough to utilize transmission optics such as MgF$_2$ or LiF. An electron beam excitation is, however, the only excitation method to oscillate the Ar$_2$* laser. We have developed a high peak power (~MW) Ar$_2$* laser using an electron beam excitation method [1,2]. The electron beam excitation method, however, requires a rather large facility, resulting in a limited repetition rate with low average power. An alternative excitation method has long been desired. Although several excitation methods have been investigated to realize practical Ar$_2$* laser, no oscillation results have been reported [3-5].

We have proposed a new excitation method to demonstrate a practical Ar$_2$* laser, where a high intensity laser-produced electrons are utilized to initiate the excimer formation kinetics. Conditions in an optical field induced ionization (OFI) plasma are strictly controlled by changing high intensity laser parameters [6,7]. We have pointed out that the OFI plasma may simulate an electron beam produced plasma. The Ar$_2$* production kinetics in an electron beam generated plasma are relatively well known and we could utilize them for the optimum Ar$_2$* production. The electron temperature and density of the electron beam produced plasma can be simulated by changing high

CP634, *Science of Superstrong Field Interactions*, edited by K. Nakajima and M. Deguchi
© 2002 American Institute of Physics 0-7354-0089-X/02/$19.00

intensity laser parameters such as focused intensity, wavelength, and polarization. A large excitation length (~40 cm) in the electron beam excitation can be reproduced by use of a hollow optical fiber installed in a high pressure Ar cell. This method has a potential to realize an Ar_2^* laser with high repetition rate and in tabletop size.

In this paper, we report on the demonstration of the Ar_2^* production initiated by using a high intensity laser-produced electrons via the OFI plasma. We observed the increase of the Ar_2^* emission intensity by use of an optical hollow fiber to guide the focused laser emission.

EXPERIMENTAL SET-UP

Our experimental apparatus consists of three elements; a high intensity laser, a high-pressure Ar cell (~10 atm) where a hollow fiber is installed, and a VUV detection system. Ti:Sapphire laser system with chirped pulse amplification produced a high intensity pulse at 780 nm. The maximum output energy was 1.5 mJ with a pulse width of 150 fs at a repetition rate of 10 Hz. The laser pulse was focused inside the pressurized Ar gas cell by using a convex lens with a focal length of 30 cm. The focused laser intensity was measured to be 10^{15} Wcm^{-2} at the focus. An optical hollow fiber with a length of 12.5 cm and a diameter of 200 μm was used to guide the focused high intensity laser pulse and thus to extend the excitation length. Time-integrated emission spectra from a plasma were detected by a VUV micro-channel plate (MCP) coupled to a VUV spectrometer. Emissions from Ar plasma were observed onto high intensity laser axis. Typical spectral resolution was better than 0.5 nm. Temporal behavior of Ar_2^* emission at 126 nm was measured using a VUV photomultiplier tube. Temporal signals were recorded by a 2 GHz digital oscilloscope. Temporal resolution of this system was about 5 ns.

EXPERIMENTAL RESULT AND DISCUSSION

A typical time-integrated emission spectrum of an Ar plasma initiated by a high intensity laser is shown in Fig. 1. In this spectrum, a focused intensity was $4x10^{13}$ Wcm^{-2} and Ar pressure was 10 atm. The linearly polarized laser beam was used. A hollow fiber was not used for this spectrum. We have observed Ar_2^* emission centered at 126 nm with a spectral bandwidth of 9 nm (FWHM). The 5th order harmonic emission of the Ti:Sapphire laser is simultaneously observed at 154 nm. Spectral modulation observed around 126 nm is caused by absorption of impurities in the gas such as nitrogen and oxygen. The Ar_2^* emission intensity at 126 nm quadratically increased with the Ar pressure (see below.). This pressure dependence of Ar_2^* emission intensity indicates that the excimer production kinetics by high intensity laser is governed by three-body association process ($Ar^* + 2Ar \rightarrow Ar_2^* + Ar$).

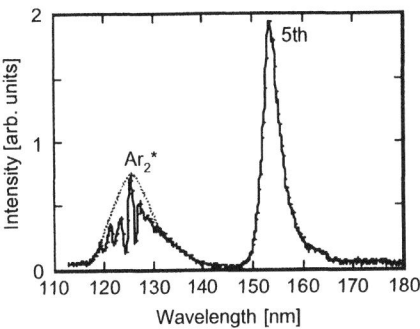

Figure 1. Emission spectrum from an Ar plasma produced by a high intensity laser.

Fig. 2 shows the temporal behavior of Ar_2^* emission at 126 nm. The focused laser intensity was 10^{15} Wcm^{-2} and Ar pressure was 10 atm. A solid line represents an experimental result. The duration of Ar_2^* emission is 18 ns (FWHM). This emission time is long enough to construct an optical cavity. On the other hand, the open circles indicate a simulated temporal behavior of Ar_2^* emission by use of electron beam simulation code [8]. The Ar_2^* production kinetics is found to be similar to that in the case of electron beam excitation, since the simulated result shows an agreement with the experiment. These experimental results suggest that a high intensity laser produced electrons via the OFI process initiated the Ar_2^* production kinetics. According to the ADK theory, the laser intensity was high enough to selectively ionize neutral Ar to Ar^+ ions [9]. A collisional relaxation process of Ar^+ produced Ar^*, a precursor of Ar_2^*. The following processes are similar those observed in an electron beam pumped plasma. Note that excimer emissions were not observed when a self breakdown plasma was produced by a Q-switched Nd:YAG laser with a nanosecond pulse width [10]. Inverse Bremsstrahlung process and following electron avalanche in a nanosecond laser-produced plasma produce highly uncontrolled plasma conditions, which is in contrast to those observed using a high intensity laser. Therefore the conditions are not fulfilled for the suitable excimer production.

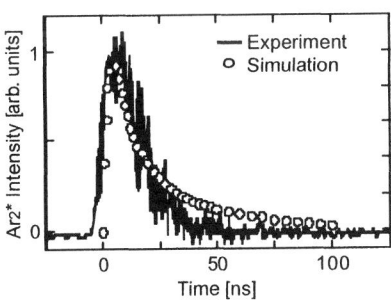

Figure 2. Time-resolved signal of Ar_2^* fluorescence at 126 nm. Solid curve represents the experimental result. Open circles are a simulated temporal behavior of Ar_2^* emission.

Fig. 3 shows the comparison of the Ar_2^* emission intensity at 126 nm with and without an optical hollow fiber. The optical hollow fiber with a length of 12.5 cm was used to guide the focused laser emission and to extend the excitation length. The inner diameter of the fiber was 200 μm. The focused laser intensity was 10^{15} Wcm^{-2} for both cases. Both Ar_2^* intensities quadratically increase with the Ar pressure. After the solid angle correction, the fiber-guided Ar_2^* intensities are approximately three times as large as those without the fiber. A visible plasma length was 4 cm in the case (b) and the fiber length was 12.5 cm in the case (a). The difference of the length may be responsible for the intensity difference. This result indicates the extension of the excitation length by use of a hollow fiber and encourages the further increase of the 126 nm intensity, finally leading to the stimulated emission.

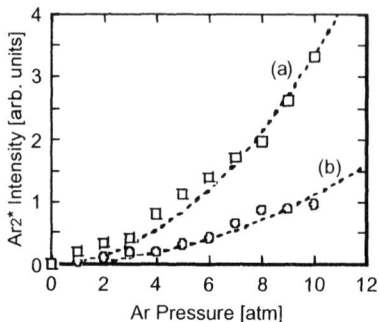

Figure 3. Comparison of Ar_2^* emission intensity at 126 nm with (a) and without (b) a hollow fiber.

SUMMARY

We have observed the Ar_2^* emission centered at 126 nm with a spectral bandwidth of 9 nm (FWHM). The Ar_2^* production kinetics suggest that the main process of the Ar_2^* production is three-body association process, since the emission intensity quadratically increases with the Ar pressure. The emission duration of Ar_2^* is 18 ns (FWHM) as predicted by kinetic simulation, which indicates the possibility of optical feedback would be possible. The OFI Ar plasma conditions are similar to those of an electron beam produced Ar plasma. The fiber guided Ar_2^* intensities are approximately three times as large as those without a fiber, which is reflected on the difference of the excited length.

REFERENCES

1. W. Sasaki, K. Kurosawa, S. Kubodera, J. Kawanaka, *J. Photopolym. Sci. Technol.* **11**, 361-366 (1998).

2. Y. Uehara, W. Sasaki, S. Kasai, S. Saito, E. Fujiwara, Y. Kato, C. Yamanaka, M. Yamanaka, K. Tsuchida, J. Fujita, *Opt. Lett* **10**, 487-489 (1985).
3. H. Ninomiya, K. Nakamura, *Opt. Commun.* **134**, 521-528 (1997).
4. T. Efthimiopoulos, B. P. Stoicheff, R. I. Thompson, *Opt. Lett.* **14**, 624 (1989).
5. H. Tanaka, A. Takahashi, T. Okada, M. Maeda, K. Uchino, T. Nishisaka, A. Sumitani, H. Mizoguchi, *Appl. Phys. B* **74**, 323-326 (2002).
6. S. Kubodera, Y. Nagata, Y. Akiyama, K. Midorikawa, M. Obara, H. Tashiro, K. Toyoda, *Phys. Rev. A* **48**, 4576-4582 (1993).
7. Y. Nagata, K. Midorikawa, S. Kubodera, M. Obara, H. Tashiro, K. Toyoda, *Phys. Rev. Lett.* **71**, 3774-3777 (1993).
8. W. Sasaki, *The Review of Laser Engineering* **13**, 912-926 (1985).
9. M. V. Ammosov, N. B. Delone, V. P. Krainov, *Sov. Phys. JETP* **64**, 1191-1194 (1987).
10. S. Kubodera, J. Kawanaka, W. Sasaki, *Opt. Commun.* **182**, 407-412 (2000).

Superradiant Amplification in Laser Produced Plasma by KrF laser

Eiichi Takahashi, Susumu Kato, Yoshiro Owadano

AIST Central2, 1-1-1 Umezono, Tsukuba, Ibaraki, 305-8568, Japan

Abstract. Superradiant amplification (SRA) in laser produced plasmas by KrF laser system for the use of large pump energy is examined by one-dimensional particle in cell simulation code. Its ultra-violet laser wavelength has many advantages for SRA. The simulation code predicts that the required signal pulse width is shorter than $\sim$ 100fs for out supposed conditions if the seed signal pulse shape is gaussian. Transient Raman scattering wavelength conversion in gaseous media can create truncated leading edge pulse. The simulation code showed the pulse can form leading edge spike by the amplification in plasmas.

INTRODUCTION

The development of the chirp pulse amplification enabled us to access ultra-intense laser up to 1PW so far. OPCPA[1] will realize wider bandwidth and is expected to generate higher intensity. However, the damage threshold of the final grating compressor still limit the total output energy.

The superradiant amplification (SRA) was proposed by Shvets, et. al. in 1998 [2]. This method uses plasmas as laser amplification media. The plasma media can handle very large laser fluence, so this is very attractive alternative scheme to generate ultra-high intensity laser pulse with large energy. It should be noted that the SRA can amplify ultrashort pulse shorter than plasma frequency. For example, for $n_e = 10^{20}$ cm^{-3}, $2\pi/\omega_p \sim$ 10fs. This is one of the merit of SRA, because the amplification process finishes before other unwanted instabilities develop in plasma.

The experiment has started at Max-Planck Institute for Quantum Optics using Ti:Sapphire laser system, ATLAS[3]. To achieve ultimate intensity, it is necessary to use pump pulses of large energy which can generated by Inertial Fusion Energy (IFE) drivers.

We have large electron beam pumped KrF laser system, Super-ASHURA[4]. And this laser has additional advantages over Ti:sapphire or Nd:glass lasers because of its ultra-violet laser wavelength as following. 1) Rayleigh range is longer assuming same focal spot diameter w, $z_R = \frac{\pi w^2}{\lambda}$. 2) Required spectral bandwidth to form ultra-short laser pulse is smaller, $\Delta\lambda = \frac{0.4\lambda^2}{c\tau_p}$. This will be important point to avoid overlapping pump and signal spectrum. 3) Gas ionization is easy due to the larger photon energy, $h\nu \sim$ 5eV [5, 6]. 4) Stimulated Raman scattering can generate longer wavelength signal seed pulse very easily.

On the other hand, the largest problem for the use of KrF laser is the generation of short

CP634, *Science of Superstrong Field Interactions*, edited by K. Nakajima and M. Deguchi

seed pulse.

In this paper, we will examine the feasibility of the superradiant amplification by electron beam pumped KrF laser using Particle in cell (PIC) simulation code.

THEORY

We will review the theory briefly [2, 7]. We assume counterpropagating pump $\vec{a}_L$ and signal $\vec{a}_S$ pulses as in Fig. 1, where $\vec{a}_S = a_S \exp\{i(-k_S z - \omega_S t)\} + c.c.$, $\vec{a}_L = a_L \exp\{i(k_L z - \omega_L t)\} + c.c.$. The dominant ponderomotive potential formed by the two

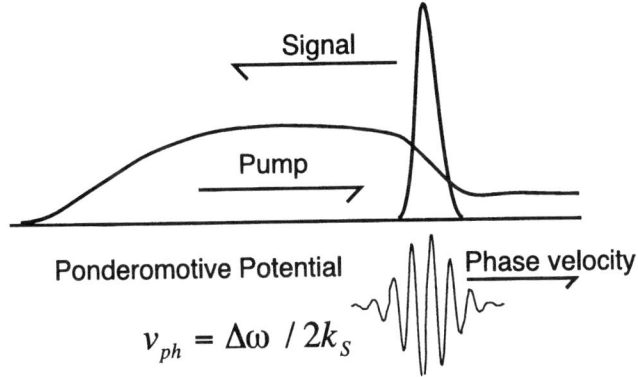

FIGURE 1. Amplification Configuration

laser interference is represented as,

$$\phi_{pond} = -mc^2 a_L a_S \cos \psi \tag{1}$$

where $\psi = (k_s + k_L)z - (\omega_L - \omega_S)t \sim 2k_s z - \Delta\omega t$. When $\omega_B^2 = 4\omega_S^2 a_L a_S > \omega_p^2$, the ponderomotive force is stronger than electrostatic force. The equation of j-th electron motion in the potential is represented simply as,

$$\ddot{\psi}_j + \omega_B^2 \sin \psi_j = 0. \tag{2}$$

Therefore, signal pulse receive both amplification and absorption alternatively in the period of ω_B. As the signal amplitude grows, the pulse width decreases with increasing of ω_B. The coherent scattering condition is automatically satisfied by the localized electron distribution determined by the potential.

In the case of stimulated Raman scattering, the three wave (pump, signal and electron plasma wave) have to satisfy phase matching condition. So, whole signal pulse grows.

PIC SIMULATION AND DISCUSSION

The results of the PIC simulation on the assumption of using KrF laser is shown in Fig. 2. The electron density is $8 \times 10^{19} \mathrm{cm}^{-3}$. $n_e/n_c \sim 0.0045$. Other parameters are noted on the figure.

As in Fig. 2(b), the growth of the electric field is proportional to the propagation distance z. This is consistent with the SRA theory $I_{signal} \propto n^2 z^2$.

Fig. 2(c) shows pulse peak structure in detail to see the relation between the potential

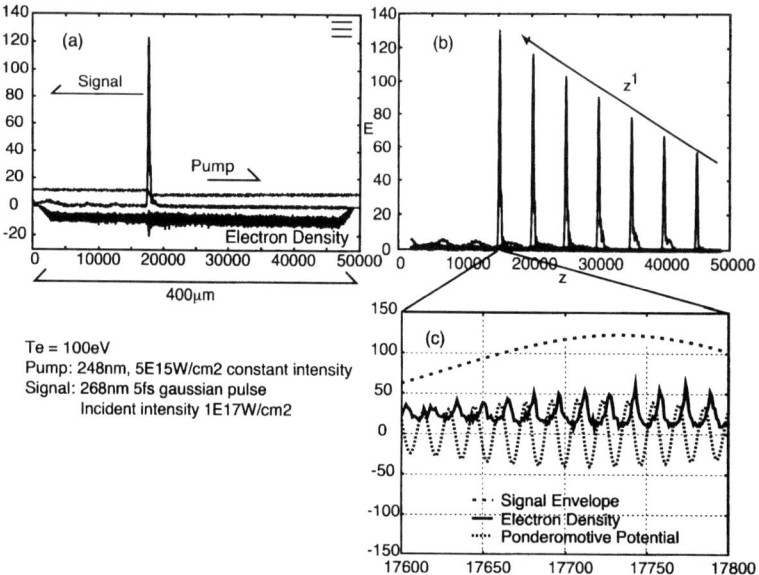

FIGURE 2. PIC Simulation for KrF laser (a) Whole simulation configuration, (b) Signal Pulse Growth, (c) the relation between the ponderomotive potential and the electron density distribution.

and the electron density. The potential is approximated by $\phi_{pond} \sim a_L a_S$. The electron density is higher in left hand side of the each ponderomotive potential cell. This is due to the direction of the potential movement is from left to right.

Fig. 3 shows the evolution of the signal pulse width and intensity for various its initial conditions. The conditions of pump intensity and plasma density are fixed ($I_L = 5 \times 10^{15} W/cm^2, n_e = 8 \times 10^{19} cm^{-3}$). We defined the pulse width as FWHM of the largest peak even for the case, whole pulse had several peaks. Basically initial points are upper left in the figure. The evolution lines are asymptotically close to the $2\pi/\omega_B$ line. We observed the conversion behavior also in weaker region than the Shvets's SRA criterion, $\omega_B^2 > \omega_p^2$.

For the longer incident signal pulse, we did not see such a behavior. This is because the long pulse has weak foot region, electron plasma wave was driven ahead and phases of the electrons are disturbed.

So, the signal pulse longer than 100fs would not likely to be SRA. KrF laser can amplify short pulse directly but the group velocity dispersion, nonlinear refractive index in the window, and the 3ω conversion from Ti:Sapphire oscillator output prevent generating

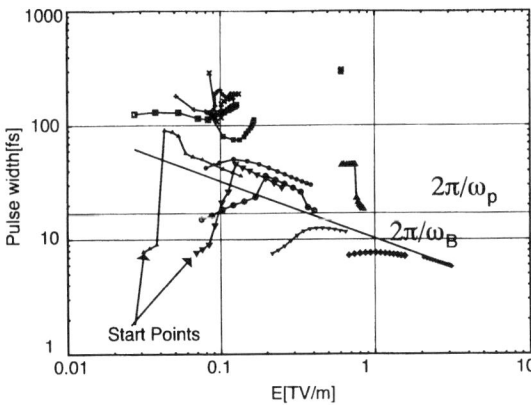

FIGURE 3. The evolution of the signal pulse width and electric field for various initial pulse width and intensity conditions. Asymptotic conversion is observed.

amplified sub-100fs pulses in KrF wavelength.

We have one solution to the problem. As in Fig. 4 (a), gaussian signal pulse of 100fs break up into several pulses. On the other hand, the truncated leading edge pulse (4 (b)) form leading edge peak. In this case, we used one-cycle sin wave to simulate the truncated leading edge pulse..

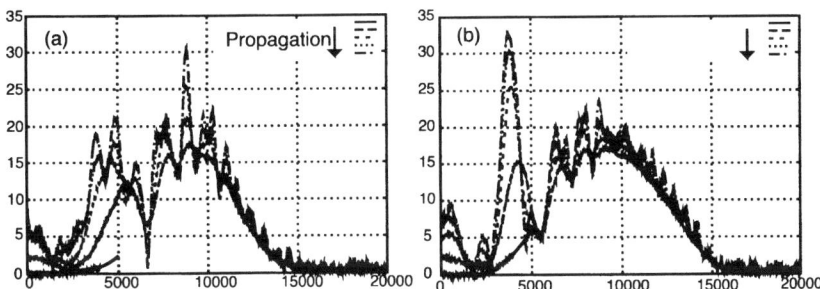

FIGURE 4. 100fs signal pulse growth for (a) Gaussian, (b) truncated leading edge pulse (sin). Pump and Signal pulse incident intensity: $1.0 \times 10^{16} W/cm^2$, Plasma density: $8 \times 10^{19}/cm^3$, Te=100eV. Propagation length $\sim 400\ \mu m$. Co-propagating coordinate.

This types of pulse can be created by transient stimulated Raman scattering in gaseous media[8]. The leading edge growth rate is independent from transverse dephasing time, T_2 of the Raman media under the transient regime during the conversion. And the conversion efficiency can be high if we use the group velocity mismatch technique.[9] The experimental investigation of this process will also be necessary.

CONCLUSION

SRA for KrF laser is examined by PIC simulation. SRA is also feasible for ultra-violet wavelength region. The signal pulse shorter than 100fs was found to be necessary for gaussian pulse shape. Transient stimulated Raman wavelength conversion is suitable not only for generating longer wavelength signal but also for having truncated leading edge. This seed pulse will create leading edge spike by the amplification in plasmas.

ACKNOWLEDGMENTS

The authors are pleased to acknowledge discussion with M. Dreher, H. Baumhacker, and Prof. K.J. Witte. A part of this study was financially supported by the Budget for Nuclear Research of the Ministry of Education, Culture, Sports, Science and Technology, based on the screening and counseling by the Atomic Energy Commission.

REFERENCES

1. I.N. Ross, P. Matousek, M. Towrie, A.J. Langley,J.L. Collier, *Opt. Comm.*, **1-3**, 125–133 (1997).
2. G. Shvets, N.J. Fisch, A. Pukhov, J. Meyer-tar-Vehn, *Phys. Rev. Lett.*, **81**, 4879–4882 (1998).
3. M. Dreher, E.Takahashi, H. Baumhacker, and K.J. Witte, "First Demonstration o Superradiant Amplificaiton in Plasma," Medeira, 2001, 28th EPS conference.
4. Y. Owadano, I. Okuda, Y. Matsumoto, I. Matsushima, E. Takahashi, E. Miura, H. Yashiro, T. Tomie, K. Kuwahara, M. Shinbo, "Super-ASHURA KrF Laser Program at ETL," in *FUSION ENERGY 1996*, IAEA, Montreal, Canada, 1996, vol. 3, pp. 215–221.
5. A.A. Offenberger, W. Blyth, A.E. Dangor, A. Djaoui, M.H. Key, A. Modena, Z. Najmudin, S.G. Preston, J.S. Wark, *Laser and Particle Beams*, **13**, 19–31 (1995).
6. Thomas M Autonsen, Jr and Zhigang Bian, *Phys. Rev. Lett.*, **82**, 3617 (1999).
7. M. Dreher, Tech. Rep. MPQ250, MPQ (2000), in German.
8. L.L. Losev, V.I. Soskov, *Opt. Comm.*, **135**, 71–76 (1997).
9. M.S. Dzhidzhoev, P.M. Mikheev, V.T. Platonenko, A.B. Savelev , *LASER PHYSICS*, **6**, 963–970 (1996).

Image of second harmonic emission generated from ponderomotively excited plasma density gradient

M. Mori, T. Shiraishi, H. Abe, S. Taki, Y. Kawada, and K. Kondo

Center for Tsukuba Advanced Research Alliance, University of Tsukuba
1-1-1 Tennoudai, Tsukuba, Ibaraki 305-8577, Japan

Abstract. We report on the observation of the second harmonic emission from underdense plasma without Raman scattering instability. Obtained second harmonic images were strongly depended on the polarization of the pump pulse. Experimental results were compared with the theoretical model of the second harmonic generation excited by ponderomotive electrons.

INTRODUCTION

The second harmonic emission from underdense plasma has been observed and explained[1-4]. In their reports, the second harmonic emission was generated ionization edge, or by the laser-plasma instabilities such as self-focusing, filamentation, or stimulated Raman scattering (SRS). In this paper, we report on the second harmonic generation from the underdense He plasma in a regime of the standard laser-wakefield illuminated by high intensity laser from 1.4×10^{17} to 5×10^{17} W/cm^2. We explained the experimentally obtained results by using the theoretical model of the second harmonic emission from the ponderomotively excited the plasma density gradient.

EXPERIMENTAL CONDITIONS

The experimental setup is shown in Fig. 1. The experiments were performed with a 10 Hz Ti:Sapphire CPA laser system (λ=800 nm). In these experiments, the pump laser pulse had a typical duration of 100 fs [full-width at half maximum (FWHM)] and energies ranging from 12 to 50 mJ. The pump pulse was directed through two $\lambda/4$ wave plates to vary the polarization. The pump pulse was focused by a fused silica planoconvex lens of f/4 in the target chamber. The spot size of the pump pulse at the focal plane was measured to be 11 μm in FWHM. Then, the peak intensity was estimated to be 5×10^{17} W/cm^2 at maximum operation. The target chamber was filled with He statically at 2 Torr. The oscillation period of the electron plasma wave (EPW) is estimated to be sufficiently longer than the pulse width of the pump pulse, thus the standard laser wakefield is formed. Therefore, neither SRS instability nor the

CP634, *Science of Superstrong Field Interactions*, edited by K. Nakajima and M. Deguchi
© 2002 American Institute of Physics 0-7354-0089-X/02/$19.00

instabilities in the self-modulated laser wakefield occur. The oscillation period of EPW was experimentally confirmed to be longer than the pulse width of the pump pulse by the frequency domain interferometric technique[5-8]. The second harmonic emission generated from the plasma was taken with a 16 bit charge-coupled-device(CCD) camera coupled with an f/4 doublet, which was set behind the interaction point. The pump pulse through the plasma was reflected behind the f/4 doublet by an IR HR mirror. By putting a band pass filter in front of the CCD, only the second harmonic image was selected. The magnification and the spatial resolution of the imaging system were estimated to be 30 times and 3 μm, respectively.

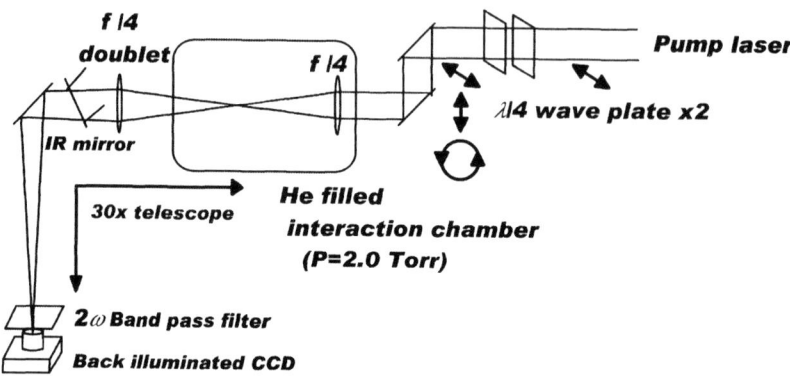

FIGURE 1. Schematic of the second harmonic experimental setup. The pump polarization was varied by using λ /4 wave plates.

RESULTS AND DISCUSSION

In Fig. 2 (a), the experimentally obtained images of the second harmonics at the focal point are shown for various pump polarizations. These images were obtained by the accumulation of 100 shots. The experimentally obtained second harmonic images apparently have a pump polarization dependence. With linearly polarized pump pulses, the images exhibited two peaks splitting along the direction of the polarization. With a circularly polarized pulse, the second harmonic image became a ring shape. The separation of the two peaks and the diameter of the ring were approximately 10 μm. The separation and diameter correspond to the spot size of the pump pulse.

They do not correspond to the size of the ionized area by tunneling ionization.

From the Ammosov-Delone-Krainov (ADK) model[9], the diameter of the fully ionized region at the focal plane is estimated to be more than 20 μm, which is two times larger than the spot diameter of the pump laser. In this ionized region, all the He atoms are fully ionized. Thus, the spatial density distribution is thought to be constant,

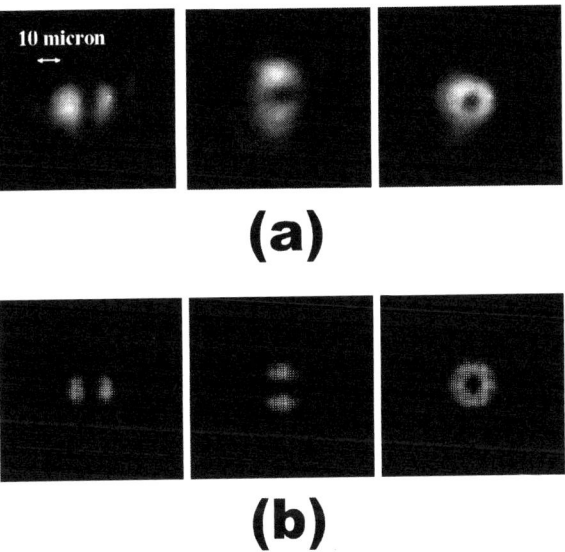

(a)

(b)

FIGURE 2. Second harmonic images for various polarizations of the pump pulse. Images in (a) were obtained in the experiment and those in (b) from the theoretical estimation. The left image was obtained with a horizontally polarized pump pulse, the middle one was with a vertically polarized pump pulse, and the right one was with a circularly polarized pump pulse.

if the ponderomotive effect of the pump pulse is neglected. However, under this experimental condition, the ponderomotive effect cannot be neglected. As is pointed out in Ref. 10, if the plasma has initial transverse density gradients prior to the arrival of the pump pulse, the second harmonics can be generated. The source current vector for the second harmonic emission $J_{2\omega}$ is proportional to $\delta n_q v$, where δn_q is the charge perturbation by the quivering electrons and v is the instantaneous velocity vector of quivering electrons. v is proportional to $a_0/\gamma \cos \omega t$, where a_0 is the normalized vector potential. δn_q is proportional to the product of $grad\ n_e \bullet v$. Then $J_{2\omega}$ becomes

$$J_{2\omega} \propto \left(\nabla n_e \bullet a_0 \right) a_0 \cos^2 \omega t / \gamma^2. \tag{1}$$

Here, the second harmonic emission was generated during laser propagation at the interaction point. Usually $grad\ n_e = 0$, then $J_{2\omega} = 0$. However, if $grad\ n_e \neq 0$, $J_{2\omega}$ becomes a finite vector. The factor of $cos^2\ \omega t$ corresponds to the electron oscillation emitting the second harmonics.

In Ref. 3, Miyazaki et al. reported that the second harmonic images were observed from the Na plasma at the intensity of $\sim 10^{12}$ W/cm². They obtained results similar to those above. In their experiment, the second harmonic emission was generated by the spatial distribution of the electron density which was caused by the spatial distribution of the laser intensity at the focus. Electrons are generated by multiphoton ionization. At 3×10^{12} W/cm², which was the maximum irradiation intensity in their report[3], the saturation of multiphoton ionization begins. The obtained image in Ref. 3 seems to be similar to ours. However, if the irradiation intensity increases sufficiently, the saturated volume around the focal point also increases. Therefore, the second harmonics should be generated from a wider area in the saturated case than in the non-saturated case. Also, Liu et al. have studied the second harmonic generation in a plasma where the plasma wavelength was longer than the pulse duration of the pump pulse[4]. In their report, the irradiation intensity is much higher than that in Ref. 3, however it reaches $\sim 10^{16}$ W/cm^2 at most. They observed the intensity of the second harmonic emission $I_{2\omega}$ scales $I_{\omega}^{1.5}$. Here, I_{ω} is the irradiation intensity of the pump pulse. They attributed this scaling to the effect of increasing partially ionized area over saturation of tunneling and/or multiphoton ionization. In our case, the same type of the second harmonic generation is thought to occur, because the irradiation intensity is much higher than theirs. However, the source current of the second harmonic emission $J_{2\omega}$ caused by the ponderomotively excited electron density gradient is much higher than that caused by the inhomogeneous electron density distribution at the partially ionized area in Refs. 3 and 4. Moreover, under our experimental conditions, the second harmonic emitting area is much smaller than the partially ionized area, which will be discussed later. Therefore, in our case, the observed second harmonic emissions are mainly from the $J_{2\omega}$ caused by the ponderomotively excited electron density gradient, not from that caused by the inhomogeneous electron density distribution at the partially ionized area. We attempted to calculate the second harmonic images by the ponderomotively excited electron density gradient. In order to estimate the second harmonic images, we calculated the spatial distribution of the electron density $n_e(r,t)$ by taking the electron dynamics into account. We solved the following equation of motion numerically.

$$\frac{d^2\mathbf{r}}{dt^2} = \kappa \nabla I(\mathbf{r},t) \tag{2}$$

Here, κ is a proportional constant. Moreover, the instantaneous laser intensity $I(r,t)$ can be given by

$$I(\mathbf{r},t) = 5 \times 10^{17} \times \exp\left\{-\left(\frac{|\mathbf{r}|^2}{r_0^2} + \frac{t^2}{t_0^2}\right)\right\} \quad W/cm^2 \tag{3}$$

where the temporal and spatial distributions were assumed to be Gaussian. In this calculation, r_0 and t_0 are 6.6 μm and 60 fs, respectively. These values were obtained from experimental results. In estimating $n_e(r,t)$, the initial spatial distribution of electron density is assumed to be constant for any spatial point. We calculate the position of an electron by solving Eq. (2) for each point in time. By this procedure, the spatial distribution of electron density can be determined at each time. $J_{2\omega}(\mathbf{r},t)$ can be obtained by $grad\ n_e(r,t)$ and $\mathbf{a_0}(\mathbf{r},t)$. The temporal intensity distribution of the second harmonic emission is given by $|J_{2\omega}|^2$. In Fig. 2(b), the time integrated $|J_{2\omega}|^2$ image is shown. The calculated images were in good agreement with the experimentally observed images. In our calculation, we did not take the longitudinal density gradient into account because under our focusing conditions, the longitudinal density gradient is negligible compared with the lateral density gradient[7,8]. Particularly for the polarization dependence, the term of $grad\ n_e \cdot a_0$ is strongly related to the spatial distribution of the second harmonic emission. In the circularly polarized pump pulse, a ring-shaped pattern was obtained because $grad\ n_e \cdot a_0$ is finite for any azimuthal direction from the center of the focus, while split patterns were obtained in the linearly polarized pump pulse. In the linear case, the direction of splitting was the direction of polarization. Moreover, at the central part of the focus, $grad\ n_e = 0$. Then we can see the dark area at the center in all cases.

Next, we will discuss the polarization of the second harmonic emission. To estimate the polarization of the second harmonic emission, the transmitted intensity through a polarizer, which was placed in front of the CCD camera, was measured as a function of the rotation angle of the polarizer. In Fig. 3, the experimental results are shown. With a linearly polarized pump pulse, the transmitted intensity apparently changed with rotation of the polarizer. In contrast, the intensity was almost constant for any angle of the polarizer with a circularly polarized pump pulse. These experimental results suggest that the polarization of the generated second harmonic emission is the same as that of the pump pulse. In the theoretical model discussed above, the polarization of the second harmonic emission is given by $J_{2\omega}$, which is proportional to a_0. This means that the polarization of the pump pulse gives that of the second harmonic emission.

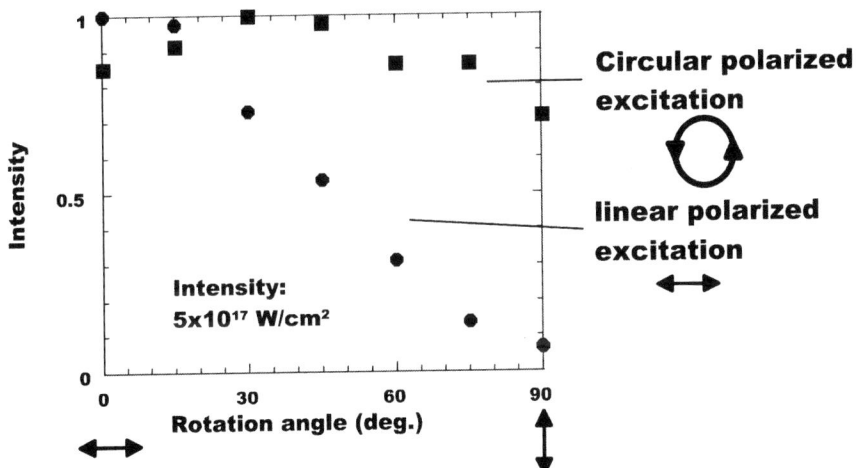

FIGURE 3. Transmitted intensity of the second harmonic emission through the polarizer was plotted as a function of the rotation angle of the polarizer. Dots were obtained with a linearly polarized pump pulse. Squares were obtained with a circularly polarized pump pulse.

As has already been pointed out in our previous report[11], the intensity of the second harmonics was proportional to the pump intensity with a power of 1.8, not 2.0, between 2.5×10^{17} W/cm^2 to 1.3×10^{18} W/cm^2. This scaling can be explained by the relativistic effect. In Ref. 11, we explained the scaling by $I_{2\omega} \propto a_0^4/\gamma^2$, which was derived by referring to Ref. 1. However, as is described above, if we take the source current vector for estimating the second harmonic emission, $I_{2\omega} \propto a_0^4/\gamma^4$. However, between 2.5×10^{17} W/cm^2 and 1.3×10^{18} W/cm^2, the intensity of the second harmonic emission is proportional to the pump intensity with a power of approximately 1.8 for both a_0^4/γ^2 and a_0^4/γ^4. In order to determine the exact intensity scaling, it is necessary to perform the experiment above 10^{19} W/cm^2. In this paper, we also paid attention to the intensity dependence of the second harmonic image. In Fig. 4, the separation of two peaks in the second harmonic image was plotted as a function of the pump intensity between 1.4×10^{17} W/cm^2 and 5×10^{17} W/cm^2. Dots were obtained from the experiment. The separation did not change significantly with increasing laser intensity. Crosses were obtained by the theory described above. They are in good agreement with each other. Therefore, the split images were thought to be caused by the electron density gradient, which was formed by ponderomotive pressure of the pump pulse. The measured second harmonic emission was not produced by the edge of the ionized region. We have also plotted the diameter of the ionized region, which was estimated by the ADK theory[9], represented by diamonds in Fig. 4. The ionized region is much wider than experimental values.

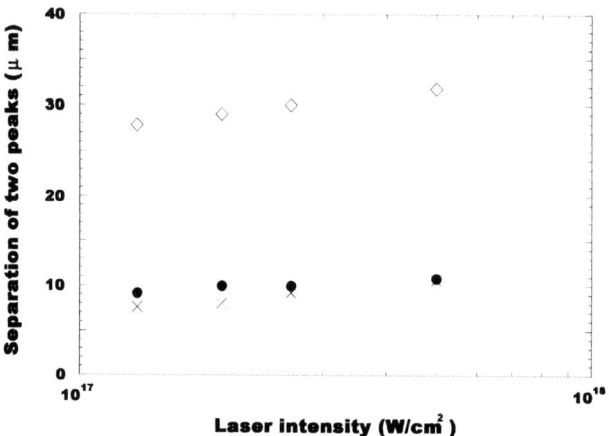

FIGURE 4. Separation of two peaks in the second harmonic image is shown as a function of the pump intensity. Dots were obtained by the experiment and crosses by the calculation. Diamonds represent the diameter of the ionized region which was estimated by the ADK model.

SUMMARY

This experiment provides the second harmonic emission from underdense plasma in a regime of the standard laser-wakefield. The second harmonic images strongly depended on the pump polarization. And the measured polarization of the generated second harmonics was the same as that of the pump pulse. These experimental results were well explained by the theoretical model, in which the plasma density gradient by ponderomotive pressure plays an important role for generating the second harmonic emission.

REFERENCES

1. Malka, V., et al., *Phys. Plasmas*, **4**, 1127-1131 (1997).
2. Krushelnick, K., et al., *Phys. Rev. Lett.*, **75**, 3681-3684 (1995).
3. Miyazaki, K., et al.., *Phys. Rev. A*, **23**, 1358-1364 (1981).
4. Liu, X., et al., *IEEE trans. Plasma Sci.*, **21**, 90-94 (1993).
5. Marques, J. R., et al., *Phys. Rev. Lett.*, **76**, 3566-3569 (1996).
6. Siders, C. W., et al., *Phys. Rev. Lett.*, **76**, 3570-3573 (1996).
7. Marques, J. R., et al., *Phys. Plasmas*, **5**, 1162-1177 (1998).
8. Takahashi, E., et al., *Phys. Rev. E*, **62**, 7247-7250 (2000).
9. Ammosov, M. V., et al., *Sov Phys. JETP*, **64**, 1191-1194 (1986).
10. Esarey, E., et al., *IEEE Trans. Plasma Sci.*, **21**, 95-104 (1993).
11. Takahashi, E., et al., *Phys. Rev. E*, **65**, 016402 (2001).

Propagation of Intense Laser Pulses in Capillary and Its Application to X-ray Laser

J.Q. Lin   H. Nakano   T. Nishikawa   T. Ozaki and K. Oguri

NTT Basic Research Laboratories, Nippon Telegraph and Telephone Corporation
3-1 Morinosato Wakamiya, Atsugi, Kanagawa 243-0198, Japan

Abstract. We investigated the guiding of femtosecond laser pulses with an intensity of $> 10^{16}$ W/cm^2 in a hollow and gas-filled capillary. We measured the laser transmission versus length of two capillaries with different inner cores in a vacuum, and also the laser propagation modes in these two capillaries. We succeeded in generating an extended plasma column that was ten times longer than the Rayleigh range by propagating high intensity femtosecond laser pulses in a gas-filled capillary. We obtained a spectral blue shift in the transmitted femtosecond laser pulses with a 5 torr nitrogen-filled capillary and used this shift to demonstrate extended plasma column formation in a gas-filled capillary. We conducted a preliminary soft X-ray laser experiment as an application of this laser-pumped gas-filled capillary.

INTRODUCTION

Many applications in physics require the propagation of intense, femtosecond laser pulses over a distance of several times the Rayleigh range (Z_R). These applications include laser particle accelerators [1], high-order harmonic generation [2] and X-ray lasers [3]. In usual cases, the absence of an additional guiding mechanism would make it difficult to generate a high–intensity field over a sufficient length with a modestly powered, ultra-short pulse laser, since the maximum vacuum focusing length is of the order of the Rayleigh range. Moreover, for intense laser pulses propagating through dense plasma, the interaction length is further limited by ionization-induced refractive defocusing. This means the effective interaction length will be less than the Rayleigh range if only natural focusing is used.

Various mechanisms have been proposed for propagating high intensity laser pulses beyond the Rayleigh range distance. The extended propagation of laser pulses in plasma beyond the Rayleigh range can be achieved with a preformed plasma channel [4]. An extended plasma column can also be realized by a laser pulse once the laser intensity is sufficiently high, and the refractive index can be modified by the laser pulse and lead to relativistic or ponderomotive self-channeling [5]. However, many applications need plasma parameters very different from those required for guiding a laser, and would benefit from a guiding structure that existed independently of the plasma.

The use of a hollow capillary seems to be such an alternative scheme. With this guiding technique, the guiding mechanism is based on Fresnel reflection occurring at the inner surface of the capillary wall. Several groups have studied ways of guiding a high intensity femtosecond laser pulse over many times the Rayleigh range of

CP634, *Science of Superstrong Field Interactions*, edited by K. Nakajima and M. Deguchi

focusing optics in a hollow capillary in a vacuum [6-8]. However, in terms of the practical application of this guiding technique, no one has yet realized a laser pulse capable of such propagation. It is essential to form an extended plasma column in a capillary. This makes it important to investigate high intensity laser pulses propagating in a capillary that contains plasma. The idea of laser pulse propagation over the Rayleigh range in a capillary containing plasma can be realized by injecting gas into a hollow capillary tube, and simultaneously employing guided high intensity laser pulses to ionize the gas over an extended region.

EXPERIMENT

We investigated high intensity laser pulses propagating through a hollow capillary under both vacuum and gas-filled conditions. The laser system we used in the experiment was a commercially available (Spectra-Physics), 10 Hz Ti:sapphire chirped pulse amplification system with a pulse duration of 100 fs and a maximum energy output of 100 mJ [9]. The linearly polarized laser pulse passed though a MgF_2 window into an evacuated chamber, and was focused by an f = 400 mm MgF_2 lens onto a capillary target. The Rayleigh range of the focusing optics was Z_R = 0.8 mm. For our experiment on laser pulse propagation through a hollow glass capillary, we fixed the capillary holder on a 5-dimensional translation stage. We measured the input and output laser energies with a calorimeter, and recorded an image of the laser mode with a CCD camera.

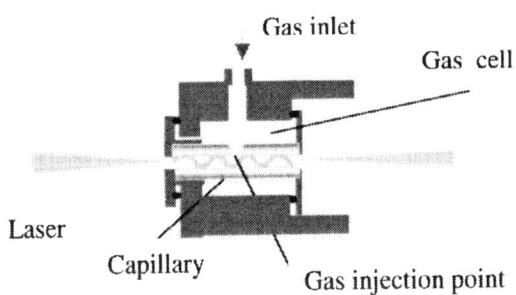

FIGURE 1. Target configuration of gas-filled capillary

Figure 1 shows the target configuration for the gas-filled capillary experiment. The capillary was cut to form an opening gas injection point, and the cut capillary was fixed in a sealed static gas cell, which we used when we filled the capillary with gas. Laser pulses were directly coupled into the capillary, and the capillary entrance simultaneously served as a differential hole between the gas-filling cell and the vacuum chamber. An output pinhole, which isolated the vacuum and gas-filling regions, was drilled in a 0.1-mm-thick stainless steel plate using a driving laser at the beginning of each experiment. The gas-filling cell was placed in a vacuum target chamber, thus allowing a femtosecond laser pulse to propagate in an almost gas-free region until it reached the capillary entrance. The present experimental set up minimizes the adverse effects of gas nonlinearity on the pulse in front of the capillary compared with the case where an uncut capillary is placed directly in the gas chamber, in which gas flows into the capillary through the capillary entrance [10]. In addition, by using this experimental set up, we can avoid the soft X-ray emission being absorbed by gas outside the capillary. Therefore, this experimental set up is particularly suitable for use with a soft X-ray laser and for high order harmonics generation research. In the case of gas-filled capillary experiment, we employed a visible range spectrometer and an X-ray transmission grating spectrometer to measure the output femtosecond laser spectra and the axial soft X-ray emission from the gas-filled capillary, respectively.

RESULTS AND DISCUSSION

We first investigated the guidance of high intensity laser pulses in a hollow capillary in a vacuum. Figure 2 shows the laser transmission versus the capillary length for two different capillary diameters used at an incident intensity of 1×10^{16} W/cm^2

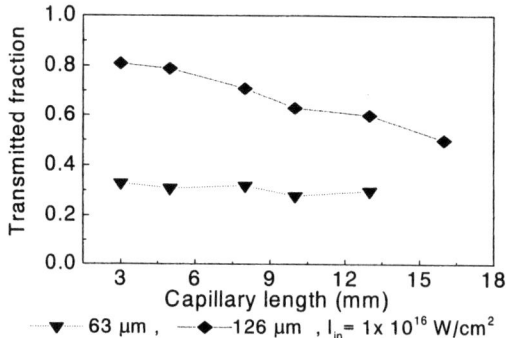

FIGURE 2. Dependence of laser transmission on capillary length in a vacuum. The dashed line is a guide for the eye.

The data in Fig. 2 show that high intensity laser pulses have propagated efficiently through a 16 mm-long capillary. Based on the fact of that Rayleigh range of the focusing optics is $Z_R = 0.8$ mm, the result shows that laser pulses propagate efficiently in capillaries with a length of up to 20 times the Rayleigh range (this experimental result was limited by capillary availability). Figure 2 also shows that the laser transmission through the two capillaries appears to decrease with capillary length. However, the transmission through an $a = 126 \, \mu m$ capillary (a is the capillary core radius) decreases much more quickly than that through an $a = 63 \, \mu m$ capillary. The origin of this difference is the different modes excited in the different core-diameter capillaries. A capillary tube with a dielectric wall is a leaky waveguide. Losses associated with refraction of the beam at the wall occur during propagation in addition to those caused by coupling the laser pulses at the entrance of the capillary tube. Specifically, there are losses caused by multiple reflections against a high order mode. The results we obtained with the analytical model show that when ω_0 / a equals 0.65 (ω_0 is the focusing spot radius), a hollow glass capillary is most favorable for EH_{11} mode excitation, and this mode has the largest damping length; when $\omega_0 << a$, the modes generated inside the capillary will be high order modes with high loss [11]. In our case, $\omega_0 / a = 0.5$ (ω_0 is about 30 μm) for an $a = 63$ μm capillary, which is close to the ideal value for EH_{11} generation. However, the value of ω_0 / a is only 0. 25 for an $a = 126$ μm capillary, which mainly excites high order modes. To provide experimental evidence for the above analysis, we measured the spatial distribution of the laser intensity at the capillary exit. Images of this spatial distribution provide information concerning the mode structure of the laser beam during its propagation. The measured modes were EH_{11}–like for the $a = 63$ μm capillary, while we observed a high order mode structure with laser pulse propagation through the $a = 126$ μm capillary.

The following part describes our experimental investigation of high intensity laser pulses propagating in a gas-filled capillary. In this experiment, the first point to be clarified was whether the gas in the capillary could seriously attenuate the laser energy when the propagated laser intensity was sufficiently high to ionize the gas. Figure 3

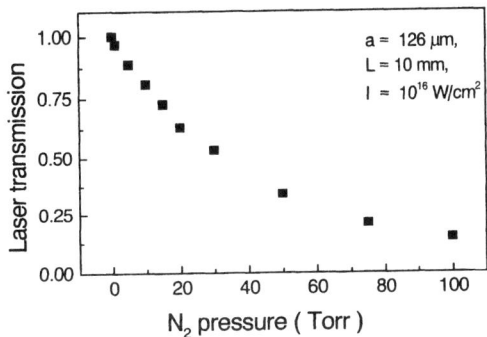

FIGURE 3. Dependence of normalized laser transmission on backing pressure of nitrogen-filled capillary. The absolute transmission is 60% at pressure of P = 0.

shows the measured normalized laser transmission as a function of the backing gas pressure at a laser intensity of $1 \times 10^{16} W/cm^2$. Generally, laser transmission decreases with increasing gas pressure. Increasing the gas density enhances not only absorption but also energy coupling to a high order transverse mode via ionization. The latter effect also reduces the laser throughput as a result of the increased propagation losses of the high order mode, and impairs the output beam quality. Self-defocusing of the incident beam in front of the capillary entrance may also reduce the transmission. Potential applications of femtosecond high intensity laser propagation in a gas-filled capillary, such as that demonstrated by the recombination mechanism soft X-ray laser experiment, usually require a very low operating gas pressure (<10 torr) [12]. We therefore focused our attention on laser transmission under low gas pressure conditions.

Figure 3 reveals that laser energy propagating through a 5 torr nitrogen-filled capillary remains at about 0.9 of the value it exhibits when propagating through the same capillary in a vacuum. Therefore, a slight reduction in the energy (intensity) of propagated laser pulses under low gas pressure will be beneficial in terms of generating an identical ionic state along their propagation path in a gas-filled capillary.

The spectral variation in output laser pulses can provide information about the laser propagation situation in a capillary. We employed the spectral change in the output femtosecond pulses from a gas-filled capillary to estimate the length of the plasma column generated in the capillary. Figure 4 shows that the laser transmitted spectrum broadens in relation to the input spectrum after passing through a 10-mm-long nitrogen-filled capillary. Interestingly, the spectral broadening shown in Fig. 4 is only towards the blue side and is significantly different from the experimental result obtained by M. Nisoli et al. for laser propagation in a nitrogen-filled capillary [13], where the broadened spectrum showed both a blue shift and a large portion of red-shift components. The essential difference between these two experiments is the laser

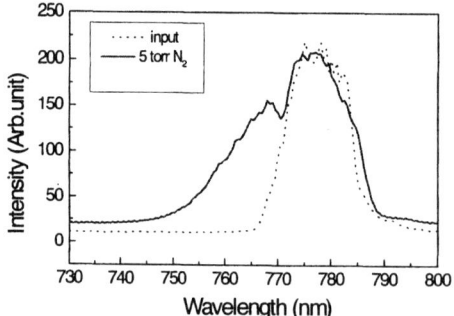

FIGURE 4. Spectral blue shift of the transmitted laser pulses after a 10-mm-long nitrogen-filled capillary, where the laser intensity is 10^{16}W/cm^2 and the radius of the capillary inner hole is 126 μm.

intensities that were employed. In Nisoli's experiment, the laser intensity was lower than the multi-photon ionization or self-focusing threshold, thus neutral electronic nonlinearity played a major role in the spectral broadening, and the self phase modulation process induced by neutral electric nonlinearity shifted the broadened spectrum towards both the blue and red sides. However, the laser intensity used in our experiment was about 10^{16}W/cm^2, for which a significant proportion of the nitrogen valence electrons was ionized around the pulse peak. Under low gas pressure, the optical field ionization process can cause a rapid increase in electron density, and hence a decrease in refractive index. This leads to the complementary effects of spatial defocusing and spectral broadening. The plasma-induced spectral shift [14] can be described as $\Delta\lambda = (\lambda_0 L/c)/(dn/dt)$, where λ_0 is the incident laser wavelength, L is the length of the gas ionization region, and n is the plasma refractive index. The plasma refractive index is given by $n = (1-Ne/Ncr)^{1/2}$, where Ne is the electron density and

Ncr is the critical density. As the ionization process is irreversible on a femtosecond time scale, this corresponds to $(dn/dt)<0$ during the whole laser pulse duration, which makes the spectrum shift only in the blue direction. The blue spectral shift under low operating gas pressure obtained above becomes obvious only when the gas ionization region L is sufficiently long. When we simply focus the laser pulses at the same intensity into a 5 torr nitrogen gas cell without using a capillary (the length of the nitrogen gas ionization region under this condition is about twice the Rayleigh range of the focusing lens), or when we guided the laser pulses in the same capillary without filling it with nitrogen gas, we were unable to observe any obvious spectral blue-shift component for either case. The above experimental results indicate that the laser ionization range in a 5 torr nitrogen-filled capillary is a distance much greater than double the Rayleigh range.

In the experiment, we simultaneously monitored the duration of the output laser pulses and found that they exhibited a slight broadening in their temporal duration. The measured pulse duration was typically about 120 fs under the above experimental conditions. From the measured laser parameters (focus spot size, pulse duration and laser transmission) at the capillary exit, the estimated output laser intensity was about 10^{15} W/cm^2. We can therefore infer that plasma is created along the whole length of the capillary ($>10Z_R$).

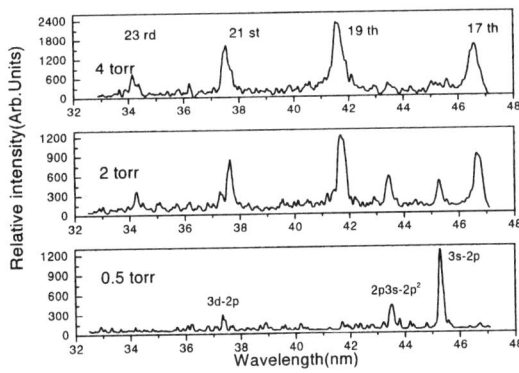

FIGURE 5. Axial soft X-ray emission was obtained from a laser-pumped nitrogen-filled capillary, where the input laser energy was 40 mJ and the pulse duration was 100fs. The capillary length and radius of inner core were 10 mm and 126 μm, respectively

By exploiting the propagation of intense femtosecond laser pulses over several times the Rayleigh range in a gas-filled capillary, we were able to carry out an experiment in which we employed a laser-pumped gas-filled capillary for recombination soft X-ray laser. Figure 5 shows the measured axial soft X-ray emission from a nitrogen-filled capillary pumped by femtosecond laser pulses. We can see clearly that the transition 3s–2p(N^{2+}) at 45.2 nm is anomalously strong at backing pressure of 0.5 torr. From our previous research [12] we know that this anomalously strong emission can be attributed to the realization of a population inversion between energy levels 3s and 2p, which results from efficient pumping into the 3s state via recombination from N^{3+} ions, and also from the collisional de-excitation of 3p or 3d states due to the relatively low electron temperature in plasma. Compared with our previous work in which a soft X-ray emission was obtained from a laser-pumped gas cell, an obvious feature of the present result is that the candidate lasing line at 45.2 nm is much stronger if we normalize the experimental data. We ascribe the enhanced soft X-ray emission to the longer plasma column formed in the present experimental set up. The other feature of the present experimental result is that high order harmonics emerge at lower gas pressure. Specifically, the high order harmonics are very strong at a backing pressure as low as 2 torr, and dominate the spectrum at a pressure of 4 torr. However, we did not observe strong high order harmonics with an operating gas pressure lower than 8 torr in our previous experiment that we conducted with a gas cell. The strong high order harmonics that emerge at a lower gas pressure with a gas-filled capillary than with a gas cell can be explained as follows. The phase mismatch between the pumping laser and the harmonics, which usually occurs while focusing laser pulses in a low pressure gas cell, is reduced due to the guiding effect of the laser pulses in the gas-filled capillary, and this enhances the generation of high order harmonics [2].

CONCLUSIONS

In conclusion, intense laser pulses with an intensity of $>10^{16} W/cm^2$ efficiently propagated over 20 times the Rayleigh range of a focusing lens in a hollow capillary in a vacuum. A plasma region with a length greater than 10 Z_R in a gas-filled capillary is generated by propagating high intensity ($>10^{16} W/cm^2$) laser pulses. An anomalously strong soft X-ray emission at 45.2 nm was obtained from a laser-pumped nitrogen-filled capillary. A scheme involving the use of a femtosecond laser to pump a gas-filled capillary shows great potential as regards the realization of soft X-ray lasers with a large gain length product.

REFERENCES

1. K. Nakajima et al., Phys. Rev. Lett. **74**, 4428(1995); A. Modena et al., IEEE Trans. Plasma Sci. **PS-24**, 289 (1996).
2. A. Rundquist et al., Science **280**, 1412(1998); Y.Tamaki, et al., Phys. Rev. Lett.**82**,1422(1999).
3. D. Korobkin et al., Phys. Rev. Lett. **81**, 1607 (1998).
4. T. Hosokai et al., Opt. Lett. **25**, 10 (2000).
5. F. Vidal et al., Phys. Rev. Lett. **77**, 1282 (1996).
6. S. Jackel et al., Opt. Lett. **20**, 1086 (1995).
7. M. Borghesi et al., Phys. Rev. E**57**, R4899 (1997).
8. C. Courtois et al., J. Opt. Soc. Am. B **17**, 864 (2000).
9. H. Nakano et al., Appl. Phys. Lett. 75, 2350 (1999).
10. P. Zhou et al., CLEO/Pacific Rim 2001 Vol. I 324. Jul. 15-19, 2001, Chiba, Japan.
11. R. Albrams et al., IEEE J. Quantum Electron **8**, 838 (1972).
12. P. Lu et al., Opt. Comm. **170**, 71-78 (1999).
13. M. Nisoli et al., Appl. Phys. B **65**, 189 (1997).
14. S. C. Rae et al., Phys. Rev. A **46**, 1084 (1992).

Demonstration of Transient Collisional Excitation X-Ray Lasers in Gases

P. Lu, T. Kawachi , M. Suzuki, K. Sukegawa, S. Namba, M. Tanaka, N. Hasegawa, R Tai, M. Kishimoto, M. Kado, K. Nagashima, H. Daido, Y. Kato, and H. Fiedorozicz[*]

Advanced Photon Research Center, Japan Atomic Energy Research Institute,
8-1 Umemidai Kizu-cho, Souraku-gun, Kyoto 619-0215 Japan
[*]Institute of Optoelectronics, Kaliskiego 2, 00-908 Warsaw, Poland

Abstract. We have demonstrated a high gain neonlike argon ion x-ray laser using a gas puff target. The elongated x-ray laser plasma column was produced by irradiating the gas puff target with line focused double picosecond laser pulses with a total energy of 9 J in a travelling-wave excitation scheme. Strong lasing was observed and a high gain coefficient of 18.7 cm^{-1} was measured on the transient collisionally excited $3p\ ^1S_0$ - $3s\ ^1P_1$ transition for neonlike argon at 46.9 nm with targets up to 0.45-cm long. Preliminary result on the nickellike xenon x-ray laser is also discussed.

I. INTRODUCTION

One of the most promising ways of developing table-top x-ray lasers for applications is a scheme based on transient collisional excitation (TCE). This scheme has the advantage that less than 10 J of laser energy from a table-top laser generating picosecond high-power laser pulses with chirped-pulse amplification (CPA) is sufficient for lasing. The first transient gain x-ray laser was demonstrated at 32.6 nm in neonlike titanium[1]. Saturated operation of transient gain x-ray lasers was demonstrated for both neonlike[2] and nickellike ions[3,4] with low pump energy. This opened the way for a practical table-top soft x-ray laser suitable for various applications.

In this paper, we present a result on transient collisional excitation x-ray lasers in which the line-shaped gain medium is produced by laser irradiation of a gas puff target. The use of a gas puff instead of a solid target offers several advantages. From the viewpoint of practical applications, the gas puff target provides a high repetition rate x-ray laser without production of target debris. In principle, the gas puff also allows better control over the density and minimizes gradients in the plasma if we can precisely characterize the interaction processes. However, it is practically difficult to control the location of gas plasma in space and time[5]. X-ray lasers using a gas puff target were first demonstrated with a 0.5 ns duration, 500 J energy laser pulse by a German-Polish collaboration[6]. In our experiment, we used two 1.5 picosecond laser pulses with a total energy of 9 J to create soft x-ray lasing in neonlike argon gas puff

CP634, *Science of Superstrong Field Interactions*, edited by K. Nakajima and M. Deguchi
© 2002 American Institute of Physics 0-7354-0089-X/02/$19.00

targets, and a transient high gain of 18.7 cm^{-1} and a narrow beam divergence of less than 3.7 mrad for the neon-like argon 3p-3s lasing line were successfully achieved. Also the nickel-like xenon lasing line at ~10 nm was clearly observed with a combination of a 600 ps prepulse followed by a 1.5 ps driving laser pulse, with a total energy of 22 J.

II. Experimental Setup

The experiment was performed on a CPA hybrid laser at the Advanced Photon Research Center of the Japan Atomic Energy Research Institute[7]. This laser provides up to 20 J of light at 1054 nm in a 1.5 ps pulse. For this work, a prepulse had to be applied before the main pulse, therefore we used the double picosecond pulse as the driving laser pulse for the neonlike argon soft x-ray lasing experiment. The separation time between the double picosecond pulses was set to be about 1.2 ns, and the energy ratio of the first pulse to the second pulse was about 1 : 6.5. The line focus with a length of 0.55 cm and a width of 20 μm was achieved by using a combination of an off-axis parabolic mirror and a spherical mirror[7]. For a typical total output energy of 9 J, the irradiances in the line focus are of about 7.0×10^{14} Wcm^{-2} for the first pulse and about 4.7×10^{15} Wcm^{-2} for the main laser pulse. Traveling wave geometry was used to irradiate the target by using a stepped mirror technique[7].

The gas puff target was formed using a solenoid valve developed at the Institute of Optoelectronics. It was equipped with a 0.6-cm-long nozzle with a 500-μm-wide slit[8]. Argon and xenon gases were used to form gas puff targets. The laser beam illuminated the gas puff target in the transverse direction with respect to the flow of gas. To achieve a high gain, a high gas density around 10^{20} cm^{-3} is required. Therefore, the laser line focus position was placed about 150 μm above the slit nozzle output. The gas backing pressure in the valve was set to be 10 bar and the valve time delay between the opening of the valve and the laser pulse was 400 μs. The maximum gas density in the interaction region was roughly estimated to be ~10^{20} cm^{-3} using the x-ray backlighting method[8]. The maximum plasma column length was set to be

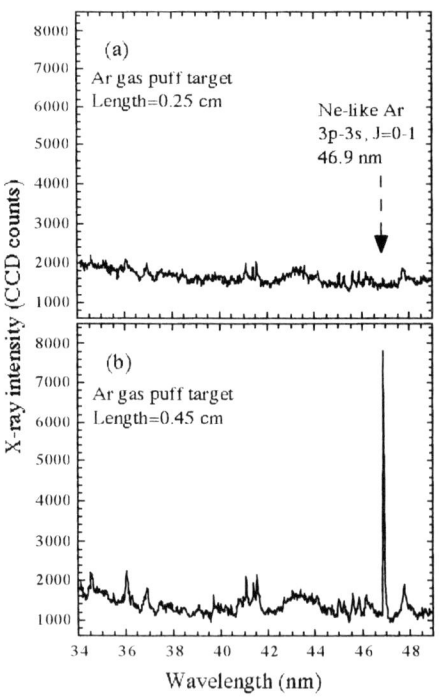

FIGURE 1. Typical on-axis emission spectra for (a) 0.25- and (b) 0.45-cm-long argon gas puff targets irradiated with a 1.2 J, 1.5 ps pulse followed by a 7.8 J, 1.5 ps pulse showing the strong collisionally excited neonlike argon $3p^1S_0 - 3s^1P_1$ laser line at 46.9 nm..

4.5 mm so that the line focus was overfilling the gas puff target in order to avoid residual absorption of the x-ray laser in the cold gas in the direction of the main detector. For the gain measurement, the length of a plasma column was changed by blocking part of the gas puff from being irradiated by means of a thin silver plate.

The main diagnostic, aligned to the axis of the line focus, was a 1200-line/mm flat-field grating spectrometer with a back-illuminated charge-coupled device (CCD). The spectrometer used a gold-coated spherical mirror collection optic to observe the on-axis x-ray emissions. An additional 2400-line/mm flat-field grating spectrometer was aligned 45° off the axis to monitor the ionization balance of the plasma column.

III. Results and Discussions

Figure 1 presents typical on-axis emission spectra for the laser-irradiated argon gas puff targets with two different plasma column lengths of 0.25 and 0.45 cm. The driving laser pulse with a combination of a ~1.2 J, 1.5 ps prepulse and a ~7.8 J, 1.5 ps main pulse was used to irradiate a target, and the total energy was 9.0 ± 1.0 J. One can clearly see from Fig. 1b that the neonlike argon $3p$ - $3s$ laser line at 46.9 nm dominates the spectrum for a gas puff target length of 0.45 cm. The x-ray laser intensity is weak but still visible in the spectrum for the length of 0.25 cm, as shown in Fig. 1a. A very large increase in the x-ray lasing intensity with increasing gas puff target lengths from 0.25 to 0.45 cm strongly indicates a very high gain for the neonlike argon 3p-3s x-ray laser. In our experiment, we did not observe the self-photopumped 3d-3p laser line at 45.1 nm, which was observed to be very strong in the experiment using a combination of a long 600 ps prepulse followed by a 6 ps driving pulse[9]. A shorter pumping pulse of 1.5 ps used in our experiment might be the reason, and our result is advantageous for possible applications.

The x-ray lasing was very stable and reproducible. Therefore, we performed a gain measurement by moving the thin silver plate to vary the plasma lengths from 0.25 to 0.45 cm. In Fig. 2, we show the intensity versus plasma column length for the neonlike argon $3p$ - $3s$ laser line. These data points were obtained under the condition of a total driving laser energy of 9.0 ± 1.0 J. Using the Linford formula[10], we obtained a high gain coefficient of 18.7 $\pm$ 1.0 cm^{-1} for the neonlike argon 3p - 3s, J = 0 - 1 transition lasing. This corresponds to a gain-length product of 8.4 for the 0.45-cm-long plasma column.

From the registered spectral distribution, we could also obtain the angular profile of the x-ray laser line and determine the beam divergence angle.

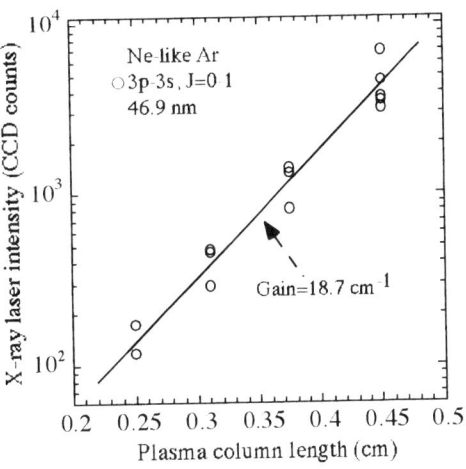

FIGURE 2. Neonlike argon $3p\ ^1S_0$ - $3s\ ^1P_1$ x-ray laser line intensity (open circles) versus length of the plasma column. A fitting (solid line) to the Linford formula[10] gives a gain coefficient of 18.7 ± 1.0 cm^{-1}.

Figure 3 shows the angular distribution of the neonlike argon $3p$ - $3s$ x-ray laser for a gas puff target of 0.45 cm. From Fig. 3, we can estimate the divergence of the x-ray laser beam to be less than 3.7 mrad. However, the angular profile is not so regular and the divergence for the central part is even less than 1 mrad. Tentatively we suppose that this is a result of modes in the amplifier. Also, this value is significantly smaller than that obtained with a combination of a long 600 ps prepulse followed by a 6 ps driving pulse, indicating a larger beam divergence of about 9 - 12 mrad[9]. The small beam divergence achieved in our experiment is mainly due to the 1.5 ps shorter pumping pulse resulting in a more transient gain in the narrower high-density gain region.

In the experiment, we also measured the ionization balance of the line-shaped argon plasma along the direction of 45^0 off the axis. Figure 4 shows the off-axis spectrum from a 0.45-cm-long gas puff target in the wavelength ranging from 1 nm to 6 nm. In Fig.4, one can clearly observe the very strong neonlike argon 3d-2p resonance lines as well as lines from higher ionization stages. The spectrum strongly indicates a nonequilibrium transient plasma in which the neonlike argon ion abundance may dominate during the ionization phase. Time-resolved spectral measurement is necessary to further clarify these processes.

To extend the scheme to shorter wavelength x-ray lasers, we performed an experiment for the nickellike xenon x-ray laser at ~10 nm with a combination

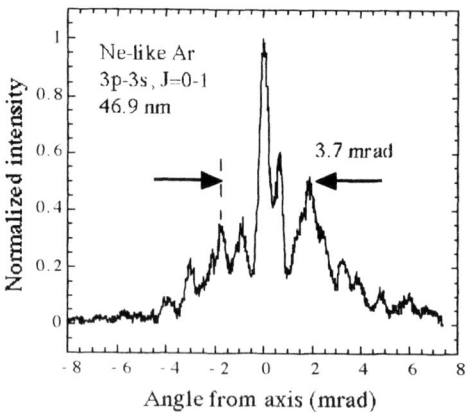

FIGURE 3. Angular distribution of the neonlike argon $3p\ {}^1S_0$ - $3s\ {}^1P_1$ x-ray laser for a 0.45-cm-long gas puff target indicating a beam divergence of less than 3.7 mrad.

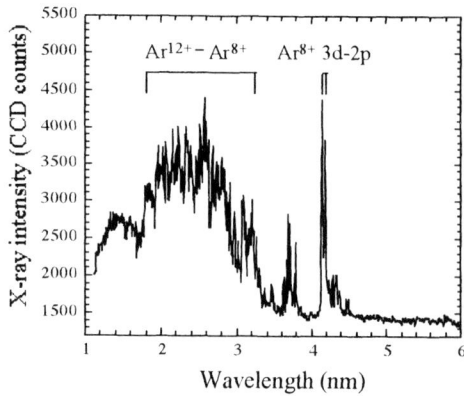

FIGURE 4. A typical 450 off-axis emission spectrum for a 0.45-cm-long argon gas puff target irradiated with a 1.2 J, 1.5 ps pulse followed by a 7.8 J, 1.5 ps pulse showing the strong neonlike argon 3d-2p resonance lines.

of a 600 ps prepulse followed by a 1.5 ps driving laser pulse, with a total energy of 22 J. A typical on-axis emission spectrum for the laser-irradiated xenon gas puff target was shown in Fig.5. The lasing line at ~10nm was clearly observed, however we failed to make it stronger for a gain measurement. We think that it is mainly due to a relative low gas density for the xenon x-ray laser, and the next experiment with a higher gas density over 10^{20} cm^{-3} is planned to obtain a high gain for the nickellike xenon TCE x-ray laser at 10 nm.

IV. Summary

In summary, we have observed large soft x-ray amplification in neonlike argon ions by irradiation of a gas puff target with two 1.5 ps driving laser pulses with ~9 J total energy in a traveling-wave geometry. Strong soft x-ray lasing on the transient collisionally excited $3p\ ^1S_0$ - $3s\ ^1P_1$ transition for neonlike argon at 46.9 nm was observed. A high gain of 18.7 cm^{-1} and a narrow beam divergence of less than 3.7 mrad were measured on the laser transition for targets up to 0.45 cm long. A weak nickellike xenon lasing line at ~10 nm was also observed with a combination of a 600 ps prepulse followed by a 1.5 ps driving laser pulse, with a total energy of 22 J.

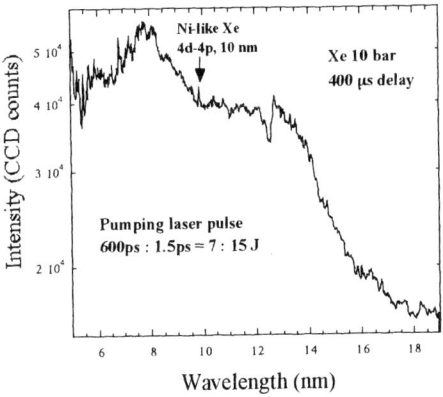

FIGURE 5 Typical on-axis emission spectra for a 0.45-cm-long xenon gas puff target irradiated with a 7 J, 600 ps pulse followed by a 15 J, 1.5 ps pulse.

ACKNOWLEDGMENTS

One of the authors (P. Lu) wishes to thank Dr. A. Nagashima for his hospitality, and Professor K. Koyama from the National Institute of Advanced Industrial Science and Technology for useful discussions.

REFERENCES

1. P. Nickles, V. N. Shlyaptsev, M. Kalachnikov, M. Schnurer, I. Will and W. Sandner, Phys. Rev. Lett. **78**, 2748-2751 (1997).
2. W. Sandner, V. N. Shlyaptsev, C. Danson, D. Neely, E. Wolfrum, J. Zhang, A. Behjat, A. Demir, G. J. Tallents, P. J. Warwick and C. L. S. Lewis, Phys. Rev. A **57**, 4778(1998).
3. J. Dunn, Y. Li, A. L. Osterheld, J. Nilsen, J. R. Hunter and S. V. Shlyaptsev, Phys. Rev. Lett. **84**, 4834(2000).
4. T. Kawachi, M. Kado, N. Hasegawa, M. Tanaka, K. Sukegawa, K. Takahashi, S. Namba, P. Lu, H.Tang, K. Nagashima, A. Sasaki, A. Nagashima, M. Koike, H. Daido and Y. Kato: Tech. Dig. CLEO/Pacific Rim 2001, Chiba, Japan, Vol.-I, pp.64-65.
5. M. Suzuki, S. Yamagami, K. Nagai, T. Norimatsu, K. Mima, Y. Murakami, T. Nakayama, W. Yu, H. Fiedorowicz, I. W. Choi and H. Daido: to be published in Proc. 2nd Int. Conf. Fusion Science & Applications, Sept. 10- 14, 2001, Kyoto, Japan.
6. H. Fiedorowicz, A. Bartnik, Y. Li, P. Lu and E. E. Fill, Phys. Rev. Lett. **76**, 415-418(1996).
7. T. Kawachi, M. Kado, M. Tanaka, N. Hasegawa, K. Nagashima, P. Lu, K. Takahashi, S. Namba, M. Koike, A. Nagashima and Y. Kato: to be published in Rev. Sci. Instrum.

8. H. Fiedorowicz, A. Bartnik, M. Szczurek, H. Daido, N. Sakaya, V. Kmetik, Y. Kato, M. Suzuki, M. Matsumura, J. Tajima, T. Nakayama and T. Wilhein: Opt. Commun. **163,** 103(1999).
9. H. Fiedorowicz, A. Bartnik, J. Dunn, R. F. Smith, J. Hunter, J. Nilsen, A. L. Osterheld and V. N. Shlyaptsev: Opt. Lett. **26,** 1403 (2001).
10. G. J. Linford, E. R. Peresini, W. R. Sooy and M. L. Spaeth: Appl. Opt. **13**, 379(1974).

Lattice dynamics of picosecond laser irradiated silicon crystal

H. Kishimura*, A. Yazaki, H. Kawano, Y. Hironaka, K. G. Nakamura and K. Kondo

*Materials and Structures Laboratory, Tokyo Institute of Technology,
4259 Nagatsuta, Midori, Yokohama 226-8503, Japan

Abstract. Direct observation of lattice dynamics of a 300 ps laser irradiated silicon crystal is performed by means of picosecond time-resolved X-ray diffraction. Change in X-ray diffraction profiles corresponds to propagation of a strain pulse inside the sample with sound velocity. The strain profiles are simulated by considering carrier dynamics and thermo-elastic treatment.

INTRODUCTION

X-ray diffraction is the most popular technique for determining crystal structure, because X-ray can interact with inner-shell electrons and are sensitive to atomic position. Ultrafast X-rays, generated by laser plasma with an intense femtosecond pulse, enable detection of structural changes as they undergo a phase transition, surface melting or recrystallization on picosecond time scale [1-6]. Real time observation using X-ray diffraction is needed to determine a transient crystal structure and transformation mechanisms under shock wave propagation. However, because of various experimental limitations, it is difficult to obtain quantitative X-ray diffraction in shocked solids [7-9]. Pump-probe technique with picosecond pulsed X-ray can apply to detect structural change under laser-induced shock compression, because the generated X-rays are completely synchronized with the driving laser. It is possible to understand transformation mechanism under shock compression in detail. In order to keep high-pressure state for a relatively long duration, long-pulse laser is required.

In this work, we studied the lattice dynamics of Si(111) under 300 ps pulsed laser irradiation by picosecond time-resolved X-ray diffraction. The picosecond pulsed X-rays are generated using laser plasma induced by femtosecond laser irradiation on metal target. Transient changes in lattice are directly observed.

EXPERIMENTAL

The laser system used was a table-top terawatt laser system (B.M. Industries, α10us). The laser pulses centered at 780 nm (E $\simeq$ 1.6 eV) are generated at 76 MHz by a Ti:sapphire laser oscillator (25 fs). The output is pulse stretched to 300 ps with energy of 400 mJ/pulse. This 300 ps pulsed beam is divided into two beams by a beam splitter (10

CP634, *Science of Superstrong Field Interactions*, edited by K. Nakajima and M. Deguchi
© 2002 American Institute of Physics 0-7354-0089-X/02/$19.00

%). One beam (10 %) is used for irradiation to Si through an optical delay line as a pump beam, and the other (90 %) is compressed to about 50 fs and used for X-ray generation. Pulsed X-rays (10 ps pulse width) are generated by focusing the 50 fs pulse on a Fe target in a vacuum chamber at laser power density of about 10^{17} W/cm^2 [10]. X-ray diffraction was performed with a characteristic $K\alpha$ line of Fe in a symmetrical Bragg diffraction geometry. The diffracted X-rays were recorded with an X-ray charge-coupled devices area image sensor. A 300 ps pump laser was irradiated to the surface of sample at a spot size of 1×4 mm^2. The sample used was a nondoped Si(111) wafer of 860 μm thickness. The sample was mounted on a motorized XY stage and moved for each laser-shot. In the experiment, two types of irradiation conditions were performed (Single shot and Multiple shots). In single shot experiment, the pump fluence was 1.0 mJ/cm^2 and XY stage moved without overlapping of the pump spot. On the other hand, in multiple shot experiment, the pump fluence was 1.2 mJ/cm^2 and laser irradiated 200 shots at approximately same spot. The diffracted signal was obtained by the accumulation of 600 data values. Details of the experimental setup are described elsewhere [5, 6].

RESULTS AND DISCUSSION

Single shot experiment

Figure 1(a) shows a typical example of the CCD image of diffracted X-rays from laser-irradiated Si(111) at a pump-probe delay time of 350 ps. The two lines are the diffracted $K\alpha_1$ (1.9360 Å) and $K\alpha_2$ (1.9399 Å) X-rays, whose diffraction angles (θ) are 17.98 and 18.02 degrees, respectively. The diffraction patterns broaden at the laser-irradiated area. Figure 1(b) shows the rocking curves from the laser-irradiated area (solid line) and the unirradiated area (dotted line). In symmetric Bragg diffraction configuration, the lattice strain can be estimated using the equation: $\Delta d/d = -\Delta\theta\cot\theta_B$, where d is the lattice spacing, $\Delta\theta$ the shift, and θ_B the Bragg angle. The rocking curve of the laser-irradiated Si(111) shows a lower-angle shifted component, which indicates the lattice expansion. The total intensity of diffraction from the irradiated area is twice the original value. This is explained by a transition from dynamical diffraction to kinematical diffraction due to an increase of strained layers.

In Fig. 2, the measured time-resolved diffraction profiles are shown as a function of the diffraction angle (rocking curves) and the time delay between X-ray and the pump pulses. Time delay $t = 0$ is defined as the time when lower-angle shifted component first appears. Time-resolved measurements were performed from -100 ps to 1000 ps at intervals of 50 ps. In the early delay time, both of the original $K\alpha$ lines broaden, while new lines appear. The new lines are broader and weaker than the original lines and deviate by approximately -0.08 degrees relative to the original Bragg angle. As the delay time increases, these new lines decrease in width, increase in intensity and merge with shifted main lines. The diffraction profile looks like a single peak at around 600 ps, and separates at longer delay time (> 800 ps). Finally the main lines deviate by approximately -0.02 degrees relative to the original Bragg angle at 1 ns.

The evolution of diffracted profiles directly depends on acoustic strain distribution.

The acoustic strain pulse is excited under laser irradiation and can be monitored by X-ray diffraction. In order to confirm a generation of strain pulse, model calculation which is described by rate and diffusion equation for carrier density N and lattice temperature T and a fluid equation was performed. Assuming that the problem is one dimensional, depth distribution of $N(x,t)$ and $T(x,t)$ depends on distance x from the surface and time t. Rocking curves were calculated using dynamical diffraction theory. The equations for N and T can be expressed as follows [11-14]:

$$\frac{\partial N}{\partial t} + \frac{N}{\tau_R} = D\frac{\partial^2 N}{\partial x^2} + \frac{\alpha(1-R)}{E}I(x,t) + \frac{\beta(1-R)^2}{2E}I(x,t)^2, \tag{1}$$

$$\frac{\partial T}{\partial t} = D_T\frac{\partial^2 T}{\partial x^2} + \frac{E_g N}{C_v \tau_R} + \left(\frac{E-E_g}{C_v}\right)\left[\left(\frac{\alpha(1-R)}{E} + \Theta N\right) + \frac{\beta(1-R)^2}{2E}I(x,t)\right]I(x,t), \tag{2}$$

$$\frac{\partial}{\partial t}I(x,t) = -[\alpha + \Theta N + \beta I(x,t)]I(x,t), \tag{3}$$

where the Auger recombination life time τ_R is represented by

$$\tau_R = \frac{1}{\gamma N^2}. \tag{4}$$

Here, D and D_T are the ambipolar and thermal diffusion coefficients, α the linear absorption coefficient, β the two-photon absorption coefficient, Θ the free-carrier absorption cross-section, R the reflectivity, $I(x,t)$ the space and time dependent intensity of the laser pulse, C_v the specific heat per unit volume and γ the Auger coefficient [15-18]. In these equation, the absorption coefficient and the specific heat are treated as a function of T and approximated by the fifth-order polynomial, fitted to the experimental data [16-18]. The equation of motion for elasticity is expressed by electronic volume effect,

$$\rho\frac{\partial^2 u}{\partial t^2} = \rho v^2\frac{\partial^2 u}{\partial x^2} + B\frac{\partial E_g}{\partial P}\frac{\partial N}{\partial x} + 3Bl\frac{\partial T}{\partial x}, \tag{5}$$

where u is the displacement, ρ the density, v the sound velocity, B the bulk modulus, l the linear thermal expansion coefficient, which depends on lattice temperature, and $\partial E_g/\partial P$ the electrical contribution for Si(111) [19, 20]. The strain can be obtained by the solution of Eq. (5) for u.

These equations were solved numerically. The laser intensity profile $I(x,t)$ is treated as Gaussian (FWHM 300 ps) centered at $t=0$. In the calculation, the following values were used[14, 21, 22]: $D = 18$ cm^2/s, $\beta = 2$ cm/GW, $\gamma = 4 \times 10^{-31}$ cm^6/s, $\partial E_g/\partial P = -2.4 \times 10^{-24}$ cm^3, $\rho = 2.3$ g cm^{-3}, $v = 9.4 \times 10^3$ m/s. The reflectivity R is set to be 0.35 for an initial value, 0.72 for $T > T_m$, where T_m is the melting temperature (1685 K) [23]. The reflectivity of $e-h$ plasma is described with the help of the Drude model. Numerically calculations were done at time between -450 ps to 1150 ps with a time step of 20 fs.

Figure 3 shows the strain profiles as a function of depth at various delay times on a gray-scale map. Significant lattice expansion is observed near the surface and small

(a)

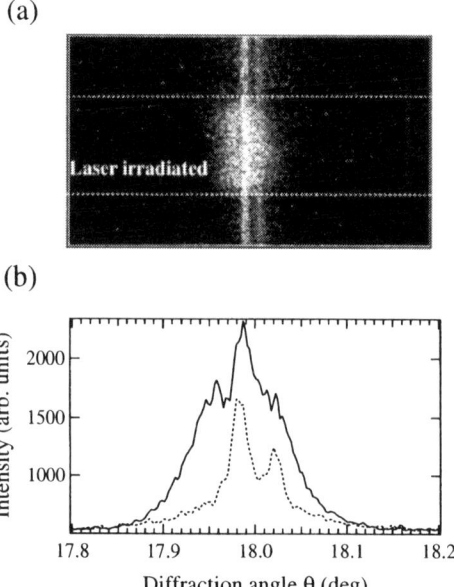

(b)

FIGURE 1. A typical example of the CCD image of diffracted X-rays from 300 ps laser irradiated Si(111) at a pump-probe delay time of 350 ps (a) and the rocking curves of X-ray diffraction (b). The solid curve is from the laser-irradiated area and the dashed curve is from unirradiated area.

compression is observed ahead of the expansion. The propagating strain pulse (acoustic phonon pulse) is observed. The speed of strain pulse propagation is 9.4×10^3 ms^{-1}, which is the sound velocity of Si(111).

The calculated rocking curves are shown as a function of diffraction angle and time in Fig. 4. The new lines appear and deviate by approximately -0.08 degrees relative to the original Bragg angle at 400 ps. As the delay time is increased, these new lines decrease in width and shifts to higher angle. Finally the main lines deviate by approximately -0.01 degrees relative to the original Bragg angle. The computational results agree approximately with the experimental data. In the calculation, gradually changes at lower-angle component can be explained.

Multiple shot experiment

The results of X-ray diffraction from Si(111) under multiple laser irradiation displayed in Fig. 5. Time-resolved measurements were performed from 0 ps to 950 ps at intervals of 50 ps. At early times, the intensities of rocking curves decreased and a new diffraction component at higher angles was observed. At later times, shifted component gradually decreased in width. The observed shift indicates lattice compression in contrast with the result of single shot experiment, although power density of both ex-

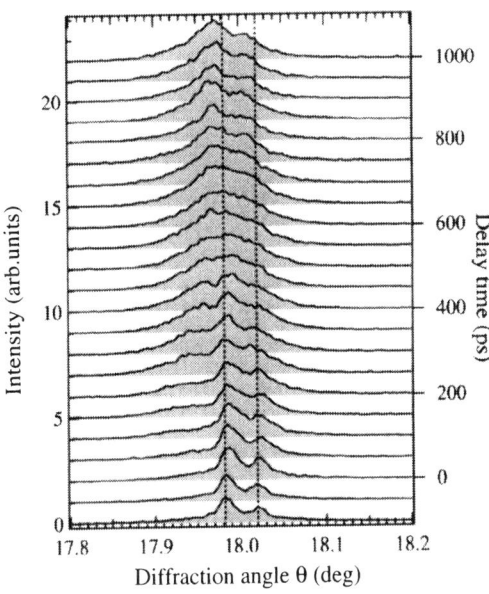

FIGURE 2. Rocking curves for 300 ps laser irradiated Si(111) for various delay time. Dashed lines indicate Bragg diffraction angle for Fe $K\alpha_1$ and Fe $K\alpha_2$ lines from the pristine Si(111).

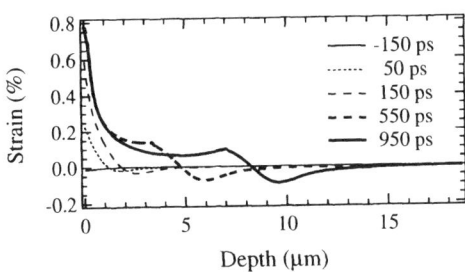

FIGURE 3. Calculated strain profiles at several delay time. Positive and negative values correspond to lattice expansion and compression, respectively.

periment were almost same. The maximum change of lattice spacing was obtained to be −0.9%. Because the lattice compression is −2.6% at approximately 5.4 GPa [24], the maximum pressure is estimated to be 1.9 GPa. We suggest that the present phenomena are shock compression due to reaction of laser ablation. It is considered that surface of irradiated area is damaged and optical properties are changed during multiple laser shot. We believe that the laser ablation occurs under multiple laser shot and the momentum transfer due to the laser ablation causes lattice compression. The time evolution of the rocking curves observes the shock wave propagation directly.

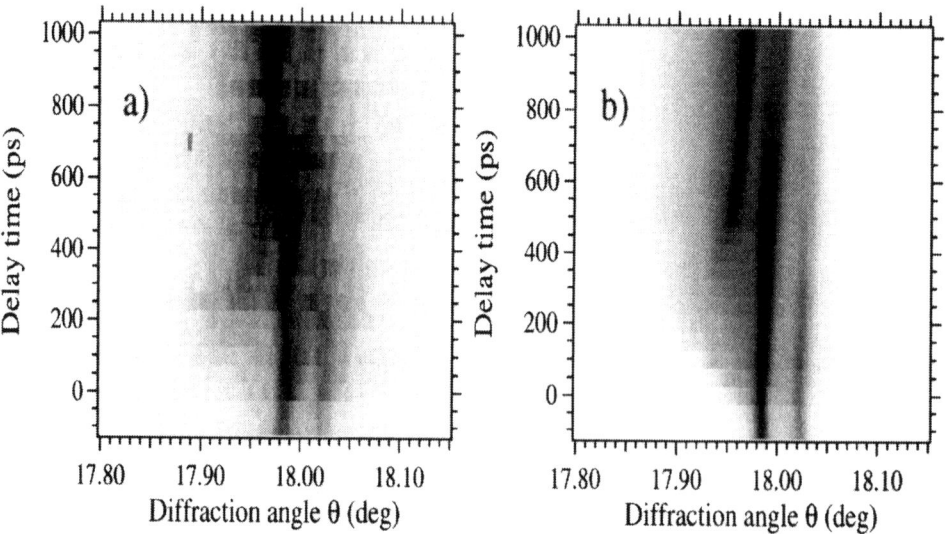

FIGURE 4. Time-resolved X-ray rocking curves for 300 ps laser irradiated Si(111) as a function of diffracted angle and of pump-prove time delay: (a) experimentally measured (b) calculated by thermoelastic treatment.

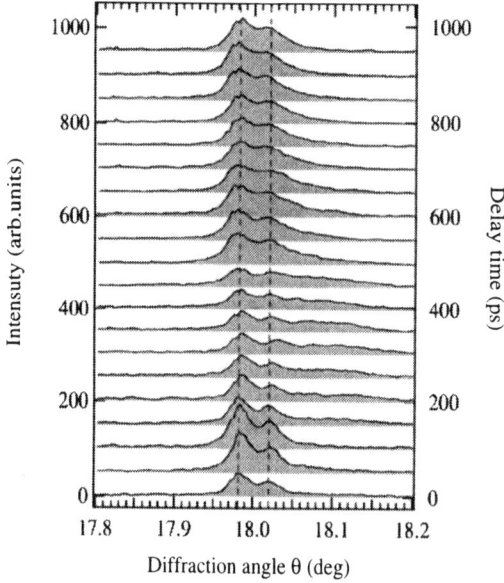

FIGURE 5. Rocking curves for Si(111) under multiple laser irradiations for various delay time. Dashed lines indicate Bragg diffraction angle for Fe $K\alpha_1$ and Fe $K\alpha_2$ lines from the pristine Si(111).

CONCLUSION

We have directly observed structural dynamics on picosecond time scale in 300 ps laser excited Si(111) crystal. The evolution of lattice motion is observed by ultrafast X-ray diffraction at a time step of 50 ps. In single shot experiment, lattice expansion was observed. Numerical simulation on carrier density and lattice temperature is performed using rate and diffusion equation under laser irradiation. Calculated rocking curves using the strain profiles represented the experimental data. On the other hand, transient lattice compression under multiple laser irradiations was observed and maximum pressure of inside silicon was 1.9 GPa. Shock-compressed state was achieved by 300 ps laser irradiation. Structuralchange in elastic-plastic transition and structural phase transition can be studied by using intense laser irradiation.

ACKNOWLEDGMENTS

The authors thank M. Hasegawa for his help in constructing the experimental setup.

REFERENCES

1. C. Rose-Petruck, R. Jimenez, T. Guo, A. Cavalleri, C. W. Siders, F. Raksi, J. A. Squier, B. C. Walker, K. R. Wilson, and C. P. J. Barty, Nature **398**, 310 (1999).
2. A. H. Chin, R. W. Schoenlein, T. E. Glover, P. Balling, W. P. Leemans, and C. V. Shank, Phys. Rev. Lett. **83**, 336 (1999).
3. D. A. Reis, M. F. DeCamp, P. H. Bucksbaum, R. Clarke, E. Dufresne, M. Hertlein, R. Merlin, R. Falcone, H. Kapteyn, M. M. Murnane, J. Larsson, Th. Missalla, and J. S. Wark, Phys. Rev. Lett. **86** 3072 (2001).
4. Y. Hironaka, T. Tange, T. Inoue, Y. Fujimoto, K. G. Nakamura, K. Kondo, and M. Yoshida, Jpn. J. Appl. Phys. **38**, 4950 (1999).
5. Y. Hironaka, A. Yazaki, F. Saito, K. G. Nakamura, K. Kondo, H. Takenaka, and M. Yoshida, Appl. Phys. Lett. **77**, 1967 (2000).
6. Y. Hironaka, A. Yazaki, F. Saito, K. G. Nakamura, and K. Kondo, Jpn. J. Appl. Phys. **39**, L984 (2000).
7. Q.Johnson, A. Mitchell, and L. Evans, Nature **231**, 310 (1971).
8. J. S. Wark, R. R. Whitlock, A. A. Hauer, J. E. Swain, and P. J. Solone, Phys. Rev. B **40**, 5705 (1989)
9. R. R. Whitlock and J. S. Wark, Phys. Rev. B **52**, 8 (1995).
10. M. Yoshida, Y. Fujimoto, Y. Hironaka, K. G. Nakamura, K. Kondo, M. Ohtani, and H. Tsunemi, Appl. Phys. Lett. **73**, 2393 (1998).
11. E. J. Yoffa, Phys. Rev. B **21**, 2415 (1980).
12. H. M. van Driel and Phys. Rev. B **35**, 8166 (1987).
13. C. Thomsen, H. T. Grahn, H. J. Maris, and J. Tauc, Phys. Rev. B **34**, 4129 (1986).
14. O. B. Wright and V. E. Gusev, Appl. Phys. Lett. **66**, 1190 (1995).
15. H. R. Shanks, P. D. Maycock, P. H. Sidles, and G. C. Danielson, Phys. Rev. **130**, 1743 (1963).
16. H. A. Weaklien and D. Redfield, J. Appl. Phys. **50**, 1491 (1979).
17. G. E. Jellison and F. A. Modine, J. Appl. Phys. **76**, 3758 (1994); Appl. Phys. Lett. **41**, 180 (1982).
18. P. D. Desai, J. Phys. Chem. Ref. Data, **15**, 967 (1986).
19. S. P. Nikanorov, Yu. A. Burenkov, and A. V. Stepanov, Sov. Phys. Solid State **13**, 2516 (1971).
20. Y. Okada and Y. Tokumaru, J. Appl. Phys. **56**, 314 (1984).
21. J. Dziewior and W. Schmid, Appl. Phys. Lett. **31**, 346 (1977).
22. M. Murayama and T. Nakayama, Phys. Rev. B **52**, 4986 (1995).
23. M. O. Lampert, J. M. Koebel, and P. Siffert, J. Appl. Phys. **52**, 4975 (1981).
24. W. H. Gust and E. B. Royce, J. Appl. Phys. **42**, 1897 (1971).

X-ray and charged particle emission from metal targets in femtosecond laser field of 10^{17} W/cm^2

Yasuaki Okano, Yoichiro Hironaka, Kazutaka G. Nakamura, and Ken-ichi Kondo

Materials and Structures Laboratory, Tokyo Institute of Technology
4259 Nagatsuta, Midori, Yokohama 226-8503, Japan

Abstract. We investigated x-ray and charged particle emissions from metal target (mainly iron target) irradiated by a femtosecond laser-pulse (52 fs, 790 nm) at power density up to 10^{17} W/cm^2. Using a time-of-flight method, kinetic energy of positive ions measured by means of a charge collector. Ion velocity distribution showed two features, which were ion-velocity cutoff and a low-velocity peak. The ion-velocity cutoff showed a strongly dependence on the incident laser energy, and the cutoff energy was comparable with ponderomotive potential in the energy of several keV. The low-velocity ion component was corresponding to ions ablated from a deep-hole (20-mm depth), and the ion energies were estimated to be a few hundred eV.

INTRODUCTION

High-energy x-rays, ions and electrons are generated via laser-matter interactions in an intense optical field above 10^{15} W/cm^2 [1]. For x-rays produced by femtosecond laser irradiation on metal targets, temporal profiles and energy distributions in keV ranges have been measured at power density of 10^{17} W/cm^2 [2,3]. These x-rays have high fluxes and extremely short pulse duration (less than tens of ps), and they have been used for time-resolved studies of molecular and crystalline dynamics [4,5]. On the other hand, there are few investigations of ions or electrons in this laser-power regime. In previous studies at relatively lower laser-power region (under 10^{15} W/cm^2), ion energy distributions of femtosecond laser plasma have been investigated [6,7]. For metal targets (Au, Cu and Al), two different ion components, consisting of high-energy (several keV) nonthermal ions and low-energy (tens of eV) ions occurring as a consequence of a thermal process following electron-lattice relaxation, have been observed at power density up to 10^{13} W/cm^2 [6].

Figure 1 shows typical energy spectra of x-rays emitted from several metal-targets by 50-fs laser irradiation at power density above 10^{16} W/cm^2. They are obtained using a direct-detection x-ray CCD camera, and characteristic x-ray emission is observed. In this work, we have studied ion and electron emission from an iron target produced by femtosecond laser irradiation (52 fs, 790 nm) where characteristic x-ray emission takes place, and surface morphology of the laser-irradiated target at power densities between 10^{16} and 10^{17} W/cm^2.

CP634, *Science of Superstrong Field Interactions*, edited by K. Nakajima and M. Deguchi
© 2002 American Institute of Physics 0-7354-0089-X/02/$19.00

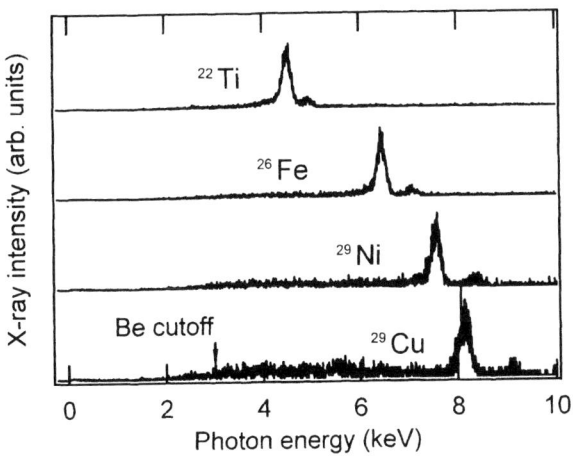

FIGURE 1. X-ray energy spectra emitted from metal-targets by 50-fs laser irradiation at power density above 10^{16} W/cm^2.

EXPERIMENTAL

The laser used in the present experiments is a tabletop terawatt laser system (B.M.industries, α10) consisting of a Ti:sapphire oscillator and amplifiers based on a chirped pulse amplification technique [2]. The amplified laser beam has a pulse duration of 52 fs (FWHM), a maximum energy of 200 mJ/pulse at a fundamental wavelength of 790 nm, and a repetition rate of 10 Hz. The pulse duration is measured by means of a single-shot autocorrelator. The contrast ratio of the main pulse to the prepulse that precedes it by 8 ns is greater than 10^6:1. In a vacuum chamber (10^{-6} Torr), the femtosecond pulse is focused on an iron target by an off-axis parabola at an incident angle of 60° with respect to the surface normal in p-polarization. A diameter of focal spot is approximately 44 µm [2]. The shape of the target is a disk of 70-mm diameter and 5-mm thick. In order to expose a fresh surface, the target is rotated and translated for each laser shot.

Charged particles from laser-produced plasma are detected by means of a charge collector consisted of an electrostatic probe in the shape of a Faraday cup. This collector is one of the simplest diagnostic tools for ions of laser plasma. The kinetic energy of positive ions is measured using a time-of-flight (TOF) method. The collector is located at a distance of 415 mm from the focal spot on the target surface at an angle of approximately 10° with respect to the surface normal. The entrance aperture in front of the collector is a hole of 3.2-mm diameter, with a resulting solid angle of 5.2×10^{-5} sr. To provide the zero-time, laser light reflected from the target surface is detected by a fast photodiode. Ion signal is amplified up to 30 dB with a high-bandwidth (350 MHz) DC voltage amplifier (Analog Modules, 353A), and is recorded by a digital-storage-oscilloscope (Tektronix, TDS684B).

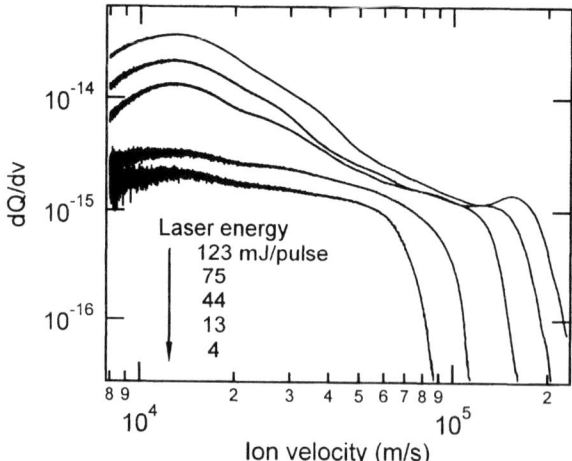

FIGURE 2. Ion velocity distributions from an iron target at laser energies between 4 and 123 mJ/pulse.

The energy spectrum of electrons is measured with a magnetic spectrometer. The spectrometer is located at specular direction for incident laser pulse. In addition, hard x-rays up to 30 keV including Kα emission are also monitored using x-ray silicon photodiodes (International Radiation Detectors, AXUV-20HE1).

RUSULTS AND DISCUSSION

Figure 2 shows typical velocity distributions for different laser energies between 4 and 123 mJ/pulse (laser power densities between 10^{16} and 10^{17} W/cm^2). Each plot is an average for 500 laser shots. They are obtained from TOF spectra of ions by the relation of

$$\frac{dQ}{dv} = \frac{t^2}{l}\frac{dQ}{dt},$$ (1)

where dQ/dv is velocity distribution of ion charge, t and l are flight time and drift distance, and dQ/dt is the time distribution of ion charge, respectively. The velocity distribution of ions shows two features, which are ion-velocity cutoff and a low-velocity peak. The velocity of the cutoff is strongly dependent on the incident laser energy and they increase in the velocity range between 6×10^4 and 2×10^5 m/s, which corresponds to the energy range between 1 and 10 keV of iron ions. As the laser energy increase, the ion charge yield of the low-velocity peak around 10^4 m/s is rapidly increase above laser energy of 44 mJ/pulse. Although the hydrogen ions emitted from adsorbates contribute to higher velocity part, these yields are negligible small compared to the total ion yield. X-ray intensity, which measured simultaneously with ion detection, increases with increasing laser energy in the present power density,

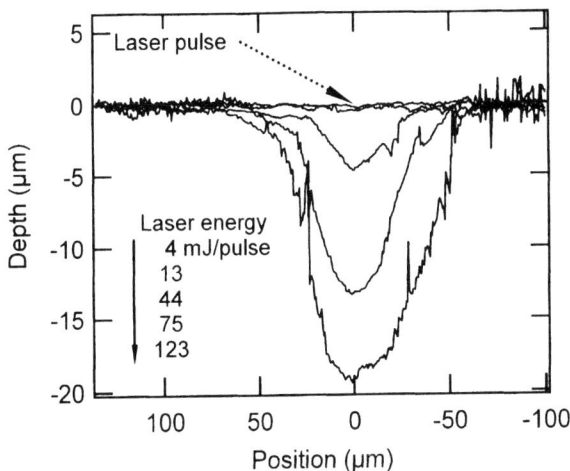

FIGURE 3. Cross-section profiles of laser-irradiated surface of an iron target at laser energies between 4 and 123 mJ/pulse.

and an average x-ray intensity at laser energy of 123 mJ/pulse is about three times higher than that of 44 mJ/pulse. Electron energy distribution is also measured at the laser energy of 170 mJ/pulse (power density of approximately 2×10^{17} W/cm^2), and the effective electron temperature is obtained more than 20 keV but the data can not be expressed by single Boltzmann-distribution.

In intense laser field, pre-plasma can be formed due to the prepulse or the envelope of the Gaussian profile of the laser beam. The laser energy is also absorbed by the electrons not only in skin-depth region on a target surface but also in this plasma. In optical fields of the order of 10^{15} W/cm^2 or greater, both ion and free electrons are fundamentally affected by ponderomotive force inherent in the strong laser light [11]. Electrons are oscillated strongly in the field of the light wave and accelerated by ponderomotive potential [12]. Ion acceleration occurs by energy transfer from the electron to ion subsystem, which occurs due to the locally induced space-charge separation fields and electron-ion collisions [8]. Ion acceleration is considered to be due to electric fields, which is caused by energetic electrons ejected from the target surface or the pre-plasma. The ponderomotive potential is described by $2\pi e^2 I (m_e c \omega^2)^{-1}$, where ω is the frequency and I is the laser-power density. The calculated ponderomotive potential is increasing linearly for incident laser power density and comparable with the ion-cutoff energy of several keV.

We have also investigated the target surface by measuring profiles of ablated holes with a laser displacement meter (KEYENCE, VF-7500). Figure 3 shows typical cross-section profiles of the ablated hole at different power densities. At incident laser energy above 44 mJ/pulse, a deep hole of the order of μm is observed and the ion charge yield of the low-velocity peak is also increasing rapidly. Therefore, low-velocity ions are corresponding to ablation, which makes these deep holes, and high-velocity ions correspond to ablation on the top surface within optical skin-depth.

These deep holes are considered to be produced as a result of indirect interactions of an intense laser pulse with a metal target, because absorption length of metal target for infrared light is limited by optical skin depth (ordinary several nm). For higher intensities enough energy is located below 1/e value extending more to the bulk. However, the laser energy at 0.1 μm is negligible small $(1/e)^{10}$ and the depth is much smaller than the hole. These deep holes have been observed in 200 fs-laser ablation of a steel plate at power density up to 5×10^{16} W/cm^2 [8]. In an interaction of an intense laser pulse with a metal target [9], the laser energy is first absorbed by free electrons through inverse bremsstrahlung. The absorption is then followed by fast electron relaxation within the electron subsystem, electron thermal diffusion, and finally, energy transfer to the lattice through electron-phonon coupling. At laser intensity of close to the ablation threshold of 10^{13} W/cm^2, two different ablation regimes in femtosecond laser ablation of metal targets have been reported by Furusawa et al. [10] and Amoruso et al. [6]. Figure 4 shows typical hole images for 44-mJ laser-pulse irradiation, and they are obtained by a field-emission scanning electron microscope (FE-SEM: Hitachi, S-4500). Melting of the target is inferred from the structure in Fig. 4 (b) and this indicates that the energy relaxation process, which mentioned above, occurred after laser irradiation. In addition, characteristic structures are observed in

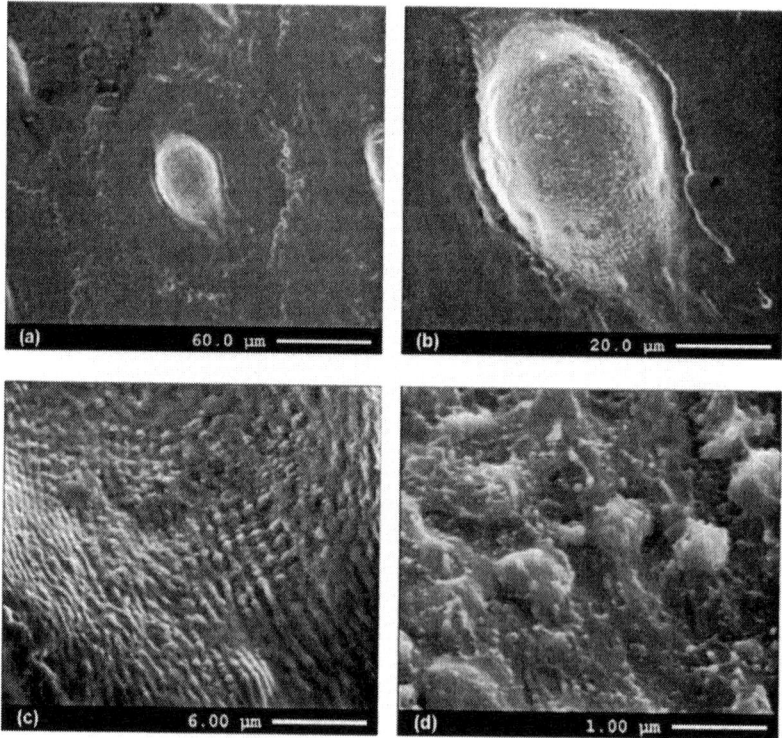

FIGURE 4. SEM images of an ablated hole on an iron target irradiated by 75-mJ-laser pulse.

258

Fig. 4 (c) and (d), however they are not suggested yet. The difference between the two regimes is attributed to the thermal-diffusion-dominant process and the skin-depth-dominant process. In addition to these processes, photon pressure [12] and ablation pressure [13] (up to several TPa), which cause shock waves into target, can be generated at the present power density region. The low-energy ion emission is considered to occur as a result of both the energy relaxation process and these mechanical processes.

ACKNOWLEDGMENTS

The authors are grateful to M. Hasegawa for constructing the experimental setup, and K. Nemoto, Y. Oishi, and T. Nayuki for valuable discussions.

REFERENCES

1. J. D. Kmetec, C. L. Gordon, III, J. J. Macklin, B. E. Lemoff, G. S. Brown, and S. E. Harris, *Phys. Rev. Lett.* **68**, 1527 (1992).
2. M. Yoshida, Y. Fujimoto, Y. Hironaka, K. G. Nakamura, K. Kondo, M. Ohtani, and H. Tsunemi, *Appl. Phys. Lett.* **73**, 2393 (1998).
3. Y. Fujimoto, Y. Hironaka, K. G. Nakamura, K. Kondo, M. Yoshida, M. Ohtani, and H. Tsunemi, *Jpn. J. Appl. Phys.* **38**, 6754 (1999).
4. C. Rose-Petruck, R. Jimenez, T. Guo, A. Cavalleri, C. W. Siders, F. Raksi, J. A. Squier, B. C. Walker, K. R. Wilson, and C. P. J. Barty, *Nature* **398**, 310 (1999).
5. Y. Hironaka, A. Yazaki, F. Saito, K. G. Nakamura, K. Kondo, H. Takenaka, and M. Yoshida, *Appl. Phys. Lett.* **77**, 1967 (2000).
6. S. Amoruso, X. Wang, C. Altucci, C. de Lisio, M. Armenante, R. Bruzzese, and R. Velotta, *Appl. Phys. Lett.* **77**, 3728 (2000).
7. P. A. VanRompay, M. Nantel, and P. P. Pronko, *Appl. Surf. Sci.* **127-129**, 1023 (1998).
8. C. Momma, B. N. Chichkov, S. Nolte, F. Alvensleben, A. Tunnermann, H. Welling, and B. Wellegehausen, *Opt. Commun.* **129**. 134 (1996).
9. S. Nolte, C. Momma, H. Jacobs, A. Tunnermann, B. N. Chichkov, B. Wellegehausen, and H. Welling, *J. Opt. Soc. Am. B* **14**, 2716 (1997).
10. K. Furusawa, K. Takahashi, H. Kumagai, K. Midorikawa, and M. Obara, *Appl. Phys. A* **69**, S359 (1999).
11. P. H. Bucksbaum, R. R. Freeman, M. Bashkansky, and T. J. McIlrath, *J. Opt. Soc. Am. B* **4**, 760 (1987).
12. S. C. Wilks, W. L. Kruer, M. Tabak, and A. B. Langdon, *Phys. Rev. Lett.* **69**, 1383 (1992).
13. A. Benuzzi, T. Lower, M. Koenig, B. Faral, D. Batani, D. Beretta, C. Danson, and D. Pepler, *Phys. Rev. E* **54**, 2162 (1996).

PHOTON ENERGY CONVERSION
OF IR FEMTOSECOND LASER PULSES INTO X-RAY PULSES
USING ELECTROLYTE AQUEOUS SOLUTIONS IN AIR

Koji Hatanaka[*], Toshifumi Miura, Hiroshi Ono, and Hiroshi Fukumura[*]

Department of Chemistry, Graduate School of Science,
Tohoku University, Sendai 980-8578, Japan

Abstract. Hard x-ray pulses were generated from aqueous solutions of electrolyte such as CsClaq and RbClaq by the irradiation of femtosecond infrared laser pulses in air. X-ray photon energy extended to 60 keV with the laser intensity of 0.6 mJ/pulse. Possible conversion mechanisms were discussed based on the results of x-ray emission spectra on laser intensity and solute concentration. A development of x-ray diffractometer utilizing a palm-top x-ray pulse source was also introduced.

INTRODUCTION

There is now a growing interest in interactions between intense laser fields and matters since they involve unsolved questions in physics and chemistry such as x-ray pulse generation. Most of studies, however, have chosen metals and gases as targets for the subject [1,2]. Reports on solutions, on the other hand, have been limited to a few such as carbon fluoride [3], copper nitrate aqueous solution [4], and others. We consider that a solution can be an appropriate target since mechanisms of the x-ray generation can be discussed by changing solute compounds, their combinations, and their concentrations, which is different from the case of metal targets. For practical applications, solution targets can be circulated by a pump and recycled, so that high stability is expected. Furthermore, x-ray pulse width may be controllable by changing solute concentrations since the lifetime of high energy electrons might be short in highly-concentrated solutions [5]. In this proceeding, x-ray pulse generation from electrolyte aqueous solutions irradiated by focused fs laser pulses in air is described. Additionally, a development of x-ray diffractometer with a palm-top x-ray pulse source is also introduced.

---

*e-mails:hatanaka@orgphys.chem.tohoku.ac.jp,fukumura@orgphys.chem.tohoku.ac.jp

CP634, *Science of Superstrong Field Interactions*, edited by K. Nakajima and M. Deguchi
© 2002 American Institute of Physics 0-7354-0089-X/02/$19.00

X-RAY PULSE GENERATION FROM AQUEOUS SOLUTIONS

Experimental

Figure 1 shows a top view of the experimental setup [6-8]. Distilled water or a highly concentrated alkali metal (Cs or Rb) chloride (> 98 %, Aldrich) aqueous solution was circulated by a pump through a flat glass nozzle of which the inner gap was about 100 μm. Those solutions are transparent to the excitation laser light at 775 nm. Refractive indexes of the solutions range from 1.333 to 1.383 depending on solute concentration and alkali metal species. Femtosecond laser pulses (Clark MXR, CPA2001, 775 nm, 130 fs, 1 kHz) were focused by an objective lens (Mitsutoyo, M Plan Apo 10X, NA = 0.28) onto a solution jet surface from the nozzle with an incident angle ~ 45 degrees to the jet surface (S polarization). Laser focus waist was estimated to be ~ 17 μm from the side view of the plasma, which is shown in Figure 1 (b). Therefore, the laser power at the focus can be calculated to be ~ 2PW/cm^2 when the laser intensity is 0.6 mJ/cm^2. X-ray emission spectra were measured with a high purity Ge solid state detector (Ge SSD, EG & G Ortec, GLP-25440-S) with a 250-μm-thick Be window. Signals from the Ge SSD were processed by a multichannel analyzer (MCA/PC98B, Laboratory Equipment Corp.). All experiments were performed under atmospheric pressure at 294 K.

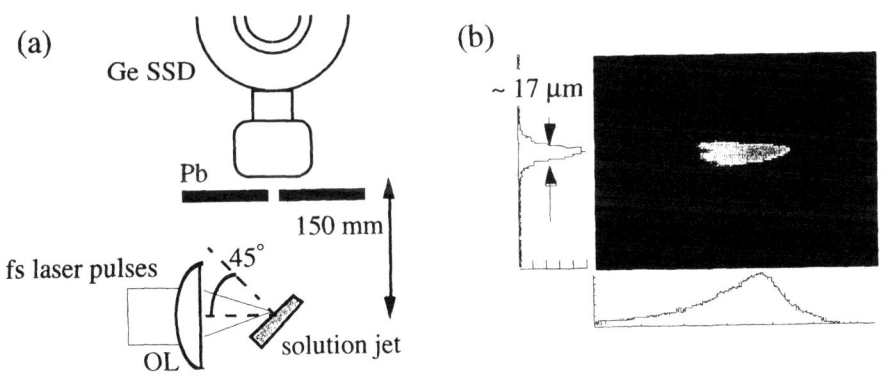

FIGURE 1. (a) A top view of experimental setup for x-ray pulse generation with solution jets and x-ray emission spectroscopy. OL: an objective lens (NA = 0.28). Pb: a lead plate (1 mm thick) with an aperture (1 mm diameter). (b) A visible side image of laser focus.

Figure 2 shows x-ray emission spectra from jets of distilled water, a CsCl aqueous solution (6.5 mol/dm^3), and a RbCl aqueous solution (6.0 mol/dm^3). The excitation laser intensity was 0.58 mJ/pulse. The spectra were uncorrected by x-ray absorption of air, the Be window, and the Ge absorbing layer (300 nm). X-ray emission intensities were normalized with their maxima. In the case of distilled water, a broad spectrum was observed with a tail up to ~ 15 keV. The intensity degradation in the lower energy region was due to the absorption effect. In the case of a CsCl aqueous solution, a similar broad spectrum with a gentler slope was observed up to ~ 40 keV. Sharp x-ray lines were also observed clearly which were assigned to $K\alpha$ (30.968 keV), $K\beta$ (34.960 keV), and $L\beta$ (4.619 keV, 4.935 keV) characteristic x-ray lines of Cs. The observation of these high-energy K lines certifies that the measurement condition is free from the pile-up effect, which is characteristic to solid state detectors. Similarly, in the case of a RbCl aqueous solution, a broad spectrum was observed with K characteristic x-ray lines of Rb (13.373 keV and 14.956 keV). Characteristic x-ray lines of Cl ($K\alpha$ = 2.6 keV, $K\beta$ = 2.8 keV) are out of the detection range. The depression observed commonly in all the spectra at ~ 11 keV are due to the Ge K absorption edge (11.1 keV). Energy conversion efficiency of laser pulse to x-ray pulse in the range 3-60 keV, in the case of a CsCl solution (6.5 mol/dm^3), was calculated to be ~ 10^{-8} under the assumption that x-ray radiation was spherically-homogeneous.

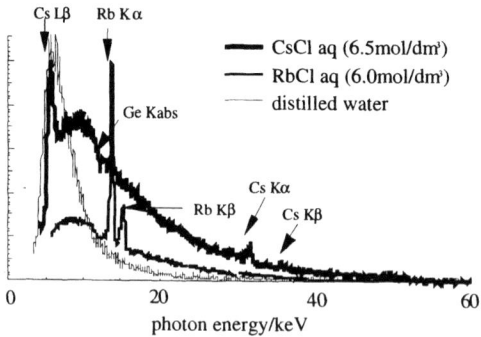

FIGURE 2. X-ray emission spectra of distilled water and alkali metal (Cs or Rb) chloride aqueous solution irradiated by focused femtosecond laser pulses (775 nm,. 130 fs, 0.58 mJ/pulse, 1 kHz). X-ray emission intensities are normalized by their maxima.

Figure 3 shows laser intensity-dependent x-ray emission spectra of a CsCl solution (6.5 mol/dm³) with a spectrum of distilled water irradiated by 0.58 mJ/pulse laser pulses. The spectra here are corrected by considering the absorption effect of air, the Be window, and the Ge absorbing layer. As the laser intensity increased, x-ray emission intensity increased, the slopes of broad spectral components became less steep, and the x-ray cut-off energy extended to higher energy region. The slopes can be analyzed quantitatively by using an equation, $I_X(E) = \exp(-E/T) \times const.$, where I_x, E, and T represents x-ray intensity, x-ray energy, and the slope parameter, respectively.

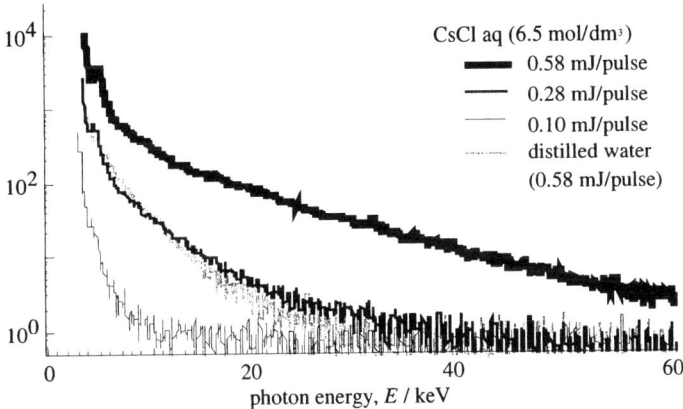

FIGURE 3. X-ray emission spectra of CsCl aqueous solution (6.5 mol/dm³) with different laser intensities and an x-ray emission spectrum of distilled water with 0.58 mJ/pulse laser intensity. The spectra are corrected by absorption effect of air, Be input window of Ge SSD, and Ge absorbing layer.

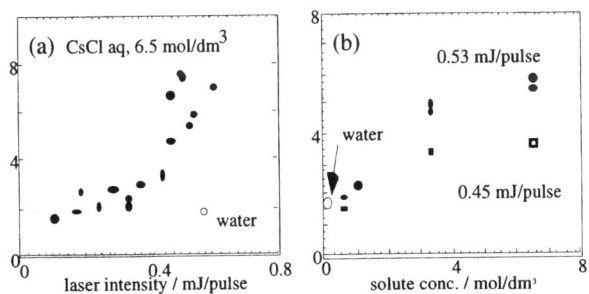

FIGURE 4. Electron temperatures as functions of excitation laser intensity (a) and solute concentration (b). Open circle represents distilled water irradiated by 0.53 mJ/pulse fs laser pulses.

Figure 4 shows the slope parameter, T, as functions of laser intensity (a) and solute concentration (b). The value of T increased gradually as the laser intensity increases. When the laser intensity exceeds ~ 0.4 mJ/pulse, the increasing slope changes suddenly to a steeper slope. On the other hand, the value of T saturates when the solute concentration increases. Furthermore, when the solute concentration is ~ 1mol/dm^3, T is almost the same with distilled water though the irradiating laser intensity is the same.

The initial ionization is induced mainly through tunneling ionization under 2 PW/cm^2 irradiation condition from the theory by Keldysh [9]. After ionized, conductive electrons are accelerated by the intense laser field through mechanisms such as inverse bremsstrahlung, stimulated Raman scattering, ponderomotive potential [10]. In this study, however, the ponderomotive potential of laser field is calculated to be only 110 eV, which is much lower than the electron temperature obtained from the slope in Figure 4. Thus we have to invoke another mechanisms such as inverse bremsstrahlung and/or stimulated Raman scattering to explain the observed high electron temperature. The sudden rise of T (Figure 4 (a)) can be due to stimulated Raman scattering and/or multiple ionization of secondary electrons. The confirmation of the mechanism details is now under consideration.

Broad X-ray emission spectra can be a result of recombination between electron and ionic species and/or bremsstrahlung. Comparing the X-ray emission spectrum of CsCl aqueous solution with that of distilled water (Figure 3), we found that X-ray intensity was higher in CsCl solution than in distilled water. This X-ray intensity enhancement by adding electrolyte with high atomic number elements is reasonable because recombination or scattering cross section is much larger in Cs$^+$ or Cl$^-$ than in H$_2$O. On the other hand, such effect of electrolyte can be observable only when the electrolyte concentration is more than 1 mol/dm^3, which is much dense from the conventional chemistry sense (The distance between Cs ions is about 1 nm when the concentration is 1 mol/dm^3.). Solute concentration may be effective to all the mechanisms such as ionization, conductive electron acceleration, and scattering and recombination for x-ray generation. Although it is difficult to clarify the mechanism on solute concentration effect at present, this is the first demonstration of solute concentration effect on x-ray pulse generation.

X-RAY DIFFRACTION WITH X-RAY PULSES

Experimental

Figure 5 (a) shows an experimental setup for x-ray diffraction with our x-ray pulse source. IR femtosecond laser pulses (775 nm, 130 fs, 0.6 mJ/pulse, 1 kHz) were focused fully by the same objective lens onto a circulated Fe_2O_3 tape target, then x-ray pulses were generated. An x-ray emission spectrum from the Fe_2O_3 target was measured by the Ge SSD, which is shown in Figure 5 (b) where two peaks are observed clearly in addition to low-intensity broad x-ray. Those peaks are assigned to Fe $K\alpha$ (0.194 nm) and $K\beta$ (0.176 nm) lines. X-ray output through a Pb collimator (the inner diameter = 0.8 mm) was used as a probe. A highly-oriented pyrolytic graphite substrate (NT-MDT, ZYH, HOPG) was chosen as a sample for x-ray diffraction. The lattice constant of the graphite plane (002) is 0.335 nm. From these values, the diffraction angle (2θ) can be calculated to be about 33 degrees for Fe $K\alpha$. Images of transmitted and diffracted x-ray from the HOPG were converted to visible images by an x-ray image intensifier (Hamamatsu Photonics, K. K., V7739P) and captured by a cooled CCD camera (Andor, DV434BV).

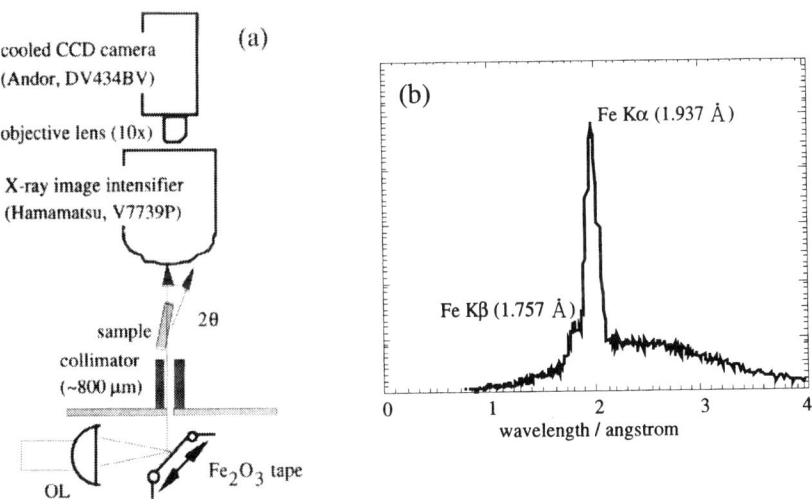

FIGURE 5. (a) An experimental setup for x-ray diffraction with x-ray pulses from a Fe_2O_3 tape target. (b) An x-ray emission spectrum from a Fe_2O_3 tape target irradiated by focused femtosecond laser pulses (775 nm, 130 fs, 0.6 mJ/pulse, 1 kHz).

A result is shown in Figure 6 (a), where the accumulation time was only 2 min. which corresponded to 1.2×10^5 shots. When the 2θ is on the Bragg angle, diffracted x-ray spots were clearly observed beside the transmitted x-ray. X-ray intensity surface plots are shown in Figure 6 (b). Diffracted x-rays of Fe $K\alpha$ and Fe $K\beta$ are clearly resolved. There is little difference in signal to noise ratios between the two accumulation times of 2 min. and 60 min. This means that time-resolved x-ray diffraction can be performed even with commercial-base laser systems in conventional laboratories. This result encourages us to proceed experiments of fs-laser-pump and x-ray–pulse-probe, which will be performed soon especially under laser ablation condition [11].

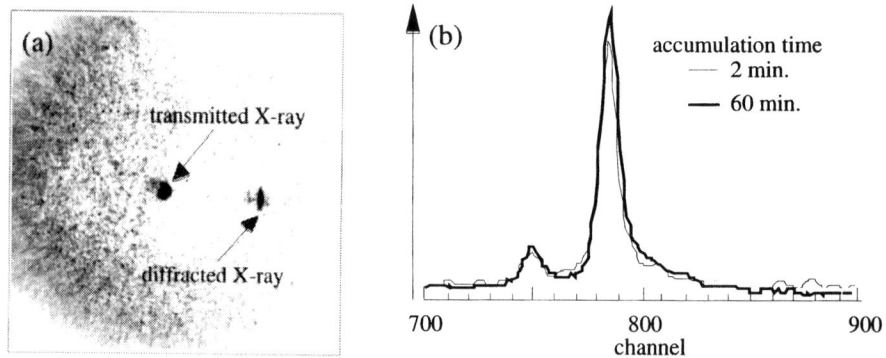

FIGURE 6. (a) An x-ray image of transmission and diffraction. (b) Diffraction patterns obtained by 2 min. and 60 min. accumulation times.

ACKNOWLEDGMENTS

The present work was supported by a Grant-in-Aid from the Ministry of Education, Science, and Culture of Japan (11355035 and 12750004). The authors are thankful to Professor Y. Udagawa at IMRAM, Tohoku University for the use of the Ge SSD, to Associate Professor T. Sekine at Department of Chemistry, Tohoku University for the use of a high voltage supplier and an amplifier, and to Andor Technology for the use of the cooled CCD camera. The authors are also thankful to Dr. Yosuke Watanabe at IMR for the great contribution on the development of x-ray diffractometer.

REFERENCES

1. Attwood, D., *Soft X-rays and Extreme Ultraviolet Radiation*, Cambridge University Press, Cambridge, 1999.

2. Hentschel, M., Kienberger, R., Spielmann, Ch., Reider, G. A., Milosevic, N., Brabec, T., Corkum, P., Heinzmann, U., Drescher, M., and Krausz, F., *Nature*, **414**, 509-513 (2001).

3. Malmqvist, L., Rymell, L., and Hertz, H. M., *Appl. Phys. Letters* **68**, 2627 (1996).

4. Tompkins, R. J., Mercer, I. P., Fettweis, M., Barnett, C. J., Klug, D. R. and Porter, G., *Rev. Sci. Instrum.*, **69**, 3113 (1998).

5. Mozumder, A., *Fundamentals of Radiation Chemistry*, Academic Press, San Diego, 1999.

6. Hatanaka, K., Miura, T., and Fukumura, H., *Appl. Phys. Letters* **80** (21), (2002), *in press*.

7. Miura, T., Hatanaka, K., Odaka, H., and Fukumura, H., *The 6ᵗʰ International Conference on Laser Ablation*, PT-24, Oct. 2001, Tsukuba, Japan

8. Hatanaka, K. and Fukumura, H., patent pending, July 31 (2001).

9. Keldysh, L. V., *Sov. Phys. JETP*, **20**, 1307 (1965).

10. Baldis, H. A., Campbell, E. M., and Kruer, W. L., *Physics of Laser Plasma*, Rubenchik, A. and Witkowski, S., Ed., Elsevier Science Publishers, North-Holland, 1991.

11. Hatanaka, K., Tsuboi, Y., Fukumura, H., and Masuhara, H., *J. Phys. Chem.*, **B 106**, 3049-3060 (2002).

X-ray generation by femtosecond laser pulses and its application to soft X-ray imaging microscope

Kenichi Ikeda[1], Hideyuki Kotaki[1,2] and Kazuhisa Nakajima[1,2,3]

[1] Graduate University for Advanced Studies,
Shonan Village, Hayama, Kanagawa, 240-0193, Japan
[2] Japan Atomic Energy Research Institute, Kizu, Kyoto, 619-0215, Japan
[3] High Energy Accelerator Research Organization, Tsukuba, Ibaraki, 305-0801, Japan

Abstract. We have developed laser-produced plasma X-ray sources using femtosecond laser pulses at 10Hz repetition rate in a table-top size in order to investigate basic mechanism of X-ray emission from laser-matter interactions and its application to a X-ray microscope. In a soft X-ray region over 5 nm wavelength, laser-plasma X-ray emission from a solid target achieved an intense flux of photons of the order of 10^{11} photons/rad per pulse with duration of a few 100 ps, which is intense enough to make a clear imaging in a short time exposure. As an application of laser-produced plasma X-ray source, we have developed a soft X-ray imaging microscope operating in the wavelength range around 14 nm. The microscope consists of a cylindrically ellipsoidal condenser mirror and a Schwarzshird objective mirror with highly-reflective multilayers. We report preliminary results of performance tests of the soft X-ray imaging microscope with a compact laser-produced plasma X-ray source.

INTRODUCTION

The present high-brightness X-ray sources have been developed as third generation synchrotron light sources based on a large-scale high energy electron storage ring and magnetic undulators. A compact, tunable, high-brightness X-ray source has basic and industrial applications in a number of fields, such as solid-state physics, material, chemical, biological and medical sciences. Particularly for biological and medical applications, such as X-ray imaging, radiography and therapy, it is essential to downscale the present synchrotron light sources into the environment of a laboratory or a hospital, keeping properties of X-ray radiations. Recently availability of compact terawatt lasers arouse a great interest in the use of lasers as a compact bright X-ray source generated by intense laser-plasma interactions, replacing conventional X-ray tubes.

We have developed laser-produced plasma X-ray sources using femtoseconds laser pulses at 10Hz repetition rate in a table-top size in order to investigate basic mechanism of X-ray emission from laser-matter interactions and its application to a X-ray microscope. In a soft X-ray region over 5 nm wavelength, X-ray emission from a solid metal target, such as Al, Cu, and W irradiated by a 130 mJ laser pulse with duration of 100 fs, achieved an intense flux of photons of the order of 10^{11} photons/rad

CP634, *Science of Superstrong Field Interactions*, edited by K. Nakajima and M. Deguchi
© 2002 American Institute of Physics 0-7354-0089-X/02/$19.00

per pulse with duration of a few 100 ps, which is intense enough to make a clear imaging in a short time exposure.

As an application of laser-produced plasma X-ray source, we have developed a soft X-ray imaging microscope operating in the wavelength range around 14 nm. The microscope consists of a grazing incidence cylindrically ellipsoidal condenser mirror and a Schwarzshird objective mirror with highly-reflective multilayers. we report preliminary results of performance tests of the soft X-ray imaging microscope with a compact laser-produced plasma X-ray source.

LASER-PRODUCED PLASMA X-RAY EMISSION FROM SOLID TARGETS

Intense ultra-short laser pulses can produce a high density plasma associating a strong electron heating generated by strong field interactions with solid state matter. As the electric field of the laser radiation exceeds the atomic electric fields above the intensities of the order of 10^{16} W/cm^2, a solid matter is ionized in a fraction of the wave oscillation through processes of the multiphoton ionization and the tunneling ionization to produce plasmas with a solid electron density of the order of 10^{23} cm^{-3}. Hence intense femtosecond laser pulses generate interactions with plasmas characterized by solid density and very steep gradients with the scale length of a few hundreds of angstroms in the plasma expansion region. In the plasma region over a length of the skin depth, typically of the order of a few hundreds of angstroms, a strong electron heating occurs in a very short time to produce the plasma heated up to temperatures of several keV at the intensity of 10^{17} W/cm^2 away from thermal equilibrium. The high density and the electron kinetic energy of these plasmas generate bright X-ray pulses with photon energy extending up to several keV through fundamental emission processes, known as bremsstrahlung, recombination, and line emissions[1].

Measurements of X-ray intensity and pulse duration

A laser-plasma X-ray source has been simply made by focusing a high peak power laser onto the surface of a taget placed in a vacuum chamber. Our laser system is a table-top size Ti:sapphire laser based on the chirped pulse amplification at the wavelength of 790 nm. This system can produce the output pulses with the pulse duration of 100 fs and the maximum pulse energy of 200 mJ at the repetition rate of 10 Hz. In order to investigate characterization of laser-produced plasma X-ray emission, the output pulses were focused by the off-axis parabolic mirror with a focal length of 150 mm. The measured focal spot profile was an elliptical shape with a horizontal diameter of about 10 μm and a vertical diameter of about 30 μm at the focal point. The peak focused laser intensity exceeds 4×10^{17} W/cm^2 for a 100 mJ pulse energy focused on the target. The solid target mounted on the translation stage was scanned by a step of 100 μm so that a fresh surface of the target was exposed at each laser shot. we measured intensity and pulse duration of X-ray radiation emitted by laser-produced plasmas for various targets with different atomic numbers (Z) using the X-ray streak

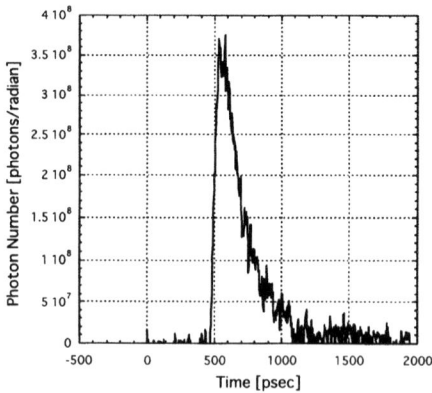

FIGURE 1. A typical temporal pulse profile of X-ray emission from plasma produced by the irradiation of a 130 mJ, 100 fs laser pulse on a Al target.

camera with 1 ps time resolution. A typical temporal pulse profile of the total X-ray emission from a Al target is shown in Figure 1.

The number of detected photons was calculated from the pulse height counts of X-ray streak devided by the photon-electron conversion effeciency of the photocathode made of a 50 μm wide and 8 mm long Au thin film. The photon intensity per shot was obtained from integrating the number of photons over the X-ray pulse duration devided by a solid angle of the photocathode mounted on the X-ray streak camera, assuming an uniform angular distribution of X-ray emission from laser-produced plasmas. The total photon intensity and the pulse duration for various solid targets irradiated by a 130 mJ laser pulse energy are summarized as a function of the atomic number Z as shown in Figure 2. It is found that the laser-irradiation on a high-Z target can produce a strong photon flux of the order of 10^{11} photons/radian per pulse.

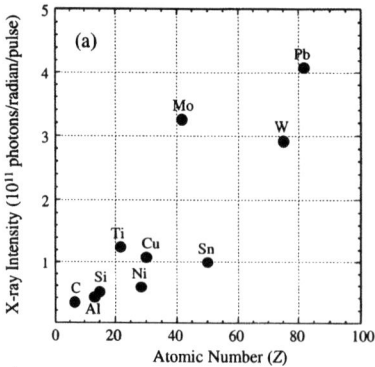
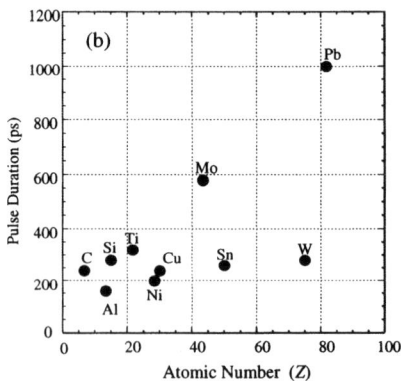

FIGURE 2. The photon intensity (a) and the pulse duration of X-ray emission from plasmas by laser irradiation on various targets as a function of the atomic number Z.

270

Measurements of X-ray spectrum

The radiation spectrum of a laser-produced plasma was measured with a flat-field normal-incident spectrograph with a platinum-coated concave grating of 1200 lines/mm blazed at 100 nm. The spectrograph equipped with a microchannel plate (MCP) detector covers the spectral range from 5 nm to 50 nm by changing the position of the 1-in. diameter MCP viewed with a charge coupled device (CCD) through an image reduction optics. The X-ray spectra of the radiation emitted by plasmas produced by Al and Cu targets are shown in Figure 3.

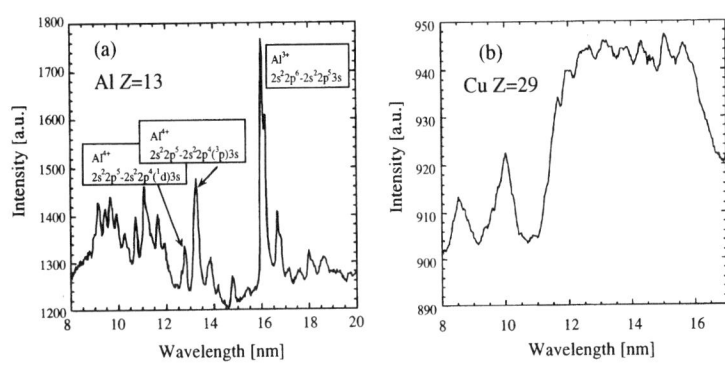

FIGURE 3. The measured X-ray spectra from plasmas produced by laser irradiation on (a) Al and (b) Cu targets.

APPLICATION TO A SOFT X-RAY IMAGING MICROSCOPE

We have developed a soft X-ray imaging microscope in a table-top size using a laser-produced plasma X-ray source as shown in Figure 4. The X-ray microscope is capable of observation of thick specimens in their natural environments, such as a living cell or biological samples in water, providing high-contrast X-ray imaging of micron-sized hydrated specimens with high penetration depth. Although the spatial resolution of an X-ray microscope is lower than that of an electron microscope, the use of X-rays as a probe can provide valuable complementary information[2, 3].

Femtosecond laser-produced plasma X-ray source

In applications to X-ray microscopy, laser-plasma X-ray sources evidently make it possible to reduce its system size and capital costs compared to the accelerator-based synchrotron radiation sources. In particular, a high brightness and a short duration of laser-plasma X-ray sources by the use of femtosecond laser irradiation can improve image quality with reduced exposure for the X-ray microscopy of micro-biological systems.

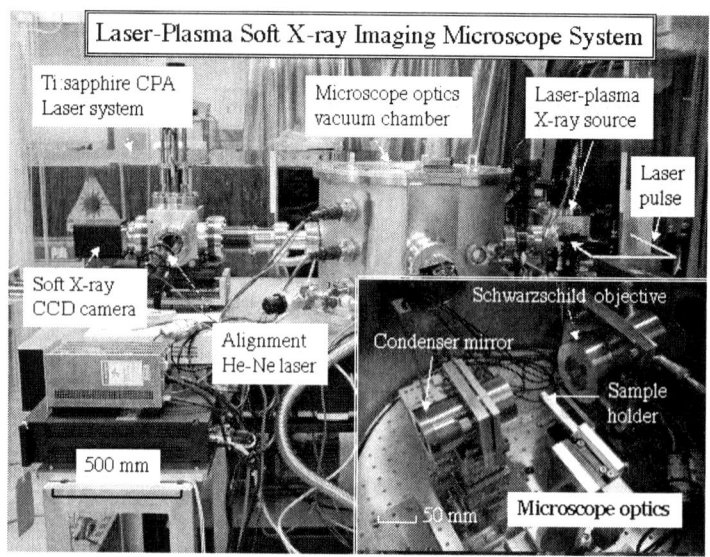

FIGURE 4. A table-top soft X-ray imaging microscope system with femtosecond laser-produced plasma X-ray source.

We constructed the target system consisting of a rotating solid target and its driving mechanisms. We used the Cu target of 3-cm diameter to generate bright soft X-rays ranging from 13 nm to 15 nm where the X-ray optics of the microscope is designed with a high efficiency. The target was irradiated by laser pulses of a 100 fs pulse duration through a 160 mm focal length lens at a 10 Hz repition rate. The laser pulse energy of less than 50 mJ was used for a typical measurement.

Microscope optics

A schematic optics of the imaging X-ray microscope is shown in Figure 5. As an ordinary optical microscope, the imaging X-ray microscope consists of a condenser and an objective. The condenser is a ellipsoidal mirror of revolution that focuses the grazing incident X-ray radiation emitted at its one focal point almost over the 4π solid angle onto the sample placed at the other focal point. The distance between two focal points is about 50 cm. The objective is a Schwarzschird optics[4] comprising two concentric spherical mirrors, obtained from NTT-AT. These objective mirrors are coated with Mo/Si multilayer coatings consisting of 40 layers with each thickness of 7.14 nm to enhance the reflectivity at a wavelenth of 13.9 nm to 73 % for normal incident X-rays[5]. Figure 6 shows a design of the Schwarzschird objective. This design can produce the magnification of 25. The numerical aperture (NA) of 0.0085 was determined by taking into account the best compromise between resolution and aberrations. With this NA, the resolution R of the microscope is expected to be R=1 μm for the wavelength λ= 13.9 nm as obtained from the Rayleigh criterion R = 0.61λ/NA.

We use a back-illuminated CCD camera as a soft X-ray imaging detector. The CCD arrays consist of 1340×1300 square pixels of 20×20 μm² with a 26×26 mm² sensitive area. A microscopic image is viewed through a 0.3 μm Be filter with the CCD arrays placed at the imaging plane 1000 mm distant from the object plane.

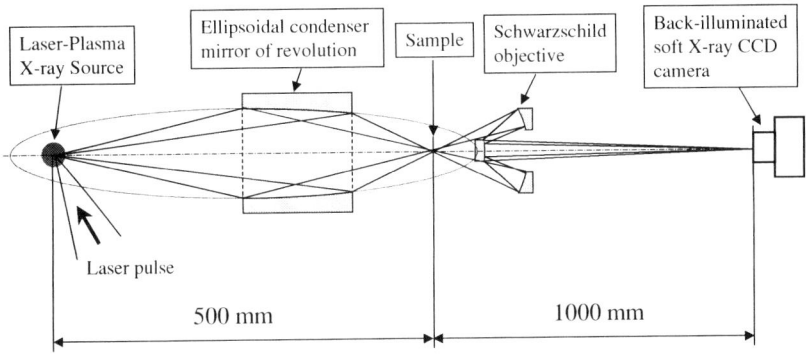

FIGURE 5. A schematic optics of the imaging soft X-ray microscope.

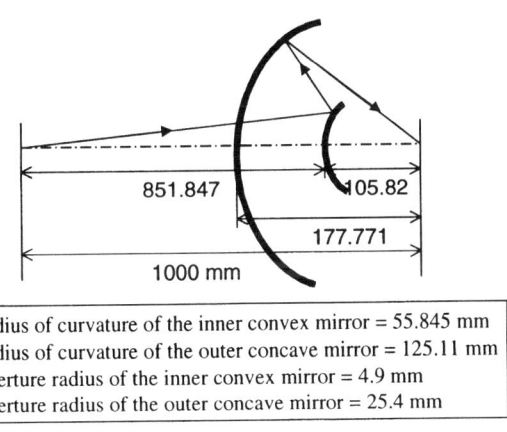

Radius of curvature of the inner convex mirror = 55.845 mm
Radius of curvature of the outer concave mirror = 125.11 mm
Aperture radius of the inner convex mirror = 4.9 mm
Aperture radius of the outer concave mirror = 25.4 mm

FIGURE 6. A design of the Schwarzschird objective for the imaging soft X-ray microscope.

273

Experiments for characterization of microscope performance

The optics alignment of the microscope was made using a visible He-Ne laser light. First, a spot size of X-ray beam at the focal point of the condenser mirror was evaluated by a knife-edge scan as shown in Figure 7. The spot size was estimated to be about 300 μm rms radius by differentiating a edge-scan profile.

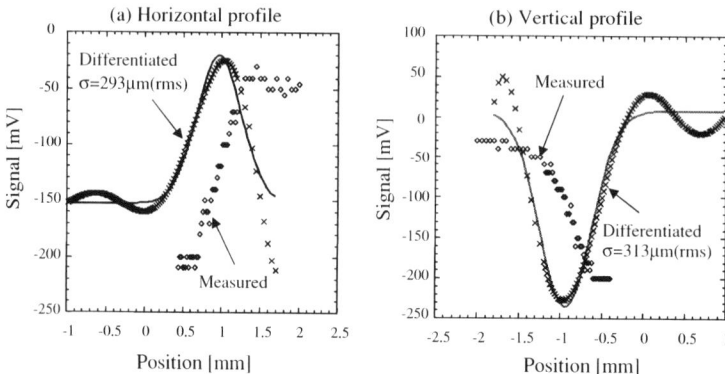

FIGURE 7. Knife-edge scan measurements of (a) a horizontal spot size and (b) a vertical spot size of the soft X-ray beam at the focal point of the condenser mirror.

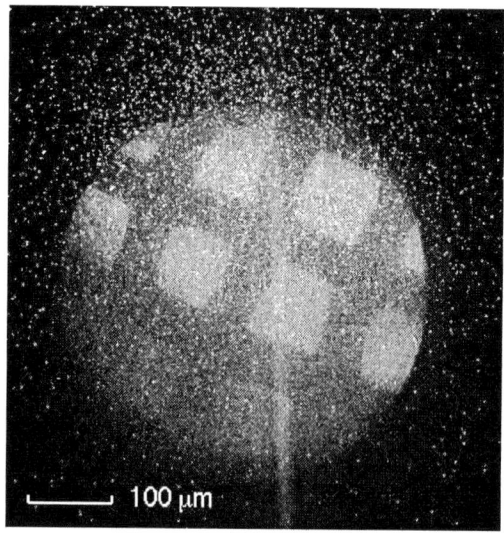

FIGURE 8. A soft X-ray microscopic image of a 150 lpi mesh (169.3 μm period).

In order to investigate a performance of the microscope, we used a mesh of 150 lpi (169.3 μm period) as a test object. A CCD image of the mesh is shown in Figure 8. The image was taken for the condition of laser irradiation of 50 mJ and about 3 second exposure time. The adjustment of imaging optics was made by changing position of the object so that a high contrast image should be obtained. From the intensity profile, spatial resolution of about 1 μm was defined by distance required to reduce the intensity from 25 % to 75 %.

CONCLUSIONS

We have investigate characteristics of laser-produced plasma X-ray emission by irradiating femtosecond laser pulses on various solid targets. It has been found that the brightness of a soft X-ray emission is strong enough to make imaging in a short exposure time.

We have developed the soft X-ray imaging microscope on a table-top for the wavelengths ranging from 13 nm to 15 nm using a bright X-ray source generated by femtosecond laser-produced plasma interactions on a Cu target. We successfully demonstrated imaging of a test mesh with spatial resolution of about 1 μm. Next we are exploring microscopy of micro-biological objects to verify capabilities of our laboratory-scale imaging soft X-ray microscope.

ACKNOWLEDGMENTS

This work is supported by the 2001th year Joint Research Project (Soken/K99-6) of Sokendai (The Graduate University for Advanced Studies).

REFERENCES

1. Giulietti, D., and Gizzi, L. A., *La Rivista del Nuovo Cimento*, 21, 1-93 (1998).
2. Aristov, V. V., and Erco A. C. , (Editors) *X-ray Microscopy IV*, Chernogolovka, Russia, 1994.
3. Methe, O., Guttmann, P., Scneide, G., Rudolph, D., Trendelenburg, M. F., and Schmahl, G., *J. Microsc.*, **188**, 125 (1997).
4. Berreman, D. W., et al., *Opt. Lett.*, **15**, 529 (1990) .
5. Takenaka, H., Ito, H., Nagai, K., Muramatsu, Y., Gullikson, E., Pererra, R. C. C., *Nucl. Instr. Methods*, **A467-468**, 341-344 (2001).

Study of ultra-high gradient wakefield excitation by intense ultrashort laser pulses in plasma

Hideyuki Kotaki[†,*], Masaki Kando[†], Takatsugu Oketa[†], Shinichi Masuda[†], James K. Koga[†], Shuji Kondo[†], Shuhei Kanazawa[†], Takashi Yokoyama[†], Toru Matoba[†] and Kazuhisa Nakajima[†,*,**]

*The Graduate University for Advanced Studies, Hayama, Kanagawa, 240-0193, Japan
†Japan Atomic Energy Research Institute, Kizu, Kyoto, 619-0215, Japan
**High Energy Accelerator Research Organization, Tsukuba, Ibaraki, 305-0801, Japan

Abstract. We investigate a laser wakefield excited by intense laser pulses, and the possibility of generating an intense bright electron source by an intense laser pulse. The coherent wakefield excited by 2 TW, 50 fs laser pulses in a gas-jet plasma around 10^{18} cm^{-3} is measured with a time-resolved frequency domain interferometer (FDI). The results show an accelerating wakefield excitation of 20 GeV/m with good coherency. This is the first time-resolved measurement of laser wakefield excitation in a gas-jet plasma. The experimental results agree with the simulation results and linear theory. The pump-probe interferometer system of FDI will be modified to the optical injection system as a relativistic electron beam injector. In 1D particle in cell simulation we obtain results of high quality intense electron beam generation.

INTRODUCTION

Laser-driven plasma accelerators have been conceived to be the next-generation particle accelerators, promising ultrahigh field particle acceleration and compact size compared with conventional accelerators[1]. In particular, it has been experimentally demonstrated that laser wake-field acceleration (LWFA) has great potential to produce ultrahigh field gradients excited by intense ultrashort laser pulses of the order of ~ 100 GeV/m[2]-[9]. The maximum energy gain has exceeded 100 MeV with an energy spread of $\sim 100\%$ due to dephasing and wavebreaking effects in the self-modulated LWFA regime, where thermal plasma electrons are accelerated[8]. The highest energy gain acceleration which exceeded 200 MeV was observed with the injection of an electron beam at an energy matched to the wakefield phase velocity in a fairly underdense plasma[9].

In order to control the wakefield for the laser-plasma experiments, we need to measure the amplitude and the phase of the wakefield. A direct measurement of the plasma density oscillation can be done by means of ultrafast time-resolved frequency domain interferometry (FDI)[10]. Several measurements have been made with FDI to demonstrate wakefield excitation by ultrashort laser pulses in an underdense plasma[9],[11]-[13]. These measurements have been done for a relatively low density plasma in a gas filled chamber using laser pulse durations around 100 fs and pump peak powers less than 1 TW. In these measurements the pump pulses were tightly focused to enhance the plasma wave excitation due to 2D effects. In a 2D dominant regime, where the pulse

CP634, *Science of Superstrong Field Interactions*, edited by K. Nakajima and M. Deguchi
© 2002 American Institute of Physics 0-7354-0089-X/02/$19.00

width is longer than a spot size, the radial wakefield is higher than the longitudinal one. Therefore a shorter pulse is preferable to generate a more 1D coherent planar wakefield in the higher resonant plasma density. We have developed a pump-probe frequency domain interferometer consisting of a 2 TW pump laser pulse with a duration of 50 fs at a wavelength of 800 nm and two frequency doubled probe pulses split from the pump pulse with a variable delay. The measurement of laser wakefields has been made in a less 2D dominated regime to suppress radial effects[14].

Recently novel schemes which use laser triggered injection of plasma electrons into the wakefield have been proposed as an optical injection. Presently there are three major schemes: nonlinear wavebreaking injection[15], transverse optical injection[16], and colliding pulse optical injection[17][18]. The colliding pulse optical injection, whose energy spread is smaller than others, uses three short laser pulses. No proof-of-principle experiment for these schemes has been yet performed because of experimental difficulties. In the optical injection schemes, provided that the probe pulses are replaced with the injection pulses, the FDI system will be modified to the optical injection system for the laser wakefield. We are planning to develop an optical injection system for LWFA to generate an ultrashort electron beam with small energy spread and low transverse emittance. We estimate the optical injection by 1D particle in cell (PIC) simulation.

Here, we present the laser wakefield excited by the intense laser pulse and the generation of the high quality relativistic electron beam. In section II the gas density measurements of the neutral gas and the laser wakefield measurement is described. In section III we present the simulation of the optical injection by using anomalous blueshift and a pump-probe mechanism, and in section IV conclusions are presented.

OBSERVATION OF LASER WAKEFIELD

Measurement of gas density distribution

The gas density was measured with a Mach-Zehnder interferometer as shown in Fig. 1. A gas-jet nozzle with an orifice diameter of 0.8 mm is placed inside a vacuum chamber evacuated to 10^{-5} Torr. The interferogram is captured with a charge-coupled-device (CCD) camera with an image intensifer. The opening duration of the gas-jet valve was 2 ms and the gate time of the CCD camera was 5 μs. The distribution of the gas density can be calculated from the phase shift.

Figure 2 shows a contour plot of the He gas density distribution at a backing pressure of 30 atm with a 1.6 ms delay. From measurements of nitrogen gas, it is found that the gas density distribution is uniform 1.5 mm from the gas-jet nozzle. Since the gas density is inversely proportional to the square of the distance from the nozzle and linearly proportional to the backing pressure, the neutral gas density of He gas is $\approx 3.6 \times 10^{17}$ cm^{-3} at the pump laser focus with a backing pressure of 10 atm.

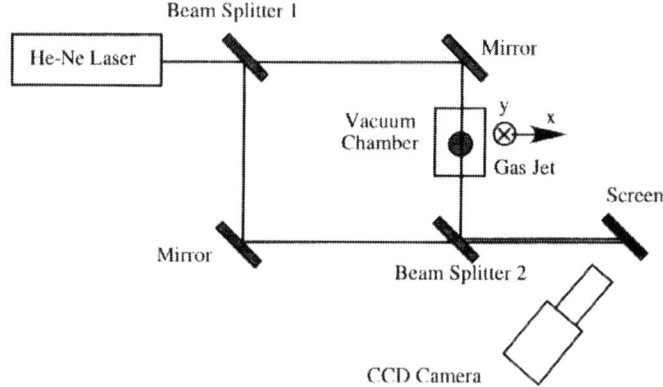

FIGURE 1. A schematic of the Mach-Zehnder interferometer for measurement of the gas density distribution of the gas-jet.

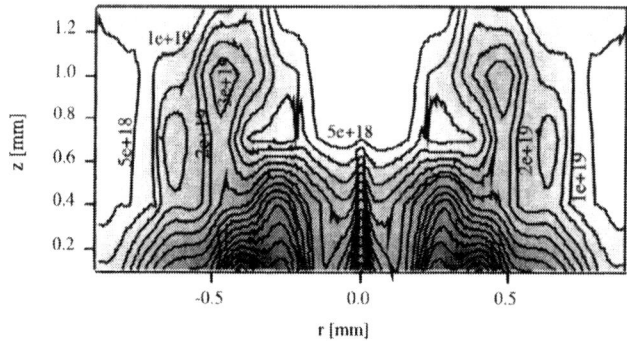

FIGURE 2. A contour map of the gas density distribution of He gas at a backing pressure of 30 atm with a delay time of 1.6 ms.

Measurement of wakefields

An intense ultrashort laser pulse produces a fully ionized plasma through optical field ionization on ultrafast time scales of the order of femtoseconds and excites electron density oscillations in the plasma behind the laser pulse. This plasma density oscillation can be measured by the frequency domain interferometry technique[10][14].

The experimental setup of the FDI is shown in Fig. 3. The multi-TW 10 Hz Ti:sapphire laser pulse at a wavelength of 800 nm with a maximum energy of 135 mJ and a duration of 52 fs (FWHM) is split into two beams. The reflected beam (80 %) is used as the pump pulse and the transmitted beam as the probe beam. The pump and probe pulses are focused by an off-axis parabolic mirror with a focal length of 15 cm. The focal spot images taken with a CCD camera show that the pump focused intensity profile is approximately a Gaussian distribution with a rms radius (at $1/e$) $\sigma_r = 8.9$ μm, while the probe beam radius is $\sigma_{rpr} = 3$ μm and centered on the pump beam. The peak pump intensity is 8.4×10^{17} W/cm^2 ($a_0 \simeq 0.6$), producing fully ionized He gas in the focal region. The probe beam is separated from the pump beam with a dichroic beam splitter after passing through the plasma and guided through an optical fiber into the spectrometer to analyze the interferograms. A part of the probe beam is sent directly to the spectrometer to make the reference fringes. The spectrometer is a Czerny-Turner type with a focal length of 1.33 m and a grating of 1800 gr/mm. The spectral resolution is 0.06 Å. The phase shift at each position was obtained by averaging 50 shots of the phase shift data to reduce pulse-to-pulse fluctuations. The gas jet was operated at a backing pressure of 10 atm for He gas. The pump pulse was focused 1.5 mm below the end of the nozzle where a fully ionized plasma was expected with an electron density of 7.2×10^{17} cm^{-3} from the neutral gas density measurements.

Figure 4 shows the results of the wakefield measurement. In the region where the probe pulse was delayed after the pump pulse, a large amplitude density perturbation excited by the pump pulse was detected as indicated by the phase shifts shown in Fig. 4 (a). From the phase shift data, the period and the amplitude of the electron density oscillation and the amplitude of the longitudinal wakefield are obtained as a function of time delay between the pump and the probe pulses. The oscillation period of the electron plasma wave is 130 ± 3.4 fs behind the pump. This plasma wave period corresponds to the electron density of $7 \times 10^{17} \pm 8.5 \times 10^{15}$ cm^{-3} in linear wakefield theory, which is in good agreement with the electron density for the fully ionized He gas expected from the neutral gas density measurements. The measured density perturbation before damping is $(\delta n/n_e)_m \simeq 0.74$ with the plasma length $L_p = 1$ mm. The maximum longitudinal wakefield of 20 GV/m is deduced. These estimates are in good agreement with the density perturbation $(\delta n/n_e)_{th} = 0.75$ and the longitudinal wakefield of 21 GV/m calculated by linear wakefield theory.

We compare the experimental and the 1D PIC simulation results. Figure 4 (b) shows the amplitude of the longitudinal wakefields and the plasma wavelength as a function of the plasma density. The solid line and the dotted line show the wakefield and wavelength from linear theory. Closed circles and triangles show the wakefields from the FDI experiment and the PIC simulation. Open circles and triangles show the plasma wavelength from the FDI experiment and the PIC simulation. The experimental results and the simulated results agree with linear theory.

GENERATION OF ELECTRON BEAMS FROM LASER-PLASMA

The wakefield excited in the gas jet has good coherency and can be used for particle accelerators. This pump-probe interferometer system will be remade into a plasma cathode laser accelerator for high quality electron beam generation. In the plasma cathode schemes, provided that the probe pulses are replaced with the injection pulses, the FDI system will be modified to the optical injection system for the laser wakefield.

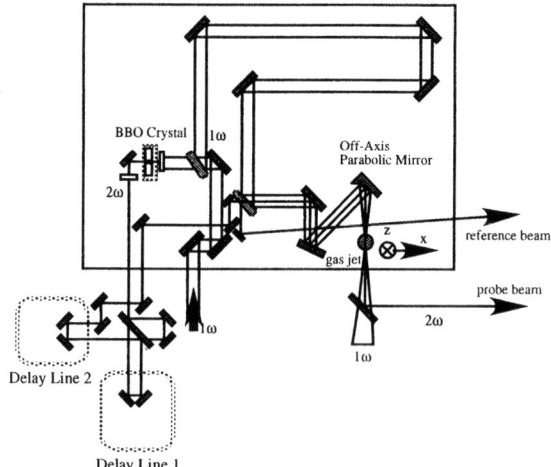

FIGURE 3. Experimental setup of the frequency domain interferometry. Delay line 1 adjusts the time delay between two collinear probe pulses and delay line 2 the time delay between the pump and the probe pulses.

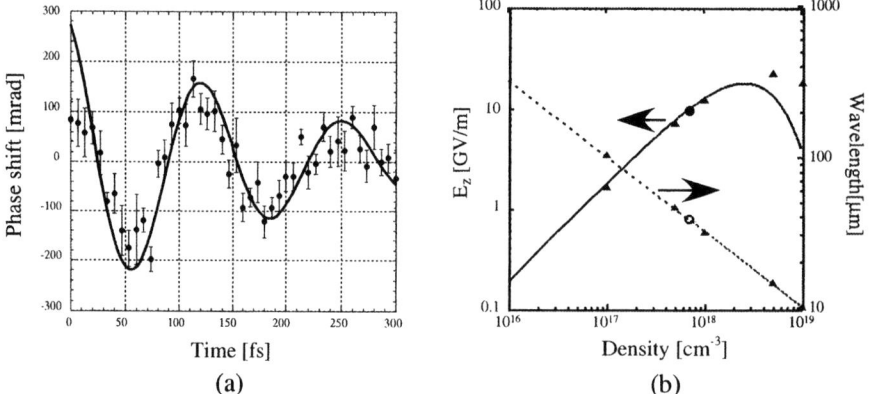

FIGURE 4. (a) The relative phase shift as a function of the time delay for the measurement of the plasma density oscillation in the gas-jet operated at a backing pressure of 10 atm. (b) The amplitude of the longitudinal wakefields and the plasma wavelength as a function of the plasma density as a function of the plasma density. The solid line and the dotted line show the wakefield and wavelength from linear theory. Closed circles and triangles show the wakefields from the FDI experiment and the PIC simulation. Open circles and triangles show the plasma wavelength from the FDI experiment and the PIC simulation.

Trapping electrons in wakefields

Three laser pulses, a pump pulse for wakefield excitation and two injection pulses for trapping the electrons in a plasma, construct a colliding optical injector[17][18]. A colliding pulse optical injection scheme for LWFA is proposed and analyzed that uses three short laser pulses: an intense pump pulse (denoted by the subscript 0), a forward going injection pulse (denoted by the subscript 1), and a backward going injection pulse (denoted by the subscript 2). The frequency, wave number, and normalized intensity are denoted by, respectively, ω_i, k_i, and a_i ($i = 0, 1, 2$). Furthermore, $\omega_1 = \omega_0 + \Delta\omega$ ($\Delta\omega \geq 0$), $\omega_2 = \omega_0$, and $\omega_0 \ll \Delta\omega \ll \omega_p$ are assumed such that $k_1 = k_0$ and $k_2 \simeq -k_0$. The laser pulse of injection 1 is the frequency up-shifted pulse by an anomalous blueshift[19]. The pump pulse generates a fast wakefield. When the injection pulses collide, they generate a slow ponderomotive beat wave with a phase velocity v_p. The phase velocity v_p is given by

$$v_p = \frac{\omega_1 - \omega_2}{k_1 - k_2} = \frac{\Delta\omega}{2k_0}. \qquad (1)$$

During the time in which the two injection pulses overlap, a two-stage acceleration process can occur; i.e., the slow beat wave injects plasma electrons into the fast wakefield for acceleration to high energies.

Electron beam generation by colliding pulse optical injection

We simulated colliding pulse optical injection for the pump pulse laser strength parameter of 1 and the injection pulses laser strength parameter of 0.3. The frequency of the forward going injection pulse ω_1 is 6% blueshifted by the anomalous blueshift[19].

Figure 5 shows the results of the colliding pulse optical injection simulation for the plasma density n_e of 7×10^{17} cm^{-3}. The graph on the left side shows γ in phase space, and the graph on the right side shows an energy spectrum. In the phase space graph, the horizontal axis shows the position of the electrons. The length of 1 mm is 200 in the horizontal axis. In the energy spectrum, the vertical axis shows the number of particles in arbitrary units, and the horizontal axis shows the normalized energy γ, where $\gamma = E/m_0c^2$ and E is the total energy of electrons. The accelerated electrons for the plasma density of 7×10^{17} cm^{-3} are shown in Fig. 6 (a). The accelerated electrons have a pulse width of 14 fs, a peak energy of 7 MeV and an energy spread $\Delta E/E$ of 7%. The accelerated electron charge becomes 26 pC when the laser spot size r is 15 μm.

The transvers βs of the accelerated electrons are shown in Fig. 6 (b). The maximum value is 0.019 and the angular spread of the electron is 19 mrad for a longitudinal β of 1. The emittance of the accelerated electrons becomes 0.28 mm mrad when the laser spot size r is 15 μm.

CONCLUSIONS

In order to investigate the laser-wakefield excited by an intense laser pulse in plasma, we study the measurement of the laser wakefield and the optical injection by the laser wakefield.

We made the first direct observation of coherent ultrahigh gradient wakefields excited by an intense ultrashort laser pulse in a gas-jet plasma with a density around 7×10^{17} cm^{-3}. The wakefield with the oscillation period of 130 fs and maximum longitudinal

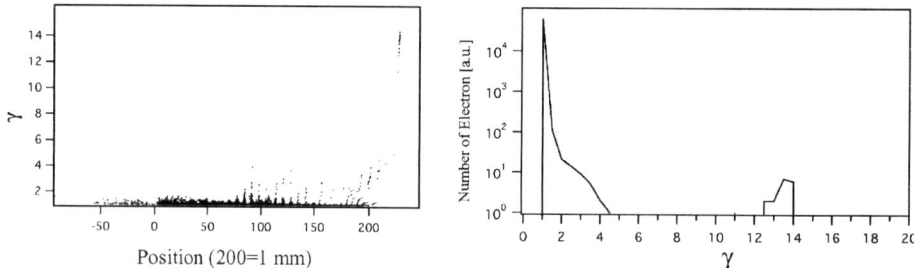

FIGURE 5. Results of the three pulse optical injection simulation for the plasma density n_e of 7×10^{17} cm^{-3} for the pump pulse laser strength parameter a_0 of 1.0 and the injection pulses laser strength parameter of 0.3. The graph on the left side are the γ in the phase space, and the graph on right side are histogram.

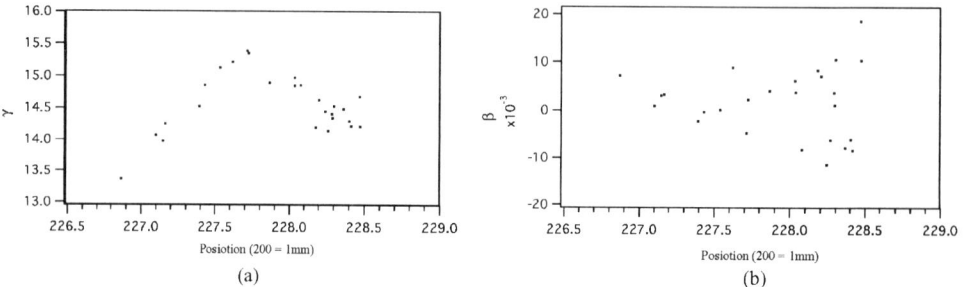

(a)   (b)

FIGURE 6. The accelerated electrons in a phase space (a) and the transverse speed of the accelerated electrons (b) of the three pulse optical injection for the plasma density n_e of 7×10^{17} cm^{-3} for the pump pulse laser strength parameter a_0 of 1.0 and the injection pulses laser strength parameter of 0.3.

amplitde of 20 GV/m was observed, showing fast damping in time from a few oscillation periods to ten periods. These measured wakefield parameters are quite consistent with the neutral gas density measurements and wakefield theory. The experimental results are compared to a 1D particle in cell (PIC) simulation. The experimental results and the simulated results agree with linear theory.

We will develop an optical injection system for laser accelerators to generate high quality electron beams using a gas jet and the pump-probe technique on a compact scale. Three laser pulses, a pump pulse for a wakefield excitation and two injection pulses for trapping the electrons in a plasma, make up a colliding optical injector. The problems of optical injection, which are the frequency shift and the timing between the 3pulses, can be solved by the FDI system and the anomalous blueshift. For the three pulse optical injection case, the energy spread is less than 10% at the energy of 7 MeV. The accelerated electrons have the pulse width of 14 fs. When we make an assumption that the laser spot radius is 15 μm, the accelerated electron charge and the emittance become 26 pC and 0.28 mm mrad, respectively.

ACKNOWLEDGMENTS

This work was supported by the JAERI-Kansai Advanced Photon Research Center. We would like to thank to K. Nagashima, N. Hasegawa, K. Sukegawa, M. Suzuki, E. Yanase, K. Yasuike, M. Nishiuchi and H. Daido for the help during the experiments.

REFERENCES

1. T. Tajima and J. M. Dawson, Phy. Rev. Lett. **43**, 267 (1979).
2. K. Nakajima, D. Fisher, T. Kawakubo, H. Nakanishi, A. Ogata, Y. Kato, Y. Kitagawa, R. Ko-dama, K. Mima, H. Shiraga, K. Suzuki, K. Yamakawa, T. Zhang, Y. Sakawa, T. Shoji, Y. Nishida, N. Yugami, M. Downer, and T. Tajima, Phy. Rev. Lett. **74**, 4428 (1995).
3. D. Umstadter, S. -Y. Chen, A. Maksimchuk, G. Mourou, and R. Wanger, Science **273**, 472 (1996).
4. D. Gordon, K. C. Tzeng, C. E. Clayton, A. E. Dangor, V. Malka, K. A. Marsh, A. Modena, W. B. Mori, P. Muggli, Z. Najmudin, D. Neely, C. Danson, and C. Joshi, Phy. Rev. Lett. **80**, 2133 (1998).
5. A. Modena, Z. Najmudin, A. E. Dangor, C. E. Clayton, K. A. Marsh C. Joshi, V. Malka, C. B. Darrow, and C. Danson, IEEE Trans. Plasma Sci., **24**, 289 (1996).
6. C. I. Moore, A. Ting, K. Krushelnick, E. Esarey, R. F. Hubbard, B. Hafizi, H. R. Burris, C. Manka, and P. Sprangle, Phy. Rev. Lett. **79**, 3909 (1997).
7. K. Nakajima, Nucl. Instr. and Meth. in Phys. Res. **A410**, 514 (1998).
8. T. E. Cowan, A. W. Hunt, T. W. Phillips, S. C. Wilks, M. D. Perry, C. Brown, W. Fountain, S. Hatchett, J. Johnson, M. H. Key, T. Parnell, D. M. Pennington, R. A. Snavely, and Y. Takahashi, Phy. Rev. Lett **84**, 903 (2000).
9. K. Nakajima, Nucl. Instr. and Meth. in Phys. Res. **A455**, 140 (2000); H. Dewa, H. Ahn, H. Harano, M. Kando, K. Kinoshita, S. Kondoh, H. Kotaki, K. Nakajima, H. Nakanishi, A. Ogata, H. Sakai, M. Uesaka, T. Ueda, T. Watanabe, K. Yoshii, Nucl. Instr. and Meth. in Phys. Res. **A410**, 357 (1998); M. Kando, H. Ahn, H. Dewa, H. Kotaki, T. Ueda, M. Uesaka, T. Watanabe, H. Nakanishi, A. Ogata and K. Nakajima, Jpn. J. Appl. Phys. **38**, 967 (1999).
10. E. Tokunaga, A. Terasaki, and T. Kobayashi, Opt. Lett. **17**, 1131 (1992).
11. C. W. Siders, S. P. Le Blanc, D. Fisher, T. Tajima, and M. C. Downer, Phys. Rev. Lett. **76**,3570 (1996); C. W. Siders, S. P. Le Blanc, A. Babine, A. Stepanov, A. Sergeev, T. Tajima, and M. C. Downer, IEEE Trans. Plasma Sci. **24**, 301 (1996).
12. J. R. Marquès, J. P. Geindre, F. Amiranoff, P. Audebert, J. C. Gauthier, A. Antonetti, and G. Grillon, Phys. Rev. Lett. **76**,3566 (1996); J. R. Marquès, F. Dorchies, P. Audebert, J. P. Geindre, F. Amiranoff, J. C. Gauthier, G. Hammoniaux, A. Antonetti, P. Chessa, P. Mori, and T. M. Antonsen Jr., Phys. Rev. Lett. **78**,3463 (1997); J. R. Marquès, F. Dorchies, F. Amiranoff, P. Audebert, J. C. Gauthier, J. P. Geindre, A. Antonetti, T. M. Antonsen Jr., P. Chessa, and P. Mori, Phys. Plasmas **5**, 1162 (1998).
13. E. Takahashi, H. Honda, E. Miura, N. Yugami, Y. Nishida, and K. Kondo, J. Phys. Soc. Jpn. **69**, 3266 (2000); E. Takahashi, H. Honda, E. Miura, N. Yugami, Y. Nishida, K. Katsura, and K. Kondo, Phys. Rev. E **62**,7247 (2000).
14. H. Kotaki, M. Kando, T. Oketa, S. Masuda, J. K. Koga, S. Kondo, S. Kanazawa, T. Yokoyama, T. Matoba, and K. Nakajima, Phys. Plasmas **9**, 1392 (2002).
15. S. Bulanov, N. Naumova, F. Pegoraro, and J. Sakai, Phys. Rev. E **58**, R5257 (1998).
16. D. Umstadter, J. K. Kim, and E. Dodd, Phy. Rev. Lett. **76**, 2073 (1996).
17. E. Esarey, R. F. Hubbard, W. P. Leemans, A. Ting, and P. Sprangle, Phys. Rev. Lett. **79**, 2682 (1997).
18. C. B. Schroeder, P. B. Lee, J. S. Wurtela, E. Esarey, and W. P. Leemans, Phy. Rev. E **59**, 6037 (1999).
19. J. K. Koga, N. Naumova, M. Kando, L. N. Tsintsadze, K. Nakajima, S. V. Bulanov, H. Dewa, H. Kotaki, and T. Tajima, Phys. Plasmas **7**, 5223 (2000).

High energy electron acceleration by stochastic acceleration mechanism

Tatsufumi Nakamura*, Susumu Kato[†] and Tomokazu Kato*

*Waseda University, 3-4-1, Ohkubo, Shinjuku-ku, Tokyo 169-8555, Japan
[†]Institute of Advanced Industrial Science and Technology, Umezono 1-1-1, Tukuba 305-8568, Japan

Abstract. A stochastic acceleration mechanism which explains the high energy electron acceleration by intense laser pulses is analyzed by Fokker-Planck approach. The Fokker-Planck equation describing is derived from the equation of motion of electrons interacting with filamented laser field. The time evolution of electron distribution function, energy distribution, effective temperature are obtained for initially rest plasma. The acceleration of 10 MeV electron beam is examined and shown to be effective.

INTRODUCTION

In recent experiments high energy electrons are observed without the existence of clear electrostatic wave where the laser pulse is split into filaments and the plasma is strongly disturbed[1,2]. The high energy electron acceleration is not explained by the conventional acceleration mechanism such as laser wakefield acceleration, beat-wave acceleration.[3,4] An acceleration mechanism under electromagnetic fields with a random force was investigated by J. Meyer-ter-Vehn and Z. M. Shen.[5] It is shown that the electrons dephased by a random force which is transverse to accelerating direction are directly accelerated by the laser field. Their theory was based on the Langevin approach, *i.e.*, the equation of motion for numbers of electrons are numerically integrated by adding a random force.

In the present paper, we investigate the stochastic acceleration mechanism by the Fokker-Planck approach. A Fokker-Planck equation is derived for the system where the laser field has multi-filament structure. The laser field is assumed to be composed of filaments of uniform electromagnetic plane waves whose phases are randomly different each other. The derived Fokker-Planck equation has nonlinear coefficients, which explains the directionality of electron acceleration towards the wave vector which is observed in experiments. Also the time evolution of the effective temperature of electrons and the ejected angle of high energy electrons are obtained from the Fokker-Planck equation.

In section 2, the Fokker-Planck equation is derived to show how the electron distribution function becomes anisotropic in momentum space. The directionality of acceleration is explained from its nonlinear coefficients. In section 3, the energy spectrum, electron temperature and ejected angle are obtained for the initially cold plasma. The acceleration of high energy electron beam is also examined. The conclusion and discussion are given in section 4.

CP634, *Science of Superstrong Field Interactions*, edited by K. Nakajima and M. Deguchi
© 2002 American Institute of Physics 0-7354-0089-X/02/$19.00

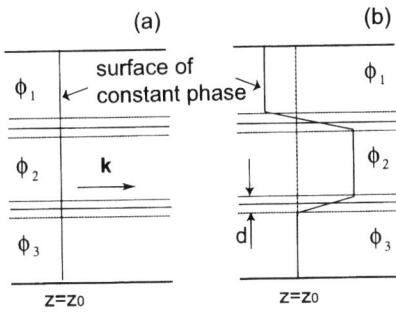

FIGURE 1. Schematic figure of the phase for (a) uniform laser field, (b) filamented laser field.

FOKKER-PLANCK EQUATION

In this paper we consider the situation where the laser field is divided into some regions in a transverse plane such as multi-filament structure. The structure is observed in the experiments of electron acceleration by intense strong laser fields, such that the laser field becomes nonuniform, *e.g.*, the laser pulse is split into filaments due to the interaction between the laser field and the plasma. The illustration of the phase of the laser field is drawn in Fig. 1. In the multi-filament structure the surface of a constant phase is modified as Fig. 1(b). For simplicity we assume $d = 0$, then the phase sharply changes at the filament edge. The phase of the laser field in each region is different, while it is uniform inside one region, *i.e.*, the laser field is consist of sectionally divided plane waves. Let us consider an electron with initial momentum u_{x0}, u_{z0}. The electron motion in a filament is written as

$$u_x = u_{x0} + a(\sin(\chi) - \sin(\chi_0)), \tag{1}$$

$$u_z = u_{z0} + u_{x0}\frac{a(\sin(\chi) - \sin(\chi_0))}{\alpha} + \frac{a^2(\sin(\chi) - \sin(\chi_0))^2}{2\alpha}, \tag{2}$$

$$\alpha = \gamma - u_z, \tag{3}$$

$$\chi = \omega t - kz. \tag{4}$$

where ω and k are the laser frequency and wave number of the electromagnetic field which corresponds to the filament. And u_{x0}, u_{z0}, and χ_0 are the initial values of u_x, u_z, and χ, respectively, and $a = eA/mc$ is the normalized vector potential. α is a constant of motion. The electron leaves the region at $t = t_1$ with $u_x = u_{x1}$ and $u_z = u_{z1}$ and enters an adjacent region. Then the electron executes again the periodic motion with u_{x1} and u_{z1} as new initial values until it moves to a different region. Therefore the electron gained $\delta u_x = u_{x1} - u_{x0}$ and $\delta u_z = u_{z1} - u_{z0}$ in moving through this region. The detail of this gain is discussed in the next paragraph. Electron experiences different phases as it enters the adjacent region with time step τ_c. The phase of the laser field in the different region is assumed to be randomly different. This allows us to treat the electron motion as a random process where $\phi = \sin\chi - \sin(\chi_0)$ is a random variable with $-2 \le \phi \le 2$. The equation of motion of an electron for n-th step is written in a same manner such as,

$$u_{x,n+1} = u_{x,n} + a\phi, \tag{5}$$

$$u_{z,n+1} = u_{z,n} + \frac{u_{x,n}a\phi}{\alpha_n} + \frac{a^2\phi^2}{2\alpha_n}, \tag{6}$$

where $\alpha_n = \gamma(u_{x,n}, u_{z,n}) - u_{z,n}$. In the time scale of $\eta \gg \tau_c$, eqs. (5) and (6) are written in differential forms as follows,

$$\frac{\partial u_x}{\partial \eta} = \frac{a\phi}{\tau_c}, \tag{7}$$

$$\frac{\partial u_z}{\partial \eta} = \frac{u_x a\phi}{\tau_c \alpha} + \frac{a^2\phi^2}{2\tau_c \alpha}. \tag{8}$$

The Fokker-Planck equation for the processes is obtained as[6],

$$\frac{\partial f_e}{\partial \eta} = \frac{a^2 D}{2\tau_c^2}\left(\frac{\partial^2 f_e}{\partial u_x^2} - 3\frac{\partial}{\partial u_z}\frac{f_e}{\alpha} + u_x^2 \frac{\partial^2}{\partial u_z^2}\frac{f_e}{\alpha^2}\right), \tag{9}$$

where f_e denotes the electron distribution function. Here the stochastic variable ϕ is assumed to be the Gaussian with $<\phi> = 0$ and $<\phi(t)\phi(t')> = D\delta(t - t')$ where $D \ll 1$. From the first term in eq. (9) it is seen that the time evolution of f_e is a normal diffusion process in u_x and estimated as $<u_x> \simeq a/\tau_c\sqrt{Dt}$. The second and third terms correspond to the drift and diffusion processes of f_e in u_z, which include $\alpha(u_x, u_z)$. In both terms $\alpha(u_x, u_z)$ plays an important role since $<u_z>$ is roughly estimated as $<u_z> \simeq (a/\tau_c\sqrt{Dt})/\alpha = <u_x>/\alpha$. It is known that $1/\alpha$ is small unless $u_x \simeq 0$, and $1/\alpha$ is proportional to u_z for $u_z \gg 1$ and $1/\alpha \simeq 0$ for $u_z \le 0$. Therefore the electron distribution drifts and diffuses in the positive u_z direction and not in negative direction. The diffusion effect in u_z is much larger than that in u_x. These facts explain the strong directionality of the acceleration towards u_z.

ACCELERATION OF ELECTRONS

The time evolution of f_e is obtained by numerically solving the Fokker-Planck equation, which is plotted in Fig. 2. The initial conditions are the Maxwell distribution with $k_B T = 5$ keV, $a = 3$, $D = 0.05$, and $\tau_c = 10$. In u_z direction, the electrons with large u_z are strongly accelerated only in positive direction. The distribution slowly diffuses in u_x. In Figs. 2 (b), (c), and (d) there are two peaks in the high energy tail, which means that the high energy electrons will be observed in a direction which is slightly deviated from the laser propagation direction. This is explained from the third term in eq. (9) such that the electrons with small u_x are slightly more accelerated than electrons with $u_x = 0$. The high energy electron beam tend to be collimated towards the direction of the wave vector. The angle exponentially decreases in time[7]. The energy distributions at

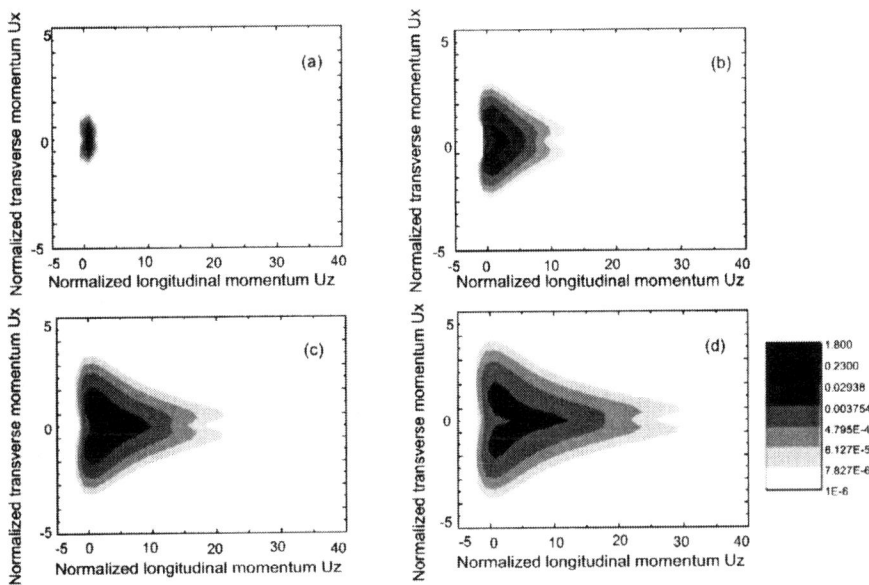

FIGURE 2. Profile of electron distribution in momentum space at (a) $\omega t = 100$, (b) $\omega t = 1000$, (c) $\omega t = 2000$, and (d) $\omega t = 3000$.

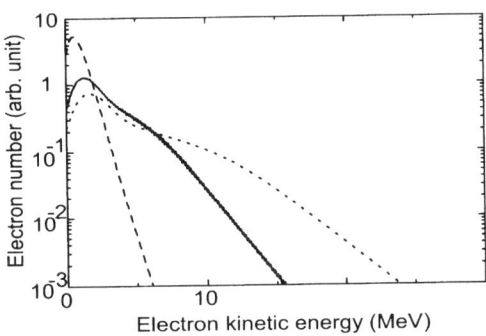

FIGURE 3. Energy distribution of accelerated electrons whose initial energy is 5 keV at $\omega t = 600$(dashed line), 1800(solid line), and 3000(dotted line).

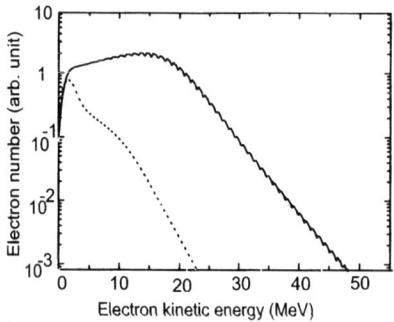

FIGURE 4. Energy distribution of accelerated electrons whose initial energy is 5 keV (dotted line) and 10 MeV (solid line) at $\omega t = 3000$.

$\omega_0 t = 600, 1800$ and 3000 are plotted in Fig. 3. It is seen that the distribution is quasi-thermal distribution and its effective energy increases up to 2.8 MeV at $\omega_0 t = 3000$ while the ponderomotive energy is 1 MeV. The acceleration of high energy electron beam is also examined. The energy distribution of electrons which are propagating in z direction with the average of 10 MeV and the thermal energy of 5 keV. The energy distribution at $\omega t = 3000$ is plotted in Fig. 4. A large numbers of electrons are accelerated up to about 17 MeV. The profile of the high energy tail for 10 MeV beam is similar to the profile for the rest plasma.

DISCUSSION AND CONCLUSION

Stochastic acceleration mechanism is analyzed by the Fokker-Planck approach. The time evolution of the electron distribution function shows the directionality of the acceleration which is the important feature of electron acceleration by the intense laser field. The effective temperature roughly scales as $T \propto t^\beta$ where $\beta \simeq 1$, while $0.5 \leq \beta \leq 1$ is obtained by Langevin approach. The effective temperature becomes higher than the temperature corresponding to the ponderomotive energy. A high energy electron beam is effectively accelerated since the electrons have the feature suitable for stochastic acceleration, $i.e.$, $p_z \gg 1$, and $p_x \ll p_z$.

In the experiment of Gahn $et\ al.$ the laser pulse is 200 fs with 1.2 TW, correspondingly $a = 1.4$, and the plasma density is $n = 4 \times 10^{20} \mathrm{cm}^{-3}$. The measured effective plasma temperature is 2.2 MeV where the observed channel length is about 400 μm. Their result is recovered in our theory by adjusting the time scale, namely $\frac{a^2 D}{2\tau_c^2}$ in eq. (9). Since the radius of the filament is approximately c/ω_p, τ_c is estimated as $\tau_c \simeq c/(\omega_p v_x)\omega_0$ where $v_x \simeq c\sqrt{u_x^2}$ is the average velocity when an electron moves to an adjacent region.

From eq. (5) the average velocity is estimated as $a\sqrt{D}$. Therefore τ_c is estimated as $\tau_c \simeq \omega_0/(\omega_p a\sqrt{D})$. The coefficient of the Fokker-Planck equation becomes $\frac{a^2 D}{2\tau_c^2} \simeq (a^2 D\omega_p/\omega_0)^2/2$. By setting the coefficient $D = 0.07$, the effective temperature at $\omega_0 t = 3100$ which corresponds to the channel length of 400μm is obtained as 2.2 MeV. In their result of angular distribution of electrons, there exist two peaks which are at $\theta = 0°$ and $\theta \simeq 5°$. The result for the second peak might be explained by our theory. In our case, the peak at $\theta = 0$ does not appear since the indirect acceleration mechanism such as wakefield acceleration is not taken into account.

The model that the laser field is divided into regions with randomly different phase is suitable for the understanding of the direct laser acceleration experiment[1,2]. It is observed that the laser pulse propagating through underdense plasma split into the filaments when its power well exceeds $P_c[\text{PW}] = 17(\omega_0^2/\omega_p^2)$ which is a critical power for self-focusing.[8] Also the plasma becomes in a turbulent state when the laser intensity is in relativistic regime. In those cases, the electron experiences the laser field with different phase during the interaction, and the stochastic acceleration plays the important role.

REFERENCE

1. Gahn, C., Tsakiris, G. D., Pukhov, A., Meyer-ter-Vehn, J., Pretzler, G., Thirolf, P., Habs, D., and Witte, K. L., Phys. Rev. Lett. **83**, 4772 (1999).
2. Koyama, K., Saito, N., and Tanimoto, M., Proceedings of the 10th International Congress on Plasma Physics, Quebec city, Canada, October 2000, p. 464.
3. Tajima, T., and Dawson, J., Phys. Rev. Lett. **43**, 267 (1979).
4. Sprangle, P., Esarey, E., and Ting, A., Phys. Rev. Lett. **64**, 2011 (1990)
5. Meyer-ter-Vehn, J., and Sheng, Z. M., Phys. Plasmas, **6**, 641 (1999).
6. Toda, M., Kubo, R., Saito, N., and Hashitsume, N., *Statistical Physics II : Nonequilibrium Statistical Mechanics, Vol. 2*, (Springer-Verlag New York, 1991), Chap. 1.
7. Nakamura, T., Kato, S., Tanimoto, M., and Kato, T., Phys. Plasmas, **9**, 1801 (2002).
8. Wang, X., Krishnan, M., Saleh, N., Wang, H., and Umstadter, D., Phys. Rev. Lett. **84**, 5324 (2000).

Prediction of Hot Electron Production by Ultraintense KrF Laser-Plasma Interactions on Solid-Density Targets

Susumu KATO*, Eiichi TAKAHASHI*, Eisuke MIURA*,
Tatsufumi NAKAMURA†, Tomokazu KATO† and Yoshiro OWADANO*

*National Institute of Advanced Industrial Science and Technology (AIST),
Tsukuba Central 2, 1-1-1 Umezono, Tsukuba, Ibaraki, 305-8568, Japan
†Department of Applied Physics, Waseda University,
3-4-1 Ohkubo, Shinjuku, Tokyo, 169-8555, Japan

Abstract. The scaling of hot electron temperature and the spectrum of electron energy by intense laser plasma interactions are reexamined from a viewpoint of the difference in laser wavelength. Laser plasma interaction such as parametric instabilities is usually determined by the $I\lambda^2$ scaling, where I and λ is the laser intensity and wavelength, respectively. However, the hot electron temperature is proportional to $(n_{cr}/n_{e0})^{1/2}[(1+a_0^2)^{1/2}-1]$ rather than $[(1+a_0^2)^{1/2}-1]$ at the interaction with overdense plasmas, where n_{e0} is a electron density of overdense plasmas and a_0 is a normalized laser intensity.

INTRODUCTION

High energy electron production by an intense ultrashort laser pulse has attracted much interest for the fast ignitor concept in inertial fusion energy (IFE) [1]. The main requirements of the fast ignitor are hole boring through the coronal plasma, penetration to high density, and generation of high energy particles. The ultra-intense irradiation experiments with infrared subpicosecond laser, e.g., Nd:glass ($\lambda = 1053$ nm) or Ti:sapphire ($\lambda = 800$ nm) lasers, the powers and focused intensities of which are exceeding 100 TW and 10^{20} W/cm^2 are possible by chirped pulse amplification (CPA) techniques [2, 3]. Namely, the classical normalized momentum of electrons $a_0 \equiv P_{osc}/mc = (I\lambda_\mu^2/1.37 \times 10^{18})^{1/2} \geq 1$, where m is the electron mass, c is the speed of light, I is the laser intensity in W/cm^2, and λ_μ is the laser wavelength in microns. On the other hand, KrF laser ($\lambda = 248$ nm) has the advantage for the fast ignitor concept that the critical density is close to core and hot electron energies are suitable since the critical density of KrF laser is ten times greater than one of infrared laser [4]. However, KrF laser irradiance intensities have been only the order of 10^{18} W/cm^2, namely $a_0 < 1$ [5, 6]. Several TW KrF laser [7] will provide irradiances exceeding 10^{19} W/cm^2 by focusing to about two times diffraction limited spot size using full vacuum propagation system.

The absorption, electron spectrum, and hot electron temperature have been usually investigated and scaled by the parameters $I\lambda^2$, n_e/n_{cr}, and L/λ [8, 9], where n_e, n_{cr}, and L are the electron density, critical density, and density scale length, respectively.

CP634, *Science of Superstrong Field Interactions*, edited by K. Nakajima and M. Deguchi
© 2002 American Institute of Physics 0-7354-0089-X/02/$19.00

Critical density absorption of the laser light converts laser energy into hot electrons with a suprathermal temperature T_{hot} approximately proportional to $\sqrt{I\lambda^2}$ for $a_0 > 1$, and $T_{hot} \sim [(1+a_0^2)^{1/2} - 1]mc^2$ at moderate density [10], where $mc^2 = 511$ keV is a electron rest mass, $T_{hot} = 160$ keV, 0.92 MeV, and 3.88 MeV for $I\lambda_\mu = 1 \times 10^{18}$, 1×10^{19}, and 1×10^{20} W/cm$^2\mu$m^2, respectively. The scaling of the hot electron temperature is supported by experiments of Nd:glass and Ti:sapphire lasers [11, 12]. On the other hand, the results from one dimensional simulation for normal incidence have shown that $T_{hot} \sim \eta(n_{cr}/n_{e0})^\alpha [(1+a_0^2)^{1/2} - 1]mc^2$ in the density region $10 < n_{e0}/n_{cr} < 100$ and the normalized intensity $4 < a_0^2 < 30$, where $\eta = 1.2 \sim 1.9$ and $\alpha = 1/2$, which weekly depend on $I\lambda^2$ and n_e/n_{cr} [9]. When $a_0 \geq 1$, $T_{hot} \propto \eta(I/n_{e0})^{1/2}$, namely the hot electron temperature is almost independent of the wavelength.

In this paper, the scaling of the hot electron temperature and the electron energy spectrum are reexamined from a viewpoint of the difference in laser wavelength using a particle-in-cell (PIC) simulation code. The difference in hot electron spectrum and flux between KrF laser and Infrared laser is clarified. The results predict the spectrum of the high energy particle generated in a KrF laser experiment at exceeding 10^{19} W/cm^2 irradiance.

PIC SIMULATION

To investigate hot electron generation for oblique incidence, we use the relativistic 1 and $\frac{2}{2}$ dimensional PIC simulation with the boost frame moving with $c\sin\theta$ parallel to the target surface, where θ is an angle of incidence [13, 14]. In the simulation, the targets are the full ionized plastic (CH) and aluminum, and the electron densities are $n_{e0} \sim 3.5 \times 10^{23}$cm^{-3} and 8.6×10^{23}cm^{-3}, respectively. The densities correspond to $n_{e0}/n_{cr} = 20$ and 48 for $\lambda = 0.25$ μm, $n_{e0}/n_{cr} = 78$ and 190 for $\lambda = 0.5$ μm, and $n_{e0}/n_{cr} = 310$ and 770 for $\lambda = 1\mu$m, respectively, where n_{cr} is the critical density. The density profile has a shape density gradient, $n_e(x) = n_{e0}$ for $x \geq 0$ and $n_e(x) = 0$ for $x < 0$. The laser intensity rises in 5 fs and remains constant after that. The irradiated intensity $I = 5 \times 10^{19}$ W/cm^2 and the angle of incidence $\theta = 30°$ (p-polarized). $a_0^2 = 2.3$, 9.2, and 36 for $\lambda = 0.25\mu$m, $\lambda = 0.5\mu$m, and $\lambda = 1\mu$m, respectively. The momentum and energy of electrons, the electromagnetic fields, and the electron and ion densities are measured after 50 fs.

Figures 1 show snapshots of the electron phase space for $n_{e0} = 3.5 \times 10^{23}$cm^{-3}. Figures 1(a-c) and 1(d-f) show for $\lambda = 0.25$ μm and 1 μm, respectively. The profiles of the electromagnetic fields and the electron and ion densities at t = 50 fs are shown in Fig.2. The results show that the electron and ion density profiles of $\lambda = 0.25$ μm are quite different from the profiles of $\lambda = 1$ μm. The gradient on the surface of the solid density plasma is still steep for $\lambda = 1$ μm. However, the gradient for $\lambda = 0.25$ μm is gentle compared with one for $\lambda = 1$ μm. In the case of $\lambda = 0.25$ μm, the electric fields penetrate into dense plasma because of the gentle density gradient. Normalized electron energy distributions are shown for the plastic ($n_{e0} = 3.5 \times 10^{23}$cm^{-3}) and aluminum ($n_{e0} = 8.6 \times 10^{23}$cm^{-3}) in Fig.3(a) and 3(b), respectively. The hot electron temperatures of the plastic ($n_{e0} = 3.5 \times 10^{23}$cm^{-3}) are 190 keV, 120 keV, and 60 keV, for 0.25

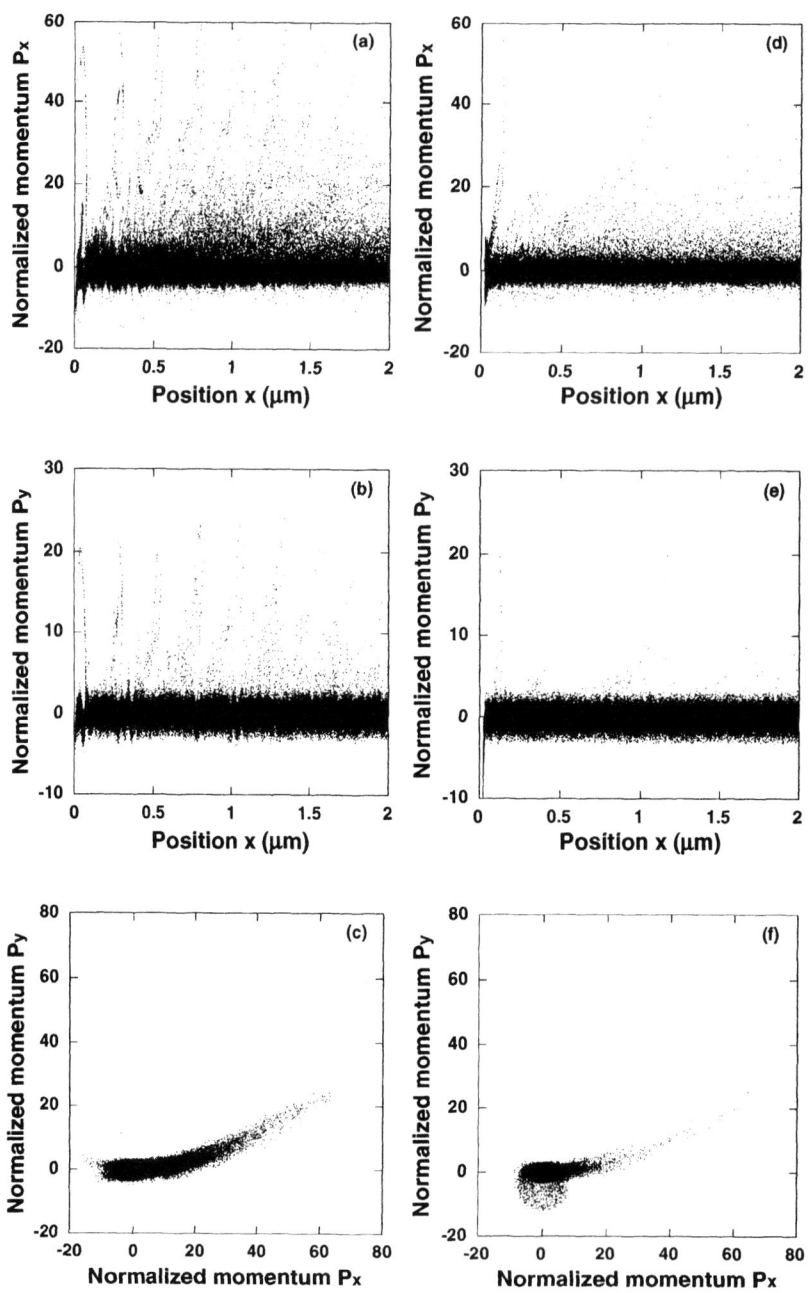

FIGURE 1. Electron phase spaces P_x vs x (a,d), P_y vs x (b,e), and P_y vs P_x (c,f) for $n_{e0} \sim 3.5 \times 10^{23} \mathrm{cm}^{-3}$ at t = 50 fs, respectively. Figures (a)-(c) and (d)-(f) are for $\lambda = 0.25$ and 1 μm, respectively.

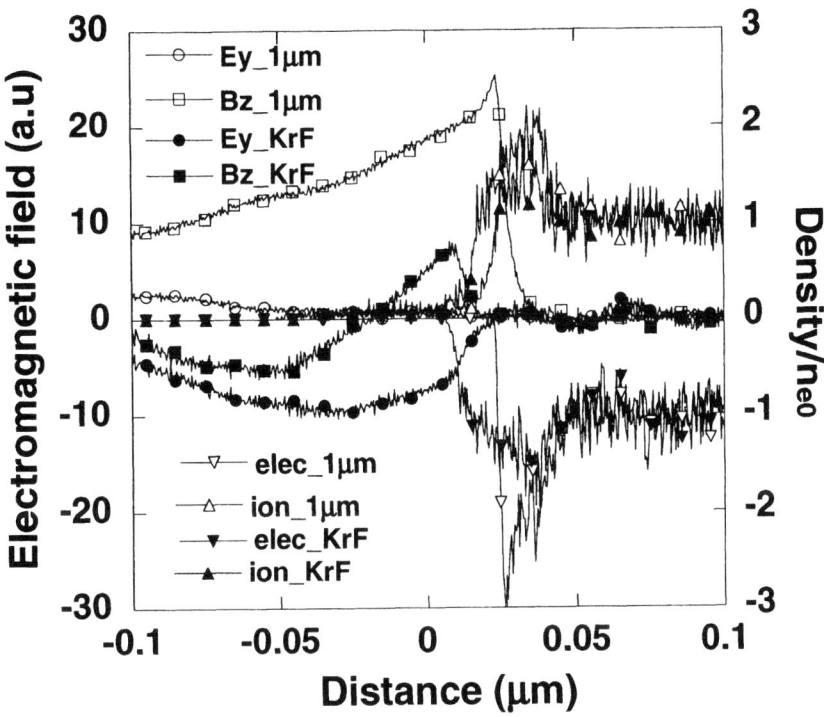

FIGURE 2. Profiles of the electromagnetic fields and the electron and ion densities at t = 50 fs, $n_{e0} = 3.5 \times 10^{23}$ cm^{-3}. The solid and open symbols are for $\lambda = 0.25$ μm and 1 μm, respectively.

μm, 0.5 μm, and 1 μm, respectively. The hot electron temperatures of the aluminum ($n_{e0} = 8.6 \times 10^{23}$ cm^{-3}) are 170 keV, 80 keV, and 40 keV for 0.25 μm, 0.5 μm, and 1 μm, respectively. Applying the simulation results of the hot electron temperature to $T_{\text{hot}} \sim \eta (n_{\text{cr}}/n_{e0})^{1/2} [(1+a_0^2)^{1/2} - 1] mc^2$, $\eta \sim 2.0$ and 2.8 for $\lambda = 0.25 \mu m$, $\eta \sim 0.95$ and 0.98 for $\lambda = 0.5 \mu m$ and $\eta \sim 0.41$ and 0.43 for $\lambda = 1 \mu m$, respectively. In one dimensional simulation for normal incidence [9], η weakly depend on both $I\lambda^2$ and n_e/n_{cr}. $\eta \propto \lambda^{-1}$ in our simulations of p-polarized oblique incidence. It is noted that η strongly depends on the incident angle, polarization, and density scale length, and also depends on n_{cr}/n_e in our simulations.

SUMMARY AND DISCUSSIONS

The hot electron temperature is scaled by $T_{\text{hot}} \sim \eta (n_{\text{cr}}/n_{e0})^{1/2} [(1+a_0^2)^{1/2} - 1] mc^2$ rather than $T_{\text{hot}} \sim [(1+a_0^2)^{1/2} - 1] mc^2$ at the interaction with overdense plasmas. In the condition of our simulations, $\eta \propto \lambda^{-1}$. The difference between our results and Wilks's results may be attributed to the polarization of the laser light. For various laser and plasma conditions, the more quantitative details of the energy and angular spectrum of

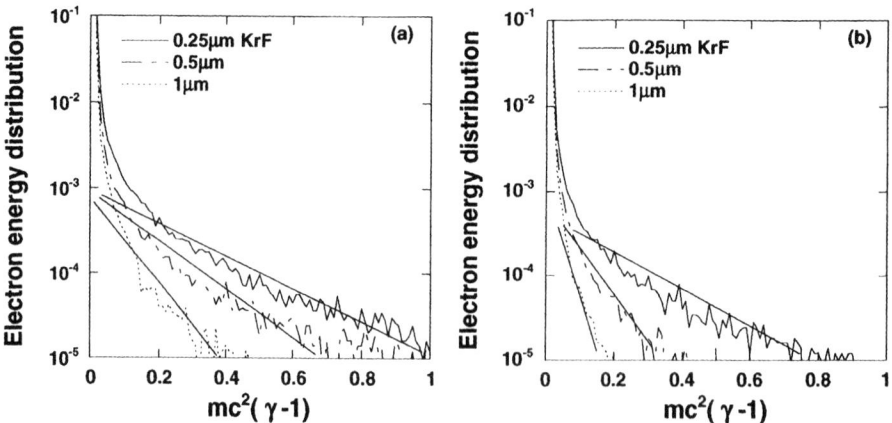

FIGURE 3. Electron energy distribution at t = 50 fs and $I = 5 \times 10^{19}$ W/cm^2 for (a) $n_{e0} = 3.5 \times 10^{23}$cm^{-3} and (b) $n_{e0} = 8.6 \times 10^{23}$cm^{-3}, respectively. The solid, dashed-doted, and dotted lines are for $\lambda = 0.25$ μm, 0.5 μm, and 1 μm, respectively. The hot electron temperatures are (a) 190, 120, and 60 keV and (b) 170, 80, and 40 keV for 0.25 μm, 0.5 μm, and 1 μm, respectively.

the fast particle and the multi-dimensional effects such as surface deformation [10] will be investigated further in the future.

Finally, we discuss the effect of preformed plasma by a prepulse of the intense short pulse laser that is an essential problem in an actual long wavelength experiment [15, 16]. We are afraid of both the absorption of the short pulse in the preformed plasma and the distance between the ablation surface and critical surface. That prevent the interaction between intense KrF laser pulse and solid density plasma. Here, we estimate a preformed plasma profile and volume by the prepulse, the intensity and pulse width of which are assumed to be 10^{12} W/cm^2 and 2 ns, respectively. Here, we estimate a preformed plasma profile and volume by the prepulse, assuming the intensity and pulse width are 10^{12} W/cm^2 and 2 ns, respectively. Total particle number and scale length by the ablation will be 2×10^{20} cm^{-2} and 40 μm. The plasma temperature will be greater than 3 keV, when the intense short pulse energy density > 2J/$\pi(10\mu$m$)^2$ in the preformed plasma and 5% of its energy is absorbed. The absorption in the preformed plasma is negligible, because the effective absorption length with 3 keV is about 50 μm. Intense laser pulse interacts with solid density plasmas, since the stand off distance is proportional to $\lambda^{14/4}$ [17] and short enough compared with wavelength by KrF laser. Therefore, KrF laser is suitable for the interaction with solid density plasmas even if the prepulses exist.

ACKNOWLEDGMENTS

A part of this study was financially supported by the Budget for Nuclear Research of the Ministry of Education, Culture, Sports, Science and Technology, based on the screening and counseling by the Atomic Energy Commission.

294

REFERENCES

1. Tabak, M., Hammer, J., Glinsky, M. E., Kruer, W. L., Wilks, S. C., Woodworth, J., Campbell, E. M., and Perry, M. D., *Phys. Plasmas*, **1**, 1629–1634 (1994).
2. Maine, P., Stricland, D., Bado, P., Pessot, M., and Mourou, G., *IEEE J. Quantum Electron.*, **24**, 398–403 (1988).
3. Perry, M. D., Pennington, D., Stuart, B. C., Tietbohl, G., Britten, J. A., Brown, C., Herman, S., Golick, B., Kartz, M., Miller, J., Powell, H. T., Vergino, M., and Yanovsky, V., *Opt. Lett.*, **24**, 160–160 (1999).
4. Shaw, M. J., Ross, I. N., Hooker, C. J., Dodson, J. M., Hirst, G. J., Lister, J. M. D., Divall, E. J., Kidd, A. K., Hancock, S., Damerell, A. R., and Wyborn, B. E., *Fus. Eng. Design*, **44**, 209–214 (1999).
5. Teubner, U., Uschmann, I., Gibbon, P., Altenbernd, D., Forster, E., Feurer, T., Theobald, W., Sauerbrey, R., Hirst, G., Key, M. H., Lister, J., and Neely, D., *Phys. Rev. E*, **54**, 4167–4177 (1996).
6. Borghesi, M., Mackinnon, A. J., Gaillard, R., Willi, O., and Riley, D., *Phys. Rev. E*, **60**, 7374–7381 (1999).
7. Owadano, Y., "Development of a high-repetition-rate electron-beam-pumped KrF laser," in *Second International Conference on Inertial Fusion Sciences and Applications, 2001, Kyoto, Japan*, 2002.
8. Lefebvre, E., and Bonnaud, G., *Phys. Rev. E*, **55**, 1011–1014 (1997).
9. Wilks, S. C., and Kruer, W. L., *IEEE J. Quantum Electron.*, **33**, 1954–1968 (1997).
10. Wilks, S. C., Kruer, W. L., Tabak, M., and Langdon, A. B., *Phys. Rev. Lett.*, **69**, 1383–1386 (1992).
11. Malka, G., and Miquel, J. L., *Phys. Rev. Lett.*, **77**, 75–78 (1996).
12. Oishi, Y., Nayuki, T., Nemoto, K., Okano, Y., Hironaka, Y., Nakamura, K. G., and Kondo, K., *Appl. Phys. Lett.*, **79**, 1234–1236 (2001).
13. Bourdier, A., *Phys. Fluids*, **26**, 1804–1807 (1983).
14. Gibbon, P., and Bell, A. R., *Phys. Rev. Lett.*, **68**, 1535–1538 (1992).
15. Yu, W., Bychenkov, V., Sentoku, Y., Yu, M. Y., Sheng, Z. M., and Mima, K., *Phys. Rev. Lett.*, **85**, 570–573 (2000).
16. Cowan, T. E., Hunt, A. W., Phillips, T. W., Wilks, S. C., Perry, M. D., Brown, C., Fountain, W., Hatchett, S., Johnson, J., Key, M. H., Parnell, T., Pennington, D. M., Snavely, R. A., and Takahashi, Y., *Phys. Rev. Lett.*, **84**, 903–906 (2000).
17. Manheimer, W. M., Colombant, D. G., and Gardner, J. H., *Phys. Fluids*, **25**, 1644–1652 (1982).

Electron Acceleration in the Moderately Underdense Plasmas

Kazuyoshi Koyama, Naoaki Saito, Eisuke Miura, and MitsumoriTanimoto

National institute of Advanced Industrial Science and Technology (AIST)
Tsukuba Central 2, Umezono 1-1-1, Tsukuba, Ibaraki, 305-8568, Japan

Abstract. The production of high-energy electron beams has been investigated by using a 1.8-TW, 100-fs laser pulse in the moderately underdense plasma of $n_e/n_c \geq 0.1$. The self-channeling was terminated by the filament structure at the length of around 1 mm, even if the large uniform density gas-target was used. The emission of MeV-electron beams correlated with the abrupt transition of the propagation mode from the self-channeling to the filament structure. The experimental results such as the power-law energy spectrum of $E^{-3.5}$ reaching the maximum energy of 2 MeV and the divergence angle of 6-8° were well explained by the stochastic direct laser acceleration model based on the random phase jump taking account of the focusing effect by the self-generated magnetic field.

INTRODUCTION

Since the super-intense and ultra-short laser pulses have a potential ability to accelerate electrons to the highly relativistic energy region in a short distance, experiments are performed at various institutes to realize compact high-energy accelerators. Important issues for utilizing compact high-energy accelerators are to clarify scaling laws of the production of high-energy electrons in various conditions as well as to keep the laser intensity high enough in a plasma for a longer distance than the Rayleigh length of $z_R = \pi r_0^2/\lambda_0$, where r_0 is the waist and λ_0 is the wavelength of the laser. If the laser power P_L is greater than the critical power of the relativistic self-focusing of $P_c[\mathrm{GW}] = 17.4(\omega_0/\omega_p)^2$, the intense laser pulse can propagate for the long distance, where ω_0 and $\omega_p = (n_e e^2/\varepsilon_0 m_e)^{1/2}$ are the laser frequency and the plasma frequency, respectively. From a point of view of producing the long self-channeling as well as getting the high electon current, the higher electron density seems to be better for the laser acceleration if harmful instabilities do not arise. Various schemes of the electron acceleration are proposed and studied; i.e. the laser wake-field acceleration (LWFA), the self-modulated LWFA, the inverse-free electron acceleration and the ponderomotive acceleration. Most of experiments were carried out in the self-modulated LWFA scheme by using the laser pulse duration of longer than 400 fs, because they were easy to excite the electron plasma wave to the large amplitude [1,2].

In case the pulse duration is shorter than 100 fs, the electron density must be in the

CP634, *Science of Superstrong Field Interactions*, edited by K. Nakajima and M. Deguchi
© 2002 American Institute of Physics 0-7354-0089-X/02/$19.00

region of a moderately underdense plasma of $n_e/n_c \geq 0.1$ for exciting the plasma wave by the LWFA, where n_c is the cutoff density. However the high electron density is favorable for the relativistic self-focusing as above mentioned, it is hard to expect to accelerate electrons in the self-modulated LWFA regime because of the insufficient growth of the plasma wave. However, the high-energy electrons are actually generated in the moderately underdense plasmas [3,4]. In this paper, we present experiments on the high-energy electron beam emission from moderately underdense plasmas as well as the theoretical consideration on the stochasticdirect laser acceleration of electrons.

EXPERIMENTS

TW-pulses delivered by a Ti:sapphire laser were focused on the supersonic pulsed gas-jets by using an F/5.5 off-axis parabolic mirror. The peak power and the duration of the laser pulse were 1.8 TW and 100 fs (FWHM), respectively. The gas jet formed a sharply bounded uniform density near the nozzle. The number density of the neutral gas was 3.4×10^{19} cm^{-3} at the reservoir pressure of 800 kPa, which corresponds to the electron density of 2.7×10^{20} cm^{-3} for Ar^{8+}-plasma. The average intensity in a focal spot of 12-15μm in a vacuum was 7×10^{17} W/cm^2, which was surrounded by a large halo of approximately 10^{15} W/cm^2. The laser pulse was focused on the surface of the gas jet. For diagnosing the laser propagation in the plasma, an ultra-fast shadowgraphy and a side scattering of the TW laser pulse were recorded. In the scattering experiment, the direction of the polarization of the TW laser beam was perpendicular to the direction of the observer. A spatial resolution of these optical diagnostic systems was 10 μm.

A typical shadowgraph of the plasma is shown in Fig. 1(a), which was taken just after the TW laser pulses passed through the gas jet. The smooth channeling around 1-

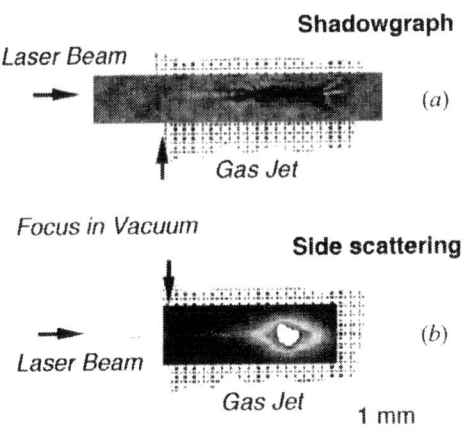

FIGURE 1. The typical shadowgraph (a) and the image of the side-scattering(b) of the plasmas produced in the gas-target of the quasi-uniform density profile.

297

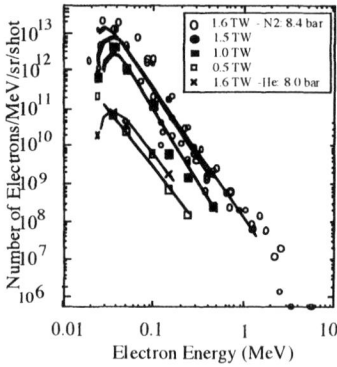

FIGURE 2. Energy spectrum of electron beams measured along the laser axis.

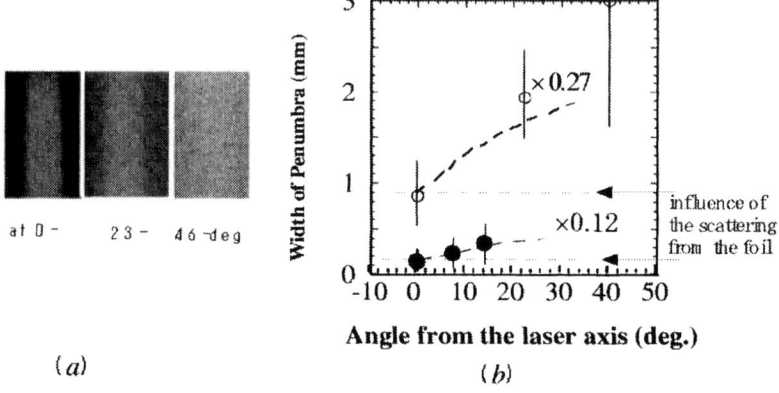

(a) (b)

FIGURE 3 (a)Electron beam shadows of SUS-bolts. (b) Measured width of penumbrae at two different configurations. Arrows indicate the influence of the scattering from the filter.

mm long was terminated by an abrupt transition of laser pulse propagation mode to a complex filamentation in the regime of the high electron density of $n_e \geq 2 \times 10^{19}$ cm^{-3} as well as the high laser power of $P_L \geq 1$ TW. A distance to the transition of the propagation mode of 1.2-0.8 mm was scaled as $P_L^{-0.4} n_e^{-0.8}$. The time-integrated side-scattering image of the TW laser beam is shown in Fig. 1(b). In the side scattered light, no shifted component from the wavelength of the laser was observed. The side-scattering image consisted of a slender, pale and straight filament and a following extremely bright region, which corresponded to the smooth channeling and the part of the complex filamentation in the shadowgraph, respectively. In case that the direction of the polarization of the TW laser was parallel to the direction of the observer, the scattered light from the slender part became very weak; the intensity of the bright part was unchanged, however. In the region of the low laser power or the low electron density, neither the transition to the complex filamentation nor the bright scattering was observed. The smooth channel of the constant diameter suggests that the intense lasepulse is guided by the relativistic self-focusing, since the laser power of 1.8 TW in the gas density of 3.4×10^{19} cm^{-3} corresponds to approximately 5 times higher than the critical power of the relativistic self-guiding (P_c) for a fully ionized helium plasma ($19 P_c$ for Ar^{8+}-plasma).

Angular distributions of high-energy electron beams were measured by using a scintillator-film detector. The energy spectrum of the electron beam was measured by using an electron energy analyzer in the forward direction along the main laser beam. The quality of the electron beam was evaluated by measuring widths of penumbrae of the thick materials placed at various angles from the laser axis. Emissions of high-energy electron beams seemed to coincied with the appearance of the transition feature of the laser pulse propagation. A divergence of the high-energy eletron beam was 6-8°FWHM. The direction of the electron beam fluctuated within the angle of 15-20°FWHM. around the optical axis of the laser from shot to shot. The maximum number of the high-energy electrons was estimated to be $(1.3 \pm 0.6) \times 10^{11}$/sr/shot from the electron current measured by using a Faraday cup. Energy spectra seem to have a power-law of $E^{-3.5}$ reaching the maximum electron energy of 2 MeV as shown in Fig.2. Slopes of spectra seem to be unchanged by the laser power. The electron-beam shadows of SUS-bolts placed at various directions and measured widths of penumbrae are shown in Fig. 3(a) and Fig. 3(b), respectively. One can deduce that the shape of the electron source after subtracting the effect of the blurring due to the electron scattering from the Al-foil which is indicated by arrows in Fig. 3(b). A radius of the source is smaller than a resolving power of 0.1 mm. The source length is estimated to be approximately 4 mm, although the plasma length is 3 mm.

DISCUSSION

In the moderately underdense plasma, it seems that electrons are hardly accelerated by the self-modulated LWFA, because the significantly large difference of phase velocities between the pump ($\approx 1.1c$) and the scattered light ($\approx 1.4c$) leads to the short dephasing length of the Raman Forward Scattering (RFS) instability of 1μm which is insufficient for exciting the plasma wave to the large amplitude, where c is the velocity of the light in the vacuum. The possibility of the direct acceleration by the oscillating electric field of the laser is also excluded, because the ponderomotive potential (the kinetic energy of the quivering of the electron) is approximately 80 keV at the laser intensity of 7×10^{17} W/cm^2 (a_0=0.65), where a_0 is the normalized vector potential of the field. Although the experiments were performed in the density region of $n_e/n_c \geq 0.2$ as well as at the weakly relativistic intensity region of $a_0 < 1$, the high-energy electron beam (>MeV) was observed coinciding with the appearance of the laser propagation changed from the smooth propagation to the complex filamentation as above mentioned. These features suggest the causality between the electron acceleration and the complex filamentation in the plasma. The merging and the splitting of the plasma filaments might cause the phase-disturbance of the oscillating field.

As one of the potential mechanisms, we have studied the stochastic direct laser acceleration that can efficiently accelerate the electrons to the substantially high energy at the weakly relativistic laser intensity. The electron motion can be analyzed by calculating the normalized vector potential of the laser field $a(\tau)$ even if the electric fields $\varepsilon(\tau)$ are discontinuous, since the vector potential is expressed by the integral of the electric filed as

$$a(\tau) = -\int_0^\tau \varepsilon(\tau)d\tau$$

, where $d\tau = dt - dz/c = (1-\beta_z)dt$, $\beta_z = v_z/c$, and z is the direction of the laser propagation.

For electrons initially at rest, there is no net energy transfer from the laser field to these electrons after leaving the laser pulse. However, if the disturbances are introduced into the laser field, electrons can gain energy from the laser field. We modeled these disturbances by the stochastic phase jumps of the coherently oscillating electromagnetic fields. The model shows the maximum kinetic energy of the electron of $E_{max} = \gamma_{max} - 1 = 2m^2 a_0^2$ in case the width of the phase jump $\Delta\phi(\tau)$ are randomly distributed, where m is the number of the phase jump. Since the physics of phase-disturbance of the oscillating field has not been understood, we assume the phase jump obeys the probability function of $p(z) = (1/\lambda)\exp(-z/\lambda)$, where λ is the mean free path of

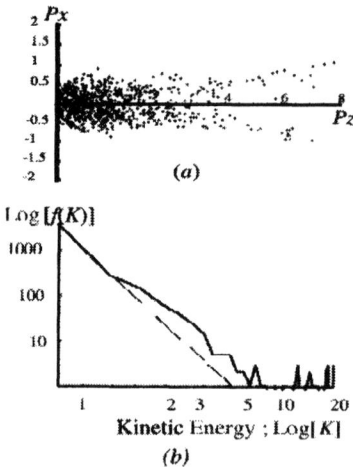

FIGURE 4. The momentum distribution (a) and the energy distribution (b) of 4000 electrons accelerated in the stochastic laser field. The average number of phase jump is $m_0 = 3$. The self-generated magnetic field is taken into account.

300

the phase jump (mean width of the phase jump). The mean number of the phase jump is defined to be $m_0 = z_p / \lambda$, where z_p is the length of the plasma filament. As shown in Fig. 4(b), the energy spectrum was well reproduced in case $m_0 = 3$, which was calculated for 4000 electrons.

The simply deduced divergence angle of the electron beam from the relation between the transverse and the longitudinal momentum gained by the direct acceleration is $\theta = \tan^{-1}(p_x/p_z) = \tan^{-1}(2/(\gamma-1))^{1/2}$. This value of $\theta = 40°$, which corresponds to the observed energy in our experiment, is mach larger than the observed divergence angle of 6-8°. In order to explain the cause of the collimated electron beam which was observed by the experiments, we supposed the electron beam was focused by the magnetic field produced by the beam. By introducing the magnetic field strength of 2MG produced by drifting electron-clouds of a relatively low energy was taken account, the beam divergence of experimentally obtained is well reproduced as shown in Fig. 4(a).

In our experiment, the fine structure followed by the abrupt transition of the propagation of the plasma was recorded by the optical shadowgraph, in spite of the longer duration of the probe pulse of 100 fs than an oscillation period of the electron plasma wave of approximately 10 fs. Hydrodynamic motion of ions is also neglgible in the duration of the probe pulse. This feature suggests that the degree of ionization in the plasma is modulated by growing hot spots in the focal spot, which might cause the phase jumps on the electron motions by merging and splitting of the filaments.

CONCLUSION

We have investigated the production of high-energy electron beams in the moderately underdense plasma of $n_e/n_c \geq 0.1$. The emission of MeV-electron beams correlated with the abrupt transition of the propagation mode from the smooth channeling to the filament structure. The self-channeling was terminated at the length of around 1 mm, even if the uniform density gas-target was used.

The experimentally observed power-law energy spectrum of $E^{-3.5}$ reaching the maximum energy of 2 MeV as well as the divergence angle of 6-8° were well explained by the stochastic direct laser acceleration model introducing the self-generated magnetic field.

ACKNOWLEDGMENTS

A part of this study was financially supported by the Budget for Nuclear Research of the MEXT, based on the screening and counseling by the Atomic Energy Commission and the Advanced Compact Accelerator Development Project supported by the MEXT.

REFERENCES

1. A.Modena, Z.Najmudin, A.E.Dangor, C.E.Clyton, K.A.Marsh, and C.Joshi, IEEE Trans. on Plasma Sci. **24** (1996) 289-295.
2. R.Wagner, S.-Y.Chen, A.Maksimchuk, and D.Umstadter, Phys.Rev.Lett. **78** (1997) 3125-3128.
3. K.Koyama, N.Saito, M.Tanimoto, "Production of MeV electrons in a gas-jet target," in *High-Power Lasers in Energy Engineering* -1999, edited by K.mima, G.L.Kulcinski and W.Hogan, Proc. of SPIE, 1999, Vol.3886, pp.76-80.
4. C.Gahn, G.D.Tsakiris, A.Pukhov, J.Meyer-ter-Vehn, G.Pretzler, P.Thirolf, D.Hobs, and K.J.Witte, Phys.Rev.Lett. **83** (1999) 4772-44775.

Ultra-Intense Laser Pulse Absorption and Fast Particles Generation at Interaction with Inhomogeneous Foil Target

A.A. Andreev[1], T. Okada[2], and S. Toraya[2]

[1]*Institute for Laser Physics, St.Petersburg, Russia*
[2]*Tokyo University of Agriculture and Technology, Tokyo, Japan*

Abstract. The absorption of a short laser pulse with duration 40 fs and the intensity more than 10^{18} W/cm^2 at the interaction with foil targets is analyzed in the theory and particle-in-cell (PIC) simulations. Initially, foil target density distribution has a smooth gradient with a variable scale-length of plasma density inhomogenity L. Laser absorbed energy transforms into the energy of fast electrons, which oscillate in the foil and partly go out from the target transforming their energy into accelerated ions. The analyses of the different mechanisms of ion acceleration in foil plasma, the influence of L and other plasma parameters on ion acceleration have been done. The angular distributions of fast electron and ion beams are calculated. The optimum laser plasma conditions for the maximum ion acceleration are found.

INTRODUCTION

Fast particles generated in laser plasma interaction can be used in many applications from technology to medicine and even for the initiation of tabletop nuclear reactions. Fast ion generation by the interaction of an ultra short high intense laser pulse with plasma has been demonstrated already in some recent theoretical[1-4] and experimental[5-9] papers, where the maximum ion energy up to 0.5 GeV is observed[10]. Different schemes of fast ion generation have been proposed in gas[10] and solid[11] targets. At the moment, the maximum conversion efficiency of laser to ion energy of fast ions has been obtained for foil targets[11]. It has been shown that energy of a laser pulse could be converted into fast ion energy with good efficiency. Some simulations[1-4] have shown that under these conditions the main mechanisms of ion acceleration are the ambipolar field from fast electrons in a sharp-gradient plasma and the Coulomb explosion. It has been shown that the main mechanism of ion acceleration from the foil rear side is the ambipolar acceleration from fast electrons accelerated by a laser field and leaving a foil to create an accelerated field. Most experimental high power lasers produce a pre-pulse, and for this reason the foil target transforms into a plasma layer with a smooth gradient. A collimated ion beam can be achieved by focusing an intense laser onto the surface of a solid film. Due to the small diameter of the laser spot and the anisotropy of the hot electron velocity distribution, the ion emittance can be comparable to or even better than that of electrostatic accelerators. Fast ions accelerate along a normal to foil surface because the ambipolar force is doing on ion along this direction. It is clear that these ions can be focused by a curve foil at some point and it has been shown in a numerical simulation but without any optimization of this process. In the present paper, we make

CP634, *Science of Superstrong Field Interactions*, edited by K. Nakajima and M. Deguchi
© 2002 American Institute of Physics 0-7354-0089-X/02/$19.00

an attempt to develop an analytical model to analyse the mechanisms of ion acceleration and, based on this model and our PIC simulations, to find the optimum foil target.

In this paper we tried to analyze some positive and negative processes for optimum impedance formation of the ion jet during its acceleration by a shaped laser pulse interacted with a curve foil target with a possibility to focus these ions at some distance.

PIC SIMULATONS

We apply a PIC method to simulate the interaction of a plasma layer with an intense ultra-short laser pulse. The method is based on the electromagnetic PIC and it is appropriate for the analysis of the dynamics of over-dense plasmas created by an arbitrary polarized, obliquely incident pulse laser. The 2D (in the rectangular Cartesian coordinate system) relativistic, electromagnetic code is used to calculate the interaction of an intense laser pulse with an over-dense plasma. The calculation with movable ions is carried out for a plasma with a variable initial density profile. Simulations are performed at wavelengths of 1 μm for a laser intensity $I > 10^{18}$ W/cm^2. The laser pulse has a Gauss shape with a duration > 20 fs. The incident angle of the pulse is $0°$. The time step is set to $0.1/\omega_p$, where ω_p is the initial plasma frequency and the spatial step $0.2c/\omega_p$. The number of electrons is 10^7 and the same for the ions. Initial electron temperature is 1 keV and ion temperature is 800 eV. The thickness of the layer varies from 1 μm to several μm.

Figure 1 shows the schematic view of the simulation model. For simplicity, the uniformity in the z-direction is assumed. The relativistic equation of motion and the Maxwell equation are solved for the components x, y, p_x, p_y, p_z and E_x, E_y and B_z

$$\partial p_j/\partial t = q_j(E + v \times B), \quad \gamma_j m_j \partial r/\partial t = p_j, \quad \partial E/\partial t = -j_j + c^2 rotB, \quad \partial p_j/\partial t = -rotE.$$

Here q_j and m_j indicate the charge and mass of a particle, respectively and j_j is the current density.

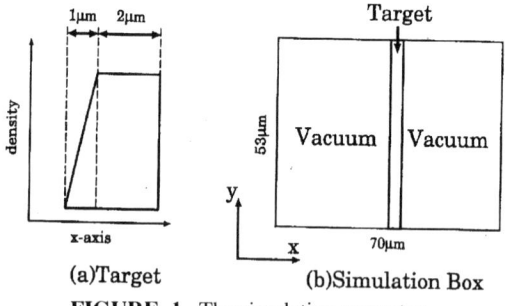

(a)Target   (b)Simulation Box

FIGURE 1. The simulation geometry.

FAST ELECTRON GENERATION

Critical for ion acceleration is the efficiency of the laser-energy conversion into a high-energy electron component, since the latter can produce the requisite strong electrostatic fields. Numerical simulations show that for $I \geq 10^{19}$ W/cm^2 intensity range, the absorption coefficient becomes independent of the angle of incidence, and the absorption is about 10 % without pre-pulse. The laser pre-pulse increases the scale of plasma inhomogenity L and the absorption coefficient η. It was shown that in the range 10^{18}-10^{20} W/cm^2 there is $\eta \propto L$. Follow[11] we will use the scaling of $\eta(L,I)$ as:

$$\eta(L,I)=(0.1+0.01L)I/(30+I)^{0.7}, \tag{1}$$

where I - radiation intensity in units 10^{18} W/cm^2, $L = L\omega_L/c$. $L \approx c_s\,t_{pl}$. Here c_s - ion sound velocity and t_{pl}.- pre-pulse duration.   In Figure 2 we see the simulation result, which confirms our model approximation.

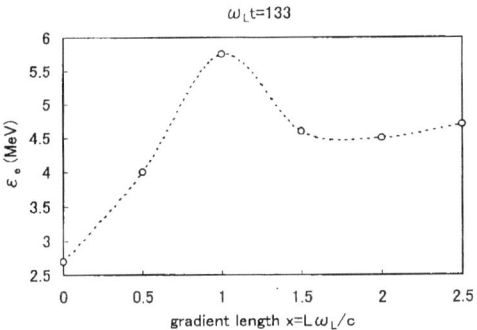

FIGURE 2.   Electron energy dependence on plasma density gradient L.

We make an estimate of the number of fast electrons accelerated by laser pulse as n_{ef} = $K_e(I)\varepsilon_L/\varepsilon_e$, where ε_L is the laser pulse energy, $K_e(I)$ is the transformation coefficient of laser energy into fast electron energy, and ε_e is the energy of an electron. It was shown that in the range 10^{18}-10^{20} W/cm^2 , $K_e(I)$ has a linear rise from 0.03 up to the η for the subpicosecond laser pulses. At a laser intensity of more than 10^{20} W/cm^2 , $K_e(I) \approx \eta$. This means that all the absorbed energy being assumed to be transferred to the motion of high-speed electrons. The average energy of an individual fast electron is obviously specified by the wave field strength inside the skin-layer

$$<\varepsilon_e> \approx m_e c^2(\sqrt{1+\eta_L I_{18}} -1) \tag{2}$$

The maximum energy of a fast electron accelerated by a laser field in a plasma corona can be calculated by this formula[5]:

$$\varepsilon_{eh} = m_e c^2(\gamma - 1) \approx m_e c^2 (p_0/mc)(\upsilon_E/c)^2, \tag{3}$$

here $v_E = eE/m\omega$, initial electron impulse $- p_0 \approx m v_E$, it determined by acceleration in laser field on λ. In Figure 3 we see the simulation results of the energy of fast electrons (solid line) and ions (dotted line), which looks like the dependence on (2). As we consider $I \geq 10^{19}$ W/cm^2 for fast electron energy we take next expression: $E_{eh} = N_{eh}\varepsilon_{eh} \approx \eta_L \varepsilon_L$ where $\varepsilon_{eh} \approx \eta_L I/cn_{eh}$, and for fast electron number: $N_{eh}=\eta_L \varepsilon_L/ \varepsilon_{eh}$. In Figure 4, we see on the front side of the foil plasma target the deepness of the electron density from the laser ponderomotive pressure and accelerated electrons from the surface are cumulated on the laser axis. On the foil rear we see the "fountain" of electrons, which go out and come back to the foil surface from the electrostatic field influence. This structure looks like a ring of electron density on the rear foil surface, which has been observed in the experiment. To understand this effect in more detail, we consider the analytical model:

On the initial stage, when electrons are separated from ions ($t<\omega_{peh}^{-1}c/v_i$), a laser beam produces on a foil the electrostatic field of the spot with the electric charge Q:

$$E = 2\pi\sigma = 2Q/R^2$$

here R is laser focal spot radius.
The time of movement in such a field of fast electron of a velocity v_e is determined by this next formula:

$$t = 2v_e m \gamma/eE = v_e m \gamma R^2/eQ$$

here γ is electron Lorentc factor.
The "fountain" effect produces a ring of non-compensated electric charge on the foil surface with internal radius R, external radius $R+l$ and height h, which are determined by the next formulas:

$$h = \gamma m v_e^2/2eE = \varepsilon_e R^2/2eQ$$
$$l = v_e (\theta - \Delta\theta)t = \theta(1-\beta_e^2)(2\varepsilon_e R^2/eQ) = \theta(2\varepsilon_e R^2/eQ\gamma^2).$$

A magnetic field from electron current produces a deviation angle:

$$\Delta\theta = ev_e H_\varphi t/cmv_e \approx \theta v_e^2/c^2$$

here $\theta< 1$ – angle of electron scattering in a foil target.
The initial can be calculated from the scattering model as follows:

$$<\theta^2> = (21MeV/\varepsilon_e)^2 (4/137)Z^2 r_0^2 n_i L_{foil} ln(183Z^{-1/3}) . \tag{4}$$

In the late stage of the interaction we see from Figure 5 an electron density perturbation because instability develops and its increment is big enough increment

$$\gamma^2 = \frac{8\pi}{3} Z \frac{m_e}{m_i} \omega^2 I_{18} kl_s$$

that at this time there is such a modulation of density. As a plasma surface at this moment has not clear curvature the electrons do not acumulate on the laser axis as before and focused jet disappear. These electrons propagate through foil with a scattering on some angle $<\theta>$, but the main part reaches the rear foil side and goes out. These electrons accelerate ions, but some part of it comes back to the foil surface as we see in Figure 4, if its energy is not enough to overcome the electrostatic barrier:

$$\varepsilon_{eh} \geq e^2 n_{eh} \, \pi L_c d (1 - 2L_c^2/d^2) \tag{5}$$

where d is laser spot size and L_c plasma density gradient.

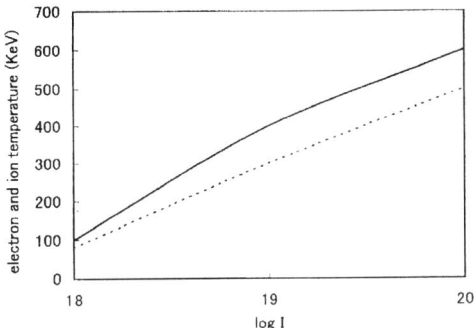

FIGURE 3. The electron and ion temperature dependence on laser intensity.

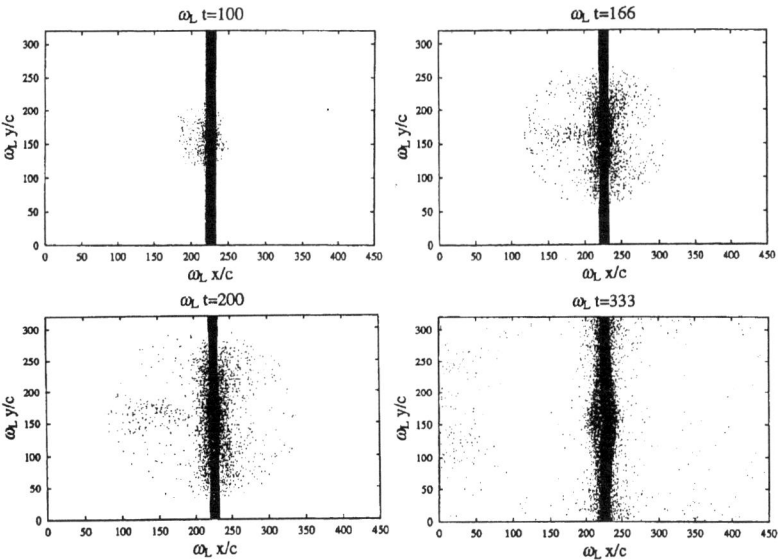

FIGURE 4. Initial stage of laser pulse interaction with foil surface.

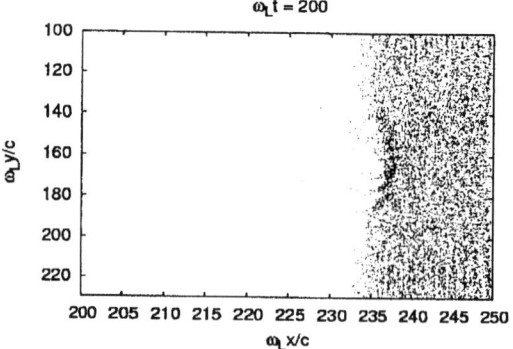

$\omega_L t = 200$

FIGURE 5. Spatial distribution of electron density. Laser beam parameters : diameter $8\,\mu$ m, $I=10^{19}$ W/cm^2, $t=40$ fs. Maximum electron density $n_e = 4n_c$, plasma density gradient $L=1\,\mu$ m, plasma slab length $2\,\mu$ m.

FAST ION PRODUCTION

It is already well known that ions are accelerated on the front side of the foil target and its rear. In the case of the front foil side ion generation and ion acceleration volume is: $V = \pi(d/2)^2 L_c$ and fast ions get electrostatic energy: $e\varphi \approx \varepsilon_{eh}$ from charge separation:

$$\varepsilon_{ih} \approx Z^* e\varphi \approx 3Z^* mc^2 (\eta_L I_{18}\, dL_c/\lambda^2)^{1/2}. \tag{6}$$

The fast proton number N_i we get from the flow of an impulse conservation low:

$$N_i m_i v_i^2 = N_{eh}\, \varepsilon_e v_{eh}^2/c^2,\ \text{as}\ v_{eh} \approx c\colon N_i \sim \eta_L \varepsilon_L/2\varepsilon_{ih}\,.$$

The average energy of fast proton is calculated from the formula :

$$<\varepsilon_p> = \frac{n_{cr}}{n_p} m_e c^2 (\sqrt{1+\eta I_{18}} - 1)$$

This dependence is close to the ion temperature dependence (dotted line) from Figure 3. The total number of fast protons, we can get from quasi-neutrality: $N_i \approx N_{eh} \approx \eta_L \varepsilon_L$ $/<\varepsilon_e>$. It is two times less than the number of ions than the above formula because in this case we have ion movement into two sides. Then the total energy of fast ions is: $E_i \approx (n_c/n_p)\, \eta_L \varepsilon_L$, where from simulations $n_c/n_p \sim 0.5$. In Figure 6, we get the ion space distribution by the PIC simulation. The rate of transversal velocity to the longitudinal one (along laser axis) decreases with the increasing of laser intensity in Figure 7. Ion acceleration on the rear side of a foil target is the most important point for jet formation. In this process we supposed that fast electrons with a concentration of n_{eh} move from ions in the distance of $r_d = v_{eh}/\omega_{ph}$. Then these electrons create the electric charged plane with the electric field:

$$E_{am}=2\pi e n_{eh}\, v_{eh}/\omega_{ph} = \text{E}_{am} = \sqrt{\pi\varepsilon_{eh}n_{eh}} \ .$$

We suppose that foil plasma has a sharp boundary because it has a short laser pulse and no time for a thermal wave to get to this side. The velocity of an ion accelerated by this ambipolar force is:$v_i = (Z/M)E_{am}t_L$, and their energy:

$$\varepsilon_{imax} = T_h\,(n_h/n_c)(m_e/m_i)\,(\omega_0 t_L)^2 \ . \tag{7}$$

From (7) we see that ion energy is proportional to the electron average energy, which means that the dependence of this energy from the plasma gradient is approximately the same as in Figure 2 (see Figure 8). The number of fast electrons we can estimate as before: $n_{eh}=\eta_L\varepsilon_L/\varepsilon_{eh}$. The number of fast protons n_i can be estimated again from the pulse stream conservation of electrons and ions and total ion energy is: $E_i = n_i\varepsilon_{ih}$ $\sim \eta_L\varepsilon_L/2$.

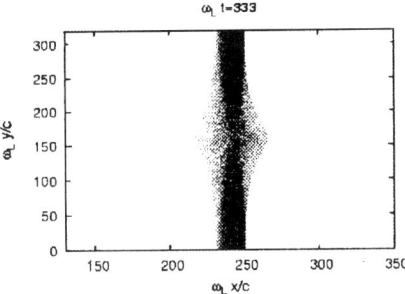

FIGURE 6. Spatial distribution of ion density at laser intensity $I=10^{19}$ W/cm^2, plasma density gradient $L=1\,\mu$ m and plasma slab length $2\,\mu$ m.

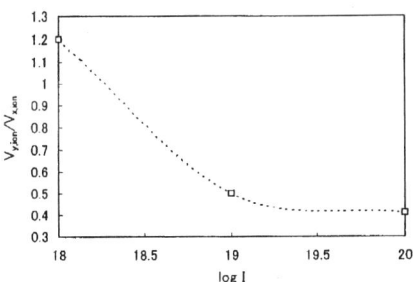

FIGURE 7. The fast ion velocity rate dependence at laser intensity.

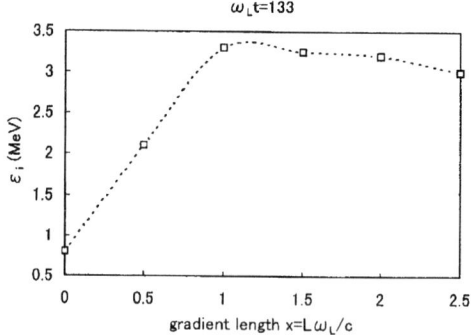

FIGURE 8. Fast ion energy dependence on plasma density gradient.

CONCLUSIONS

Focused before the foil target electron bunch is produced by a curve surface of plasma, a critical density from a ponderomotive laser beam pressure at an early stage of the process occurs. After instability developed the surface is modulated and the bunch disappeared. A laser beam produces a beam of fast electrons and fast ions from the rear surface of the foil target. The average energy of accelerated electrons and ions depends on the laser intensity as a square root. The angle of velocity of the ion at a maximum energy decreases on the laser intensity approximately as a square root. The increase of scale-length of plasma in-homogeneity on some interval increases the laser absorption, particle energy and decreases the angle distribution of the generated ion bunch from the rear side of a foil target.

REFERENCES

1. Wilks S.C., *Phys. Fluids B* **5** , 2603-2608 (1993).
2. Lefebvre E., Bonnaud G., *Phys. Rev. E* **55**, 1011-1014 (1997).
3. Andreev A.A., Platonov K.Yu., *Physics Report* **24**, 26 (1998).
4. Hatchet S.P., Brown C.G., Cowan T.E. *et al.*, *Phys. Plasmas* **7**, 2076-2082 (2000).
5. Wei Yu., Bychenkov V., Sentoku Y. *et al.*, *Phys. Rev. Lett.* **85**, 570-573 (2000).
6. Gurevich A.B., Mescherkin A.P., *ZhETF* **80**, 1810 (1981).
7. Maksimchuk A., Gu S., Flippo K. *et al.*, *Phys. Rev. Lett.* **84**, 4108-4111 (2000).
8. Clark E.L., Krushelnick K., Davies J.R. *et al.*, *Phy. Rev. Lett.* **84**, 670-673 (2000).
9. Krushelnick K., Clark E.L., Zepf M. *et al.*, *Phys. Plasmas* **7**, 2055-2061 (2000).
10. Wilks S.C. Langdon A.B., Cowan T.E. *et al.*, *Phys. Plasmas* **8**, 542-549 (2001).
11. Andreev A.A., Platonov K.Yu., *Laser and Particle Beams* **18**, 81-86 (2000).

Fast Electron Production by Irradiation of Ultrashort Laser Pulses on Copper

Yuji Oishi*, Takuya Nayuki*, Takashi Fujii*, Koshichi Nemoto*,
Tsutomu Kayoiji†, Yasuaki Okano¶, Yoichiro Hironaka¶,
Kazutaka G. Nakamura¶, and Ken-ichi Kondo¶

*Central Research Institute of Electric Power Industry, 2-11-1, Iwado kita, Komae-shi,
Tokyo 201-8511, Japan
†Interdisciplinary Graduate School of Science and Engineering, Tokyo Institute of Technology, 4259,
Nagatsuta-cho, Midori-ku, Yokohama 226-8502, Japan
¶Materials and Structures Laboratory, Tokyo Institute of Technology, 4259 Nagatsuta-cho,
Midori-ku, Yokohama 226-8502, Japan

Abstract. Fast electrons were produced by laser irradiation of 43-fs, 2.7×10^{18} W/cm^2 on a 30 μ m thick copper target. The energy spectra of the electrons were directly measured using a magnetic spectrometer with an imaging plate. The typical temperature was 350 keV for 15° irradiation and was found to be close to the ponderomotive potential at the intensity 2.7×10^{18} W/cm^2. The energy spectra of high-energy photons, which are expected to be produced from the electrons, were also calculated.

INTRODUCTION

The chirped pulse amplification technique opened new fields of laser interaction physics including high energy particle generation such as electrons, x-ray photons, ions, neutrons, and positrons[1]. Among them, the electron generation is the most basic phenomena because other particles are generated by the interaction of electrons with matters. Although relativistic electrons are obtained using a subpicosecond pulse laser, when shorter pulses are used, a smaller energy laser can obtain the intensity which is necessary for production of relativistic electrons. The result is that laser systems become compact and high repetition rate irradiation is possible.

Recently, several groups reported the generation of fast electrons using pulse durations of several ten femtoseconds. Gahn *et al.*[2] estimated 0.9 MeV as the temperature of electrons which were produced by a 130-fs pulse of intensity higher than 10^{18} W/cm^2 when an artificial prepulse of an order of 10^{16} W/cm^2 was combined. Without an artificial prepulse, Schwoerer *et al.*[3] measured the angular variation of the hard x-ray spectrum at 5×10^{18} W/cm^2 using a 60-fs laser, and estimated an electron temperature of 0.7 MeV in the specular direction and 0.3 MeV in the laser forward direction. However, the electron temperatures of these experiments using shorter pulses were not directly estimated, but estimated from photon spectra.

CP634, *Science of Superstrong Field Interactions*, edited by K. Nakajima and M. Deguchi
© 2002 American Institute of Physics 0-7354-0089-X/02/$19.00

Here, we report on results of direct measurements of fast electrons penetrating through copper film targets using 43 fs, 90 mJ pulse without an artificial prepulse [4]. The energy spectra of electrons were measured using a magnetic spectrometer with an imaging plate (IP), which is used for electron microscopy and x-ray imaging. In addition, spectra of high-energy photons originating from the electrons were simulated by GEANT4 [5].

EXPERIMENTAL SETUP

The diagram of the experimental setup is shown in Fig.1. The experiments were performed using a 4 TW Ti:Sapphire laser at the Materials and Structures Lab. in Tokyo Institute of Technology [6], which is based on Chirped Pulse Amplification technique and able to deliver up to 200 mJ, 43 fs pulse duration at the wavelength of 780nm with a repetition rate of 10 Hz. The contrast ratio of the main pulse to the undesirable small prepulse that precedes it by 8 ns is grater than 10^6. In this study, the delivered energy to the vacuum chamber was between 85 and 100 mJ. The p-polarized laser beam was incident at 15 and 45 $^\circ$ to the normal of the target and focused down to a spot size of 5 μm $\times$ 20 μm in FWHM, with an f/3 ($f_{effective}$=152.4 mm) off-axis parabolic mirror. Therefore, the maximum intensity was estimated to be $\sim 2.7 \times 10^{18}$W/cm^2. Tape-like targets (30 μm thick) made of copper were irradiated and its surface was translated for every shot. The surface displacements of the target caused by tape movement was measured to be less than ± 10 μm which was small enough compared with the collimated range of 90 μm. The copper target was chosen because this was the well-known material in the previous experiments for X-ray generation. The energy of electron with the projectile range of 30 μm is 120 keV, therefore, electrons higher than 120 keV was expected to be able to penetrate through the copper target from laser irradiation side to backside and to be captured by an electron detector.

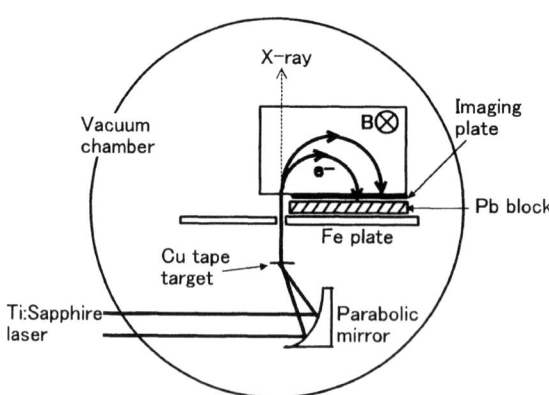

FIGURE 1. Experimental setup

The electron energy spectrum was measured with a magnetic electron spectrometer located 13 cm from the backside surface of the target. The energy of electron deflected by 180° was derived from the two dimensional calculation of electrons trajectories using a map of the measured magnetic field, which was about 1.3 kG. Because of mechanical structure of a tape drive equipment, only electrons emitted to the target normal direction were detected by the spectrometer. The slit width of the spectrometer entrance was 5 mm, so an energy resolution is calculated to be about ±50 keV. The IP [Fuji Film FDL-UR-V] were used to monitor the electron flux and were covered with 15 μm thick aluminum film as a light shield. A lead block was used to shield the imaging plates from X-ray generated by laser produced plasma. The electron flux was calibrated using the sensitivity data provided by the manufacturer in the energy range between 50 keV and 1 MeV. For the range higher than 1 MeV, the electron flux was corrected by a linear extrapolation of the sensitivity data. The IP sensitivity has a maximum value at 150 keV and gradually decreases to about 0.18 of the maximum value at 1 MeV.

We obtained very close signal intensities per one shot with 100 times different shot numbers, therefore, the response of an imaging plate was linear enough and the deviation of imaging plates did not cause any significant errors to the temperature measurements. The background noise obtained by reversing the magnetic field was negligible.

RESULTS AND DISCUSSIONS

Figure 2 shows an example of an electron image recorded on the IP. Regions where electrons reached are dark. From this figure, it is evident that relativistic electrons with energies more than 1.0 MeV were generated. The energy spectra of electrons for laser irradiation at 15° and 45° are shown in Fig. 3. These spectra were obtained with 20 shots in both cases.

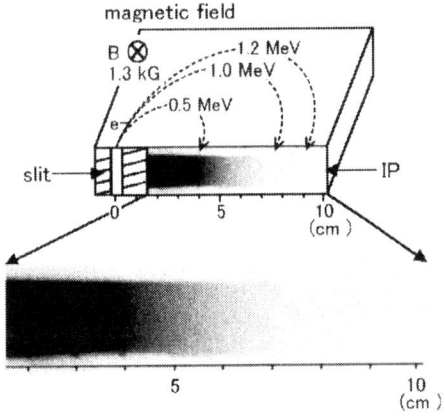

FIGURE 2. Electron trajectories and electron image recorded on the imaging plate attached to spectrometer.

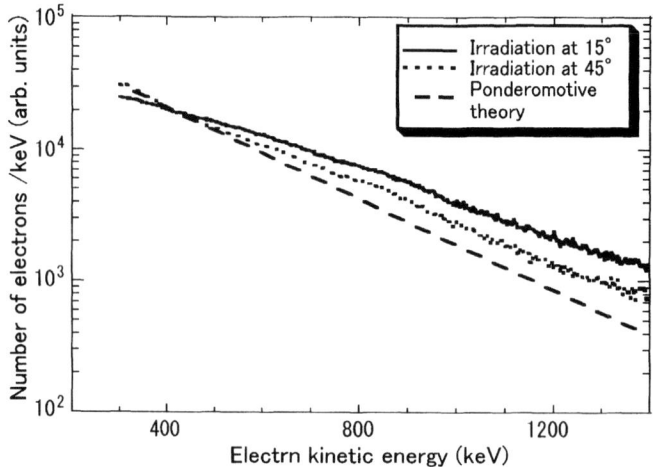

FIGURE 3. Spectra of fast electrons penetrate the 30 μm thick copper target measured at normal direction to the target surface for laser irradiation at 15° and 45° . The Broken line shows ponderomotive scaling.

The electron temperatures were estimated from fitting to the Boltzmann distribution in the range between 500 keV and 1 MeV because in the low energy region around 200 keV, the sensitivity of the IP changes significantly. The estimated temperatures were 350 keV for the irradiation at 15° and 300 keV for the irradiation at 45° . The broken line shows ponderomotive scaling described by $k_B T_h = m_0 c^2 \left(\sqrt{1 + I\lambda^2 \left[W/cm^2 \cdot \mu m^2 \right] / 1.38 \times 10^{18}} - 1 \right)$ [7]. Although the laser intensities of these experiments were on the border of the relativistic region where the ponderomotive force is dominant, the obtained electron temperatures were close to the ponderomotive potential of 250 keV. We observed fast electrons for the irradiation at 15° , which was not expected to cause significant resonance absorption. Therefore, the main acceleration mechanism should be due to the ponderomotive force. Figure 4 shows the relation between laser intensities and electron temperatures. Our result is marked as an asterisk, which is almost on the curve of the poderomotive scaling.

These fast electrons can produce high-energy photons by bremsstrahlung, and the estimation of photon energy and yield is important for photon applications. The photon spectrum for the 30 μm thick copper target was calculated using a simulation program GEANT4 and the results are shown in Fig. 5. In this calculation, x-rays were generated by electrons which were assumed to be accelerated to a high temperature of 350 keV in the direction normal to the target surface and the cut-off electron energy was assumed to be 3 MeV. The estimated temperature of the x-ray spectrum was 260 keV. In Schwoere's experiments,[10] a laser pulse at an intensity of 5.0×10^{18} W/cm^2 with 60 fs duration was irradiated on a 1 mm thick tantalum target and x-ray temperature was estimated at 200 keV. We also calculated a photon spectrum for a 1 mm thick tantalum target, by which efficient x-ray production can be expected. The calculated temperature was 250 keV, which was close to the results of Schwoerer *et al.*

[3] and the estimated intensity of photons was ten times larger than that for 30 μ m thick copper target, as indicated in Fig.5.

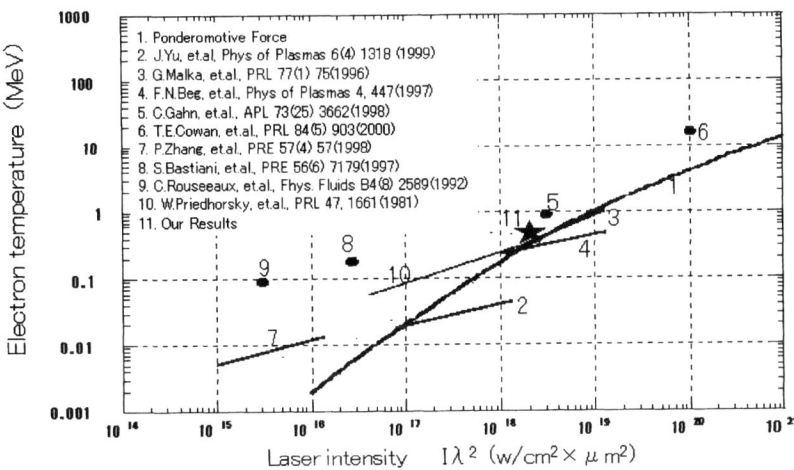

FIGURE 4. Relation between laser intensities and electron temperatures. Our result is marked as an asterisk.

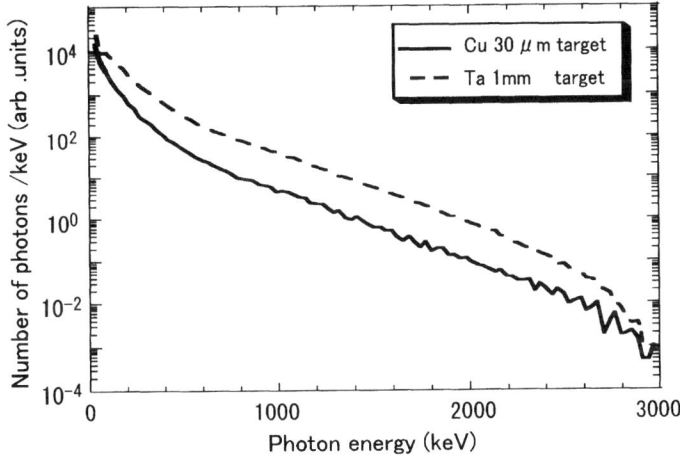

FIGURE 5. Calculated energy spectra of photons which are generated by the interaction of electrons with 30 μ m thick copper target and 1mm thick tantalum target.

SUMMARY

In summary, fast electrons were produced by laser irradiation of 43-fs, 2.7×10^{18} W/cm^2 on a 30 μ m thick copper target and the energy spectra were directly measured using a magnetic spectrometer with an imaging plate. The typical temperature was 350 keV for 15° irradiation and was found to be close to the ponderomotive potential at the intensity 2.7×10^{18} W/cm^2. With ultrashort pulses, particles with energy corresponding to the ponderomotive potential can be produced by a small laser. Higher temperatures may be possible with an artificial prepulse, and a small and high-repetition ultrashort laser can be used for studies of high-energy particle generation.

ACKNOWLEDGMENTS

Authors would like to thank T. Fukuchi, S. Akita, S. Sasaki of Central Research Institute of Electric Power Industry and T. Yamazaki of The Institute of Physical and Chemical Research for their support and helpful discussions, and M. Hasegawa of Tokyo Institute of Technology for experimental support. The sensitivity data of the imaging plate was courtesy of Fuji film co. We also thank T. Suzuki of Univ. of Tokyo for useful advice for simulation using GEANT.

REFERENCES

1. Mourou, G., and Umstadter, D., *Phys. Fluids B* **4**, 2315 (1992); Umstadter, D., *Phys. Plasmas* **8**, 1774 (2001); Wilks, S., C., Kruer, W., L., Tabak, M., and Langdon, A., B., *Phys. Rev. Lett.* **69**, 1383 (1992);Cowan, T., E., Hunt, A., W., Phillips, T., W., Wilks, S., C., Perry, M., D., Brown, C.,Fountain, W., Hatchett, S., Johnson, J., Key, M., H., Parnell, T., Pennington, D., M., Snavely, R., A., and Takahashi, Y., *Phys. Rev. Lett.* **84**, 903 (2000); Kmetec, J., D., Gordon III, C., L., Macklin, J., J., Lemoff, B.E., Brown, G., S., and Harris, S., E., *Phys. Rev. Lett.* **68**, 1527 (1992); Nemoto, K., Maksimchuk, A., Banerjee, S., Flippo, K., Mourou, G., Umstadter, D., and Bychenkov, V., Yu., *Appl. Phys. Lett* **78**, 595 (2001); Norreys, P.A., Fews, A., P., Begg, F., N., Bell, A., R., Dangor, A., E., Lee-P, P., Nelson, M., B., Schmidt, H., Tatarakis, M., and Cablel, M., D., P., *Plasma Phys. Contr. Fus.* **40**, 175 (1998); Ditmire, T., Zweiback, J., Yanovsky, V., P., Cowan, T., Hays, E., G., and Wharton, K., B., *Nature* **398**, 489 (1999); Gahn, C., Tsakiris, G., D., Pretzler, G., Witte, K., J., Delfin, D., Wahlstrom, C.-G. and Habs, D., *Appl. Phys. Lett.* **77**, 2662 (2000).
2. Gahn, C., Pretzler, G., Saemann, A., Tsakiris, G., D., Witte, K., J., Gassmann, D., Schatz, T., Schramm, U., Thirolf, P., and Habs, D., *Appl. Phys. Lett.* **73**, 3662 (1998).
3. Schwoerer, H., Gibbon, P., Dusterer, R. Behrens, R., Ziener, C., Reich,C., and Sauerbrey, R., *Phys. Rev. Lett.* **86**, 2317(2001).
4. Oishi, Y., Nayuki, T., Nemoto, K., Okano, Y., Hironaka, Y., Nakamura, K., G., and Kondo, K., *Appl. Phys. Letters* **79**, 1234 (2001).
5. CERN Program Library at http://geant4.web.cern.ch/geant4/
6. Yoshida, M., Fujimoto, Y., Hironaka, Y., Nakamura, G., K., Kondo, K., Ohtani, M., and Tsunemi, H., *Appl. Phys. Letters* **73**, 2393 (1998).
7. Malka, G., and Miquel, J., L., *Phys. Rev. Lett.* **77**, 75 (1996).

Identification of Ions Generated by Ultrashort Laser Pulses using Thomson Spectrometer

Yuji Oishi*, Takuya Nayuki*, Takashi Fujii*, Koshichi Nemoto*,
Tsutomu Kayoiji[†], Kazuhiko Horioka[†], Yasuaki Okano[¶],
Yoichiro Hironaka[¶], Kazutaka G. Nakamura[¶], and Ken-ichi Kondo[¶]

*Central Research Institute of Electric Power Industry, 2-11-1, Iwado kita, Komae-shi,
Tokyo 201-8511, Japan

[†]Interdisciplinary Graduate School of Science and Engineering, Tokyo Institute of Technology, 4259,
Nagatsuta-cho, Midori-ku, Yokohama 226-8502, Japan

[¶]Materials and Structures Laboratory, Tokyo Institute of Technology, 4259 Nagatsuta-cho,
Midori-ku, Yokohama 226-8502, Japan

Abstract. Fast ions generated by irradiation of laser pulses of width 55 fs, intensity 8.6×10^{18} W/cm^2 on 5 μ m thickness copper film were measured by use of a Thomson mass spectrometer. From the spectgram, ions ejected from the target surface which was opposite side of the laser irradiation were determined to be protons. Copper ions were not observed. From the enegy measurements using Mylar filter method, the maximum proton energy was estimated more than 650 keV.

INTRODUCTION

High-energy particle generation by T-cubed laser is an attractive current issue and many studies are reported [1,2]. However, most of their pulse durations are subpicoseconds, and few investigations have been conducted with pulse durations of several ten femtoseconds. Using shorter pulses, we can generate high-energy particles with a smaller energy, so that the laser system becomes more compact and irradiation at higher repetition rate is possible. Thus, it is important to clear whether high-energy particles are accelerated in several ten femtoseconds.

Previously, we have measured the energy spectra of fast electrons produced by ultrashort, high-intensity laser pulses using a magnetic spectrometer with an imaging plate [3]. Relativistic electrons with energies more than 1 MeV were observed for irradiation on 30 μ m thick copper film by pulses of width 43 fs, intensity 2.7×10^{18} W/cm^2, repetition rate 10 Hz without artificial prepulses. The typical temperature of electrons was estimated to be 350 keV and was found to be close to the ponderomotive potential.

Here, we report the measurement of energetic ions from copper film target by use of the same laser system. Identification of generated ions was conducted by a Thomson mass spectrometer. The maximum energy of ions was also estimated using Mylar filter method.

CP634, *Science of Superstrong Field Interactions*, edited by K. Nakajima and M. Deguchi
© 2002 American Institute of Physics 0-7354-0089-X/02/$19.00

EXPERIMENTAL SETUP

Figure 1(a) shows a schematic drawing of the experimental setup. The experiments were performed using a 4 TW Ti:Sapphire laser, which is based on the CPA technique and is able to deliver up to 200 mJ, 43-fs pulses at the wavelength of 780 nm with a repetition rate of 10 Hz [4]. The contrast ratio of the main pulse to the undesirable small prepulse that precedes it by 8 ns is greater than 10^6. In this study, the delivered energy to the vacuum chamber was 185 mJ. The p-polarized laser beam was incident at the angle of $30°$ relative to the perpendicular to the target surface and was focused down to a spot size of 5 μm × 10 μm in FWHM with an f/3 (f $_{eff}$=152.4 mm) off-axis parabolic mirror. The maximum intensity was estimated to be $\sim 8.6 \times 10^{18}$ W/cm^2.

Thin copper film, of which thickness was 5 μm, was used as a target. The copper film was translated for every shot so that new fresh surface was irradiated. Ions ejected from the target surface which was opposite side of the laser irradiation were collimated by two 500 μm pinholes and passed through a Thomson mass spectrometer. A Thomson mass spectrometer analyzes ions into different energy and ratio of mass to charge. Ions with the same charge to mass ratio reach the same parabola on the detector plane with higher energy ions closer to the center. The schematic diagram of our Thomson mass spectrometer is shown in Fig. 1(b). When a high voltage was applied to the electrodes, parallel electric and magnetic fields existed between the electrodes, deflecting ions according to their energies and ratios of mass to charge. The electric and magnetic fields were 3.3×10^5 V/m and 5.1 k Gauss, respectively. The deflection length by the fields was 4.5 cm. A nuclear track plate as known CR-39 located 10 cm from the center of the deflector was used as the ion detector. The CR-39 is sensitive to ions with energies greater than 100 keV/nucleon. The tracks on the CR-39 were observed with an optical microscope and a CCD camera after etching for 6 hours in 6.25 N NaOH solution.

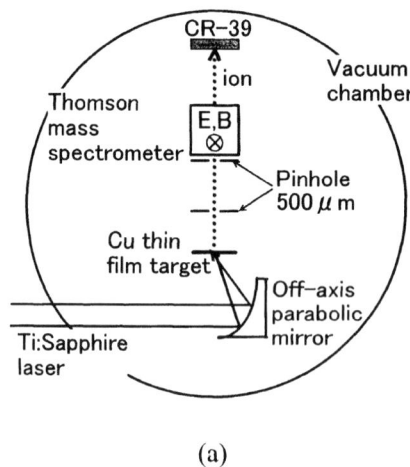

(a)

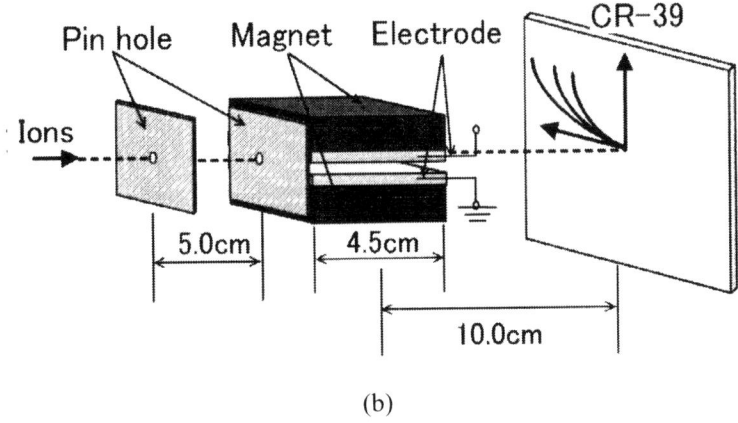

(b)

FIGURE 1. (a) Experimental setup. (b) Thomson mass spectrometer

RESULTS AND DISCUSSIONS

Figure 2 shows the experimental results of the mass spectrograms. The magnetic and the electric fields deflected ions toward horizontal and vertical directions, respectively. A thin black curve is observed in the left hand side of Fig. 2(a). The magnified image of this curve is shown in Fig. 2(b). From this figure, a black curve is determined to be a Thomson parabola, because (i) the curve consists of ion tracks, (ii) the width of the curve was about 500 μm, which corresponded with the diameters of the pin holes, and (iii) the experimental curve was very close to the theoretical proton parabola, which is shown as a broken line in Fig. 2(a). The origin of the protons was derived from contaminants on the target surface. Parabolas for copper ions were not observed, although the target was made of copper.

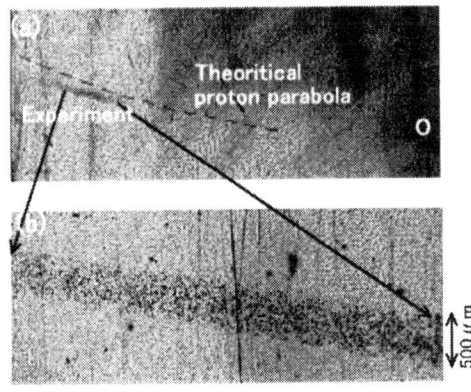

FIGURE 2. (a) Thomson spectgram for proton. (b) Magnified image of (a)

Figure 3 shows the energy spectrum of fast protons estimated by the counting of track numbers in the parabola. The counting of the number was conducted at intervals of 500 μ m in the horizontal direction and the numbers of protons were translated into the number per keV per sr. Proton intensity gradually decreases with increasing kinetic energy and drops sharply around 210 keV. The maximum energy was estimated to 250 keV from the spectrum. The total number of protons was 1.1×10^9 /sr/shot.

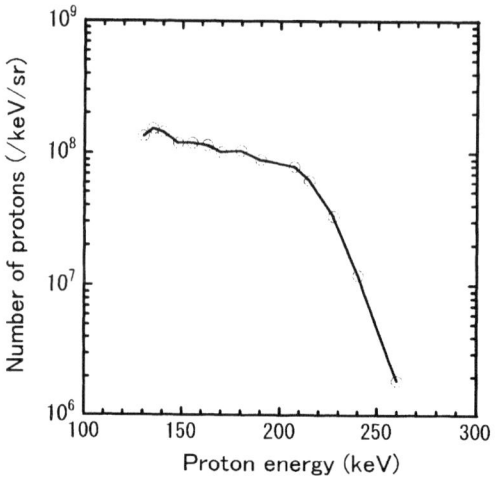

FIGURE 3. Spectrum of fast protons.

In another experiment, we measured the maximum energy of protons using steps of Mylar filter instead of a Thomson mass spectrometer. Figure 4(a) shows the experimental setup. The CR-39 was covered with seven steps of Mylar filter of thickness from 2.5 to 17.5 μ m at 2.5 μ m intervals. The spatial distribution of the thickness of Mylar filter looking from the ion beam axis direction is shown in Fig. 4(b). Figure 4(c) shows the image of protons penetrated the steps of Mylar filter. Regions where protons reached are dark and where protons could not reach are light. From the figure, it is evident that fast protons with energies more than 650 keV were generated, while protons with energies more than 780 keV were not generated. The cutoff energies by the filter were calculated using a program *PSTAR* [5].

Maximum energies of protons measured by the Thomson mass spectrometer and measured using Mylar filter method are different, and both values are less than those reported before [6-11]. Protons with energies of several MeV are reported for laser intensities of order 10^{18} W/cm^2 and 10 MeV are expected for our laser intensity of 8.6×10^{18} W/cm^2, according to experimental equation $E_{p\,max} = 1.2 \times 10^{-2} \times I^{0.313}$ [keV] suggested by F. N. Beg *et al* [6]. Thus, it seems that protons are not so accelerated with pulse durations of several ten femtoseconds as to be with subpicosecond durations, although laser intensities are same order. However, our values were the preliminary results and higher energies can be obtained by optimizations. Further experiments and optimizations are needed to discuss in more detail.

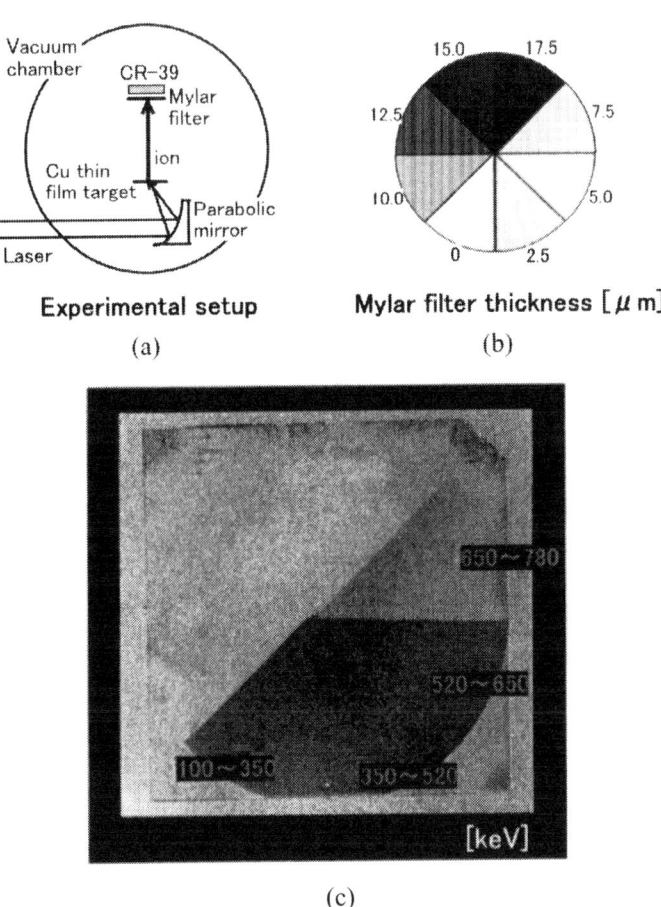

FIGURE 4. (a) Experimental setup for measurement of proton energy with Mylar filter. (b) Spatial distribution of thickness of Mylar filter looking from the ion beam axis direction. (c) Image of protons penetrated the steps of Mylar filter. Protons reached in dark regions.

SUMMARY

In summary, fast ions were generated by irradiation of ultrashort (55fs) laser pulses on copper film. From the Thomson mass spectgram, ions ejected from the copper target surface which was opposite side of the laser irradiation were found to be protons. Copper ions were not observed. From the energy measurements using Mylar filter method, the maximum proton energy was estimated more than 650 keV. However, this value was the preliminary result and higher energies can be obtained by optimizations.

ACKNOWLEDGMENTS

Authors would like to thank T. Fukuchi, S. Akita, S. Sasaki of Central Research Institute of Electric Power Industry and T. Yamazaki of The Institute of Physical and Chemical Research for their support and helpful discussions, and M. Hasegawa of Tokyo Institute of Technology for experimental support.

REFERENCES

1. Mourou, G., and Umstadter, D., *Phys. Fluids B* **4**, 2315 (1992).
2. Umstadter, D., *Phys. Plasmas* **8**, 1774 (2001), and references therein.
3. Oishi, Y., Nayuki, T., Nemoto, K., Okano, Y., Hironaka, Y., Nakamura, K., G., and Kondo, K., *Appl. Phys. Letters* **79**, 1234 (2001).
4. Yoshida, M., Fujimoto, Y., Hironaka, Y., Nakamura, G., K., Kondo, K., Ohtani, M., and Tsunemi, H., *Appl. Phys. Letters* **73**, 2393 (1998).
5. PSTAR Program at http://physics.nist.gov/PhysRefData/Star/Text/PSTAR.html
6. Beg., F., N., Bell, A., R., Dangor, A., E., Danson, C., N., Fews, A., P., Glinsky, M., E., Hammel, B., A., Lee, P., Norreys, P., A., and Tatarakis, M, *Phys. Plasmas* **4**, 447 (1997).
7. Maksimchuk, A., Gu, S., Flippo, K., Umstadter, D., and Bychenkov, V., Yu., *Phys. Rev. Lett.* **84**, 4108 (2000).
8. Nemoto, K., Maksimchuk, A., Banerjee, S., Flippo, K., Mourou, G., Umstadter, D., and Bychenkov, V.,Yu., *Appl. Phys. Lett* **78**, 595 (2001).
9. Clark, E., L., Krushelnick, K., Zepf, M., Beg, F., N., Tatarakis, M., Machacek, A., Santala, M., I., K., Watts, I., Norreys, P., A., and Dangor, A., E., *Phys. Rev. Lett.* **85**, 1654, (2000).
10. Hatchett, S., P., Brown, C., G., Cowan, T., E., Henry, E., A., Johnson, J., S., Key, M., H., Koch, J., A., Langdon, A., B., Lasinski, B., F., Lee, R., W., Mackinnon, A., J., Pennington, D., M., Perry, M., D., Phillips, T., W., Roth, M., Sangster, T., C., Singh, M., S., Snavely, R., A., Stoyer, M., K., Wilks, S., C., Yasuike, K., *Phys. Plasmas* **7**, 2076 (2000).
11. Murakami, Y., Kitagawa, Y., Sentoku, Y., Mori, M., Kodama, R., Tanaka, K., A., Mima, K., Yamanaka, T., *Phys. Plasmas* **8**, 4138 (2001).

Electron-Positron Pair Production by Ultra-Intense Lasers

K. Nakashima*, T. E. Cowan* and H. Takabe†

*General Atomics, San Diego, CA 92122 USA
†Institute of Laser Engineering, Osaka University, Osaka, 565-0871 JAPAN

Abstract. Pair production by ultra-intense lasers are studied by use of numerical code based on Fokker Planck equation. Using ultra-intense laser with the intensity more than 10^{19}W/cm^2, electron-positron pairs are produced from hot electrons accelerated by the laser field. The number of positrons reaches 10^{-4} t the number of hot electrons. To analyze the physical processes of laser pair production, a numerical code based on relativistic Fokker-Planck equation are developed. The results from this simulation agree with the theoretical studies and experimental studies. The positron yield is 10^{10} when the laser intensity 10^{20}W/cm^2, laser energy 280J, wave length 1μm. The number of pair production increases at 10^{19}W/cm^2, and starts to saturate at 10^{21}W/cm^2.

INTRODUCTION

The pair production by ultra-intense lasers are shown in this paper. A positron is an anti-matter of an electron, which has the mass, 0.511MeV, exactly same quantity of an electron. To produce an electron-positron pair, more than 1MeV energy is needed. Ultra-intense laser can load this energy into plasmas.

The energy of electrons quivering in laser electro-magnetic field is expressed as follows[1],

$$\varepsilon = mc^2 \left\{ 1 + \frac{1}{2} \left(\frac{eA}{mc} \right)^2 \right\}, \qquad (1)$$

where ε is the energy of electron, m is the electron rest mass, c is the velocity of light, e is the electron charge, and A is the electro-magnetic potential. When a laser intensity exceeds the 10^{19}W/cm^2, the electrons have relativistic energy, that is, $\varepsilon > mc^2$. The electron energy reaches about 7MeV with an laser intensity 10^{21}W/cm^2. This energy is enough to produce an electron-positron pair.

There are many electron acceleration mechanism not only quivering but beat wave, wake field, and self-modulated wake field accelerations *etc.*[2–4] The energy of electrons reaches higher energy region using these mechanisms. These effects increase the number of pair productions.

The well-known plasma simulation method is particle in cell method (PIC). The studies using PIC method have brought many remarkable results[5, 6]. But PIC simulation cannot treat electro-magnetic waves with very short wave length like x-ray or γ-ray because the mesh width on numerical space has the order of Debye length. Therefore we directly solved the transport equation for electrons, photons and positrons.

CP634, *Science of Superstrong Field Interactions*, edited by K. Nakajima and M. Deguchi
© 2002 American Institute of Physics 0-7354-0089-X/02/$19.00

PAIR PRODUCTION BY ULTRA-INTENSE LASERS

It was noticed that intense lasers with more than 10^{19}W/cm^2 is needed to produce electron-positron pairs in the previous section. These high intensity laser have been realized using CPA method[7, 8] from early '90. So the experimental study is just started recently.

There are mainly two processes to produce electron-positron pairs from laser energy. One is to produce them directly from laser energy. The potential energy within the Compton length of electron becomes comparable to two times of the electron rest mass.

$$eE\lambda_C = 2mc^2 \qquad (2)$$

The laser intensity which satisfies this equation is about 10^{28}W/cm^2. If the laser is a plane wave, the pair production does not occur because of the violation of momentum conservation. Bunkin *et al.* estimated pair production in case that a Gaussian laser is focused with 45° cone angle in the vacuum[9]. This laser has a curved wave surface. In this case, the laser power to produce pairs is 10^{19}W. This laser intensity at the beam waist is almost same with 10^{28}W/cm^2. However, it is too high intensity to construct the lasers with recent technology.

Another process is pair production from high energy electrons which generate in the laser-plasma interactions. The threshold intensity relatively low in comparison with direct pair production. The threshold is 10^{19}W/cm^2[10, 11]. In addition, there are two pair production processes from high energy electrons. One is an interaction between an electron and a nucleus (Trident pair production).

$$e^- + Z \to e^+ + 2e^- + Z, \qquad (3)$$

where e^- is electron, e^+ is positron, and Z is nucleus. The other process is the interaction between Bremsstrahlung photon and nucleus (Photo pair production).

$$\gamma + Z \to e^+ + e^- + Z, \qquad (4)$$

Liang *et al.* showed that Trident process becomes dominant when a target thickness is about a few micron, and derived positron production rate[12]. The positron yield reaches 10^{-4} times of the number of produced hot electrons.

At the same time, Gryaznykh *et al.* estimated the positron yield by ultra-intense lasers independently. they showed that photo pair production becomes dominant when the target thickness is a few milli-meters[13]. One can find the dominant pair production process alters if the target thickness changes from the order of micron to milli-meter. The cross point is at 20μm[14].

There are also several studies on pair production by ultra-intense lasers in Refs[15–19].

So far theoretical studies are more advanced than experimental studies, but the intense lasers with capability of pair production can be built using CPA technology now. So a few pair production experiments were done. Cowan *et al.* carried out this first experiment using NOVA laser system in Lawrence Livermore National Laboratory[20]. In this experiment, the number of positrons estimated are 10^{-4} of hot electrons.

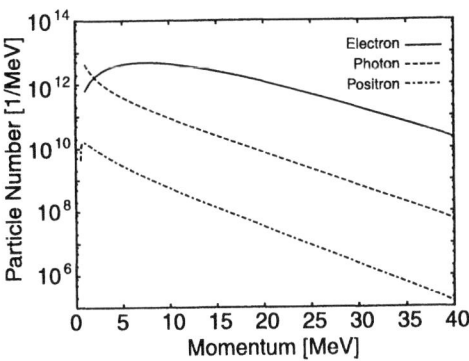

FIGURE 1. The electron, photon, and positron momentum spectrums are plotted. The feature of each spectrum is shown around 1-5MeV. The electron number is about 10^{14}, the photon number is about 10^{12}, the positron number is about 10^{10}. The thickness of gold foil is $125\mu m$, the laser intensity is 10^{20}W/cm^2.

Other experiment performed by Gahn *et al.* showed that continuous positron source can be realized using ultra-intense lasers. They chose He gas as a target of laser irradiation. many hot electrons are successively produced from the interaction region. Electrons were guided to Pb target, then hot electrons are converted to electron-positron pairs.

The other experiment was quite different from previous ones. In this experiment, laser photons are converted to pairs directory. At first, the head-on collisions between laser and 47GeV electrons were occurred. High energy γ-ray with 29GeV were created in this interaction. This γ-ray proceeds in the laser field, then γ-ray and multi-photon of laser interaction was realized.

$$\gamma + n\omega \rightarrow e^+ + e^-, \tag{5}$$

where ω is a photon of laser. The energy of laser photon is several eV. Multi-photon interaction is important experiment in the point of non-perturbation theory of quantum electrodynamics.

As we said above, the pair production by ultra-intense lasers are hot topics in view of theory and experiment. But the spatial distribution, energy distribution, and higher order quantum dynamical processes have never studied yet. We are developing a numerical simulation code based on Relativistic Fokker-Planck equation to investigate these unknown problems.

NUMERICAL SIMULATION BASED ON RELATIVISTIC FOKKER-PLANCK EQUATION

Electrons generated by ultra-intense lasers have the energy up to several MeV. These electrons proceed in a plasma or solid target with the velocity of light. The electron with MeV energy feels the interaction of Coulomb scattering, bremsstrahlung, and pair production in the target. Especially Coulomb scattering is a long range interaction, so

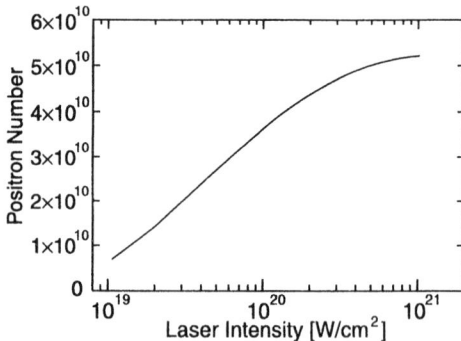

FIGURE 2. Total number of created positrons versus intensity of ultra-intense laser are plotted. The positron yield increases dramatically around 10^{19}W/cm^2, and saturates around 10^{21}W/cm^2. Total energy of the laser is fixed at 280J. The thickness of gold foil is 125μm.

this force must be treated statistically. Fokker-Planck equation is the equation derived from this statistical view. We reduced Relativistic Fokker-Planck equation[21, 22] as follows under the condition that the velocity of back ground electron and ions are small enough in comparison with hot electrons,

$$\frac{\partial f}{\partial t} + \frac{p\mu}{m\gamma}\frac{\partial f}{\partial x} = Y_e n_e m \frac{1}{p^2}\frac{\partial}{\partial p}(\gamma^2 f) + \frac{1}{2}(Y_e n_e + Y_i n_i)\frac{m}{p^3}\frac{\partial}{\partial \mu}\left\{\gamma(1-\mu^2)\frac{\partial f}{\partial \mu}\right\}, \quad (6)$$

$$Y_e = 4\pi\left(\frac{e^2}{4\pi\varepsilon_0}\right)^2 \ln\Gamma_{ee}, \quad Y_i = 4\pi\left(\frac{Ze^2}{4\pi\varepsilon_0}\right)^2 \ln\Gamma_{ei},$$

where $n_{e,i}$ is electron and ion density, p is the hot electron momentum, $\Gamma_{ee,ei}$ is the Coulomb logarithm, $\mu = \cos\theta$, $f(x,p,\theta)$ is the distribution function of hot electron. This equation has 1-D real space and 2-D momentum space with spherical symmetry. In addition to Coulomb scattering, Bremsstrahlung and pair production are set in the numerical code[14].

Target condition in the simulations are used in the parameters of the experiment of Cowan *et al.*[20] The target is gold. The thickness of the target is 125μm. The intensity of laser is 10^{20}W/cm^2. We initially put the hot electrons with Maxwellian. The temperature of hot electrons is given by Ponderomotive potential[23]. That is, this simulation shows the hot electron beam transport produced by ultra-intense lasers. We got hot electron, photon and positron spectrum from this simulation(FIGURE.1). The results show that the number of positron is 10^{10}, and the ratio of each particles are, electron : photon : positron = 1 : 10^{-1} : 10^{-4}. These results agree with the experiment.

We derive the relation between laser intensity and the number of electrons(FIGURE.2). The theoretical studies said that there is a threshold around 10^{19}W/cm^2. Our simulation shows the rapid increasing of positrons at 10^{19}W/cm^2. One of the remarkable results is the saturation of the positron yield at 10^{21}W/cm^2.

326

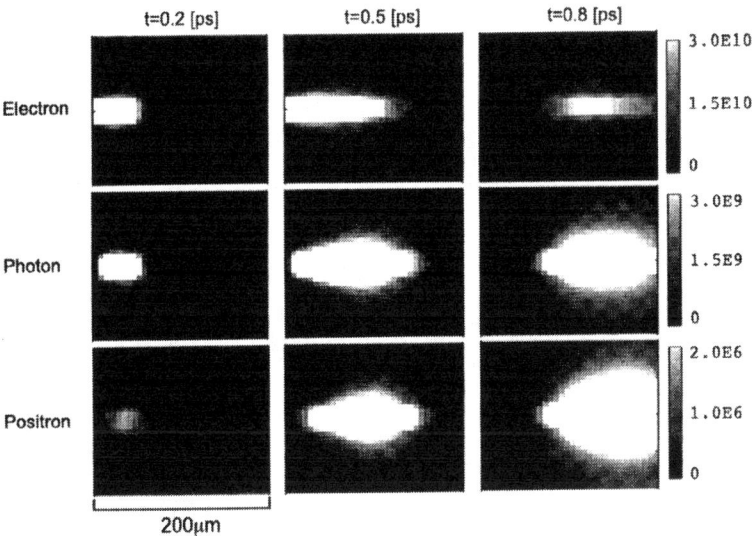

FIGURE 3. Initial temperature of hot electron is 4MeV. This temperature is the same temperature of electrons produced by ultraintense laser with 10^{20}W/cm^2. The electrons are injected from the left side of simulation area.

We are also developing higher dimensional simulation code. That has 2-D real space and 3-D momentum space. The basic equation is as follows,

$$\frac{\partial f}{\partial t} + \frac{p}{m\gamma}\left\{\frac{\mu}{r}\frac{\partial(rf)}{\partial r} + \xi\frac{\partial f}{\partial z} - \frac{1}{r}\frac{\partial(\eta f)}{\partial\phi}\right\}$$
$$= Y_e n_e m_e \frac{1}{p^2}\frac{\partial}{\partial p}(\gamma^2 f)$$
$$+ \frac{1}{2}(Y_e n_e + Y_i n_i)\frac{m}{p^3}\left\{\frac{\partial}{\partial\mu}\left[(1-\mu^2)\frac{\partial f}{\partial\mu}\right] + \frac{1}{(1-\mu^2)}\frac{\partial^2 f}{\partial\phi^2}\right\} \qquad (7)$$

One of the results from this 2-D simulation is shown in FIGURE.3. Photons go long distance (mean free path) in the target before creating pairs. So the density peak of positrons goes on ahead in comparison with the density peak of electrons. Other results are now being analyzed.

CONCLUSION

The pair production by ultra-intense laser is considered in this paper. The threshold of pair production is around 10^{19}W/cm^2 according to the theoretical studies. This was confirmed from our simulation.

Recently, some experimental studies were carried out. The experimental study by Cowan *et al.* showed the electron and positrons spectrum. The ratio of electrons and positrons is 10^{-4}. The results of our study agrees with it.

Then, we are developing the 2-D simulation code to investigate the spatial distribution and electro-magnetic effects. One of the results shows the positron density peak is disagree with the electron density peak.

We continue to study the electron-positron dynamics to apply particle sources, relativistic plasma, and astrophysical plasma research.

REFERENCES

1. Schmidt, G., and Wilcox, T., *Phys. Rev. Lett.*, **31**, 1380 (1973).
2. Kitagawa, Y., Matsumoto, T., Minamihata, T., Sawai, K., Matsuo, K., Mima, K., Nishihara, K., Azechi, H., Tanaka, K. A., Takabe, H., , and Nakai, S., *Phys. Rev. Lett.*, **68**, 48 (1992).
3. Modena, A., Najmudin, Z., Dangor, A. E., Clayton, C. E., Marsh, K. A., Joshi, C., Malka, V., Darrow, C. B., Danson, C., Neely, D., and Walsh, F. N., *Nature*, **377** (1995).
4. Nakajima, K., Fisher, D., Kawakubo, T., Nakanishi, H., Ogata, A., Kato, Y., Kitagawa, Y., Kodama, R., Mima, K., Shiraga, H., Suzuki, K., Yamakawa, K., Zhang, T., Sakawa, Y., Shoji, T., Nishida, Y., Yugami, N., Downer, M., and Tajima, T., *Phys. Rev. Lett.*, **74**, 4428 (1995).
5. Sentoku, Y., Mima, K., Shen, Z. M., Kaw, P., Nishihawa, K., and Nishikawa, K., *Phys. Rev. E*, **56**, 046408 (2002).
6. Pukhov, A., and ter vehn, J. M., *appl. Phys. B*, **74** (2002).
7. Strickland, D., and Mourou, G., *Opt. Commun.*, **56**, 219 (1985).
8. Perry, M. D., and Mourou, G., *Science*, **264** (1994).
9. Bunkin, F. V., and Tugov, I. I., *Sov. Phys. -Dokl.*, **14**, 678 (1970).
10. Bunkin, F. V., and Kazakov, A. E., *Sov. Phys. -Dokl.*, **15**, 758 (1971).
11. Shearer, J. W., Garrison, J., Wong, J., and Swain, J. E., *Phys. Rev. A*, **8**, 1582 (1973).
12. Liang, E. P., Wilks, S. C., and Tabak, M., *Phys. Rev. Lett.*, **81**, 4887 (1998).
13. Gryaznykh, D. A., *Phys. Atom. Nucl.*, **61**, 394 (1998).
14. Nakashima, K., and Takabe, T., *Phys. Plasmas*, **9**, 1505 (2002).
15. Hora, H., *Laser Int. Rel. Plasma Phenom.*, **3B** (1974).
16. Seely, J. F., *Laser Int. Rel. Plasma Phenom.*, **3B** (1974).
17. Brezin, E., and Itzykson, C., *Phys. Rev. E*, **2**, 1191 (1970).
18. Becker, W., *Laser and Partcle Beams*, **9**, 603 (1991).
19. Berezhiani, V. I., Tskhakaya, D. D., and Shukla, P. K., *Phys. Rev. A*, **46**, 6608 (1992).
20. Cowan, T. E., Perry, M. D., Key, M. H., Ditmire, T. R., Hatchett, S. P., Henry, E. A., Moody, J. D., Moran, M. J., Pennington, D. M., Phillips, T. W., Sangster, T. C., Sefcik, J. A., Singh, M. S., Snavery, R. A., Stoyer, M. A., Wilks, S. C., Young, P. E., Takahashi, Y., Dong, B., Fountain, W., Parnell, T., Johnson, J., Hunt, A. W., and Kühl, T., *Laser Part. Beams*, **17**, 773 (1999).
21. Braams, B. J., and Karney, C. F., *Phys. Fluids*, **B1**, 1355 (1989).
22. Braams, B. J., and Karney, C. F. F., *Phys. Rev. Lett.*, **59**, 1817 (1987).
23. Wilks, S. C., Kruer, W. L., Tabak, M., and Langdon, A. B., *Phys. Rev. Lett.*, **69**, 1383 (1992).

Neutral Beam Generation
by Laser Irradiation of Thin Foils

Y. Wada*, T. Kubota* and A. Ogata*

*Graduate School of Advanced Sciences of Matter, Hiroshima University
1-3-1 Kagamiyama, Higashi-Hiroshima 739-8530, Japan

Abstract. We irradiated thin ($< 10 \ \mu$m) plastic and metal foils by a 1 TW, 50 fs, T^3 laser. The foil surface was located at $\pi/4$ to the laser injection. Particle beams were obtained at both sides of the foil with respect to the laser injection. The beam at the rear side, flowing to the direction perpendicular to the foil surface, was insensitive to the magnetic field. Measurements using the track detector CR39 tell that it has small divergence (~ 200 mrad) and suggest that their particles could have high energies (up to 1 MeV). We therefore conclude that these particles make a neutral beam. To the contrary, ions were obtained at the same side of the foil with respect to the laser injection. They also have energies up to ~ 1 MeV, but larger angler distribution.

INTRODUCTION

High energy (> 100 keV) ion generations using high-energy ($\sim$ kJ) and long-pulse ($\sim$ ns) lasers with plasmas were reported in 1970s $\sim$ 1980s [1]. Recent experiments have used high-power (> 10 TW) and short-pulse (< 1 ps) lasers with solid targets to generate ions with much higher energies (> 1 MeV) [2, 3, 4]. Many simulation studies have been made on this phenomenon [5, 6, 7]. These experiments and simulations indicate that lasers with intensities over 10^{17} W cm^{-2} can generate MeV ions. High-energy electrons play an important role in the high-energy ion generation. Two mechanisms proposed for the high energy electron generation are "vacuum heating" [8] and "$\mathbf{J} \times \mathbf{B}$ ponderomotive acceleration" [9]. In the vacuum heating, p-polarized laser pulses push and pull electrons with quiver velocity across the boundary between vacuum and solid. In the $\mathbf{J} \times \mathbf{B}$ ponderomotive acceleration, ponderomotive force of the laser pushes electrons toward the laser direction. In both cases, pulled or pushed electrons form a cloud of electrons between the plasma and vacuum interface. Ions pulled by this cloud are accelerated to several MeV.

In this paper, we report similar experiments to irradiate thin ($< 10 \ \mu$m) plastic and metal foils by a laser with smaller power (1 TW) and shorter pulse width (50 fs). Our motivation had been to develop an ion source of an ion accelerator using a T^3 laser. Though ions were obtained at the same side of the foil, neutral particles were obtained at the rear surface of the foil with respect to the laser injection. As far as we know, this is the first observation of the neutral particles by the interaction of a high power laser with a thin foil target.

In the next two sections, we describe the experimental apparatus and the experimental results. The following section discusses the results. The final section gives conclusions.

CP634, *Science of Superstrong Field Interactions*, edited by K. Nakajima and M. Deguchi
© 2002 American Institute of Physics 0-7354-0089-X/02/$19.00

EXPERIMENTAL APPARATUS

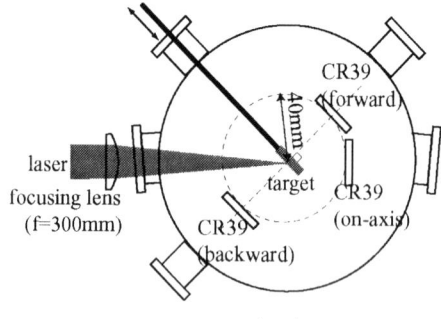

target chamber

FIGURE 1. Experimental setup. CR39 plates are located at forward, backward and on-axis directions.

FIGURE 1 shows the experimental setup. The experiment was performed with 1 TW (50 mJ, 50 fs) Ti:Sapphire laser, 800 nm in wavelength and 10 Hz in pulse frequency. A main pulse is accompanied with a pre-pulse, whose power is $\sim 1/2000$ of the main. An f=300 mm lens located outside of a vacuum chamber focused the laser to a target in the chamber. Diameter and depth of the chamber are 400 mm and 200 mm, respectively. The surface of the target foil was located at $\pi/4$ to the laser injection. The laser intensity on the target was $\sim 4 \times 10^{16}$ Wcm^{-2}. Three types of materials were tried as the target foils; Mylar ($C_{10}H_8O_4$)$_n$, polypropylene (C_3H_6)$_n$ and aluminum (Al). Their thickness was mostly less than 10 μm. Once penetrated by a laser pulse, a target foil was bored, so the foil frame was moved after each laser shot so that the laser pulse always irradiates a virgin surface. Typical vacuum in the chamber was $\sim 10^{-3}$ Pa.

A CR39 was used to detect generated particles. It is a track detector sensitive only to ion (and neutron generated by recoil protons or carbons) [10]. An energetic ion cuts chemical bonds of the CR39. Chemical etching enlarges and exposes these radiation damages so that we can observe the resultant pits by using a microscope. Typical etching used 7N NaOH solutions at 70°C for 5 hours, but other combinations of etching conditions were also tried. About $5 \sim 20$ μm of the CR39 surface was scraped away in this operation. The size of each pit enables us to estimate the energy of the particle, if the particles are identified and the size of a pit is calibrated to the beam energy beforehand. We had done so by using proton beams ($0.5 \sim 2.4$ MeV) of a Van de Graff accelerator in Hiroshima University.

A magnetic energy analyzer combined with a CR39 plate was used to measure energies of the charged particles. The analyzer consists of a pair of dipole magnets behind a 200 μm slit. The computer code MAFIA[1] was used to estimate the tracks of the charged particles under the magnetic field. This setup was designed to detect down to 50 keV protons, 150 keV C^{6+} ions and 300 keV Al^{13+} ions. FIGURE 2 shows experimental setup for the energy analysis and the estimation of particle deflection. We also measured

---

[1] See http://www.cst.de

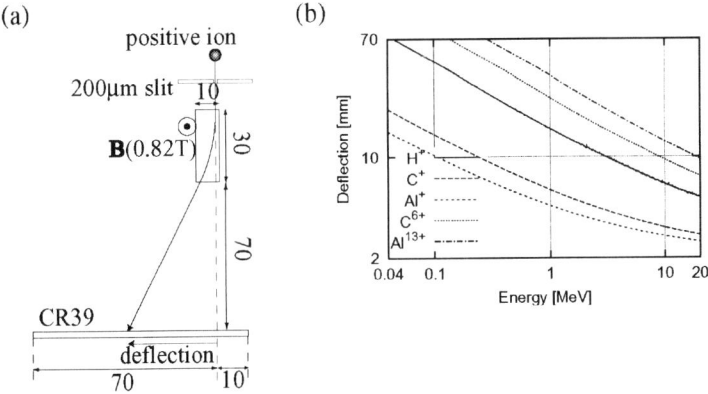

FIGURE 2. (a): Experimental setup of energy analysis. (b): Estimation of particle deflection by MAFIA.

spectrum of the scattered laser light.

EXPERIMENTAL RESULTS

CR39 Measurement of Particle Distributions

In order to measure particle distributions, we placed the CR39 plates in both forward and backward directions across the target, and also in the direction on the laser axis, as shown in FIGURE.1. In the forward and backward directions the CR39 plates accept particles generated in the normal direction against the target. In the on-axis direction, the CR39 plates accept particles on the laser axis. Each CR39 plate is located with distance of 40 mm away from the interaction point. FIGURE 3 shows a photograph taken at the moment of the laser irradiation, from the direction parallel to the laser axis and perpendicular to the foil surface, Al in 3 μm-thick in this case. It shows a fine but bright trace to the forward direction, besides the backward laser reflection.

The target irradiation left many particle tracks on the CR39 detectors in the forward and the backward directions, while few tracks in the on-axis direction. In the forward direction, the tracks were concentrated within angular divergence of $\sim$ 200 mrad. On the other hand, tracks in the backward direction had a larger anguler distribution ($> \pi/2$).

FIGURE 4 shows the number densities of etched pits in the forward(a) and the backward(b) directions. In the forward direction, the number of pits increases as the foil thickness decreases in each material. Aluminum targets got more pits than plastics. The total number of the pits was $\sim 1 \times 10^5$ at the most in the plastic targets and $\sim 6 \times 10^5$ in Al targets. In the backward direction, the number of pits had weak but opposite dependence on the foil thickness; i.e., the thicker targets give the more particles. The total number of the pits was $\sim 1 \times 10^6$ at the most in the plastic targets, greater than the case of the forward direction, and $\sim 6 \times 10^5$ in Al targets.

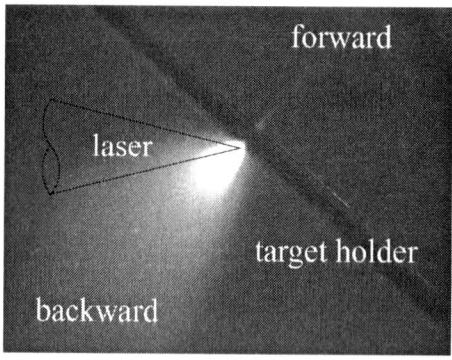

FIGURE 3. A photograph at the laser irradiation taken from the direction parallel to the laser axis and perpendicular to the 3 μm thick Al target foil surface

(a)   (b)

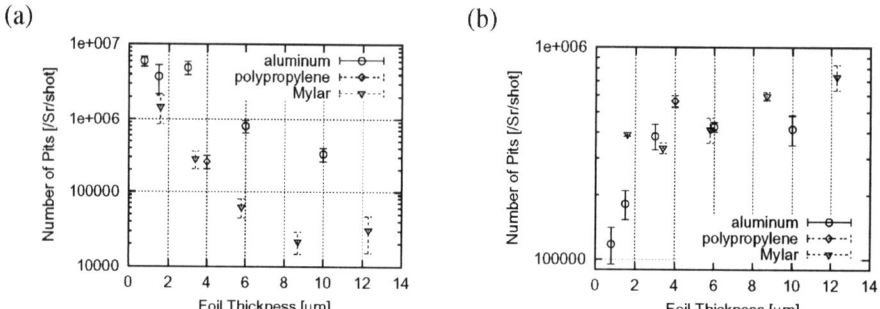

FIGURE 4. The relation between the number density of etched pits and target thickness. (a): The forward direction. (b): The backward direction. Thickness of each target is 0.8 μm, 1.5 μm, 3 μm, 6 μm, 10 μm in Al, 1.6 μm, 3.4 μm, 5.8 μm, 8.7 μm, 12.3 μm in Mylar and 4 μm in polypropylene.

Particle Energy Measured by a Magnetic Energy Analyzer

FIGURE 5 shows the distribution of etched pits in the forward direction in the setup of FIGURE 2, caused by 10 laser shots irradiation onto an Al foil in 3 μm-thick. Distributions of the etched pits with and without the magnetic field coincided within statistical errors. Similar results were obtained in plastic targets. Because these particles are insensitive to the magnetic field, we have to conclude that they are neutral. On the other hands, particles in the backward direction were deflected to a certain degree by the magnetic field. FIGURE 6 shows their energy spectrum caused by the irradiation of 4 μm-thick polypropylene, under assumption that all the particles were protons. Maximum proton energy then became ~ 1 MeV.

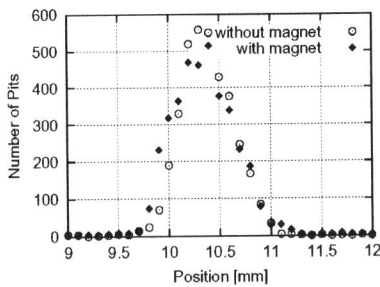

FIGURE 5. Deflections of etched pits caused by the magnetic energy analyzer. White circles represent the distribution without magnetic field. Filled diamonds represent the distribution with magnetic field. Target was Al in 3 μm-thick. Note : X-axis shows not deflections (FIGURE 2) but positions on the CR39 plate.

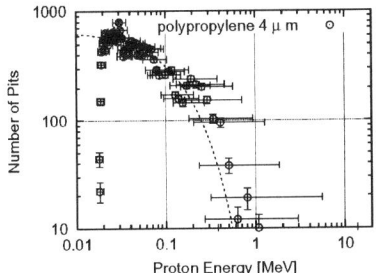

FIGURE 6. The energy spectrum estimated from the magnetic deflection in backward direction. Target was polypropylene in 4 μm-thick. The reduced curve is $\sim \exp[-E[MeV]/0.13]$.

DISCUSSIONS

Particles in Backward Direction

The particles in the backward direction were deflected by the magnetic field. Their energies range from ~ 50 keV up to 1 MeV. These results are consistent with the hitherto experiments [1, 3] using lasers of $\sim 10^{16} - 10^{17}$ Wcm^{-2}, except that the maximum energy of 1 MeV is somewhat high. This may be because the assumption is wrong that all the particles were protons. Generated particles can include carbon ions (in plastic targets) or aluminum ions (in Al targets). Carbon ions or Al ions with mass larger than proton are caused smaller deflection than protons with the same energy/nucleon. We are going to introduce a Thomson parabola to identify the charges of the particles.

Particles in Forward Direction

The particles in the forward direction are neutrals. We should discuss here the mechanism of their generation, but we have to refrain from it, because we do not have enough data. Only clues we have at the present are traces left on CR39 plates.

The effect of the neutral atoms onto the CR39 detectors has not been reported so far. However, studies of CR39 response to D-T neutrons report that neutrons with energies between 0.1 and 1 MeV knock out hydrogen atoms consisting CR39 and these hydrogen atoms leave tracks an it[11]. The conversion efficiency from neutrons to hydrogen atom is very low, $\sim 10^{-4}$ pits/n.

We cannot tell the components of the neutrals at present. They can be hydrogens, carbons, oxygens, and/or metals. We can neither tell whether the neutrals form clusters or not. The sizes of pits etched on the CR39 plates are various. Some are smaller than, some are same with, and some are larger than those created by accelerator protons with energy in the $0.1 \sim 1$ MeV range. These pits distribute homogeneously. Covering a CR39 by an up-to 3 μm-thick Al foil, we found that particles penetrated it leaving fairly large ($\sim 5\mu$m in diameter) holes. If the sizes of these holes were comparable to the particle size, the particles could be clusters, although we can attribute these sizes to some heat process after the penetration.

It is a standard technique to derive particle energies from pit sizes left on the CR39, if we know what the particles are. What if we apply this technique assuming that protons have made the pits? Comparing their sizes with those caused by high-energy accelerator protons, we depicted energy spectra of particles in the forward direction as shown in FIGURE 7 (a). Their energies would range up to 1 MeV and their temperatures would range between 200 and 500 keV.

We have made another experiment to study the ability of penetration of particles, putting Al filters with in front of CR39 plates. The filter thickness was 0.8 μm, 1.5 μm, 3 μm or 6 μm. Dependence of the number of pits on the filter thickness is shown in FIGURE 7 (b). Some particles penetrated 6 μm thick Al. If the particles were protons, they would have kinetic energies over 600 keV and if Al^{13+} ions they were over 18 MeV. The SRIM code[2] was used in these calculations of penetration.

The particles in the forward direction could be neutrons. However, we do not have any conditions favorable for production of neutrons [12]. Assuming that the neutral particles were atoms or molecules, we tried to strip electrons from them using carbon foils, 30 to 100 nm in thick, locating them in front of the magnetic analyzer. Such stripping foils have been used for the purpose of charge exchange in ionaccelerators[13]. The results were unsuccessful. We only observed that the particles were scattered without regard to the magnetic field. Improvement of the vacuum ($\sim 5 \times 10^{-5}$ Pa) at the target brought little change.

---

[2] The program code can be downloaded from http://www.srim.org/

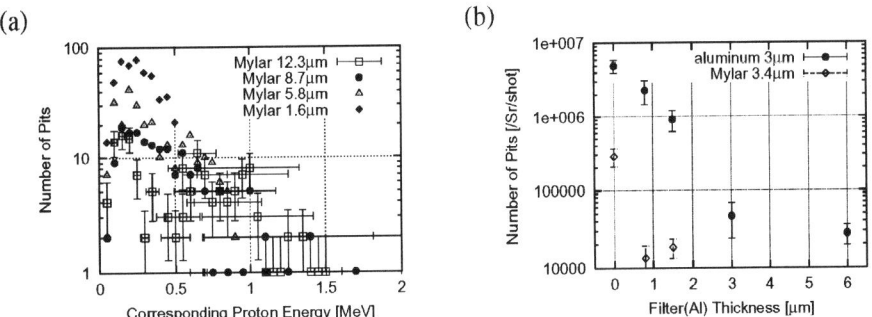

FIGURE 7. (a) : Energy spectra of particles in the forward direction under the assumption that all the particles were protons. The targets were Mylars with various thicknesses. Error bars are given only for the case of 12.3 μm thickness, but other data points also have similar errors. (b) : Dependence of the number of pits on the filter(Al) thickness. The target was 3.4 μm-thick Mylar or 3 μm-thick aluminum.

Further Experiments

Some experiments are under way and some are planned with the purpose of identifying the particles and clarifying their mechanism of generation. In order to identify the charged state of the particles in the backward direction, a Thomson parabola has been introduced.

Energies of the particles in the forward direction can be measured by the time-of-flight method. We have constructed a larger vacuum chamber to carry out in with a better resolution. The most probable origin of the neutrals is the recombination of ions and electrons. An optical spectrometer has been introduced study of the recombination radiation, or radiation from excited ions after the recombination.

Dependence of the neutral particle generation on the laser intensity has been also studied. In order to increase the laser intensity, a new off-axis parabola mirror will be introduced in a new vacuum chamber. The gratings of a compressor of the CPA will be changed to a dielectric type in a future.

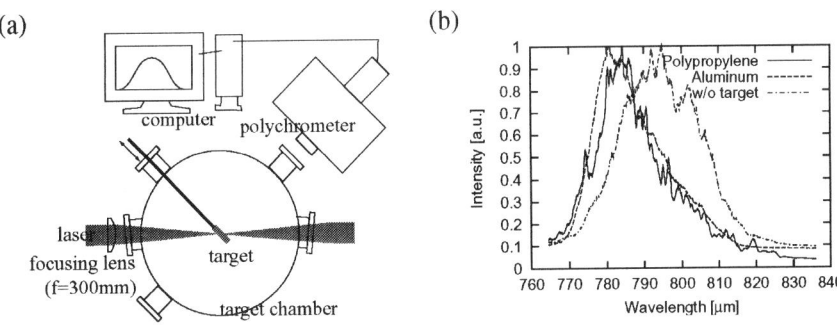

FIGURE 8. Measurement of spectrum of the scattered laser light. (a): Experimental setup. (b): Typical spectrum of the laser light. The target was either a polypropylene foil in 4 μm-thick aluminum foil in 3 μm-thick.

We have measured spectrum of the scattered laser light using the optical spectrometer.

FIGURE 8 shows typical spectrum of the scattered light(b) together with the experimental setup(a). Some blue shifts were observed when the laser irradiated targets. We have some comments on the blue shift of the scattered laser light. We cannot tell if this blue shift is due to the same mechanism observed in past experiments [14, 15] due to interaction between an underdense plasma and a laser light.

CONCLUSIONS

We have detected neutral beams in the interaction between high intensity ($<$ 10^{17} Wcm^{-2}) laser and thin foil. The neutral beams were detected at the rear side of the foil with respect to the laser injection. They were concentrated in an angle of $\sim$ 200 mrad, and possibly had fairly high energies. Ions were detected at the same side of the foil with respect to the laser injection. They have larger angular distribution and energies up to $\sim$ 1 MeV if they were protons. We have to refrain from telling what are the components of the neutrals and what is the mechanism to generate them at present.

ACKNOWLEDGMENTS

We thank Fumitaka Nishiyama, for providing us plastic foils, Ryoko Katsube and Akio Seiki for calibration of CR39 detector, Yuji Tsuchimoto for software development for CR39 data processing, Koji Matsukado and Daisuke Fukuta for their help in experiments, Osamu Kamigaito, Naruhiko Sakamoto and Makoto Tobiyama for running the MAFIA code. This work was supported by Nuclear Research Promotion Program of JAERI, Advanced Compact Accelerator Development Project of MEXT and NIRS.

REFERENCES

1. S. J. Gitomer, et al., *Phys. Fluids*, **29**, 2679–2688 (1986).
2. E. L. Clark, et al., *Phys. Rev. Lett.*, **84**, 670–673 (2000).
3. A. Maksimchuk, et al., *Phys. Rev. Lett.*, **84**, 4108–4111 (2000).
4. R. A. Snavely, et al., *Phys. Rev. Lett.*, **85**, 2945–2948 (2000).
5. A. Zhidkov, et al., *Phys. Rev. E*, **61**, R2224–R2227 (2000).
6. S. C. Wilks, et al., *Phys. Plasmas*, **8**, 542–549 (2001).
7. A. Pukhov, *Phys. Rev. Lett.*, **86**, 3562–3565 (2001).
8. F. Brunel, *Phys. Rev. Lett.*, **59**, 52–55 (1987).
9. W. L. Kruer, and K. Estabrook, *Phys. Fulids*, **28**, 430–432 (1985).
10. B. G. Cartwright, E. K. Shirk, and P. B. Price, *Nucl. Instr. and Meth.*, **153**, 457–460 (1978).
11. K. Oda, et al., *Bulletin of Kobe Univ. Mercantile Marine*, **36**, 175–183 (1988), in Japanese.
12. L. Disdier, et al., *Phys. Rev. Lett.*, **82**, 1454–1457 (1999).
13. I. Yamane, and H. Yamaguchi, *Nucl. Instr. and Meth. A*, **254**, 225–228 (1986).
14. W. M. Wood, et al., *Phys. Rev. Lett.*, **67**, 3523–3526 (1991).
15. R. L. Savage, et al., *Phys. Rev. Lett.*, **68**, 946–949 (1992).

Field Polarization Effect On Electron Dynamics In Strong Laser Field

L. Shao, Y. K. Ho*, P. X. Wang, Q. Kong, Y. Q. Yuan, N. Cao, J. Pang, Y. J. Xie

*Institute of Modern Physics, Fudan University,Shanghai 200433, China; *Corresponding author. E-email: hoyk@fudan.ac.cn ; Fax: 86-21-65643815*

Abstract. In this paper we present a comparison between the effects on electron dynamics by a linearly polarized and a circularly polarized field of an intense laser beam. Special attention is given to the vacuum laser acceleration scheme, also known as capture and acceleration scenario (CAS). It has been found that CAS phenomenon can occur in both polarized fields but there are differences. The CAS of a circularly polarized field exhibits axisymmetric feature, whereas CAS of a linearly polarized field can only be observed when electrons are injected nearly along the polarization plane. Given the same laser parameters (intensity, beam width), the maximum energy gained by the electrons from linearly polarized field is higher than that from circularly polarized field. Physical explanations based on the space distributions of the field amplitudes are described. We hope this study provides significant help to those designing experimental setup to verify CAS as well as to those developing laser accelerators of the future.

INTRODUCTION

The rapid development of intense laser technology in recent years has stimulated many frontier research areas in both applied and fundamental physics [1-3].

Among these new areas we have studied the feasibility of laser-driven electron acceleration in vacuum and have obtained meaningful results[4]. We found a free electron can be captured by the laser field and accelerated violently with an acceleration gradient in excess of $1 GeV/cm$. We call this process 'capture and acceleration scenario' (CAS), and present a physical explanation to this acceleration scheme[5].

In all our previous work the laser beam was assumed to be linearly polarized and the electrons were injected along the electric polarization plane of the electromagnetic field. The main tasks of the present investigation are to study the effect of the field polarization state on the CAS.

This study has great interest to experimentalists designing setup to verify CAS or implementing the laser accelerator.

CP634, *Science of Superstrong Field Interactions*, edited by K. Nakajima and M. Deguchi
© 2002 American Institute of Physics 0-7354-0089-X/02/$19.00

CALCULATION MODEL

The schematic geometry of the electron scattering by laser beam used in the calculation is shown in Fig.1. The laser beam adopted is in the lowest-order Hermit-Gaussian (0, 0) mode[6] expressed analytically in the paraxial approximation and propagates along the z axis, which is the laser axis. The incidence angle between the incident direction of the electron and the propagation direction of the laser is defined as θ. Four-dimensional vector configurations of energy-momentum $(\gamma_i, P_{xi}, P_{yi}, P_{zi})$ and $(\gamma_f, P_{xf}, P_{yf}, P_{zf})$ are used to specify the incoming and outgoing states of electrons respectively. For a linearly polarized (LP) field, we assume that the electric component of the laser field is polarized along the x-axis. The angle between the projection on the $x-y$ plane of the electron's incident direction and the x-axis is defined as the polarization azimuth angle α.

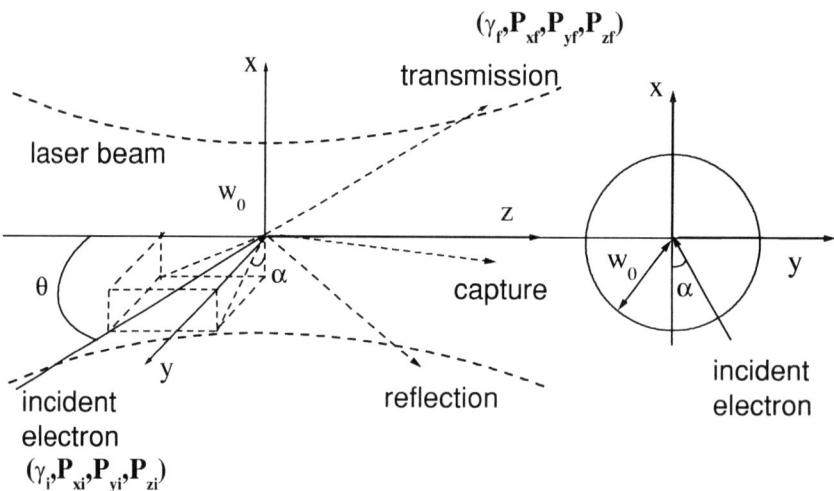

FIGURE 1. Schematic geometry of electron scattering by a laser beam. The laser propagates along the z-axis, W_0 is the laser beam width at the waist. The free electron enters from the minus-x side. The four-dimensional vectors $(\gamma_i, P_{xi}, P_{yi}, P_{zi})$ and $(\gamma_f, P_{xf}, P_{yf}, P_{zf})$ denote the incident and outgoing state of the electron, respectively. The Lorentz factor γ designates the electron energy and the momentum P is normalized in units of $m_e c$. Angle θ is the electron incident angle, with $\tan\theta = P_{ti}/P_{zi}$ (P_{ti} denotes the transverse component of P_i). Angle α denotes the polarization azimuth angle. Style for Figure Captions. Center this text if it doesn't run for more than one line.

As for the circularly polarized (CP) field, the electromagnetic field components of the laser in the normal paraxial approximation are expressed in the following equations:

$$E_x = E_0 \frac{w_0}{w(z)} \exp\left(-\frac{x^2+y^2}{w(z)^2}\right) \exp[\varphi],$$ (1)

$$E_y = E_0 \frac{w_0}{w(z)} \exp\left(-\frac{x^2+y^2}{w(z)^2}\right) \exp\left[\varphi+\frac{\pi}{2}\right],$$ (2)

$$E_z = (\frac{i}{k})(\frac{\partial E_x}{\partial x} + \frac{\partial E_y}{\partial y}),$$ (3)

$$\mathbf{B} = -(\frac{i}{\omega})\nabla \times \mathbf{E},$$ (4)

$$\varphi = i\left(kz - \omega t - \phi(z) - \phi_0 + \frac{k(x^2+y^2)}{2R(z)}\right).$$ (5)

Where w_0 is the radius of the beam width at focus, $w(z) = w_0\left[1+(z/Z_R)^2\right]^{1/2}$, $R(z) = z\left[1+(Z_R/z)^2\right]$, $\phi(z) = \tan^{-1}(z/Z_R)$, $Z_R = kw_0^2/2$ is the Rayleigh length.

And if an LP field and a CP field have the same field strength a_0, ($a_0 = eE_0/m_e c\omega$ is a dimensionless parameter measuring the laser intensity, with $-e$ and m_e being the electron charge and mass respectively, ω the angular frequency of the electromagnetic wave and c the speed of light in vacuum), the energy flow densities I of the two fields are not the same. The latter is two times of the former:

$$I_L = 1.366 \times 10^{18} (\frac{a_0}{\lambda})^2, \text{ and } I_C = 2.73 \times 10^{18} (\frac{a_0}{\lambda})^2.$$ (6)

for the LP and CP field respectively. In equation (6) λ is the wave length of the laser of unit μm and the unit of I is W/cm^2

RESULTS AND DISCUSSION

First we examine the dependence of the electron dynamics on the polarization azimuth angle α for LP fields. Figure 2 shows the electron final energy γ_f versus the initial phase ϕ_0 of the field at different α values. Various values of ϕ_0 correspond electrons impinge on the laser beam at different delay-time. It can be seen from this figure that with α being increased γ_f decreases quickly and the capture plateau becomes narrower. This means for LP field we have to inject electrons into an azimuthal angle as smaller as possible to achieve the CAS mode. Figure 3 presents a comparison of γ_{fm} (the maximum value of γ_f as the initial laser phase ϕ_0 is varied over the range 0 to 2π) versus α between the LP field and CP field. The laser intensities for the both fields are $a_0 = 100$. This figure illustrates two properties. One is that the electron dynamics in LP fields are strongly axis-asymmetric, whereas in CP fields they are axisymmetric. The other is that the maximum energy gained by electrons from the CP field is a bit lower than that from the LP field.

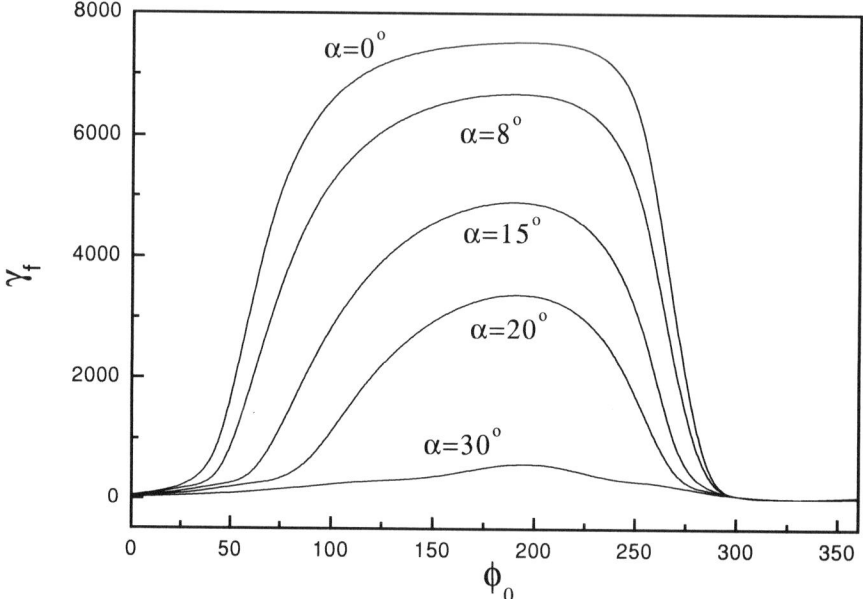

FIGURE 2. The final electron energy γ_f versus the initial phase ϕ_0 of the field at different α values for LP fields with $a_0 = 100$. Other parameters used are $kw_0 = 170$, $\tan 6 = 0.1$ and $P_{ri} = 3$.

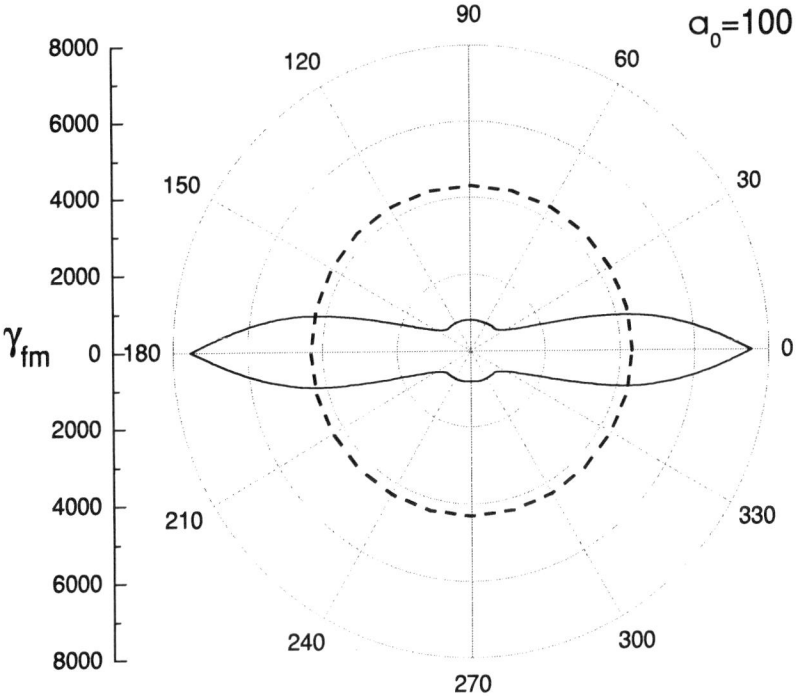

FIGURE 3. A comparison between the effects of LP and CP field on the maximum final energy γ_{fm} vs. α. Other parameters used are the same as in Fig. 2. The solid line and the dashed line denote the LP and CP field respectively.

The physical explanation to the first point is according to the distribution of the longitudinal component of the electric field, which plays a major role in accelerating electron in longitudinal direction. Figure 4(a) and (b) show the amplitude distributions on the focus plane of LP and CP laser beam respectively. It can be seen obviously that the longitudinal component of the electric field distributes only along the polarization plane of the transverse component of the electric field for LP field, but it has a same distribution both along the polarization plane and normal to the polarization plane for CP field. Since the amplitude of E_z is not axisymmetric in LP field, it could be well-understood that the electron dynamics in LP field exhibit axis-asymmetric characteristics. In contrast to that, the whole CP field amplitudes, including the longitudinal component amplitude, are symmetric with respect to the azimuth angle α. Therefore, the electron dynamics in CP field should be axisymmetric.

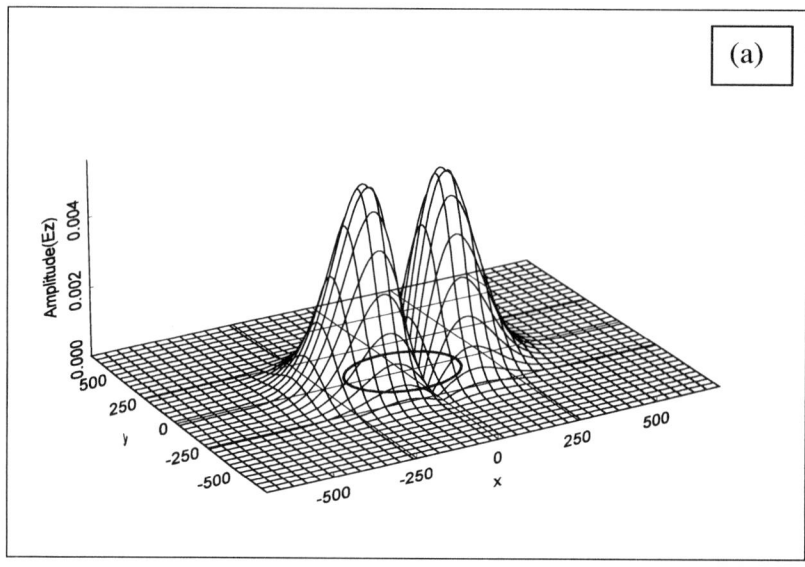

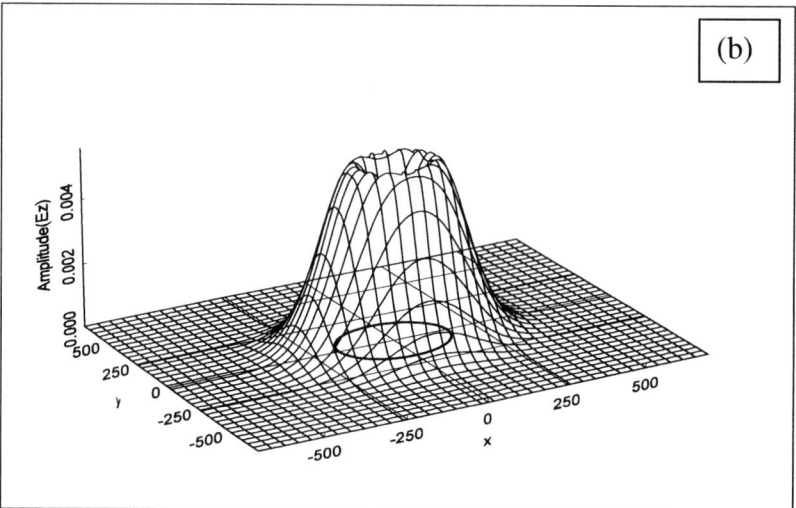

FIGURE 4. The distribution of the longitudinal electric field amplitude at the $z = 0$ plane with $a_0 = 100$, (a) is for LP field and (b) is for CP field. The bold circle on the bottom plane represents the laser beam width on the focused plane.

To the second property of the Fig. 3, it can be explained as follows. In the CP field, in addition to the $-eE_x$ force, which is the same for both fields with the same a_0, there is another transverse force $-eE_y$ for CP only. This force will pull the electrons away from the laser axis which is the strong region of the electric longitudinal component and thus the effectiveness of the acceleration from the longitudinal electric force is decreased. Accordingly the maximum energy gained by electrons from the LP field is larger than that from the CP field.

CONCLUSION

The results of this paper can be summarized as follows.
1. CAS can emerge in both the LP field and CP field provided the laser intensities are high enough.
2. CAS of the LP field exhibits highly axis-asymmetric feature. Only in a small range of the polarization azimuth angle will the incident electrons be substantially accelerated. It is in direct contrast to that in CP field where axisymmetric characteristics is the rule.
3. CAS phenomenon for LP field (with $\alpha = 0$) is more prominent than that of the CP field.

ACKNOWLEDGMENTS

The authors would like to thank A. M. Sessler and E. Esarey for helpful discussion, and C. M. Fou for carefully reading of this manuscript. This work is partly supported by the National Natural Science Foundation of China under Contracts No. 19984001 and No. 10076002, National High-Tech ICF Committee in China, Engineering-Physics Research Institute Foundation of China, National Key Basic Research Special Foundation (NKBRSF) under grant No. G1999075200, and Zhonglu fellowship.

REFERENCES

1. D.Strickland and G.Mourou, *Opt. Commun.* **56**, 219 (1985); M. D. Perry and G. A. Mourou, *Science* **264**, 917 (1994); N. Blanchot *et al.*, *Opt. Lett.* **20**, 395 (1995); G. A. Mourou, C. P. Barty, and M. D. Perry, *Phys. Today* **51**, Jan., 22 (1998).
2. A. Modena, Z. Najmudln, A. E. Dangor, C. E. Clyton, K. A. Marsh, C. Joshl, V. Malka, C. B.Darrow, C. Danson, D. Neely, and F. N. Walsh, *Nature (London)* **377**, 606 (1995); P. Sprangle, E. Esarey, and J. Krall, *Phys. Plasmas* **3**, 2183 (1996); D. Umstadter, S.-Y. Chen, A. Maksimchuk, G. Mourou, and R. Wagner, *Science* **273**, 472 (1996); C. I. Moore, A. Ting, S. J. McNaught, J. Qiu, H. R. Burris, and P. Sprangle, *Phys. Rev. Lett.* **82**, 1688 (1999); K. Krushelnick, E. L. Clark, Z. Najmudin, M. Salvati, M. I. K. Santala, M. Tatarakis, and A. E. Dangor, *ibid.* **83**, 737 (1999); P. Sprangle, E. Esarey, A. Ting and g. Joyce, *Appl. Phys. Lett.* **53**, 2146 (1988).
3. J.Zhang *et al.*, *Phys. Rev. Lett.* **76**, 3865 (1997); G. Malka, E. Lefebvre and J. L. Miquel, *Phys. Rev. Lett.* **78**, 3314 (1997); K. T. McDonald, *ibid.* **80**, 1352 (1998); P. H. Bucksbaum, M. Bashkansky and T. J. McIlrath, *Phys. Rev. Lett.* **58**, 349 (1987); P. Monot, T. Auguste, L. A. Lompr, G. Mainfrary, and C. Manus, *ibid.* **70**, 1232 (1993); C. I. Moore, J. P. Knauer and D. D. Meyerhofer,

Phys. Rev. Lett. **74**, 2439 (1995); N.Yugami, K.Kikuta and Y.Nishida, *Phys. Rev. Lett.* **76**, 1653 (1996).

4. Y.K.Ho, J.X.Wang, L.Feng, W. Scheid and H.Hora, *Phys. Lett.* **A220**, 189 (1996); J.X.Wang *et. al.*, *Phys. Rev.* **E58**, 6575 (1998); L. J. Zhu, Y. K. Ho, J. X. Wang and Q. Kong, *Phys. Lett.* **A248**, 319 (1998).

5. P. X. Wang, Y. K. Ho, *et. Al.*, *Appl. Phys. Lett.*, **78**, 2253 (2001); *J. of Appl. Phys.* **91,** 856 (2002).

6. L. W. Davis, *Phys. Rev.* **A19**, 1177 (1979); J. P. Barton and D. R. Alexander, J. *Appl. Phys.* **66**, 2800 (1989).

7. G. V. Stupakov and M. S. Zolotorev, *Phys. Rev. Lett.* **86**, 5274 (2001).

QUANTUM ASPECT OF SUPER STRONG FIELD INTERACTION

HIROSHI TAKAHASHI

Brookhaven National Laboratory
Upton New York, 11973

Abstract. Quantum aspects of the interactions between the high-intensity laser and electron plasma has been studied from the point of view of coherent-state formalism In this paper, I offer a sound theoretical foundation based on quantum-mechanical and coherent-state-formalism for analyzing the interactions between the high-intensity laser and electron plasma in a many electron system. Using first the Two Times Green Function method, I derived a quantum theory for a free electron laser when the laser intensity is not very high. Then, the dispersion relation for the free electron high-intensity laser can be formulated using the classical laser field. To provide the sound foundation using the classical field, I discuss the coherent-photon state. For analyzing many electron systems, I describe the use of the atomic-coherent state, which provide proper description of the superradiant transition , the atomic laser, and the production of positrons by a very strong laser.

INTRODUCTION

The classical formalism for an electromagnetic field was used to advance a theory for the free electron laser. However, because the electron beam employed in the free electron laser was not refined very well in the past, its formulation by quantum - mechanical formalism was not considered seriously. Recently, the electron beam has been refined, it is desirable to formulate a theory that rests on a good foundation. I derived a quantum theory for a free-electron laser of not too high an intensity r using the Two Times Green function formalism [1] that Matsubara derived for solid physics theory. I compared the dispersion relationship that I obtained with Kwan s classical theory [2]. However, when the laser intensity becomes high. multi-photon reactions play an important role, so that the highly correlated Green function has to be taken into account. And a large number of simultaneous equations have to be solved; these become more complicated as the laser 1 s intensity increases. Instead of solving each highly correlated Green function, I proposed describing the many photon field by the classical approximation field. This approach greatly simplified the formulation

To obtain a proper foundation for the classical approximation of the laser field, others have suggested incorporating the coherent-state expansion for an electro-magnetic field [3] . Although coherent state expansion is over-complete, it offers a well-defined mathematical basis for exploring the relation between the quantum

CP634, *Science of Superstrong Field Interactions,* edited by K. Nakajima and M. Deguchi
© 2002 American Institute of Physics 0-7354-0089-X/02/$19.00

formula and the classical formula, and provides the methodology for treating more complex systems.

In a previous paper[4], I discussed on the classical field and applied coherent-state expansion to the photon field, but the many electrons involved in the theory of the free electron were not treated as the coherent state. By taking into account the multiple spin-state for the assembly of electrons in the same way as for the many photons, a quantum free-electron theory based on the electron-coherence state can be derived in a simple way as the classical photon field, and the super-radiant transition can be described.

Considering the two levels of spin up and spin down states for single electron, the state of assembly of the electrons can be expressed by the so-called atomic coherent state, and the free electron theory can be simplified using both the coherent photon- and atomic-states. This electron-coherent theory simplifies the formulation of parameters, such as the super-radiant transition by using the so-called Dicke Ⅎ s super-radiant states. This use of the atomic-coherent states can be generalized for the analysis of the atomic laser by a simple extension of the theory.

THE FORMALISM OF DISPERSION RELATION OF FREE ELECTRON

When a relativistic electron is moving in a electro-magnetic field and a magnetic field B , the Hamiltonian for such a system may be written as

$$H = \sum_i (c\,\alpha_i \cdot (p_i - eA/c) + \beta_i mc^2) + H_r + H_{ee} + H_s,$$

where A is the sum of the vector potential, A_s of the B-field and the vector potential, A_r, of the electromagnetic wave; $A = A_s + A_r$, m and C are the mass of the electrons and the speed of light, respectively, while α and β are the Dirac matrix operators, and the suffix stands for the ith particle. H_r is the Hamiltonian for the electromagnetic wave. H_{ee} is the Hamiltonian for the interaction between electrons , and Hs is the Hamiltonian for the static magnetic field. The last Hamiltonian, the constant of motion, is not incorporated in the dynamics of the system; it can thus be neglected in the following derivation.

To solve this Hamiltonian system by taking into account the quantum aspect, we use a second quantization formalism in which the Hamiltonian is expressed by the creation and annihilation operator of photons and electrons, The dispersion relation of laser photon, is derived from the equation of motion for the two-time Green function of the photon by taking its time derivative. In this successive derivation, the highly correlated Green functions of the photon and electron are created. The resulting equation of motion becomes a simultaneous equation with large number of highly correlated Green functions for a high intensity laser.

In our previous paper, we obtained the dispersion relation by assuming that the laser intensity was not high enough, so that the highly correlated Green function of the photon and electron can be decoupled, and it results to closing the simultaneous equation in the small order , and we got the dispersion relation as .

$$
\left[h^2(\omega^2 - \omega_{q,\lambda}^2) - h\omega_{q,\lambda} \{ \sum_{\substack{k_3 k_2 \\ \sigma_1 \sigma_2}} M_{k_1 k_2 q} (h\omega + T_{k_1 \sigma_1} - T_{k_2 \sigma_2})^{-1} \sum_k [n_{k_2 \sigma_2} H(k_1, k_2, k, \sigma_1) - n_{k_1 \sigma_1} H(k_1, k_2, k, \sigma_1)] \right.
$$

$$
x(-(4\pi e^2/k^2)[1 + (4\pi e^2/k^2) \sum_{k k_4 \sigma_3} \frac{(n_{k_2 \sigma_2} - n_{k_2 \sigma_{21}})}{(h\omega + T_{k_1 \sigma_1} - T_{k_2 \sigma_2})} | H^*(k_3, k_4, k, \sigma_3) |^2]^{-1}
$$

$$
\sum_{k_{31} k_4 \sigma_3} H^*(k_3, k_4, -k, \sigma_3) M_{k_4 k_3 q} \left(\frac{(n_{k_3 \sigma_3} - n_{k_4 \sigma_{31}})}{(h\omega + T_{k_3 \sigma_3} - T_{k_4 \sigma_3})} \right)
$$

$$
\sum_{\substack{k_3 k_4 \\ \sigma_1 \sigma_2}} | M_{k_1 k_2 q} |^2 \left(\frac{(n_{k_2 \sigma_2} - n_{k_2 \sigma_{21}})}{(h\omega + T_{k_1 \sigma_1} - T_{k_2 \sigma_2})} \right) \left] G_{q\lambda}(\omega) \right. = h^2 \omega_{q\lambda}/\pi
$$

After few manipulations , we finally obtained the dispersion relation as

$$
\omega^2 - q^2 c^2 - \frac{\omega_p^2}{2\gamma} \left[\frac{\left(hq^2/m\gamma \right)^2 / 4}{\left(\omega \cdot q V_b \right)^2 - \left(hq^2/m\gamma \right)^2 / 4} \right]
$$

$$
= \frac{\omega_p^2 \omega_{cc}^2}{4\gamma^3} \sum_{K=\pm K_0} \left(\frac{q+K}{k} \right)^2 \left[\left(\omega - (q+K) \bullet V_b \right)^2 - \frac{1}{4} \left(\frac{h(q+K)^2}{m\gamma} \right)^2 + \frac{\omega_p^2}{\gamma} \right]^{-1}
$$

where $\omega_p^2 = 4\pi e^2 n_0 / m$ is the plasma frequency.

This can be compared with the one obtained by Kwan using classical formalism as

$$
\left[\omega^2 - q^2 c^2 - \frac{\omega_P^2}{\gamma} \right] = \frac{1}{2} \frac{\omega_P^2 \, \omega_{ce}^2}{\gamma^5} \frac{(q + K_0)^2}{K_0^2}
$$

$$
x \left[(\omega - (q + K_0) \bullet V_b)^2 - \frac{\omega_P^2}{\gamma^3} (1 + 3(q + K_0)^2 \gamma_D^2) \right]^{-1}
$$

where $\lambda_D^2 = T/4\pi n_0 e^2$ and T is the temperature of the electron beam.

DISPERSION FORMULA FOR THE HIGH INTENSITY LASER

To obtain the dispersion relation for a high intensity laser, a large number of simultaneous equation has to be solved; hence, it is too complicated to be resolved the problem analytically. Therefore, I treated the laser field as the classical field , and then the dispersion formula for $G_{q,\lambda}(\omega)$ is obtained as (see Reference [6])

$$\Big[h^2(\omega^2 - \omega_{q,\lambda}^2) - h\omega_{q,\lambda} \{ \Pi_s \sum_{L_s = \infty}^{\infty} J_{L_s}(Z_s) \exp[-iL_s(\theta_s - \Delta_s)] \} \sum_{M_s = \infty}^{\infty} J_{M_s}(Z_s) \exp[iM_s(\theta_s - \Delta_s)]$$

$$x \sum_{\substack{k_1,k_2 \\ \sigma_1\sigma_2}} M_{\sigma_1\sigma_2,\lambda}^{a k_1 k_2 q} (h\omega + T_{k_1\sigma_1} - T_{k_2\sigma_2} - \sum_s (L_s - M_s)\omega_s)^{-1} \sum_k [n_{k_2\sigma_2} H^a(k_1,k_2,k,\sigma_1) - n_{k_1\sigma_1} H^a(k_1,k_2,k,\sigma_1)]$$

$$x(-(4\pi e^2/k^2)[1 + (4\pi e^2/k^2) \sum_{kk_4\sigma_3} \frac{(n_{k_2\sigma_2} - n_{k_2\sigma_{21}})}{(h\omega + T_{k_1\sigma_1} - T_{k_2\sigma_2} - \sum_s (L_s - M_s)\omega_s)} \Big| H^{a^*}(k_3,k_4,k,\sigma_3) \Big|]^{-1}$$

$$\sum_{k_{31}k_4\sigma_3} H^{a^*}(k_3,k_4,-k,\sigma_3) M_{\sigma_{31}\sigma_2,\lambda}^{a k_4 k_3 q} (\frac{(n_{k_3\sigma_3} - n_{k_3\sigma_{31}})}{(h\omega + T_{k_3\sigma_3} - T_{k_4\sigma_3} - \sum_s (L_s - M_s)\omega_s)}) \Big|$$

$$\sum_{\substack{k_3 k_4 \\ \sigma_1\sigma_2}} \Big| M_{\sigma_1\sigma_2,\lambda}^{a k_1 k_2 q} \Big|^2 (\frac{(n_{k_2\sigma_2} - n_{k_2\sigma_{21}})}{(h\omega + T_{k_1\sigma_1} - T_{k_2\sigma_2} - \sum_s (L_s - M_s)\omega_s)}) \Big] G_{q\lambda}(\omega) = h^2 \omega_{q\lambda}/\pi$$

Although, we obtained a very simple dispersion relation for a high-intensity laser case using the classical field approximation, the justification of its use must be clarified. This can be done using the coherent state which gives the sound foundation of the classical formula.

COHERENT , SQUEEZE, AND SUPER-RADIANT THEORIES

The coherent state of photon is defined as

$$|\alpha\rangle = \exp(-|\alpha|^2/2) \sum_{n=0}^{\infty} \alpha^n / \sqrt{n!} / |n\rangle$$

Here

$$|n\rangle = 1/\sqrt{n!}(a^+)^n|\varphi_0\rangle \quad |\varphi_0\rangle = |vacume\ state\rangle$$

is the eigen state of th number operator $N = a^+ a$ containing light quanta. For our consideration it is useful to split a+ and a into a sum of Hermitian operators i.e.

$$a = (u + ip)/\sqrt{(2h)} \quad a^+ = (u - ip)/\sqrt{(2h)}$$

Coherent state $|\alpha\rangle$ is the eigen state of the non-Hermitian annihilation operator a with the complex eigen value value $\alpha = (u + ip)/\sqrt{(2h)}$ If this complex eigen value α, which labels the coherent states runs over the whole complex plane, the coherent states becomes over-complete for the Hilbert space. Such an over-complete sets of the coherent states can not be used for our consideration because they are linearly dependent. However, Bargmann al.[7] and Perelomov[8] proved that subset of the over complete sub-state form a complete set.

This subset is given by

$$\{ \; |\alpha\rangle: \alpha = \sqrt{\pi}(l+im) \; ; \; l = 0, \pm 1, \pm 2, \pm 3, \pm 4, \ldots\ldots \}$$

This fact was originally stated by von Neumann, without proof. Therefore these states are called von Neumann lattice coherent state(VNLCS). This coherent state is over complete state and the some dicretalization is needed to reduces the over-completeness to the conventional completeness. Here the uncertainty of the quantum description is comes in.

This discretization introduces the uncertainty principle of quantum formulation, the delicacy of the quantum effect on the theory such as the super radiant state and the squeeze states which is prominent in the quantum theory of the laser can be studied.

—The one of advantage of using these coherent state description is the fact that many modes in the EM field can be formulated without difficulty. Furthermore another transition associated with the super-radiant. The squeezed states which might be created by the free electron laser is described by the u or p in the Eq.(5.3).

ATOMIC COHERENT STATES

In the above formalism the electron system is assumed to be an assembly of single-level particles neglecting the spin dependence of magnetic field. When the magnetic field is taken into account, the single electron state is split into two states of spin-up and spin-down. The excitation and de-excitation of the spin dependent electrons by laser field play role of important factor for the analysis. The wave function of electron plasma is expressed by function of the spatial and spin, but when the wave length of the laser is larger than the distance between electrons, the wave function of assembly of the electrons can be described by the total spin which is the sum of each spin. For the analysis of many electron states, the use of the atomic coherent state of Bloch state can simplify the formalism. similar to the coherent state of photon $|\alpha\rangle$ as the many photon analysis. The atomic coherent state of Bloch state $|\theta,\varphi\rangle$ is defined as

$$|\theta,\varphi\rangle \equiv R_{\theta,\varphi} \; |\text{-J}\rangle,$$

where $R_{\theta,\varphi} = \exp[-i \; \theta \; J_n] = \exp[-i \; (J_x \sin\varphi - J_y \cos \varphi)] = \exp[\zeta J_+ - \zeta^* J_-]$ and $\zeta = \_$ $\theta \exp[-i\varphi]$.

When a strong laser field irradiates the electron plasma, the spin state of the electrons will be excited from the down state to the up state and the Dick⊥s super-radiant state will be created. These analyses can be carried out using the Bloch state. When the positively charged particles are in the plasma, the angular momentum will participate in the equation of motion of the whole system, and various electro-magnetic field will be excited including the Landau level (in the case of strong magnetic plasma). The coherent state of the plasma state can be described as the spin and angular variable, and classical formula will be obtained with a proper quantum foundation.

COHERENT STATE FOR FERMIONS

Toyoda et al [4] gave a sound foundation for the classical formalism for the high-intensity photon field. However, in their formulation, classical approximation is applied to the radiation field, assuming that the electron state is not subjected to such an approximation. To use the coherent state for electrons, the formalism developed by Ohnuki et al [5] should be applied which describes the coherent state of fermion particles

$$|(\xi)_n\rangle = \exp(-\sum_{j=1}^{n} \xi_j a_j^+)|0\rangle$$

where ξs are the Grassmann numbers instead of the complex number α for a boson field.

CONCLUSION

In the above formulation, we used the atomic coherent state to describe the many electron states. When a very strong laser field irradiates electron plasma, very many exotic states will be created. The straightforward application of the Green function method becomes very complex, but it might be simplified by using Group theoretical approaches.

Recent simulation of very strong laser irradiation to plasma shows the vortex type and soliton like excitation. By taking a group theoretical approach, such as, L-S coupling, these excitation can be analyzed using the classical approach for these coherent states. I would like to mentioned that the coherent state is related with Gauge field. The group theory provides a proper foundation for the coherent state theory, and not only the strong laser theory, many exotic gauge theories give simple descriptions for complicated reactions. When the laser field is extremely strong, positrons might be created, in order to analyze this case, the formalism of using four- components Dirac wave function is required instead of two components. And the coherent state can be described by the group theory in high dimensions To formulate these process, the real quantum description provides fundamental foundation for this analysis.

REFERENCES

[1] Hiroshi Takahashi, "Theory of the free electron laser". Physica 123 C (1984) 225-237 North-Holland Amsterdam

[2] T. Kwan., J.M.Dawson, and A.T. Lin. Phys. Rev. 16A (1977)

[3] Hiroshi Takahashi, "Interaction between a Plasma and a Strong Electromagnetic Wave",
Physica 98C (1980) 313-324

[4] Tadashi Toyoda and Karl Widermuth, "Charged Schroedinger particle in a c-number radiation field", Phy. Rev. 10D, p2391 (1980)

[5] Yosio Ohnuki and Taro Kashiwa,"Coherent State of Fermi Operators and the Path Integral", Prog. Theo. Physics, 60. 548, (1978)

[6] F.T.Arecchi, E.Courtens, R. Gilmore and H. Thomas, "Atomic coherent states in Quantum Optics", Phys. Rev A, 6, 2211, (1972).

[7] V.Bargmann, P.Butera, L. Giradello and J.R.Klauder, "On The Completeness of the Coherent States", Rep. Math.Phys. 2, 221, (1971).

[8] A.M. Perelomov, "On the Completeness of Coherent States", Theor. Mat. Fiz. 6, 213, (1971).

SHONAN Lectures

Nonadiabatic Transitions and Laser Control of Molecular Processes

Hiroki Nakamura

Department of Theoretical Studies, Institute for Molecular Science,
Department of Functional Molecular Science,
The Graduate University for Advanced Studies,
Myodaiji, Okazaki 444-8585, Japan

Abstract. Nonadiabatic transitions play crucial roles in various dynamic processes in physics, chemistry, and biology. This is true also for laser control of molecular dynamics. In this lecture, I will first explain the importance of nonadiabatic transitions together with the basic theories and then demonstrate how various molecular processes can be controlled by manipulating lasers. Actually, by controlling nonadiabatic transitions among dressed states, we can control various molecular processes.

WHAT IS NONADIABATIC TRANSITION?

Nonadiabatic transition is a transition between adiabatic states caused by a variation of the adiabatic parameter. This is a highly multi-disciplinary concept, being an origin of the mutability of this world [1]. Potential energy surface crossing is the most typical example, governing various dynamic processes, and is described by the time-independent theory of nonadiabatic transition. On the other hand, transitions induced by external time-dependent fields are treated by the time-dependent theory of nonadiabatic transition.

The most basic model was first discussed by Landau, Zener, and Stueckelberg independently in 1932. The Landau-Zener formula or the Landau-Zener- Stueckelberg theory is well known. This theory has, unfortunately, many defects and cannot work well in the physically and chemically important regions. After 60 years, we have successfully obtained a complete set of solutions which can be directly applied to practical problems (see the references in [1]). This is called Zhu-Nakamura theory.

FLOQUET STATE REPRESENTATION AND NONADIABATIC DYNAMICS

Molecular processes in a time-dependent laser field can be described by the following Schroedinger equation:

$$i\frac{d\Psi}{dt} = H\Psi = [H_0 + \mu\epsilon(t)\cos(\omega(t)t + \phi(t))]\Psi, \tag{1}$$

CP634, *Science of Superstrong Field Interactions*, edited by K. Nakajima and M. Deguchi
© 2002 American Institute of Physics 0-7354-0089-X/02/$19.00

where μ, $\epsilon(t)$, $\omega(t)$, and $\phi(t)$ are the dipole operator, laser intensity, laser frequency, and phase, respectively. Here we use the adiabatic approximation, assuming that the time dependence of laser parameters $X = \{\epsilon(t), \omega(t), \phi(t)\}$ is weaker than the time-dependence of laser itself, and divide the time derivative as

$$\frac{d}{dt} = (\frac{\partial}{\partial t})_X + \dot{X}\frac{\partial}{\partial X}. \tag{2}$$

First, we solve the eigenvalue problem at fixed X,

$$i(\frac{\partial}{\partial t})_X \Phi_\alpha = H\Phi_\alpha. \tag{3}$$

After taking a time average over the laser period T, we have

$$\det|H_F - \lambda I| = 0, \tag{4}$$

where λ_α is called quasi-energy, and the Floquet Hamiltonian matrix is given by

$$H^F_{\alpha n, \beta m} = (E_\alpha - m\hbar\omega)\delta_{m,n}\delta_{\alpha,\beta} - (\delta_{m,n+1} + \delta_{m,n-1})\mu_{\alpha,\beta}\epsilon/2, \tag{5}$$

where α and β designate the molecular state and $\mu_{\alpha,\beta}$ represents the transition dipole moment. The eigenstates obtained from the above equation are called dressed states. They are shifted up or down from the original molecular states by the photon energy, $m\hbar\omega$, and deformed by the laser-molecule dipole coupling. This is called dressed-state- or Floquet-state-representation (see for instance, [2]). The nonadiabatic transitiopns among the dressed states are induced by the time-variations of the laser parameters X, and are described by

$$\Psi = \sum_\alpha a_\alpha(X)\Phi_\alpha(t : X) \tag{6}$$

$$\text{with} \tag{7}$$

$$i\frac{d}{dt}a_\alpha(X) = -i\sum_\beta a_\beta(X)\langle\langle\Phi_\alpha|\dot{X}\frac{\partial}{\partial X}|\Phi_\beta\rangle\rangle, \tag{8}$$

where

$$\langle\langle\cdots\rangle\rangle = \frac{1}{T}\int_0^T dt <\cdots>. \tag{9}$$

If the laser is not very strong and do not induce violent multi-photon processes, we can use the two-state picture described by the following Hamiltonian:

$$H = \begin{pmatrix} E_1 + \hbar\omega, & -\mu\epsilon(t)/2 \\ -\mu\epsilon(t)/2, & E_2 \end{pmatrix}. \tag{10}$$

Various theories of time-dependent nonadiabatic transition can be applied in the dressed state representation in order to control molecular processes [1].

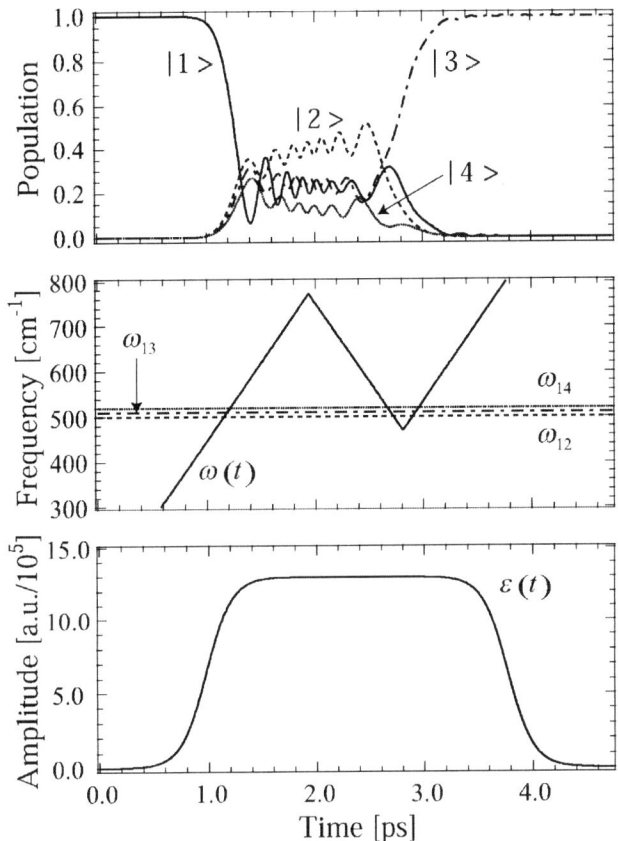

FIGURE 1. Complete excitation from $|1>$ to $|3>$ by one and half period of frequency chirping in the four-level model. Time variations of the population (upper part), laser frequency (middle part), and laser envelope (bottom part) are shown. Reprinted in part with permission from [4]. Copyright(2002) American Chemical Society.

VARIOUS IDEAS OF LASER CONTROL OF MOLECULAR PROCESSES

Various ideas of laser control have been proposed so far such as coherent control, adiabatic rapid passage, chirped pulse, pump-dump, pulse-shape driven, STIRAP, and π−pulse method. Nice summaries can be found in the references [3, 4]. Beacuse of the shortage of space, further description is not provided here. The reader should refere to these books.

OUR NEW IDEAS

We have been proposing two different methods, usage of the complete reflection phenomenon and control by periodic sweeping of laser parameters [1, 5, 6, 7, 8]. The complete reflection phenomenon is a peculiar phenomenon in the case of time-independent
nonadiabatic tunneling (NT) type of transition, in which the two diabatic potential curves
as a function of spacial coordinate cross with opposite signs of slopes and create a potential barrier in the lower adiabatic potential. By shifting up an attractive ground potential
surface by cw laser and creating NT type crossings with a repulsive excited potential,
we can control molecular dissociation by properly adjusting the laser frequency and creating the complete reflection condition where we want to stop the dissociation. This idea
was demonstrated to work well in the case of one- and two-dimensional models [5]. The
second idea is to control the nonadiabatic transition at curve crossing between dressed
states by sweeping the laser parameter, for instance the frequency, and realizing the con-
structive interference for the desirable process [6, 7, 8]. The control conditions can be
formulated by using the theory of time-dependent nonadiabatic transition. This scheme
has been extended so that the periodic sweeping of laser parameters can be replaced by a
sequence of linearly chirped pulses [9]. This is expected to be realized relatively easily,
since linear chirping can now be rather commonly used in experiments. This second idea
is a generalization of chirping, $\pi-$pulse, and adiabatic rapid passage, and can be applied
not only with laser but also with any other time-dependent external field.

NUMERICAL EXAMPLES

Various numerical demonstrations based on our ideas have been published in the references cited above. Among them, the selective excitation among closely lying multi-
levels by using the periodical sweeping of laser frequency is shown here [8]. The level
1 is the ground level and the other three, 2-4, are the excited states. The level spacings
and transion dipole moments are assumed to be $\omega_{12} = 500 \text{cm}^{-1}$, $\omega_{23} = \omega_{34} = 10 \text{cm}^{-1}$,
and $\mu_{12} = \mu_{13} = \mu_{14} = 1.0 a.u.$.Figures 1 demonstrate the complete excitation to the level
$|3 >$ with use of one and half period of sweeping. The transition time is about 3ps, which
is close to the limit determined by the uncertainty principle. The peak intensity of the
laser pulse is 0.5917GW/cm^2. The laser frequency is chirped linearly as follows:

$$
\omega(t) = \begin{cases}
\omega_{14} + c(t - t_0) & (t \leq t_1), \\
\omega_{14} - c(t - t_2) & (t_1 < t < t_3), \\
\omega_{14} + c(t - t_4) & (t > t_3),
\end{cases} \tag{11}
$$

where $t_1 = t_0 + \Delta t_1, t_2 = t_0 + 2\Delta_1, t_3 = t_2 + \omega_{24}/c + \Delta t_2, t_4 = 2t_3 - t_2, c =$
$346.6 \text{cm}^{-1}/\text{ps}, \Delta_1 = 726.3 \text{fs}$, and $\Delta_{t2} = 90.08 \text{fs}$. The pulse shape is a combination
of hyperbolic tangent functions.

Figures 2 depict the similar results by $\pi-$pulse with a similar short pulse duration as
that in Figure 1. The pulse shape is a hyperbolic-secant function.The peak intensity is
0.05GW/cm^2. Since the pulse duration is short and thus the band-width is broad, the

358

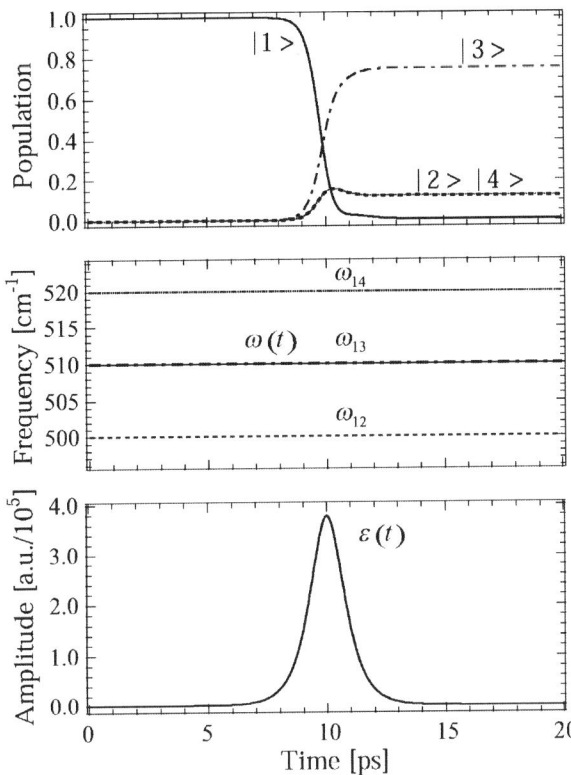

FIGURE 2. Excitation from $|1>$ to $|3>$ by π-pulse with a short time duration for the same system as in Figure 1. Reprinted in part with permission from [4]. Copyright(2002) American Chemical Society.

transitions to $|2>$ and $|4>$ cannot be avoided and the complete excitation to $|3>$ cannot be attained as in Fig.1.

In order to achieve the complete transfer, one order of magnitude longer time is required, as shown in Fig.3. The peak intensity in this case is $0.005\mathrm{GW/cm^2}$. The adiabatic rapid passage (ARP) certainly requires further longer time, since two ARP processes, $|1>$ to $|2>$ and $|2>$ to $|3>$, are required to make the selective excitation from $|1>$ to $|3>$. Thus, the present scheme would be quite useful and effective for the systems with fast relaxation.

FUTURE PERSPECTIVES

For the experimental realization of our idea of periodic sweeping, it would probably be better to use a sequence of linearly chirped pulses in stead of the quadratic chirping. This

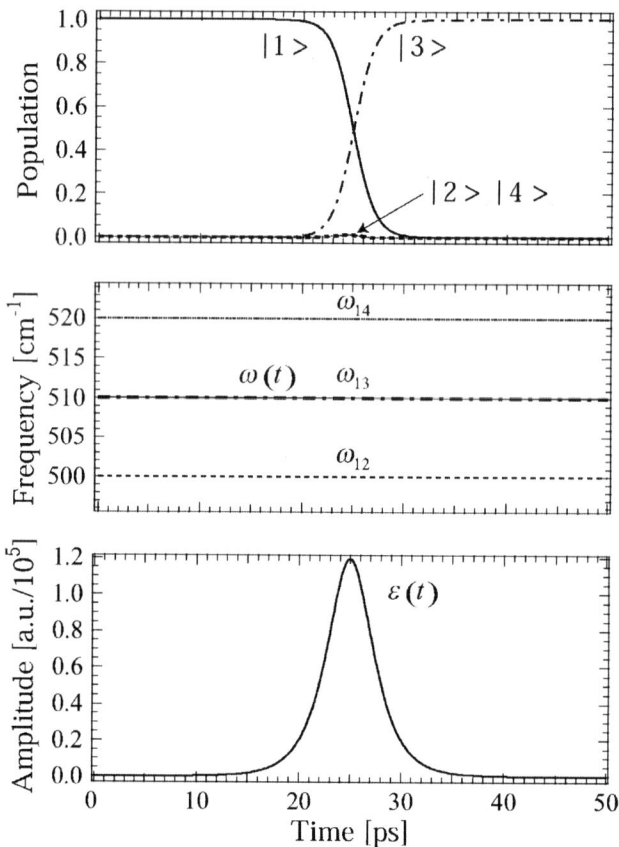

FIGURE 3. The same as Figure 2 with a long time duration to achieve the complete excitation. Reprinted in part with permission from [4]. Copyright(2002) American Chemical Society.

is actually possible as discussed in Ref.[9]. Another problem we have to overcome is the feasibility of applications to multi-dimensional systems. This is true for both complete reflection and periodic sweeping. As for the periodic sweeping, a formulation based on the semiclassical propagation and the semiclassical theory of nonadiabatic transition would be useful.

ACKNOWLEDGMENTS

The works presented here have been carried out by the collaboration with Drs.Y.Teranishi and K.Nagaya. The partial financial support by a research grant No. 10440179 from the Ministry of Education, Culture, Sports, Science, and Technology of Japan is also acknowledged.

REFERENCES

1. H.Nakamura, *Nonadiabatic Transition: Concepts, Basic Theories, and Applications* (World Scientific and Imperial College Press, Singapore, 2002).
2. S.I.Chu, *Advances in Multi-photon processes and Spectroscopy*, Vol.2 (World Scientific, Singapore, 1986).
3. S.A. Rice and M. Zhao, *Optical Control of Molecular Dynamics* (John Wilcy and Sons, New York, 2000).
4. *Laser Control and Manipulation of Molecules* edited by A. Bandrauk, R.J. Gordon, and Y. Fujimura (Amer. Chem. Soc., 2002).
5. K. Nagaya, Y. Teranishi and H. Nakamura, *J. Chem. Phys.* **113**, 6197 (2000).
6. Y. Teranishi and H. Nakamura, *Phys. Rev. Lett.* **81**, 2032 (1998).
7. Y. Teranishi and H. Nakamura, *J. Chem. Phys.* **111**, 1415 (1999).
8. K.Nagaya, Y. Teranishi and H. Nakamura, in Ref. [4].
9. K. Nagaya, Y. Teranishi and H. Nakamura, *J. Chem. Phys.* (in press).

Formation Dynamics of Exciton Domain Pattern and Anisotropy of Interactions in Photoinduced Structural Phase Transitions

Ryotaro Yabuki and Keiichiro Nasu

Institute of Materials Structure Science, KEK,
Graduate University for Advanced Study,
1-1, Oho, Tsukuba, 305-0801, Japan

Abstract. We theoretically study nonlinear nonequilibrium real-time dynamics of domain pattern formation by photo-generated excitons in a 2-D insulating crystal, and search for successful conditions of exciton proliferation, which finally can result in macroscopic photoinduced structural phase transitions. Our model is a strongly coupled many-exciton Einstein-phonon system, interacting with a reservoir, which consists of the acoustic phonons and the radiation field. Using this model, we numerically calculate early-stage time-evolution dynamics of photo-generated exciton-domains, full-quantum mechanically. It is found that the spatial anisotropy of inter-exciton interactions is essential for a large size exciton-domain nucleation of the early stage, and this anisotropy also makes tunneling-type slow exciton proliferations successful even in the retarded stage, as compared with the isotropic cases.

INTRODUCTION

In recent years, there discovered many new insulating solids, which, being shined by only a few visible photons, become pregnant with a macroscopic excited domain that has new structural and electronic orders quite different from the starting ground state. This phenomenon is called "photoinduced phase transition"(PIPT).[1,2] According to recent development of laser spectroscopy, quite novel and noticeable properties of this PIPT are now discovered in various insulating solids.[2,3,4] Especially, the efficiency of the photoinduced phase transition is proved to depend quite nonlinearly on frequency and intensity of exciting light.[2,3,4] While, it is also clearly shown that the resultant photoinduced phase is different from any equilibrium phases realized in the relevant material, such as a low temperature phase or high temperature ones of this material.[2,5]

However, as for the real-time dynamics of these nonequilibrium phase transition phenomena, there are still many theoretical and experimental problems, which are left unclarified. For this reason, in the present paper, we will be concerned with the real time dynamics of domain pattern formations, which finally can result in the macroscopic PIPT in a 2-D crystal. Taking a strongly coupled many-exciton Einstein-phonon system in a 2-D insulating crystal as our theoretical model system, we numerically calculate spatio-temporal evolutions of photo-generated excitons, their proliferations and domain pattern formations, full-quantum-mechanically.

CP634, *Science of Superstrong Field Interactions*, edited by K. Nakajima and M. Deguchi

As for the way of light excitation, we assume the case of pulse excitation composed of several successive ones with an equal time interval. In order to clarify the elementary process of exciton proliferation, we, at first, will show the time evolution of the total exciton number and the total energy, owing to only a single pulse excitation. After that, we will show how the whole exciton system develops during and after the several successive pulse excitations. We will show that there are two stages of exciton proliferation. One is the stage, which we call the early stage, and it is equal to the time region wherein the successive pulse excitation is going on. Another is the retarded stage, wherein the sequential pulse excitation is turned off. In the former stage, the total number of excitons drastically increases by making use of large excess phonon (or vibronic) energy just given by the light pulse. While, in the retarded stage, the whole exciton system reaches some local adiabatic potential energy minimum, and hence the total number of excitons only gradually increases, by the tunneling effect through various adiabatic potential energy barriers.

We will conclude that the spatial pattern of exciton domain formed in the early stage, sensitively inherits the spatial anisotropy of inter-exciton interactions. It will be also shown that this anisotropy makes the aforementioned tunneling quite efficient, and the proliferation quite successful in the retarded stage, as compared with the cases wherein the inter-exciton interaction is isotropic.

MANY-EXCITON PHONON COUPLED SYSTEM

Let us now define our relevant system composed of many excitons coupling strongly with Einstein phonons. The total Hamiltonian ($\equiv H_s$) of our model system is written as ($\hbar = 1$)

$$H = H_s + H_r + H_{sr}, \tag{1}$$

where H_s denotes the strongly coupled exciton-phonon system, which is given by

$$
\begin{aligned}
H_s = & E\sum_l B_l^+ B_l + \sum_{l \neq l'} T(|l - l'|) B_l^+ B_{l'} \\
& + \omega \sum_l b_l^+ b_l - \sqrt{\omega S} \sum_l B_l^+ B_l (b_l^+ + b_l) \\
& + \sum_{l \neq l'} G(|l - l'|) B_l^+ B_l (B_{l'}^+ + B_{l'}) + \sum_{l > l'} V(|l - l'|) B_l^+ B_l B_{l'}^+ B_{l'} .
\end{aligned}
\tag{2}
$$

Here E is the energy of an exciton, B_l^+ and b_l^+ are creation operators of an exciton and a phonon, respectively, at a lattice site specified by a position vector l in a 2-D square lattice. T is the exciton transfer which operates from site l' to l, ω is the energy of the Einstein-phonon, and S is an exciton-phonon coupling constant. G and V are the third- and the fourth-order anharmonic inter-exciton interactions, respectively, which come from a long range nature of Coulomb interaction among electrons and holes constituting these excitons.

H_r represents a reservoir composed of the radiation field and the acoustic phonons, which are linearly coupling with the exciton and the Einstein phonon fields through H_{sr}. Consequently, various relaxation channels can occur in our relevant system, such as vibrational relaxations, radiative and nonradiative decays of excitons, and they contribute to stabilize resultant photoinduced phases.

There may be various cases which can be described by these parameters T, V and G. However, in order to make our later discussions simple and clear, we focus only on two typical cases of these parameters, that is, an anisotropic case and an isotropic one as shown in TABL 2. All the used parameters for these cases are listed in TABLE.1 and 2, and spatial extensions of T, V and G in the 2-D square lattice are illustrated in Fig.1, in which the same notation is used as that of TABLE 2.

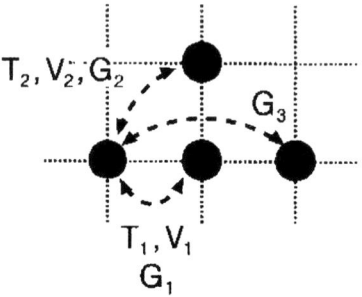

FIGURE 1. Spatial extensions of T, V and G in the 2-D square lattice. Parameter values are listed in TABLE.2.

TABLE 1. Common parameters in eq. (2)

ω	0.10 eV
E/ω	7.80
S/ω	6.45

TABLE 2. Parameters for inter-exciton interactions used in eq. (2) and in Figure 1.

	Anisotropic case	Isotropic case
$(V_1, V_2)/\omega$	(-1.45, 0.63)	(-0.9, 0.0)
$(T_1, T_2)/\omega$	(-1.0, 0.5)	(-1.0, 0.0)
$(G_1, G_2, G_3)/\omega$	(0.30, -0.03, 0.15)	(0.30, 0.0, 0.0)

The isotropic case is the most standard one since it includes the interactions only between neighboring two sites. In this model, two excitons at neighboring two sites attract each other, and it tends to make an exciton-cluster. Throughout the present paper, the occupation of a single site by more than one exciton is excluded from the beginning. From eq.(2), we can easily see that the photo-excited excitons can proliferate through the third-order anharmonicity G. In the anisotropic case, on the other hand, the interactions between neighboring sites and that between next neighboring sites are assumed to be opposite in their signs. Such an anisotropy mainly comes from the nonlocal natures of Wannier functions of the electron and the hole constituting the exciton. These Wannier functions are usually extending over many sites from their central sites, and are oscillating from site to site. For this reason, we can expect various spatial anisotropies for T, V and G. However, in the present paper, we will not be concerned with their microscopic origins. We will treat them only phenomenologically, and compare aforementioned two typical cases in connection with the domain pattern formation and the proliferation. In order to perform practical calculations, we derive the master equation under Markov approximation for the reservoir. We also tacitly assume that our system is a three dimensional one with a layer structure, whose one layer is just this 2-D square lattice with only a weak inter-layer interaction. The total number of excitons in one layer is assumed to be restricted within 100 or so, because of this weak inter-layer interaction.

METHOD AND APPROXIMATIONS

At first, we introduce a set of basis states with n_l (= 1 or 0) exciton and m_l (= 0, 1, 2,...) phonons at each lattice site l, as

$$\prod_l [(B_l^+ U_l)^{n_l} \frac{(b_l^+)^{m_l}}{\sqrt{m_l!}}] | 0 >, \quad U_l \equiv e^{-\sqrt{S/\omega}(b_l - b_l^+)}, | 0 > \equiv \text{exciton-phonon vacuum,}$$

(3)

where U_l denotes the operator of phonon displacement, which appears or disappears, according to the presence or absence of an exciton at site l.[6] Even if we have used this basis set, however, we still have serious difficulty, since the total number of excitons changes from 0 to about 100, while m_l also changes from 0 to about S/ω, almost independently at each site. Thus the direct calculation of this time evolution leads to too large dimensional ones. In order to overcome this numerical difficulty, we derive a new iterative method for the exciton proliferation. Its basic idea was developed by Mizouchi [7] only for the 1-D case. However, in the case of the present 2-D system, we have to extend this theory so that we can describe the problem of spatial pattern formation, which was absent in the 1-D case.

Our new iterative method is as follows. We focus only on the most forwardly expanding part (the most front) of the exciton domain boundary, wherein an exciton with the excess energy coming from the photo-excitation is always included. This most front is searched by try and error method, so that it will be the most efficiently growing part of the domain boundary. The contribution from other excitons not in this front is approximated by a mean field. As proliferation proceeds by using the excess energy, the position of this front also moves. For the practical reason mentioned before, the size of this front can not be so large. As schematically shown in the left part of Fig.2(B), we take the shaded 4 lattice sites as this front. This front (the 4 sites) is our relevant system, within which we calculate excitons, Einstein phonons and their interactions, full-quantum-mechanically, as well as various damping and decay channels mentioned before.

Here, we define aliases of our exciton states; (1) "Mother exciton", denoted by the black circle in Fig.2(B). It is in the front, and has an excess phonon or vibronic energy inherited from the light. (2) "Frozen exciton", denoted by the shaded circle in Fig.2(B). It is in the outside of the front, and is always in the zero phonon state $m_l = 0$ defined by eq.(3)

As the proliferation proceeds, the total number of exciton in the front increases from 1 to 2, as shown in Fig.2 (B). At this stage, we reconstruct a new front just as schematically shown in the right hand side of Fig.2(B). The mother exciton is now frozen, and the new exciton becomes a new mother exciton. This new mother inherits the excess energy, which has now somewhat decreased from the initial excess energy, because of the dampings or the relaxations mentioned before. In this reconstruction, the site with the largest exciton density within the front, is taken as the site wherein the new mother is. We call this reconstruction procedure the "generation crossover". While, the new 4 sites (new front) of the new generation is chosen by try and error method, so that they will be the most efficiently growing part around the new mother. The total energy in the system is conserved before and after this generation crossover. We iterate this procedure, until we can get a large domain. Thus, using this method, we can numerically calculate the temporal evolution of a large system involving many excitons and phonons. It should be noted that this approximation is valid only when the excitons are rather localized, $S \gg |T_1|$.

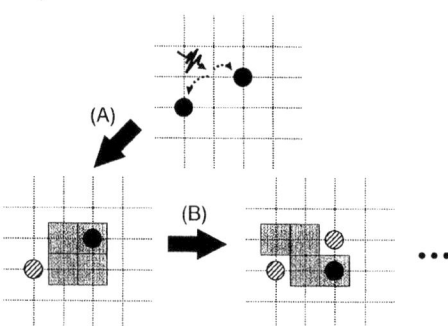

FIGURE 2. Iterative procedure in the 2-D lattice. (A) Photo-generated two excitons (for example) are replaced by a mother exciton (black circle) and a frozen one (shaded circle). (B) The shaded 4 sites in the left part of (B) denotes the front. The shaded 4 sits in the right part of (B) is the new front of the next generation.

FORMATION OF DOMAIN PATTERN AND ANISOTROPY

The pattern formation characteristics are well known in the studies for the diffusion-limited-aggregation phenomena. [8] For example, an anisotropy of surface tension forces a resultant cluster pattern to be a rod like one. Generally speaking, an anisotropy in the elementary process of the growth always brings some characteristic patterns of the resultant cluster. In the case of our present PIPT, the inter-exciton interactions are the main origin of exciton proliferation. Thus, we can expect that the domain pattern will sensitively reflect the anisotropy or the isotropy of those interactions. In some case, we can expect that even the success or the failure of the PIPT itself will also be dominated by the absence or the presence of the anisotropy. For these reasons, we have chosen the two typical cases mentioned in TABLE 2. Keeping these points in the mind, let us see the results of the numerical calculations, performed by using the model and the method given in previous sections.

RESULTS AND DISCUSSION

At first, we have shown in Figs.3 (A) and (B), the temporal evolution of the total exciton number and the total energy, owing to only one photon pulse injection at time zero. This pulse is assumed to be strong enough to generate two excitons at once, as shown in Fig.2(A) and Fig.3(A). Our process is the literal nonlinear one, and as already shown by Mizouchi [7], a single exciton alone can not results in efficient proliferations. Moreover, even if two excitons are simultaneously created at the beginning, the resultant proliferation is also shown to sensitively depend on the initial inter-exciton distance, and this situation is called the initial condition sensitivity.[7] For this reason, in the present study, the inter-exciton distance is chosen to make the subsequent proliferation most efficient under the one-pulse excitation condition.

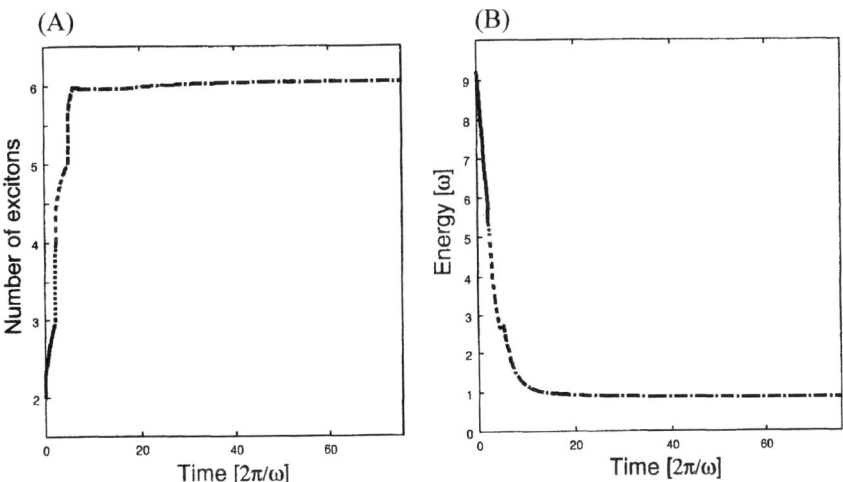

FIGURE 3. Temporal evolution of total exciton number (A) and total energy (B) after only one photon pulse injection, at time zero. The parameters of the anisotropic case are used.

From Fig.3(A), we can clearly see that there are two stages in the exciton proliferation process. One is the early stage just after the pulse excitation, wherein the number of excitons increases very drastically from 2 to 6 or so. This rapid increase is easily seen to occur by making use of the excess vibronic energy just donated from the photon pulse, since the total energy also rapidly decreases at the same time, as shown in Fig.3 (B). After this rapid process, the retarded stage starts. In this stage, the exciton number only gradually increases, and the total energy also decreases gradually. The whole exciton system has now reached some local adiabatic potential energy minimum, and hence the total number of excitons only gradually increases, by the tunneling effect through various adiabatic potential energy barriers.

In the next, let us proceed to the case of multi-pulse excitation by six successive ones with an equal time interval, which is 100 oscillation periods of the Einstein phonon, as shown in Fig.4. The center of mass position of the two excitons generated by each photon pulse is randomly determined within the 2-D (15×15) lattice. We can

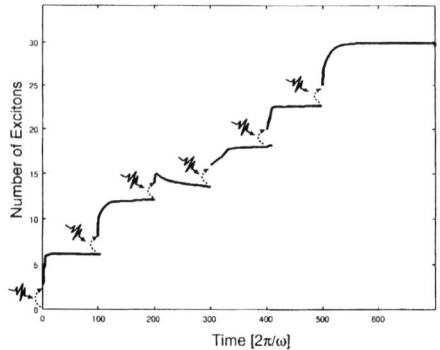

FIGURE 4. Temporal evolution of total exciton number by 6 photon pulses injection at random sites. The wavy arrow represents a photon pulse. The parameters of the anisotropic case are used.

easily see from Fig.4 that the 100 period time interval is long enough for each early stage associated with each pulse to finish. However, these early stages do not result in the same proliferation, since they are influenced by the frozen excitons created by the preceding pulse excitations. This effect due to the frozen excitons in the out side of the front is taken into account by the mean field approximation mentioned before.

As inferred from Fig.4, the effect of this successive six pulses excitation looks like saturated after 600 period. Hence, we can call that this time region from zero to 600 period the elongated early stage, since this stage is just after the successive six pulses excitation. However, after this six-pulse excitation, the domain still gradually grows, and this slow growth continues up to about 100000 period. This is the literal retarded stage relative to the aforementioned elongated early stage, and the slow growth is just due to the tunneling process, explained before.

Let us now proceed to the effects of the anisotropy of the inter-exciton interactions, as compared with the isotropic one. Figure.5 shows the characteristic exciton domain pattern obtained by using the anisotropic interactions in the elongated early stage, that is, just after 600 period in Fig.4, and the shaded circle at each lattice site denotes the exciton in the 2-D (15×15) lattice. We can clearly see that the domain pattern has an "island", namy "capes" and many "peninsulas" stretched outside, with a "strait", "bays" and "gulfs" in between. These characteristics reflect the anisotropy of the inter-exciton interactions. Here, we should note that the interactions between neighboring

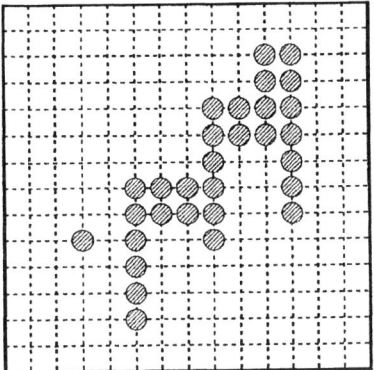

FIGURE 5. Domain pattern formed in the elongated early stage (600 periods) by anisotropic inter-exciton interactions shown in Fig.1 and TABLE 2. The shaded circle is an exciton in the 2-D (15×15) lattice. 6-photon pulses are injected at random.

sites and that between next neighboring sites are opposite in their signs.(TABLE.2) Therefore, these interactions bring aforementioned structures to the domain pattern.

Next, let us proceed to the tunneling type slow proliferation in the retarded stage, that is from 600 period to 100000 period in Fig.4. As is shown in Fig.6, the strait, bays and gulfs are now filled up by newly grown excitons, which are denoted by the black

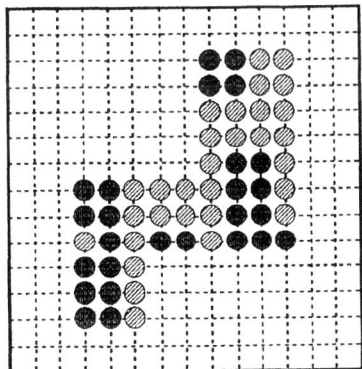

FIGURE 6. Domain pattern at 100000 period after the successive 6 photon pulses injections, under the anisotropic interactions shown in Fig.1and TABLE 2. The 2-D(15×15) lattice is used. The black circle represents the exciton newly grown in the retarded stage. The shaded circles are same as that of Fig.5.

circles. Thus, the characteristic pattern is lost by the tunneling process. That is, the pattern peculiar to the anisotropy of interactions, appears only in the early stage. However, this tunneling process itself is the result of the anisotropy. In the present case, 30-excitons are generated in the elongated early stage, and 26-excitons are added in the retarded stage. Thus the PIPT in this case is successful.

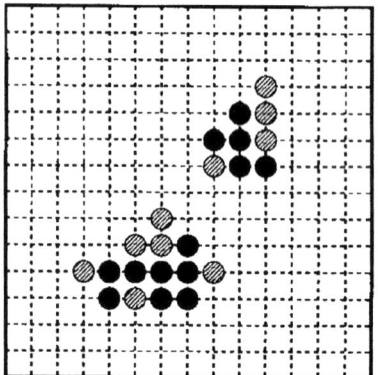

FIGURE 7. Domain pattern at 100000 period after the successive 6 photon pulses injections, under the isotropic interactions shown in Fig.1and TABLE 2. The 2-D(15×15) lattice is used. The black circle represents the exciton newly grown in the retarded stage. The shaded circles are excitons generated in the elongated early stage(just after 600 period in Fig.4).

Let us proceed to the isotropic case. Figure.7 shows a domain pattern obtained by using the isotropic interactions with parameter values shown in TABLE.2. The calculations are performed by keeping the same conditions except for the inter-exciton interactions shown in TABLE.1 and 2. The sites and the intervals of photon pulse injections are also same as that of the above two calculations for anisotropic case. In this isotropic case, however, only some small block type patterns are formed. Thus the PIPT in this case is not successful. Comparing these results, we can clearly see that the anisotropic interactions makes the proliferation successful, and the resultant larger domain formation possible.

CONCLUSION

In the present work, the early time relaxation process of photoexcited state with an excess phonon energy has been simulated by full-quantum-mechanical calculations. We have introduced the iterative method to overcome numerical difficulties, and to clarify how the exciton proliferation proceeds. Using this method, numerical calculations involving a large number of excitons and phonons are executed. Numerical results have indicated that the pattern of the exciton domain grown in the early stage sensitively reflects the anisotropy of the inter-exciton interactions, and that the successful proliferation is also realized by this anisotropic interactions.

To extend the front size to more than 4 sites is our future problem.

REFERENCES

1. Nasu, K., *"Relaxation of Excited States and Photo-induced Structural Phase transitions"*, Springer-Verlag, Berlin, 1997, pp.3-16.
2. Nasu, K., Huai, P., and Mizouchi, H., *J.P.CM*, **13**, R693 (2001).
3. Koshihara, S., Takahashi, Y., Sakai, H., Tokura, Y., and Luty, T., *J. Phys. Chem.,* **B103**, 2592 (1999).
4. Ogawa, Y., Koshihara, S., Koshino, K., Ogawa, T., Urano, C., and Takagi, H., *Phys. Rev. Lett.* **84**, 3181 (2000).
5. Huai, P., and Nasu, K., *J.Phys. Soc. Jpn.* **71**, 1182(2002).
6. Cho, K., and Toyozawa, Y., *J. Phys. Soc. Jpn.* **30**, 1555 (1971).
7. Mizouchi, H., and Nasu, K., *J. Phys. Soc. Jpn.* **70**, 2175 (2001).
8. Ball, R., Brady, R., Rossi, G., and Thompson, B., *Phys. Rev. Lett.* **55**, 1406 (1985).

Introduction of Physics related to ultra-short-pulse laser-plasma

Hitoki Yoneda

Institute for Laser Science, University of Electro-communications
Chofugaoka, Chofushi, Tokyo 182-8585, Japan

Abstract. This lecture deals with the basic physics of the laser-plasma interaction. Specially, the subjects related to the experiments of solid-state target with ultra-short-pulse lasers are chosen. Since these interaction processes are very complex and it needs a lot of time to overview whole of this area, only simple and basic physical models are used for discussion in this article. The common sense with normal solid-states and/or optics is also introduced for understanding optical response for high density plasma.

1.INTRODUCTION

Recently, chirped pulse amplification has made it possible to achieve high intensity electromagnetic field in the laboratory. Even with commercial base lasers, their output power can exceed multi TW level. When these laser lights are focused, very high concentration of optical field can be obtained easily. The materials having the extreme condition are achieved with these lasers so that these research tools has opened new research areas in nonlinear optics, plasma physics, and new material processing. However, at the same time, the physics inside these interaction areas becomes more complex and user of these lasers, sometimes, escape from consideration of details physics and/or creation of models by themselves.

In this lecture, the basic idea for understanding the laser-plasma interaction is introduced for beginner scientist of this field. In this article, I'll show some common sense between these interactions of high optical field and solid-states physics and normal optical physics.

2.BASIC PROCESS IN LASER PLASMA INTERACTION

When the laser light illuminates the solid-state materials, thin surface layer of the target is heated, melted and evaporated. Finally, plasma is created on the surface. In this stage, electrical field of laser heats up material with Joule heating mode. After creation of the plasma, energy of the laser is absorbed with electrons in the plasma. The absorbed energy drives thermal conduction wave into cold target material, and the plasma expansion toward vacuum area with normal direction to the surface. This energy is also used for excitation and ionization of ions. These ions will loss the energy with emitting radiation in late stage. Changing density and temperature and

CP634, *Science of Superstrong Field Interactions*, edited by K. Nakajima and M. Deguchi
© 2002 American Institute of Physics 0-7354-0089-X/02/$19.00

changing ionization state and/or the populations of excitation levels cause change of the optical properties (n(ω), k(ω)) of the plasmas. Then, the laser propagation is modified due to this change of the optical constant. Figure 1 shows relation of above-mentioned processes. This feedback rout, sometimes, produces strong non-linearity in the laser plasma interaction. These complexly also put difficulty in prediction of the whole phenomena.

To remove from this difficulty, some models use the assumption of 'quasi-steady state' for this interaction. For example, with nanosecond laser pulse, several self-similar solutions [1] has been proposed for analysis of thermal expansion and or heat waves. In these models, laser light interacts with the plasma having given profile of density and temperature, which has slowly changed in time.   In the case of interaction with ultra-short-pulse lasers, since pulse duration of laser is so small that some plasma parameters are frozen and are ignored during the interaction time. Among the processes shown in figure 1, for example, hydrodynamics and some of ionization process are slower than pulse duration of femtosecond lasers. Even relaxation time from electron system and ionic one, sometimes, become slower than pulse duration. In this case, initial parameters of plasma are kept during laser heating, and it is enough to consider only some dominant processes. Therefore, laser light wave can interact with plasmas having quasi-constant optical parameters.

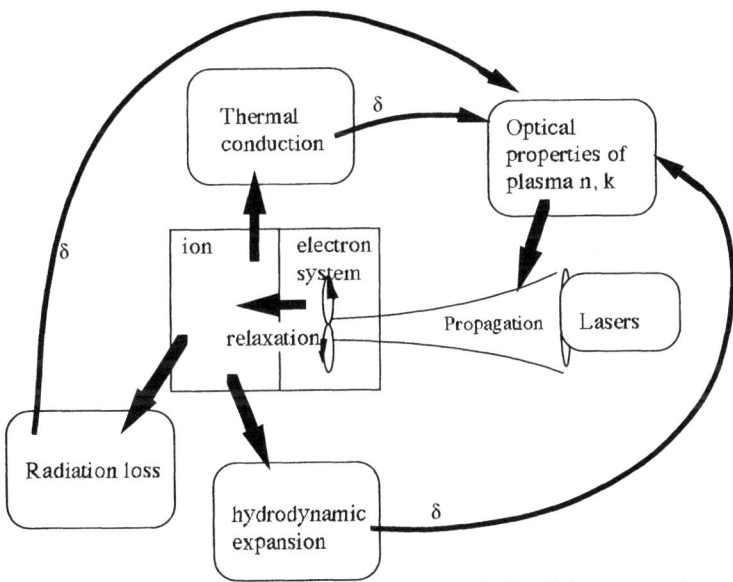

FIGURE 1. The process inside laser-plasma interaction. These 'feedback' loops, sometimes, result in strong nonlinear feature of laser plasma interaction.

In usual treatment in optics, light propagation is considered with following Maxwell equations.

$$\nabla \times \mathbf{E} = \frac{i\omega}{c}\mathbf{B}, \quad \nabla \times \mathbf{B} = \frac{4\pi}{c}\sigma\mathbf{E} - \frac{i\omega}{c}\mathbf{E} \qquad (1)$$

In these equations, a harmonic light wave ($\mathbf{E} = \mathbf{E}(x)\exp[-i\omega t]$) was considered as incident laser light. Only σ is a material parameter. To estimate this value, the free response of the free electrons under the harmonic wave is considered. They are quivering with electric field of laser $\dfrac{\partial \mathbf{u}_e}{\partial t} = -\dfrac{e}{m_e}\mathbf{E}(x)\exp[-i\omega t]$ and induced current density becomes $\mathbf{J} = -n_e(x)e\mathbf{u}_e$. According to simple ohm's law, conductivity of plasma is represented with $\sigma = i\dfrac{\omega_{pe}^2}{4\pi\omega}$ because $\dfrac{\partial \mathbf{J}}{\partial t} = -n_e e\dfrac{\partial \mathbf{u}_e}{\partial t} = \dfrac{e^2 n_e}{m_e}\mathbf{E} = \dfrac{\omega_{pe}^2}{4\pi}\mathbf{E}$

and $\mathbf{J} = \sigma\mathbf{E} = i\dfrac{\omega_{pe}^2}{4\pi\omega}\mathbf{E}$.

Including this discussion, eq.(1) rewrite a wave equation such as,

$$\nabla^2\mathbf{E} - \nabla(\nabla\cdot\mathbf{E}) + \dfrac{\omega^2}{c^2}\varepsilon\mathbf{E} = 0 \qquad (\because \varepsilon = 1 - \dfrac{\omega_{pe}^2}{\omega^2}). \qquad (2)$$

If variation of the dielectric constant ε is slow in space, the second term of left-hand side can be neglected. This assumption maybe reconsider carefully in the case of ultra-short-pulse laser interaction with matter because its density scale length of plasma is, sometimes, shorter than wavelength of laser. It is also treat this thin layered plasma carefully since it changes absorbance and mechanism of absorption, easily. Dispersion relation of light in plasma is deduced from eq.(2), that is $\omega^2 = \omega_{pe}^2 + k^2 c^2$. Figure 3 shows this relation. Any elastic and non-elastic scattering will be discussed on this plot. (for example, Stimulated Brillouin scattering process is denoted with vectors)

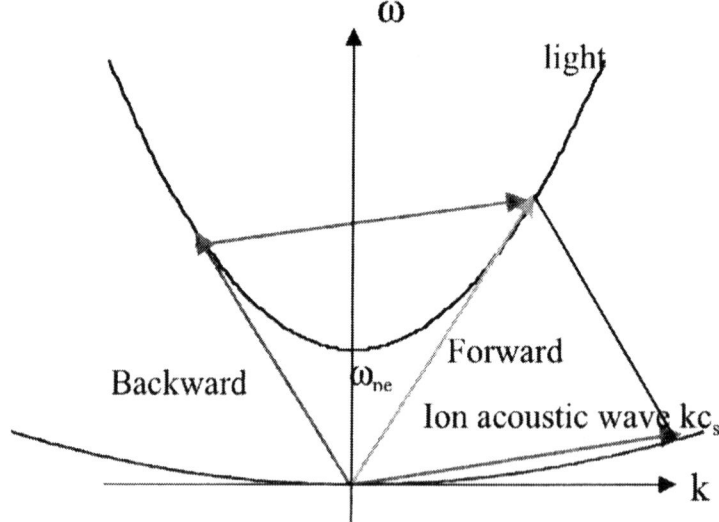

FIGURE 2. Dispersion relations in laser plasma interaction. The light has $\omega^2 = \omega_{pe}^2 + k^2 c^2$ and the other waves have the other formula. (For help of understanding, that of ion acoustic wave is also plotted in this figure.) Phase matching condition between two waves can be treated on this figure.

To proceed the discussion of propagation problem in plasma, linear density gradient $(n_e = n_c \frac{z}{L})$ are considered in following part. Dielectric constant of the plasma becomes

$$\varepsilon(\omega, z) = 1 - \frac{\omega_{pe}^2}{\omega^2} = 1 - \frac{n_e}{n_c} \quad (3).$$

The wave equation then, rewrites $\nabla^2 E + \frac{\omega^2}{c^2}\left(1 - \frac{z}{L}\right)E = 0.$ The solution of this equation [2] is already known as mathematics subject and that is combination with two Airy functions of Ai(x) and Bi(x) (shown in Figure 3).

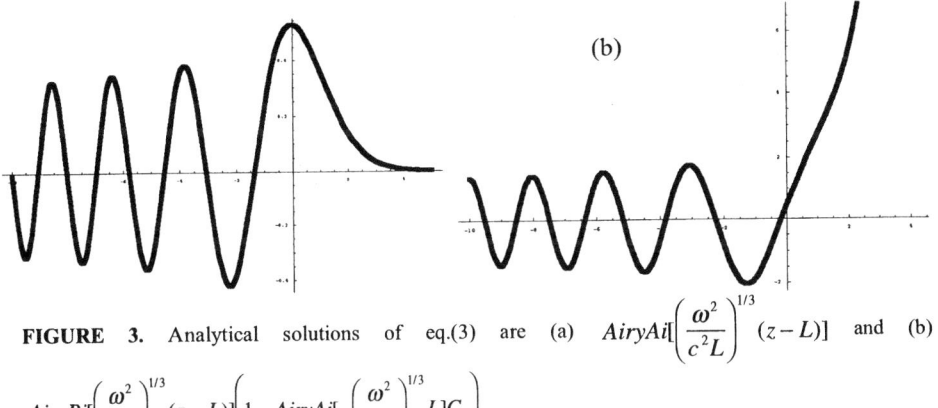

(b)

FIGURE 3. Analytical solutions of eq.(3) are (a) $AiryAi\left[\left(\frac{\omega^2}{c^2 L}\right)^{1/3}(z - L)\right]$ and (b)

$$\frac{AiryBi\left[\left(\frac{\omega^2}{c^2 L}\right)^{1/3}(z - L)\right]\left[1 - AiryAi\left[-\left(\frac{\omega^2}{c^2 L}\right)^{1/3} L\right]C_2\right]}{AiryBi\left[-\left(\frac{\omega^2}{c^2 L}\right)^{1/3} L\right]}.$$

As shown in Fig.3, one of them (b) is not correct physical meaning so that only Ai(x) function can be solution of this subject.

When $x \ll -1$,

$$Ai[x] \approx \frac{1}{(-x)^{1/4}} \cos[\frac{2}{3}(-x)^{3/2} - \frac{\pi}{4}] \text{ and}$$

cosine function can be rewrite with a pair of exponential functions

$(e^{i\frac{2}{3}(\omega L/c)(1-z/L)^{3/2} - i\frac{\pi}{4}}$ and

$e^{-i\frac{2}{3}(\omega L/c)(1-z/L)^{3/2} + i\frac{\pi}{4}}$). This denoted that the solution is made with both of the forward and reflected backward light waves.

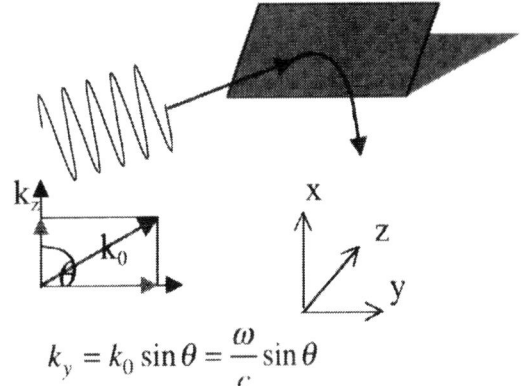

$$k_y = k_0 \sin\theta = \frac{\omega}{c}\sin\theta$$

FIGURE 4. Oblique incidence of laser light

If the incident laser light enters the target with oblique condition, analysis of ray trajectory becomes important. Again, we start from the wave equation (eq.2) with two orthogonal vector components for E and B. When the laser light has s-polarization, and axis notation is chosen in the manner of Fig.4, this equation can be solved with variable separation method. If we can rewrite $E_x(y,z) = F(y)G(z)$, the equation becomes

$$-\frac{1}{F(y)}\frac{\partial^2}{\partial y^2}F(y) = \frac{1}{G(z)}\left(\frac{\partial^2}{\partial z^2}G(z) + \frac{\omega^2}{c^2}\varepsilon(z)G(z)\right) = const. \tag{4}.$$

The first equation produces a solution of harmonic wave for y-direction so that the light can propagate with constant k_y ($=k_0\sin\theta$) vector during the interaction for this direction. The second one rewrites with above k_y.

$$\frac{\partial^2}{\partial z^2}G(z) + \frac{\omega^2}{c^2}\left(\varepsilon(z) - \sin^2\theta\right)G(z) = 0 \tag{5}$$

With similarity with eq. (2), it is easy to image that the laser light can penetrate into the position of $n_e = n_c\left(1 - \sin^2\theta\right)$. The ray of this laser trajectory is calculated with integral of dz/dy with its path, therefore,

$$\frac{dz}{dy} = \frac{k_z}{k_y} = \frac{\sqrt{1 - \frac{n_e}{n_c} - \sin^2\theta}}{\sin\theta} \tag{6}.$$

Typical results of this analysis are shown in Figure 5. It is important to consider the adaptability of this assumption of variable separation. It is much more clear when this solution is compared with ray tracing in a graded index fiber. In the latter case, variable separation can't be used so that its solution is different from Fig.5.

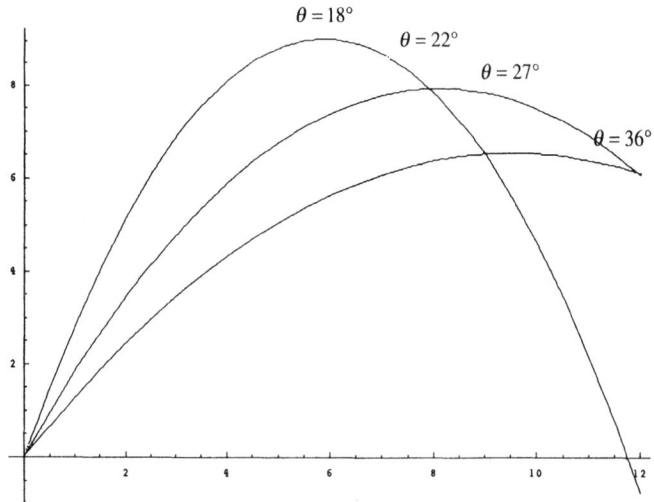

FIGURE 5 Ray tracing of oblique incident laser in plasma

3. ABSORPTION PROCESS IN PLASMA AND COMPARISON WITH OPTICAL RESPONSE OF METALS

There are several absorption mechanisms in the laser-plasma interaction. Among them, collisional absorption and resonance absorption are dominant processes at moderate intensity condition. From dispersion relation of Fig.2 and ray tracing of Fig.3, the laser wave can penetrate into the point having density of $n_c(1-\sin^2\theta)$. However, beyond this point, the electrical field can penetrate into higher density region with evanescent-like mode(see Fig.2(a)). This component can be coupled with electron-plasma wave at the critical density point ($n_e=n_c$). With this resonance, the amplitude of electrical field at this point becomes huge value and any small dissipation can absorb the laser energy, effectively. This is explanation of resonant absorption.
About the collisional absorption, we can start eq. (1) and (2) with ohm's law. When the laser light enter the plasma, the electrons are driven by its electric field. If some resistive force can be assumed in the plasma, equation of motion of the electrons is

$$\frac{\partial \mathbf{u}}{\partial t} = -\frac{e}{m_e}\mathbf{E} - v_{ei}\mathbf{u} \quad (7)$$

When the incident laser is harmonic wave, the solution of this equation is

$$u[t] = \frac{E_0 \exp[i\omega t]}{v_{ei} + i\omega}\frac{e}{m_e} + C_1 \exp[-v_{ei}t] \quad (8).$$

Therefore, conductivity of the plasma is written with $\sigma = i\dfrac{\omega_{pe}^2}{4\pi(\omega + iv_{ei})}$ and its

dielectric constant is with $\varepsilon = 1 - \dfrac{\omega_{pe}^2}{\omega(\omega + iv_{ei})}$. With the same procedure in Sec.2,

dispersion relation is

$$\omega^2 = \omega_{pe}^2\left(1 - i\frac{v_{ei}}{\omega}\right) + k^2c^2. \quad (9)$$

Therefore, this electrical field damped in the resistive plasma and the coefficient is

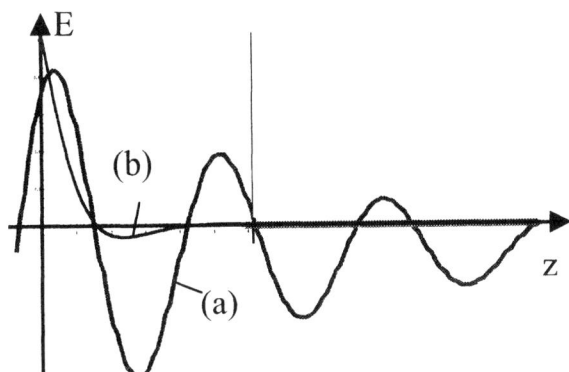

FIGURE 6. Damping properties of optical electric field in the plasma (a) (see manuscript for its assumption) and normal metals (b).

$\gamma = \dfrac{\nu_{ei}}{2}\dfrac{\omega_{pe}^2}{\omega^2}$. In this discussion, we used an assumption of smaller damping coefficient.

Therefore, whole solution of this wave is damping oscillation with ω frequency. (see Fig.6 (a))

It is very interest to compare this absorption and the interaction with normal metals. As well-known, for interaction of light with metals the similar Drude models can be used. The incident light also penetrates into metal with evanescent mode (skin depth).

The dispersion relation of metals [3] is $\hat{k}^2 c^2 = \omega^2 \mu\left(\varepsilon + i\dfrac{4\pi\sigma}{\omega}\right)$. Since any metals at

normal condition (at room temperature) has larger conductivity ($\dfrac{4\pi\sigma}{\omega\varepsilon} \gg 1$), k has to

be square root of pure imaginary value. It means

$$\hat{k}\dfrac{\omega}{c}\sqrt{i\dfrac{4\pi\sigma\mu}{\omega}} = \dfrac{\omega}{c}\sqrt{\dfrac{4\pi\sigma\mu}{\omega}}\left(\dfrac{1}{\sqrt{2}} + \dfrac{1}{\sqrt{2}}i\right) \quad (10).$$

Therefore, it can be concluded that the 'propagation wave' inside metals has strongly damping feature and its oscillating frequency and its damping coefficient are equal each other. (see Fig.6 (b))

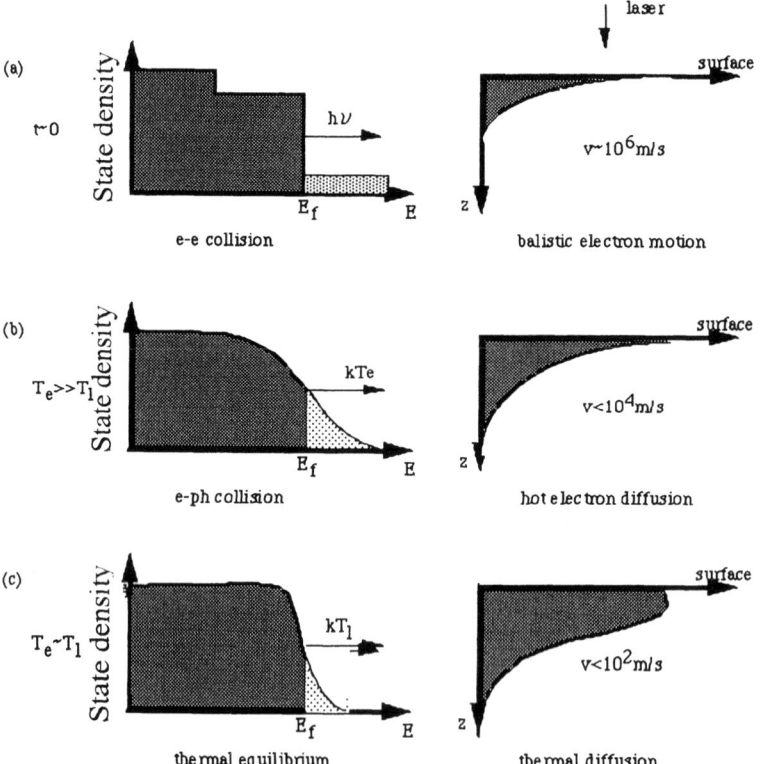

FIGURE 7. Post process of ultra-short-pulse laser heated materials

If this normal metal is illuminated with ultra-short-pulse laser, it will be heated up to so called plasma temperature in the meanwhile. During this heating process, both curves in Fig.6 will be identical. How does it change? These comparisons and questions will produce one of the important evidence when people create common sense between solid-state physics and plasma physics.

4. ENERGY RELAXATION PROCESS AFTER ABSORPTION

As mentioned before, the absorption process and laser plasma interaction itself, sometimes, are able to separate other physical process. So that, after energy absorption in electron system, relaxation, conduction, and hydrodynamics expansion will occurred as a post process in some case of ultra-short-pulse laser experiments (Fig.7). Among them, they may separate each other in some condition. Typical time scales of each process are τ_{ee}, $\tau_{e\text{-ph}}$, τ_{ei} respectively. If we can assume $\tau_{ee} \ll \tau_{e\text{-ph}}, \ll \tau_{ei}$, the discussion of this post process will be simplified. More explanation will be appeared in future.

REFERENCES

1. For example, references in text book " Physics of laser driven plasmas" edited by H.Hora, A Wiley-Interscience Pub., NewYork 1981 Chapter 5 p.58.
2. W. Kruer, "The Physics of Laser Plasma Interactions", Addison-Wesley, Redwood city 1988 p.32-35.
3. See textbook of M.Born and E.Wolf, "6th edition Principle of Optics", Pergamon Press Oxford 1980 Chapter XIII p.611-664.

Relativistic Interaction of Laser Pulses with Plasmas

S. V. Bulanov

General Physics Institute of Russian Academy of Sciences,
Vavilov St. 38, Moscow 119991, Russia
bulanov@fpl.gpi.ru

Abstract. The lecture presents an introduction to the theory of the interaction of relativistically strong, ultra-short laser pulses with plasmas. The laser interaction with underdense plasmas and its application to the problem of charged particle acceleration, of laser frequency upshifting, of relativistic self focusing, and of the generation of a quasistatic magnetic field is discussed. The properties of nonlinear coherent structures such as relativistic solitons and vortices and the production of high harmonics are discussed.

Over the last few years we have witnessed extremely fast progress in laser technology. The laser intensity I has increased by two orders of magnitude each couple of years and has now reached the value of $\geq 10^{21} W/cm^2$ in the radiation emitted by petawatt lasers [1]. The electric field of these pulses is of the order of $\approx 10^{12} V/cm$ and significantly exceeds the inter-atomic field. Such a large electric field fully ionizes the matter with which it interacts and can force the electrons in the plasma to oscillate with relativistic energy. In these regimes the specific features of the nonlinear dynamics of collisionless plasmas and their interaction with the electromagnetic waves become very important and attractive for theoretical studies [2]. On the other hand, the increasing interest in the problems of the interaction of relativistically strong laser radiation with plasmas finds various broad applications in the development of new concepts of compact laser-based accelerators of charged particles [3], powerful ultra-fast X-ray sources and controlled nuclear fusion in the framework of the Fast Ignition Concept [4]. They are also connected with the problems of the propagation of relativistically strong electromagnetic waves in space plasmas and with the mechanisms of acceleration of cosmic rays [5] and with the problems of high energy physics [6]. Particle acceleration by an ultra-intense laser pulse interacting with plasma also has practical applications to laser induced nuclear reactions [7], ion injection into conventional accelerators, and hadrontherapy in medicine [8]. When a petawatt laser pulse interacts with matter, conditions can be produced that were imagined to occur only in astrophysical objects. This opens the way for experimental plasma astrophysics to study the properties of matter under these extreme conditions [9].

In the relativistic range of amplitudes of the laser radiation, when its intensity is above $1.3 \times 10^{18} W/cm^2$, the ratio the electron quiver velocity in the laser field $v_E = eE/m_e\omega$ to the speed of light (v_E/c) becomes close to one in the

CP634, *Science of Superstrong Field Interactions*, edited by K. Nakajima and M. Deguchi
© 2002 American Institute of Physics 0-7354-0089-X/02/$19.00

case of the laser light with wavelength $1 \mu m$. In particular, this means that the magnetic part of the Lorentz force $e \mathbf{v} \times \mathbf{B} / c$ becomes as important as the electric part $e \mathbf{E}$. Relativistic effects modify the nonlinear processes that are known in the limit of moderate radiation amplitudes and make it possible for essentially new nonlinear phenomena to occur.

One of the most attractive applications of ultra-short super-intense laser pulses is connected with the development of new methods of accelerating charged particles. It seems that presently operating accelerators of charged particles has approached their maximum reasonable size. These were evident as far back as the fifties when it was proposed the use of collective electric fields excited in a plasma (collective methods of acceleration) to accelerate charged particles. The generation of high energy particles, both electrons and ions, when strong electromagnetic radiation interacts with plasma is a well known basic phenomenon. However, it is necessary to find the plasma and radiation parameters that optimize this process.

Among the wide variety of methods for generating a regular electric field in plasmas with strong laser radiation, the most attractive at the present time is the scheme [3] of the laser wake field accelerator (LWFA). In this method a strong Langmuir wave is excited in the plasma and the charged particles are accelerated by the electric field of this wave. The advantages of Langmuir waves for charged particle acceleration, emphasized in Ref. [10] are related to the fact that the electric field in the wave is longitudinal and that its frequency ω_{pe} does not depend on the wave vector. This means that the group velocity of a Langmuir wave in a cold plasma, $\partial \omega_{pe} / \partial k$ is equal to zero, i.e., the wave stays in the plasma for a long time at the place where it is generated and can be used for charged particle acceleration. A plasma wave with wave-number $k_p = \omega_{pe} / c$ has a relativistic phase velocity, and the charged particles at resonance can be accelerated up to ultra-relativistic energies.

Another fascinating suggestion for using ultra short laser pulses is related to the scheme which aims at achieving fusion conditions with reduced laser pulse drive energy [4]. In the standard scheme of inertial confinement fusion (ICF) a pellet containing high pressure deuterium-tritium (DT) fuel is made to achieve a large isoentropic compression by ensuring a high degree of symmetrical irradiation of the driving nanosecond laser pulse. A significant amount of laser energy is needed in order to create with converging shock waves at the center of the target a hot spark with the required high temperature. Then a burn wave propagates through the fuel via alpha particle heating. The amount of additional energy needed in order to reach high gain depends on the degree of spherical convergence, which can be spoiled by the development of the Raleigh-Taylor instability. On the contrary, in the fast ignition scheme it is suggested that a multi-terawatt laser pulse drills a channel towards the center in a pre-compressed target. Then, a picosecond petawatt pulse accelerates electrons up to multi-MeV energy values. The relativistic electrons heat a small portion of the fuel before it can disassemble. A number of fundamental questions must be addressed in this scheme, such as the channel drilling and the transport of the laser pulse energy, the acceleration of fast particles and their interaction with dense plasma. An other proposed scheme of Fast Ignition involves the laser accelerated ions to ignite the precompressed thermonuclear target.

The laser radiation is characterized by its incidence angle on the plasma, θ and by its polarization, which can be circular, elliptic or linear. In the case of linear polarization, for oblique incidence one distinguishes s- and p-polarized waves. We consider a laser pulse of length l, with amplitude of the electric field E_0, and carrier frequency ω_0. In describing the laser pulse interaction with plasmas, it is convenient to use dimensionless units in order to characterize the plasma density and the laser pulse amplitude.

The plasma density can be described by a dimensionless parameter defined as the ratio of the density n to the critical density n_{cr}. In a plasma with density equal to the critical density, the carrier frequency of the laser radiation ω_0 is equal to the Langmuir frequency $\omega_{pe} = \sqrt{4\pi n e^2 / m_e}$, i.e., $\omega_0 / \omega_{pe} = (n_{cr} / n)^{1/2} = 1$. When $\omega_0 / \omega_{pe} > 1$ the plasma is described as "underdense". Such plasma is transparent to the laser radiation. The group velocity of the electromagnetic wave is close to the speed of light. A plasma where $\omega_0 / \omega_{pe} < 1$ is described as "overdense". The electromagnetic radiation cannot penetrate deeply into an overdense plasma. It penetrates into the plasma as far as the evanescence length, which is of the order of $d_e = c / \omega_{pe}$, where d_e is the collisionless skin depth. However, for relativistic intensities, the collisionless skin depth changes (the refractive index changes) and an overdense plasma can become transparent.

When the dimensionless ratio $a = eE_0 / m_e \omega_0 c$ is much smaller than unity, the quiver velocity, $v_E = eE / m_e \omega_0$, is small compared with the speed of light in vacuum, c (and the quiver radius $r_E = eE / m_e \omega_0^2$ is much shorter than the wavelength of the laser radiation, $\lambda = 2\pi c / \omega_0$) and we can neglect relativistic effects and use classical mechanics in order to describe the interaction of the laser light with charged particles. In the opposite limit, when $a \gg 1$, we must describe the laser matter interaction within the framework of the relativistic theory.

First of all the relativistic effects modify qualitatively the charged particle dynamics in the field of the electromagnetic wave. From the exact solution of the equations of motion of a charged particle in a planar electromagnetic wave [11] it follows that the transverse component of the generalized momentum is constant, $\mathbf{p}_\perp = (e/c)\mathbf{A}_\perp(x - ct) = \text{constant}$, and the energy and the longitudinal component of the momentum are related as $m_e c^2 \gamma - p_\parallel c = m_e c^2 \sqrt{1 + (p_\parallel / m_e c)^2 + (p_\perp / m_e c)^2} - p_\parallel c = \text{constant}$. In the reference frame, where the charged particle was at rest before the interaction with laser pulse, the particle kinetic energy $K = m_e c^2 (\gamma - 1)$ and momentum p are given by the expressions: $K = m_e c^2 |\mathbf{a}_\perp|^2 / 2$, $p_\parallel = m_e c |\mathbf{a}_\perp|^2 / 2$, and $\mathbf{p}_\perp = m_e c \mathbf{a}_\perp$. The acceleration length, which the particle must pass to acquire the energy, is equal to $\xi_\perp = a_0 \lambda$ in the transverse direction and it is $\xi_\parallel = a_0^2 \lambda / 2$ in the longitudinal direction. We see that in order to get this energy the particle must remain in the laser field in the region that has the size in the transverse direction not less than $a_0 \lambda$ and the region size in the longitudinal direction must not be less than $a_0^2 \lambda / 2$. Here $\mathbf{a}_\perp(x - ct) = e\mathbf{A}_\perp(x - ct) / m_e c$. We see that for $a > 2$ the particle acquires a relativistic energy and the longitudinal component of its momentum is larger than the transverse component.

We shall call the laser radiation with $a >> 1$ superintense or relativistically strong. In this case we come into range of the relativistic optics [12]. The dimensionless amplitude of the electromagnetic wave can also be expressed, via the radiation intensity I and the wavelength λ, as: $a_0 = (I / 1.35 \times 10^{18} W / cm^2)^{1/2} (\lambda / 1 \mu m)$. Since the radiation of the petawatt lasers, focused into a spot with diameter of the order of the radiation wavelength, reaches a magnitude about $1.35 \times 10^{21} W / cm^2$, we see that the dimensionless amplitude a in this case is well above unity, $a_0 \approx 40$. We see that the value of the amplitude of the petawatt laser radiation written in dimensionless units becomes larger than $(m_p / m_e)^{1/2}$, which means that the physical processes due to the nonlinear ion dynamics come into play [2,13].

Probably the most impressive nonlinear phenomenon in underdense plasma is the self-focusing of the laser radiation. The self-focusing, discovered by G. A. Askar'yan in 1962 [14], appears due to the nonlinear change of the refractive index of the medium in the region where a high intensity electromagnetic wave propagates. In the multi-terawatt or petawatt laser pulse - plasma interaction self focusing appears due to the relativistic increase in the electron mass and to the plasma density redistribution under the action of the ponderomotive force. The threshold (critical) power for relativistic self-focusing is [15] $P_c \geq 2m_e^2 c^5 \omega_0^2 / e^2 \omega_{pe}^2 \approx 17(\omega_0 / \omega_{pe})^2 GW$. When the laser power far exceeds the critical power, the laser beam can split into separate filaments.

Superintense laser radiation in a plasma is subject to a host of instabilities. The fastest is the stimulated Raman scattering (SRS) instability, which develops on the electron time-scale. The stimulated forward Raman scattering (SFRS) leads to the self-modulation of the laser pulse with modulation length of the order of $2\pi / k_p = 2\pi c / \omega_{pe}$ [16]. Self-modulation of the laser pulse is of great importance for the LWFA, where the laser pulse excites a longitudinal electric field, which in turn accelerates electrons up to high energies [3].

For ultrashort superintense laser pulses, the stimulated backward Raman scattering (SBRS) instability is the first to develop. SBRS leads to the erosion of the amplitude profile at the leading edge of the pulse and to the formation of a steep laser front similar to a shock in a time of the order of $\approx \omega_{pe}^{-1}(\omega_0 / \omega_{pe})^2$. Such a shock front generates a wake field, which accelerates the electrons in the plasma, thus leading to fast depletion of the laser-pulse energy ($\tau_p(\omega_0 / \omega_{pe})^2$) [17,18] and to the induced focusing of the laser light behind the front [19].

The electrons accelerated inside a self focused laser pulse produce electric currents in the plasma and a quasistatic magnetic field associated with them. The attraction of the electric currents leads to the redistribution of the fast electrons. This in turn changes the refractive index because, due to the relativistic increase of the electron mass, the effective plasma frequency is smallest in the regions with the highest concentration of fast electrons. This process causes high intensity laser radiation to interact magnetically in plasmas, makes the laser light filaments merge and provides a mechanism for transporting the laser energy over long distances [20].

As a consequence of the equation in a plasma dominated by the electron dynamics the quasistatic magnetic field is associated with electron fluid vortices. In this case the vorticity is $\nabla \times \mathbf{B} = 4\pi ne\mathbf{v}/c$. Both the electrostatic wake fields and the magnetic field vortices stay in the plasma much longer than the laser pulse because of their very low propagation velocity. The distance between the vortices is comparable to, or in their final stage even larger than, the collisionless skin depth. The vortex row moves as a whole in the direction of the laser pulse propagation with a velocity much smaller than the pulse group velocity. The velocity of the vortex row decreases with increasing distance between the vortex chains that form the row. The vortex system evolution is described by the Hasegawa-Mima equation [21]. In Ref. [22] it was shown that symmetric vortex raw is always unstable, but the antisymmetric raw can be stable similarly to the Von Karman vortex raw in Eulerian fluids [23].

In inhomogeneous plasma the vortices move along the lines of constant density according to the Hasegawa-Mima equation. This is a consequence of the Eurthels theorem. When the laser pulse interacts with the finite length plasma slab, the magnetic field advected by the electron vortices appears at the rear side of the slab in the form of a dipole. Depending on the length scale of the plasma inhomogeneity the vortex pair either propagates in the longitudinal direction in a weakly inhomogeneous case or it breaks into two separated vortices in the case of a steep plasma boundary, and each of them moves in the opposite direction in the transverse direction.

Sharply focalized bunches of ultrarelativistic electrons, accelerated during the breaking of the wake-wave [24], can generate a strong magnetic field which propagates together with the bunch and which is due to the magnetic component of the Lienard-Wiechert potential of the electron bunch [25].

Recently in the interaction of a laser pulse with a plasma, a novel phenomenon has been identified [26,27] that directly involves the expansion of a cavity filled with electromagnetic energy, its interaction and coalescence with other cavities in the plasma. This phenomenon is related to the long time evolution of the relativistic subcycle electromagnetic solitons [28-31] that are produced when an ultrashort ultraintense laser pulse propagates in an underdense plasma. Relativistic solitons consist of electron density cavities inside which an electromagnetic field oscillates coherently with a frequency below the local plasma frequency and a spatial structure corresponding to half a cycle. Recently, they have been identified in 3D PIC simulations [32].

In inhomogeneous plasmas, in contrast to the electron vortices, which move across a density gradient, solitons move along the density gradient towards the lowest density level. When a soliton approaches the plasma-vacuum boundary some critical density level, it radiates its energy in the form of a low-frequency short electromagnetic burst [33]. On an ion timescale due to the ion acceleration caused by the time-averaged electrostatic field inside the soliton it evolves into the post-soliton, which is a slowly expanding bubble in the plasma. The typical energy of fast ions is of the order of $m_e c^2 a$, where a is the amplitude of the trapped electromagnetic field.

In an overdense plasma with $\omega_0 / \omega_{pe} < 1$ one can never neglect the plasma inhomogeneity. The laser radiation penetrates the plasma to just over the evanescence length which is of the order of the collisionless skin depth. When the plasma has a sharp boundary the laser plasma interaction takes place at the plasma-vacuum interface. At a sharp plasma-vacuum interface the laser radiation can easily extract the electrons from the plasma and accelerate them towards the vacuum region. This process, known as "vacuum heating of the electrons" [34], provides an effective mechanism of anomalous absorption of the laser light.

When the plasma has a smooth density distribution, the Langmuir frequency is a function of the coordinates $\omega_{pe} = \omega_{pe}(r)$, and the processes that occur in the vicinity of the critical surface, in the region of the plasma resonance where $\omega_0 = \omega_{pe}(r)$, play a key role. In the plasma resonance region the electromagnetic wave excites resonantly an electric field with very high amplitude, localized in the narrow region where the laser radiation is absorbed and fast particles are generated [36].

The ponderomotive pressure of the pulse causes a plasma density redistribution in the transverse direction that changes the refractive index and allows the laser pulse to penetrate into an overdense plasma. This is the basic idea of the hole boring and of the laser energy transport in the fast ignition concept [4].

When the laser radiation interacts with a thin slab of high density plasma, this leads to effective generation of fast ions at both the front side and the rear side of the target. The ions can be focused by tailoring the target. In this case, the generated fast ions appear in the form of highly collimated beams.

As it has been mention above, effective ion acceleration requires laser radiation in the petawatt power range. In the simplest scheme of ion acceleration, the laser pulse interacts with a thin foil. When multiterawatt laser radiation interacts with a foil, matter is ionized in an interval shorter than a single optic oscillation period of the laser radiation, producing in such way collisionless plasma. Under the action of the laser radiation the electrons are expelled from a region on the foil with transverse size of the order of the diameter of the focal spot. For femtosecond long laser pulses with multiterawatt power, the typical time scale of the hydrodynamic expansion of a micron plasma slab is much longer than the laser pulse duration. Under these conditions ions remain at rest, which results in the formation of a positively charged layer of ions. However after a time interval equal to or longer than the inverse of the ion Langmuir frequency $\omega^{-1}_{pi} = (4\pi n_0 Z_i^2 e^2 / m_i)^{-1/2}$ the ion layer explodes because of the Coulomb repulsion of electric charges with equal sign.

However if we compare the form of the energy spectra of the accelerated ions observed in the experiments or obtained in numerical simulations with the narrow energy spectrum required for various applications in controlled nuclear fusion, for the injectors, and in hadrontherapy, we see that the energy spectra of the laser accelerated ions are at present quite far from the required ones. It was found that the ion spectra below a maximum energy can be approximated by a quasi-thermal distribution with an effective temperature several times smaller than the maximal ion energy.

In order to improve the proton beam quality one can cut the beam into beamlets with a narrow spread in energy space. However, in this case the efficiency of the laser energy transformation into the energy of fast particles decreases significantly, and, what is more important, the number of fast particles decreases. A different approach which seems more promising is connected with the use of multi-layer targets. In this scheme a thin foil is used as a target and its rear surface is coated with a thin hydrogen layer. When the ultra short laser pulse irradiates the target, the heavy atoms are partly ionized and the ionized electrons abandon the foil, generating an electric field due to charge separation. Because of their inertia the heavy ions remain at rest, while the lighter protons are accelerated [37].

Recently a new type of target for laser-matter interaction has appeared which consists of a gas made up of clusters, which are relatively small pieces of a solid material. A very efficient absorption of the laser energy interacting with the clusters and the formation of very high temperature underdense plasmas have been demonstrated in Ref. [38]. Such high temperature plasmas make table top fusion experiments possible.

The process of the laser plasma interaction is accompanied by the high harmonic generation. The main mechanism of the high harmonic generation in the range of the relativistic intensity of the laser radiation is due to the reflection of the electromagnetic wave at the "oscillating mirror"[35]. The oscillating mirror appears due to the nonlinear motion of the electrons in the narrow region near the plasma boundary. The high harmonic generation efficiency becomes much higher when the laser pulse propagates inside a narrow channel [39]. If the channel has thin walls, it makes possible to generate the high harmonics in the form of ultra short coherent wave packet.

The complexity of the laser-plasma interaction due to the high dimensionality of the problem, to the lack of symmetry and to the importance of nonlinear and kinetic effects makes analytical methods unable to provide a detailed description. On the other hand, powerful methods for investigating the laser-plasma interaction have become available through the advent of modern supercomputers and the developments of applied mathematics. In the case of ultra-short relativistically strong laser pulses, simulations with 3D Particle In Cell codes provide an unique opportunity for describing adequately the nonlinear dynamics of laser plasmas, including nonlinear wave breaking, the acceleration of charged particles up to high energy and the generation of coherent nonlinear structures such the relativistic solitons and vortices.

REFERENCES

1. Mourou, G., Barty, C., and Perry, M., *Physics Today* **51**, 22-28 (1998).
2. Bulanov, S.V., et al., "Relativistic Interaction of Laser Pulses with Plasmas" in *Reviews of Plasma Physics*. Vol. 22, edited by V.D. Shafranov, New York. Kluwer Academic / Plenum Publishers, 2001, pp. 227-335.
3. Tajima, T., and Dawson, J.M., *Phys. Rev. Lett.* **43**, 267-270 (1979); Nakajima, K., et al., *Phys. Rev. Lett.* **74**, 4428-4431 (1995); Modena, A., et al., *Nature* **337**, 606-608 (1995).
4. Tabak, M., et al., *Phys. Plasmas* **1**, 1626-1634 (1994); Roth, M., Cowan, T., Key, M., Hatchett, S., Brown, C., Fountain, W., Johnson, J., Pennington, Snavely, D., Wilks, S., Yasuike, K., Ruhl, H., Pegoraro, F., Bulanov, S., Campbell, M., Perry, M., and Powell, H., *Phys. Rev. Lett.* **86**, 436-439 (2001).

5. Berezinskii, V.S., Bulanov, S.V., Ginzburg, V.L., Dogiel, V.A., Ptuskin, V.S., *Astrophysics of Cosmic Rays*, Amsterdam, North Holland Publ. Elsevier Science Publ. 1990.

6. Tajima, T., and Mourou, G., *Phys. Rev. ST* **5**, 031301-40 (2002).

7. Bychenkov, V.Yu., Sentoku, Y., Bulanov, S.V., Mima, K., Mourou, G., and Tolokonnikov, S.V., *JETP Lett.* **74**, 586-589 (2001).

8. Bulanov, S.V., and Khoroshkov, V.S., *Plasma Phys. Rep.* **28**, 453-456 (2002); Bulanov, S.V., Esirkepov, T.Zh., Kuznetsov, A.V., Khoroshkov, V.S., and Pegoraro, F., *Physics Letters A* **299**, 240–247 (2002).

9. Farley, D.R., et al., *Phys. Rev. Lett.* **83**, 1982-1985 (1999).

10. Dawson, J. M., *Plasma Phys. Controlled Fusion* **34**, 2039-2044 (1992)

11. Landau, L.D., and Lifshits, L.M., *The Classical Theory of Field,* Oxford. Pergamon Press. 1984.

12. Mourou, G., et al., *Plasma Phys. Rep.* **28**, 12-27 (2002)

13. Bulanov, S.V., et al., *Plasma Phys. Rep.* **25**, 701-714 (1999).

14. Askar'yan, G.A., *Sov. Phys. JETP* **15**, 8-12 (1962).

15. Max, C., Arons, J., and Langdon, A., *Phys. Rev. Lett.* **33**, 209-212 (1974).

16. Antonsen, T.M., and Mora, P., *Phys. Rev. Lett.* **69**, 2204-2207 (1992).

17. Bulanov, S.V., Inovenkov, I.N., Kirsanov, V.I., Naumova, N.M., and Sakharov, A.S., *Physics of Fluids* B **4**, 1935-1942 (1992).

18. Mori, W.B. et al., *Phys. Rev. Lett.* **72**, 1482-1485 (1994); Barr, H.C., et al, *Phys. Rev. Lett.* **81**, 2910-2913 (1998).

19. Bulanov, S.V., and Sakharov, A.S., *JETP Lett.* **54**, 203-206 (1991).

20. Askar'yan, G.A., Bulanov, S.V., Pegoraro, F., and Pukhov, A.M., *JETP Lett.* **60**, 251-254 (1994).

21. Hasegawa, A., and Mima, K., *Phys. Rev. Lett.* **39**, 205-208 (1977)

22. Bulanov, S.V., Esirkepov, T. Zh., Lontano, M., Pegoraro, F., and Pukhov, A. M., *Phys. Rev. Lett.* **76**, 3562-3565 (1996).

23. Lamb, H., *Hydrodynamics*, Cambridge. Cambridge University Press, 1932.

24. Bulanov, S., Pegoraro, F., Pukhov, A., and Sakharov, A., *Phys. Rev. Lett.* **78**, 4205-4208 (1997); Bulanov, S., Naumova, N., Pegoraro, F., and Sakai, J.-I., *Phys. Rev. E* **58**, R5257-5260 (1998).

25. Liseikina, T.V., et al., *Phys Rev E* **60**, 5991-5997 (1999)

26. Naumova, N.M. et al., *Phys. Rev. Lett.*, **87** 185004-1-185004-1 (2001).

27. Borghesi, M. et al., *Phys. Rev. Lett.* **88**, 135002-1-135002-4 (2002).

28. Esirkepov, T.Z., Kamenets, F.F., Bulanov, S.V., and Naumova, N.M., *JETP Lett.* **68**, 36-41 (1998)

29. Bulanov, S.V., Esirkepov, T.Zh, Naumova, N.M., Pegoraro, F., and Vshivkov, V.A., *Phys. Rev. Lett.* **82**, 3440-3443 (1999); Bulanov, S.V., Califano, Esirkepov, T.Zh., Mima, K., Naumova, N.M., Nishihara, K., Pegoraro, F., Sentoku, Y., and Vshivkov, V. A., *J. Plasma Fusion Research* **75**, No. 5-CD, 506-509 (1999).

30. Farina, D., Lontano, M., and Bulanov, S.V., *Phys. Rev. E* **62**, 4146-4151 (2000).

31. Farina, D., and Bulanov, S.V., *Phys. Rev. Lett.* **86**, 5289-5292 (2001).

32. Esirkepov, T.Zh., et al., *Phys. Rev. Lett.* (2002) submitted for publication

33. Sentoku, Y., et al., *Phys. Rev. Lett.* **83**, 3434-3437 (1999).

34. Brunel, F., *Phys. Rev. Lett.* **59**, 52-55 (1987).

35. Bulanov, S.V., Esirkepov, T.Zh, Naumova, N., and Pegoraro, F., *Phys. Plasmas* **1**, 745-757 (1994).

36. Bulanov, S.V., Kovrizhnykh, L.M., and Sakharov, A.S., *Physics Reports* **186**, 1-50 (1990).

37. Esirkepov, T.Zh., et al., *Phys. Rev. Lett.* (2002) submitted for publication.

38. Ditmire, T., et al., *Nature* **386**, 54-55 (1997).

39. Bulanov, S.V., Kamenets, F., Pegoraro, F., and Pukhov, A., *Physics Letters A* **195**, 84-87 (1994).

Review and Comparison of Particle-in-Cell and Vlasov Simulation methods with application to relativistic self-focusing

James Koga

Advanced Photon Research Center, JAERI, Kyoto-fu, 619-0215, Japan

Abstract. In this paper we present a review and comparison of Particle-in-Cell and Vlasov methods for plasma simulation with applications to relativistic self-focusing of high intensity laser pulses in plasmas.

INTRODUCTION

Plasma phenomena occur throughout nature and are the result of the complex nature of the collective interaction of many charged particles. Simulation of plasmas on large scale computers has become an invaluable tool in analyzing various aspects of plasma behavior. In particular for laser plasma interaction it has become the dominant means of explaining results of experiments using high intensity short pulse lasers. In this paper we will be discussing two types of plasma simulation techniques. They are particle and Vlasov simulation techniques.

PARTICLE SIMULATION

In the real world individual charged particles in a plasma are coupled to each other via electromagnetic fields $(\vec{E}, \vec{B})$. Particles are accelerated by the electromagnetic fields via the Lorentz force equation:

$$\frac{d\vec{p}}{dt} = q(\vec{E} + \frac{\vec{v}}{c} \times \vec{B}) \tag{1}$$

$$\frac{d\vec{x}}{dt} = \frac{\vec{p}}{\gamma m} \tag{2}$$

where $\vec{x}$ is the position, $\vec{p}$ is the momentum, q is the charge, γ is the relativistic factor, and m is the mass of the particle. In particle simulation a large number of simulation particles are advanced by these equations.

Typically a finite differencing scheme is used to advance the particles which was developed by Boris [1]. First, we rewrite equation 1 in the form:

$$\frac{d\vec{u}}{dt} = \frac{q}{m}(\vec{E} + \frac{\vec{u} \times \vec{B}}{\gamma c}) \tag{3}$$

CP634, *Science of Superstrong Field Interactions*, edited by K. Nakajima and M. Deguchi
© 2002 American Institute of Physics 0-7354-0089-X/02/$19.00

where $\vec{u} = \gamma \vec{v}$. Finite differencing this equation we get:

$$\frac{\vec{u}^{n+\frac{1}{2}} - \vec{u}^{n-\frac{1}{2}}}{\Delta t} = \frac{q}{m}\left(\vec{E}^n + \frac{\vec{u}^{n+\frac{1}{2}} + \vec{u}^{n-\frac{1}{2}}}{2\gamma^n c} \times \vec{B}^n\right) \tag{4}$$

where the superscript n refers to whole time steps and $n \pm \frac{1}{2}$ refers to fractional time steps. This equation contains both $\vec{E}$ and $\vec{B}$. One can eliminate $\vec{E}$ by expressing introducing the following variables:

$$\vec{u}^{n-\frac{1}{2}} = \vec{u}^- - \frac{q}{2m}\vec{E}^n \Delta t \quad \vec{u}^{n+\frac{1}{2}} = \vec{u}^+ + \frac{q}{2m}\vec{E}^n \Delta t \tag{5}$$

where $\gamma'^2 = 1 + (\frac{\vec{u}^-}{c})^2$. Rewriting equation 3 we get :

$$\frac{\vec{u}^+ - \vec{u}^-}{\Delta t} = \frac{q}{2\gamma^n mc}(\vec{u}^+ + \vec{u}^-) \times \vec{B}^n \tag{6}$$

where

$$\vec{u}^+ = \begin{pmatrix} u_x^+ \\ u_y^+ \\ u_z^+ \end{pmatrix} \quad \vec{u}^- = \begin{pmatrix} u_x^- \\ u_y^- \\ u_z^- \end{pmatrix}. \tag{7}$$

Equation 6 represents a matrix equation which can be inverted to get:

$$\vec{u}^+ = \frac{1}{1 + \Omega^2 \Delta t^2} \Omega \cdot \vec{u}^- \tag{8}$$

$$\Omega = \begin{pmatrix} 1 + (\Omega_x^2 - \Omega_y^2 - \Omega_z^2)\Delta t^2 & 2\Omega_x\Omega_y\Delta t^2 + 2\Omega_z\Delta t & 2\Omega_z\Omega_x\Delta t^2 - 2\Omega_y\Delta t \\ 2\Omega_x\Omega_y\Delta t^2 - 2\Omega_z\Delta t & 1 + (-\Omega_x^2 + \Omega_y^2 - \Omega_z^2)\Delta t^2 & 2\Omega_y\Omega_z\Delta t^2 + 2\Omega_x\Delta t \\ 2\Omega_z\Omega_x\Delta t^2 + 2\Omega_y\Delta t & 2\Omega_y\Omega_z\Delta t^2 - 2\Omega_x\Delta t & 1 + (-\Omega_x^2 - \Omega_y^2 + \Omega_z^2)\Delta t^2 \end{pmatrix}, \tag{9}$$

where $\Omega^2 = \Omega_x^2 + \Omega_y^2 + \Omega_z^2$, $\Omega_x = \frac{qB_x}{2mc\gamma^n}$, $\Omega_y = \frac{qB_y}{2mc\gamma^n}$, and $\Omega_z = \frac{qB_z}{2mc\gamma^n}$. Using equations 5 and 8 we can get $\vec{u}^{n+\frac{1}{2}}$. Finite differencing equation 2 we get:

$$\vec{x}^{n+1} = \vec{x}^n + \frac{\Delta t}{\gamma^{n+\frac{1}{2}}} \vec{u}^{n+\frac{1}{2}} \tag{10}$$

where $(\gamma^{n+\frac{1}{2}})^2 = 1 + (\frac{\vec{u}^{n+\frac{1}{2}}}{c})^2$ which can be used to advance particle positions.

One way the electromagnetic fields used in Equation 1 can be calculated is to calculate the contribution from other particles in the plasma via the Lienard-Wiechert fields [2]:

$$\vec{E}(\vec{x},t) = e\left[\frac{\vec{n} - \vec{\beta}}{\gamma^2(1 - \vec{\beta} \cdot \vec{n})^3 R^2}\right]_{ret} + \frac{e}{c}\left[\frac{\vec{n} \times \{(\vec{n} - \vec{\beta}) \times \dot{\vec{\beta}}\}}{(1 - \vec{\beta} \cdot \vec{n})^3 R}\right]_{ret} \quad \vec{B}(\vec{x},t) = \left[\vec{n} \times \vec{E}\right]_{ret} \tag{11}$$

where $\vec{E}(\vec{x},t)$ and $\vec{B}(\vec{x},t)$ are the electric and magnetic fields, respectively, generated by charged particles other than the particle which feels the field. Here, *ret* refers to the time in the past where the trajectory of the other particle intersects with the light cone of the particle which sees the fields. $\vec{n}$ is the unit normal vector between the particle and other particle's past position, β and γ are the usual relativistic factors, and $\dot{\vec{\beta}}$ is $d\vec{\beta}/dt$ which is the usual acceleration divided by c.

From a computational point of view it can be seen that if there are N particles which interact via the Lienard-Wiechert fields then N^2 interactions must be calculated. Resultingly, the amount of computation increases rapidly with particle number so only a limited number of particles can be calculated in a reasonable time even using supercomputers.

One way of getting around the amount of computation required from direct particle-particle interaction simulations is to compute the electromagnetic fields on a finite number of grids. This method is called Particle-in-Cell (PIC). Many excellent references can be found describing this method [3, 4, 5, 6] so in this section we will only cover briefly the essential details of the method. In the PIC method there are still particles, however, the field through which they interact is calculated using grids on which Maxwell's equations are solved:

$$\nabla \cdot \vec{E} = 4\pi\rho \qquad \nabla \cdot \vec{B} = 0 \qquad (12)$$

$$\nabla \times \vec{E} = -\frac{1}{c}\frac{\partial \vec{B}}{\partial t} \quad \nabla \times \vec{B} = \frac{4\pi}{c}\vec{J} + \frac{1}{c}\frac{\partial \vec{E}}{\partial t} \qquad (13)$$

where ρ refers to the charge density and $\vec{J}$ refers to the current density. The charge and current density are accumulated on the grid from the particles. By using grids instead of calculating direct interactions the number of calculations for a N particle system goes as [6]: $M \ln M + bN$ where M is the number of grids and b is a constant. The increase in computation only goes as roughly N as opposed to N^2 for particle-particle simulations. This makes possible calculations of the interaction of many particles through simulation. The charged particles are coupled to eachother via the grid.

There are several ways to solve Maxwell's equation on a uniform grid. They include Fast Fourier Transforms [5] and implicit finite difference schemes [4]. We will describe in more detail an explicit finite difference scheme [3] which is more suitable for implementation on massively parallel computers where local solutions are optimal for speed. Rearranging Maxwell's equations where equations 12 are taken as initial conditions and finite differencing each component of the electric field (E_x, E_y, E_z) for a two dimensional grid we get:

$$\frac{E_{x_{j+\frac{1}{2},k}}^{n+1} - E_{x_{j+\frac{1}{2},k}}^{n}}{\Delta t} = c\frac{B_{z_{j+\frac{1}{2},k+\frac{1}{2}}}^{n+\frac{1}{2}} - B_{z_{j+\frac{1}{2},k-\frac{1}{2}}}^{n+\frac{1}{2}}}{\Delta y} - 4\pi J_{x_{j+\frac{1}{2},k}}^{n+\frac{1}{2}} \qquad (14)$$

$$\frac{E_{y_{j,k+\frac{1}{2}}}^{n+1} - E_{y_{j,k+\frac{1}{2}}}^{n}}{\Delta t} = -c\frac{B_{z_{j+\frac{1}{2},k+\frac{1}{2}}}^{n+\frac{1}{2}} - B_{z_{j-\frac{1}{2},k+\frac{1}{2}}}^{n+\frac{1}{2}}}{\Delta x} - 4\pi J_{y_{j,k+\frac{1}{2}}}^{n+\frac{1}{2}} \qquad (15)$$

390

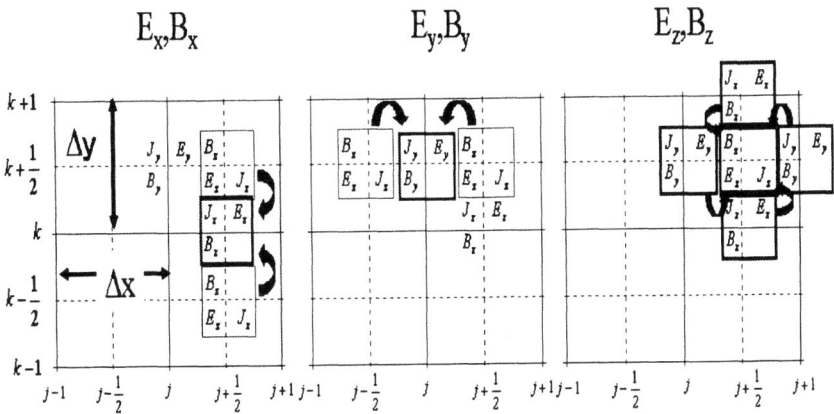

E_x, B_x E_y, B_y E_z, B_z

FIGURE 1. The finite difference positions of the fields on a uniform two dimensional grid is shown.

$$\frac{E_z{}_{j+\frac{1}{2},k+\frac{1}{2}}^{n+1} - E_z{}_{j+\frac{1}{2},k+\frac{1}{2}}^{n}}{\Delta t} = c\frac{B_y{}_{j+1,k+\frac{1}{2}}^{n+\frac{1}{2}} - B_y{}_{j,k+\frac{1}{2}}^{n+\frac{1}{2}}}{\Delta x} - c\frac{B_x{}_{j+\frac{1}{2},k+1}^{n+\frac{1}{2}} - B_x{}_{j+\frac{1}{2},k}^{n+\frac{1}{2}}}{\Delta y}$$
$$-4\pi J_z{}_{j+\frac{1}{2},k}^{n+\frac{1}{2}} \qquad (16)$$

and for the magnetic field (B_x, B_y, B_z) we get

$$\frac{B_x{}_{j+\frac{1}{2},k}^{n+\frac{1}{2}} - B_x{}_{j+\frac{1}{2},k}^{n-\frac{1}{2}}}{\Delta t} = -c\frac{E_z{}_{j+\frac{1}{2},k+\frac{1}{2}}^{n} - E_z{}_{j+\frac{1}{2},k-\frac{1}{2}}^{n}}{\Delta y} \qquad (17)$$

$$\frac{B_y{}_{j,k+\frac{1}{2}}^{n+\frac{1}{2}} - B_y{}_{j,k+\frac{1}{2}}^{n-\frac{1}{2}}}{\Delta t} = c\frac{E_z{}_{j+\frac{1}{2},k+\frac{1}{2}}^{n} - E_z{}_{j-\frac{1}{2},k+\frac{1}{2}}^{n}}{\Delta x} \qquad (18)$$

$$\frac{B_z{}_{j+\frac{1}{2},k+\frac{1}{2}}^{n+\frac{1}{2}} - B_z{}_{j+\frac{1}{2},k+\frac{1}{2}}^{n-\frac{1}{2}}}{\Delta t} = -c\left(\frac{E_y{}_{j+1,k+\frac{1}{2}}^{n} - E_y{}_{j,k+\frac{1}{2}}^{n}}{\Delta x} - \frac{E_x{}_{j+\frac{1}{2},k+1}^{n} - E_x{}_{j+\frac{1}{2},k}^{n}}{\Delta y}\right). \quad (19)$$

In figure 1 we show the sequence of calculation for the fields on a uniform two dimensional grid. Note that the $\vec{E}$ and $\vec{B}$ fields are offset from one another by half time steps and half a grid cell. This finite differencing scheme is stable as long as the Courant condition is satisfied for the simulation time step Δt. In the case of two dimensions the condition is[3]: $c\Delta t < \Delta x/\sqrt{2}$ assuming $\Delta x = \Delta y$ where Δx and Δy are the grid sizes in the x and y direction, respectively.

The current terms (J_x, J_y, J_z) in equations 14, 15, and 16 are calculated by accumulating the current contributions from the simulation particles onto the grid. By appropriately accumulating current on the grid one can maintain charge conservation without having to recalculate equation 12. The technique is fairly detailed so we refer the reader to the reference [7]. Figure 2 shows the collection of current in the simplest

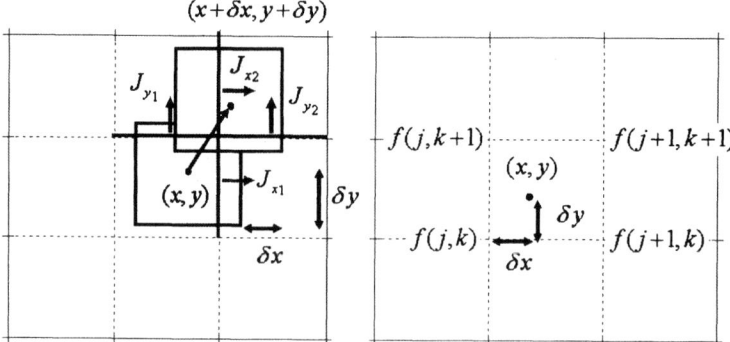

FIGURE 2. The collection of current from a particle onto a uniform two dimensional grid (left) and the interpolation of the fields on the grid to the particle (right) are shown. (x, y) is the initial particle position and $(x + \delta x, y + \delta y)$ is the final position

case where four cell boundaries are crossed by the particle. The currents are calculated as: $J_{x1} = \delta x(\frac{1}{2} - y - \frac{1}{2}\delta y)$, $J_{x2} = \delta x(\frac{1}{2} + y + \frac{1}{2}\delta y)$, $J_{y1} = \delta y(\frac{1}{2} - x - \frac{1}{2}\delta x)$, and $J_{y2} = \delta y(\frac{1}{2} + x + \frac{1}{2}\delta x)$, where δx and δy refer to the change in the particle position in one time step in the x and y directions, respectively. There are more complicated crossings of 7 and 10 boundaries which are described in [7].

Once the new fields have been calculated on the uniform grid, they need to be interpolated to the particle position. This is done by an area weighting scheme [3] which is shown in figure 2: $f(x,y) = f(j,k)(1 - \delta x)(1 - \delta y) + f(j+1,k)\delta x(1 - \delta y) + f(j,k+1)\delta y(1 - \delta x) + f(j+1,k+1)\delta y\delta x$ where f represents the field quantity being interpolated to the particle position.

In addition to the various proceedures described above additional constraints are placed on the simulation due to numerical instabilities. One instability is the thermal instability. If the temperature of the plasma particles is not high enough then the plasma will numerically heat up unless the following condition is met [3]: $\frac{\lambda_D}{\Delta x} \geq 0.3$ where $\lambda_D = \sqrt{kT/4\pi n_0 e^2}$ is the Debye length, T is the temperature of the simulation particles, and n_0 is the plasma density.

For an example of the application of the PIC method see the article in these proceedings dealing with the study of proton acceleration and relativistic self-focusing by this author [8].

VLASOV SIMULATION

In this section we describe Vlasov simulation methods. In the previous section we talked about combinations of simulation particles and grids to model plasma behavior. In the case of Vlasov simulation only grids are used to model the plasma. The advantage of

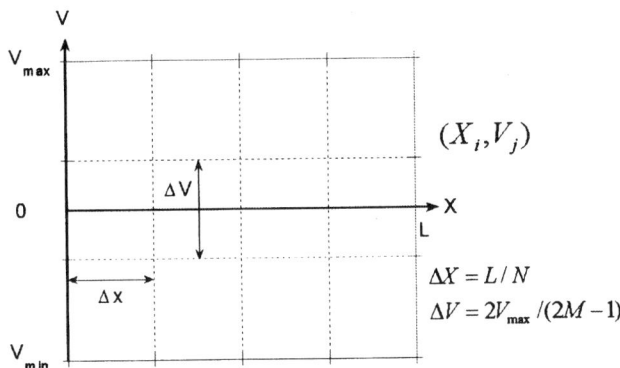

(X_i, V_j)

$\Delta X = L/N$

$\Delta V = 2V_{max}/(2M-1)$

FIGURE 3. The grid on which the Vlasov equations are solved. The grid has $2M+1$ cells in velocity V with indicies $j = -M, -M+1, ..., -1, 0, 1, ..., M-1, M$ and N cells in space X with indicies $i = 1, 2, ..., N$

this method is that it is very accurate. The noise level is very low. Since we deal only with grids parallelization on massively parallel computers is fairly straightforward. The disadvantage of this technique is that large amounts of computer memory are needed and different types of numerical instabilities occur. In this section we describe numerical solution of the Vlasov equation in one dimension using electrostatic fields:

$$\frac{\partial f}{\partial t} + v\frac{\partial f}{\partial x} - E\frac{\partial f}{\partial v} = 0 \tag{20}$$

$$\frac{\partial E}{\partial x} = 1 - \int_{-\infty}^{\infty} f \, dv \tag{21}$$

where f is the distribution function $f(x,v,t)$, x is the position, v is the velocity, t is the time, and E is the electrostatic field $E(x,t)$. The following normalization is used: $\Delta x \to \lambda_D$, $\Delta t \to \omega_p = \sqrt{4\pi n_0 e^2/m}$, $v \to \lambda_D \omega_p$. Equations 20 and 21 are solved on a uniform grid which is shown in figure 3.

Equation 20 is a hyperbolic equation so we can use the cubic interpolation spline technique (CIP)[9, 10]. In addition to increase accuracy we use differential algebra (DA) which allows one to calculate derivatives algebraically, see [11]. This combination is called the DA-CIP scheme[12]. In the following section we will briefly describe this method. The reader is referred to [12] for further details.

The general form of the equations which we are solving can be written in the form:

$$\frac{Df}{Dt} = \frac{\partial f}{\partial t} + \sum_\alpha u_\alpha \frac{\partial f}{\partial r_\alpha} = g \quad \frac{Dr_\alpha}{Dt} = u_\alpha \tag{22}$$

where $\alpha = (x,y,z)$, $\vec{r} = (r_x, r_y, r_z)$, $\vec{u} = (u_x, u_y, u_z)$, and $g(\vec{r}, f, \partial f/\partial \vec{r}, t)$ is a forcing term. The equation for the advance of the derivatives can be written in the form:

$$\frac{D\partial_\alpha f}{Dt} = \frac{\partial g}{\partial r_\alpha} - \sum_\beta (\frac{\partial u_\beta}{\partial r_\alpha} \frac{\partial}{\partial r_\beta}) f = g_\alpha \tag{23}$$

Equations 20 and 21 can be expressed in Lagrange form as:

$$\frac{Df}{Dt} = 0 \quad \frac{Dx}{Dt} = v \quad \frac{Dv}{Dt} = -E(x,t) \tag{24}$$

with the derivatives expressed as:

$$\frac{D\partial_x f}{Dt} = \frac{\partial E}{\partial x} \partial_v f \quad \frac{D\partial_v f}{Dt} = -\partial_x f. \tag{25}$$

These equations can be expressed in a more compact form as:

$$\frac{D\vec{q}}{Dt} = \vec{G}(\vec{q}) \tag{26}$$

where $\vec{q} = (x, v, f, \partial_x f, \partial_v f)$ and $\vec{G} = (v, -E, 0, (\partial E/\partial x)\partial_v f, -\partial_x f)$.

To calculate the time advance of equation 26 one can Taylor series expand the equation:

$$\vec{q}^*(t + \Delta t) = \vec{q}(t) + \frac{D\vec{q}(t)}{Dt}\Delta t + \frac{1}{2}\frac{D^2\vec{q}(t)}{Dt^2}\Delta t^2 + \dots \tag{27}$$

$$= \vec{q}(t) + \vec{G}(\vec{q}, t)\Delta t + \frac{1}{2}(\frac{\partial \vec{G}(\vec{q}, t)}{\partial t} + \frac{\partial \vec{G}(\vec{q}, t)}{\partial \vec{q}}\vec{G}(\vec{q}, t))\Delta t^2 + \dots \tag{28}$$

where Δt is the time step size. We can calculate this equation via a second order Runge-Kutta integration scheme:

$$\vec{q}^*(t + \Delta t) = \vec{q}(t) + \frac{\Delta t}{2}(\vec{h}_1 + \vec{h}_2) \tag{29}$$

$$\vec{h}_1 = \vec{G}(\vec{q}) \quad \vec{h}_2 = \vec{G}(\vec{q} + \vec{h}_1 \Delta t) \tag{30}$$

In the first step of the calculation we calculate $\vec{h}_1 = \vec{G}(\vec{q})$. In order to determine this we need to calculate the electric field E. We know that: $E = -\frac{\partial \phi}{\partial x}$ where ϕ is the scalar potential so that after finite differencing we get $E_i = \frac{\phi_{i+1} - \phi_{i-1}}{2\Delta X_i}$. In addition, we can get $\partial_x E_i = -\frac{(\phi_{i+1} - 2\phi_i + \phi_{i-1})}{\Delta X_i^2}$ where the indicies i are the same as figure 3. Equation 21 can then be written in the form:

$$\frac{\phi_{i+1} - 2\phi_i + \phi_{i-1}}{\Delta X_i^2} = \int_{-\infty}^{\infty} f(X_i, v, t)dv - 1 \cong \sum_{j=-j_{max}}^{j_{max}-1} \int_{V_j}^{V_{j+1}} F_{i,j}(X_i, v, t)dv - 1 \tag{31}$$

$$= \sum_{j=-j_{max}}^{j_{max}-1} \Delta V_i(f_{i,i} + f_{i,j+1})/2 - 1. \tag{32}$$

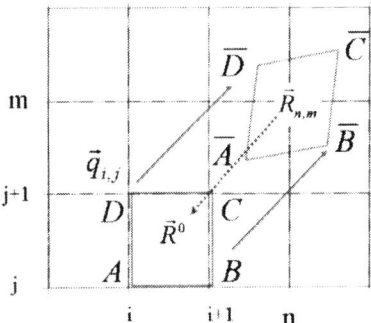

FIGURE 4. Each cell moves when $\vec{q}(t)$ is advanced.

This equation represents a tridiagonal matrix which can quickly be solved by the Thomas algorithm. Once the potential ϕ has been solved the electric field can be calculated. In equation 31 the two dimensional cubic interpolation function was used $F_{i,j}$ for $f(x,v)$ for $(x,v) \in ([X_i, X_{i+1}], [V_j, V_{j+1}])$ where

$$F_{i,j}(x,y) = \sum_{n=0}^{3} \sum_{m=0}^{3} c_{n,m} \tilde{x}^n \tilde{v}^m$$

$$\tilde{x}^n = \frac{(x-X_i)}{\Delta X_i} \quad \tilde{v}^m = \frac{(v-V_j)}{\Delta V_j}$$

(33)

where some of the coefficients c_{nm} are:

$$c_{1,1} = f_{i,j} \quad c_{1,2} = \partial_v f_{i,j} \Delta V_j$$
$$c_{2,1} = \partial_x f_{i,j} \Delta X_i \quad c_{2,2} = \partial_{xv} f_{i,j} \Delta X_i \Delta V_j.$$

All the coeffiecents can be found in [12].

The second step involves calculating $\vec{h}_2 = \vec{G}(\bar{q})$ where $\bar{q} = \vec{q} + \vec{h}_1 \Delta t$ is the time advanced $\vec{q}$ intermediate state. To calculate this we need the electric field at the advanced time: $E(x,t)|_{x=\bar{x}}$. This requires the distribution function f at the advanced time: $f(x,v,t+\Delta t)$ constructed from the intermediate states $\bar{f}, \partial_x \bar{f}$ and $\partial_v \bar{f}$. When $\vec{q}(t)$ is advanced in time each cell position also moves as seen in figure 4. We need to reconstruct the distribution function from these new cell positions. This can be done by using the cubic interpolation function in equation 33. Replacing the old values, $f_{i,j}, \partial_x f_{i,j}, \partial_v f_{i,j}$ with the intermediate values, $\bar{f}_{i,j}, \partial_x \bar{f}_{i,j}, \partial_v \bar{f}_{i,j}$ in the definitions of the coefficients c_{nm}. The interpolation function $\bar{F}_{i,j}(x,v)$ satisfies:

$$\bar{F}_{i,j}(X_i, V_j) = \bar{f}_{i,j} \quad \partial_x \bar{F}_{i,j}(X_i, V_j) = \partial_x \bar{f}_{i,j} \quad \partial_v \bar{F}_{i,j}(X_i, V_j) = \partial_v \bar{f}_{i,j} \quad (34)$$

$$\bar{F}_{i,j}(X_{i+1}, V_j) = \bar{f}_{i+1,j} \quad \partial_x \bar{F}_{i,j}(X_{i+1}, V_j) = \partial_x \bar{f}_{i+1,j} \quad \partial_v \bar{F}_{i,j}(X_{i+1}, V_j) = \partial_v \bar{f}_{i+1,j} \quad (35)$$

$$\bar{F}_{i,j}(X_i, V_{j+1}) = \bar{f}_{i,j+1} \quad \partial_x \bar{F}_{i,j}(X_i, V_{j+1}) = \partial_x \bar{f}_{i,j+1} \quad \partial_v \bar{F}_{i,j}(X_i, V_{j+1}) = \partial_v \bar{f}_{i,j+1} \quad (36)$$

$$\bar{F}_{i,j}(X_{i+1}, V_{j+1}) = \bar{f}_{i+1,j+1} \quad \partial_x \bar{F}_{i,j}(X_{i+1}, V_{j+1}) = \partial_x \bar{f}_{i+1,j+1} \quad \partial_v \bar{F}_{i,j}(X_{i+1}, V_{j+1}) = \partial_v \bar{f}_{i+1,j+1} \quad (37)$$

By using this function we can determine the values of the grid points within each new cell $\overline{ABCD}$ in figure 4. Let $\vec{R}_{nm} = (X_n, V_m)$ represent the grid point in $\overline{ABCD}$. In order to calculate the value at this point we use the cubic function. However, it is a function of the old cell positions in $ABCD$. This can be resolved by finding the mapping between the new cell and old cell. To find the position $\vec{R}^0 = (X^0, V^0)$ in $ABCD$ corresponding to $\vec{R}_{nm}$ we assume a linear transformation between the two cells of the form:

$$\vec{R}^0 = T_{i,j}^{-1}(\vec{R}_{n,m} - \bar{\vec{r}}_{i,j}) + \vec{R}_{i,j} \tag{38}$$

$$\bar{\vec{r}}_{i+1,j} - \bar{\vec{r}}_{i,j} = T_{i,j}(\vec{R}_{i+1,j} - \vec{R}_{i,j}) \quad \bar{\vec{r}}_{i,j+1} - \bar{\vec{r}}_{i,j} = T_{i,j}(\vec{R}_{i,j+1} - \vec{R}_{i,j}) \tag{39}$$

where $\bar{\vec{r}}_{i,j}$ and $R_{i,j}$ are the new and old grid positions, repectively, and $T_{i,j}$ is a linear transformation matrix defined by equation 39. Once we know this transformation we can write:

$$\bar{f}_{n,m} = \bar{F}_{i,j}(X^0, V^0) \quad \partial_x \bar{f}_{n,m} = \frac{\partial}{\partial x}\bar{F}_{i,j}(X^0, V^0) \quad \partial_v \bar{f}_{n,m} = \frac{\partial}{\partial v}\bar{F}_{i,j}(X^0, V^0) \tag{40}$$

This is done for all the grid cells to get $f(x, v, t + \Delta t)$. Once this is done h_2 can be determined and used in equation 29 to get $\vec{q}^*(t + \Delta t)$ which is the time advanced grid. We repeat this whole process for each time step until the desired number of time steps is reached.

COMPARISON AND CONCLUSION

Figure 5 shows results for PIC (left) and Vlasov (right) simulations with initial conditions at the top and final results at the bottom. The simulation run is for the two stream instability where initially oppositly flowing electron beams are unstable and merge to form a vortex in x-v phase space. The parameters of each simulation are somewhat close to eachother. It can be seen that there are fluctuations in the distribution function for the PIC simulation whereas in the Vlasov simulation there is none. Each simulation converges to a single vortex. In the case of the Vlasov simulation the distribution is unchanging after some time. However, the PIC simulation is still evolving in time. It will be of further study to determine which type of simulation is closer to reality over long time scales and over what time scales each type of simulation can be useful.

ACKNOWLEDGMENTS

I especially would like to acknowledge Takayuki Utsumi for his development of the techniques described in the section on Vlasov simulation. I would like to thank Kazuhisa Nakajima for inviting me to give a review talk concerning plasma simulation and the students who attended my talk asking many thought provoking questions.

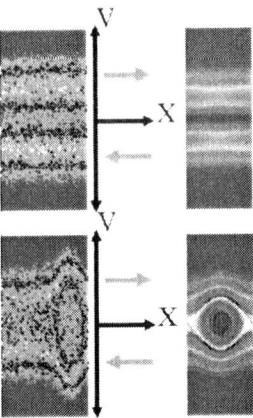

FIGURE 5. Comparison of PIC results (left) and the Vlasov results (right) are shown for the two-stream instability where the initial condtions are at the top and the final states are at the bottom.

REFERENCES

1. Boris, J., "Relativistic plasma simulation-optimization of a hybrid code", in *Proceedings of the 4th Conference on Numerical Simulation of Plasmas*, Naval Research Laboratory, Washington, D. C., 1970, pp. 3–67.
2. Jackson, J. D., *Classical Electrodynamics*, John Wiley and Sons, Inc., New York, 1975.
3. Birdsall, C. K., and Langdon, A. B., *Plasma Physics via Computer Simulation*, McGraw-Hill Book Company, New York, 1985.
4. Hockney, R. W., and Eastwood, J. W., *Computer Simulation Using Particles*, Adam Hilger, Bristol, 1988.
5. Dawson, J. M., *Rev. Mod. Phys.*, **55**, 403–447 (1983).
6. Tajima, T., *Computational Plasma Physics: With Applications to Fusion and Astrophysics*, Addison-Wesley Publishing Company, Inc., Redwood City, 1989.
7. Villasenor, J., and Buneman, O., *Computer Physics Communications*, **69**, 306–316 (1992).
8. Koga, J., Nakajima, K., Yamagiwa, M., and Zhidkov, A., *these proceedings* (2002).
9. Yabe, T., and Aoki, T., *Computer Physics Communications*, **66**, 219–232 (1991).
10. Yabe, T., Ishikawa, T., and Wang, P. Y., *Computer Physics Communications*, **66**, 233–242 (1991).
11. Berz, M., *Particle Accelerators*, **24**, 109–124 (1989).
12. Utsumi, T., Kunugi, T., and Koga, J., *Computer Physics Communications*, **108**, 159–179 (1998).

Particle Acceleration with Strong Field Interactions

Kazuhisa Nakajima

Graduate University for Advanced Studies,
Shonan Village, Hayama, Kanagawa, 240-0193, Japan
Japan Atomic Energy Research Institute, Kizu, Kyoto, 619-0215, Japan
High Energy Accelerator Research Organization, Tsukuba, Ibaraki, 305-0801, Japan

Abstract. A number of concepts of particle acceleration by laser fields have been proposed almost since the beginning of the laser evolution. The recent tremendous progress of ultraintense lasers has created new concepts of high energy particle generation and acceleration due to strong field interactions with matter. In a state of the art of ultra-intense lasers, the highest laser intensities reach a TeV range in terms of the ponderomotive energy exerted on matter. In this paper, the laser-driven particle acceleration concepts are reviewed on basic mechanism of strong field-particle beam and plasma interactions. New concepts of super high energy particle acceleration mechanism based on super-strong laser-plasma interactions are presented.

INTRODUCTION

A number of concepts of particle acceleration by laser fields have been proposed almost since the beginning of the laser evolution. Recently great advances of ultraintense ultrashort pulse lasers have brought about tremendous experimental and theoretical progress in maturity of laser-driven particle accelerator concepts. In particular relativistic electron acceleration mechanism due to ultraintense laser-plasma interaction can be expected to make a compact high energy accelerator. In this context there is a great interest growing in the world-wide community of laser, plasma and accelerator physics. Although an accelerating field limit of conventional accelerators increases as wavelength and pulse duration of driving electric fields decrease, material destruction due to surface heating limits accelerating fields to around 1 GV/m even for optical wavelengths less than 10 μm at a short pulse duration of a few psec[1]. In vacuum, intense focused optical fields can exceed a few TV/m. Such ultrahigh fields have evolved a great deal of particle acceleration concepts by ultraintense laser interaction with particle beams and plasmas.

A novel particle acceleration concept was proposed by Tajima and Dawson[2], which utilizes plasma waves excited by intense laser beam interactions with plasmas for particle acceleration, known as laser-plasma accelerators. In particular recently there has been a great experimental progress on the laser wakefield acceleration (LWFA) of electrons since the first ultrahigh gradient acceleration experiment made by Nakajima et al. [3]. Recent experiments have successfully demonstrated that the self-modulated LWFA mechanism is capable of generating ultrahigh accelerating gradient of $\sim$ 100 GeV/m[4]. In the self-modulated LWFA, however, the maximum energy gain has been limited at most to 100 MeV with energy spread of $\sim$ 100 % because of dephasing and wavebreak-

CP634, *Science of Superstrong Field Interactions*, edited by K. Nakajima and M. Deguchi
© 2002 American Institute of Physics 0-7354-0089-X/02/$19.00

ing effects in plasmas where thermal plasma electrons are accelerated. The first high energy gain acceleration exceeding 200 MeV has been observed with the injection of an electron beam at an energy matched to the wakefield phase velocity in a fairly underdense plasma[5].

The ponderomotive motion of charged particles in the field of non uniform electromagnetic wave is under investigation for a long time, being initiated by the P. L. Kapitza and P. A. M. Dirac [6]. A new concept of laser ponderomotive acceleration named as "Dirac accelerator" is presented. Particles are accelerated due to direct strong field interactions with laser pulses of which a group velocity is less than the vacuum speed of light. In this regime, the energy gain of particles is determined by the group velocity and does not depend on laser intensity. The laser intensity determines a probability for the particle to be reflected from a leading edge of the pulse, and, thus, the number of particles captured in the acceleration. Here we present some examples of applications of a "Dirac accelerator" concept to superhigh energy particle accelerators with GeV to TeV energies.

The peak amplitude of the transverse electric field of a linearly polarized laser pulse is given by

$$E_L[\text{TV/m}] \simeq 2.7 \times 10^{-9} I^{1/2}[\text{W/cm}^2] \cong 3.2 a_0/\lambda_0[\mu\text{m}], \tag{1}$$

where I is the laser intensity, λ_0 is the laser wavelength, and a_0 is the laser strength parameter defined by $a_0 \equiv eA_0/m_e c^2$ in terms of the peak amplitude of the laser vetor potential A_0 and the electron rest energy $m_e c^2$. Using the laser peak intensity $I = cE_L^2/8\pi = ck^2A_0^2/8\pi$, the laser strength parameter is given by

$$a_0 = (2e^2\lambda_0^2 I/\pi m_e^2 c^5)^{1/2} \cong 0.85 \times 10^{-9} \lambda_0[\mu\text{m}] I^{1/2}[\text{W/cm}^2]. \tag{2}$$

Physically a_0 is equal to the normalized momentum of the electron quiver motion in the laser field. In a strong laser field, an electron absorbs energy and momentum from the wave to cause mass shift from m_e to $m_e\gamma_L$, where $\gamma_L = (1 + a_0^2/2)^{1/2}$. The effective potential of the electron inside the laser field for $a_0 << 1$ is

$$U_{eff} = m_e c^2 (1 + a_0^2/2)^{1/2} \approx m_e c^2 + m_e c^2 a_0^2/4 \tag{3}$$

In the nonrelativistic regime, the ponderomotive force or the field gradient force can be defined as

$$\mathbf{F} = -\nabla U_{eff} \approx -m_e c^2 a_0^2/4 \tag{4}$$

The laser pulse propagating in underdense plasmas expels plasma electrons exerted by the ponderomotive force to excite plasma waves. In this regime particle acceleration with strong laser fields is mainly attributed to the laser wakefield acceleration mechanism. In the highly relativistic regime for $a_0 >> 1$, if the initial mometum of a free electron is smaller than the quiver momentum, the electron is reflected from the laser pulse. The reflection of particles from the laser pulse moving in plasmas results in their acceleration. This mechanism is the basis of the "Dirac accelerator".

LASER ACCELERATION IN VACUUM

Lawson-Woodward theorem

Particle acceleration in vacuum can eliminate the difficulties associated with gas and plasmas where the accelerating field is limited due to the gas breakdown and the plasma wave-breaking besides suffering from beam-gas and plasma collisions and laser-plasma instabilities[7]. A major shortcoming of laser-vacuum acceleration is attributed to the phase velocity of the electric field in the accelerated direction of particles greater than the vacuum light velocity c for a focused laser beam.

Assuming the propagation of a nearly plane wave, the electric field of the Hermite-Gaussian TEM$_{lm}$ mode is given by[8]

$$E_{l,m}(x,y,z) = E_0 \frac{r_0}{w(z)} H_l\left(\frac{\sqrt{2}x}{w(z)}\right) H_m\left(\frac{\sqrt{2}y}{w(z)}\right) \exp\left[-\frac{r^2}{w^2(z)}\right] \exp(i\psi), \qquad (5)$$

where the phase of the transverse field E_{lm} is

$$\psi = kz - \omega t + zr^2/(Z_R w^2(z)) - (l+m+1)\tan^{-1}(z/Z_R) + \phi_0, \qquad (6)$$

E_0 is the maximum field amplitude, $w(z) = r_0[1 + (z/Z_R)^2]$ is the laser spot radius, r_0 is the minimum spot radius at focus, and $Z_R = \pi r_0^2/\lambda_0$ is the Rayleigh length, i.e. the distance over which the spot size expands to $\sqrt{2}r_0$, $\lambda = 2\pi/k$ is the wavelength, $\omega = ck$ is the frequency, k is the wavenumber, $r = \sqrt{x^2+y^2}$, H_l is the Hermite polynomial of order l, $H_0(\xi) = 1$, $H_1(\xi) = 2\xi$, $H_2(\xi) = 4\xi^2 - 2, \cdots$, and ϕ_0 is a constant.

Since a Gaussian laser field propagating in the z direction is transversely bounded, the finite longitudinal component of the electric field is obtained from $\nabla \cdot \mathbf{E} = 0$, as $E_z = -(1/ik)\nabla_\perp \cdot \mathbf{E}_\perp$ that can accelerate electrons in the z direction. The phase velocity of the electric field in the propagation direction is obtained from $d\psi/dt = 0$. Along the propagation axis $r = 0$, the phase velocity is

$$v_{ph} = c\left[1 - \frac{l+m+1}{kZ_R(1+z^2/Z_R^2)}\right]^{-1} > c. \qquad (7)$$

Therefore relativistic electrons with the longitudinal velocity $v_z \simeq c$ will slip in phase with respect to the accelerating field E_z and eventually decelerate. Acceleration occurs over a slippage distance Z_s, defined by $kZ_s|v_{ph} - v_z| \simeq \pi$, which gives $Z_s \simeq \pi Z_R$. If an highly relativistic electron interacts over an infinite region ($z = -\infty$ to ∞), a net energy gain is zero resulting from deceleration canceling out acceleration. This is pointed out by the Lawson-Woodward theorem assuming that the region of interaction is infinite, no static electric or magnetic field is present, and nonlinear effects (e.g., ponderomotive, and radiation reaction forces) are neglected[9]. In order to make a nonzero net energy gain, one or more of these assumptions must be violated.

Laser Beat Wave Accelerator

The laser beat wave accelerator is based on the nonlinear ponderomotive force[10]. Two laser beams of different wavelengths, λ_1 and λ_2, are copropagated in the presense of an injected electron beam. Properly phased electrons travelling along the same axis as the two laser beams undergo an axial acceleration from the beat term in the $\mathbf{v} \times \mathbf{B}$ force. The total laser field is represented by the vector potential,

$$\mathbf{A}(z,r,t) = \mathbf{A}_1(z,r,t) + \mathbf{A}_2(z,r,t), \tag{8}$$

where $\mathbf{A}_1$ and $\mathbf{A}_2$ represent laser 1 and 2, respectively, and the circularly polarized laser fields are given by

$$\mathbf{A}_i(z,r,t) = A_{0i}\frac{r_{0i}}{w_i(z)} \exp\left[-\frac{r^2}{w_i^2(z)}\right] [\cos(\psi_i)\mathbf{e}_x + \sin(\psi_i)\mathbf{e}_y], \tag{9}$$

where $i = 1,2$ denotes the laser beam, $w_i(z) = r_{0i}(1 + z^2/Z_{Ri}^2)^{1/2}$, $Z_{Ri} = \pi r_{0i}/\lambda_i$, and

$$\psi_i = k_i z - \omega_i t + r^2(z/Z_{Ri})/w_i^2(z) - \tan^{-1}(z/Z_{Ri}) + \phi_{0i}. \tag{10}$$

The relativistic Lorentz force equation can be written as

$$d\mathbf{u}/dt = \partial \mathbf{a}/\partial t - (c\mathbf{u}/\gamma) \times (\nabla \times \mathbf{a}), \tag{11}$$

where $\mathbf{u} = \mathbf{p}/m_e c$ is the normalized electron momentum $\gamma = (1 + u^2)^{1/2}$ is the Lorentz factor, and where $\mathbf{a} = \mathbf{a}_1 + \mathbf{a}_2$, $\mathbf{a}_{1,2} = |e|\mathbf{A}_{1,2}/m_e c^2$ are the normalized vector potentials. The electron energy equation is given by

$$d\gamma/dt = (\mathbf{u}/\gamma) \cdot \partial \mathbf{a}/\partial t. \tag{12}$$

In the one dimensional limit, i.e. $\lambda_i/r_{0i} \ll 1$, the transverse canonical momentum is approximately conserved, i.e. $d/dt(\mathbf{u}_\perp - \mathbf{a}_\perp) = 0$. Hence using $\mathbf{u}_\perp = \mathbf{a}_\perp$, the energy equation is described by

$$d\gamma/dt = (1/2\gamma)\partial a_\perp^2/\partial t + (u_z/\gamma)\partial a_z/\partial t, \tag{13}$$

where $a_\perp^2 = \mathbf{a}_\perp \cdot \mathbf{a}_\perp$ is given by

$$a_\perp^2 = \hat{a}_1^2 + \hat{a}_2^2 + 2\hat{a}_1\hat{a}_2\cos(\psi_2 - \psi_1), \tag{14}$$

and $\hat{a}_i = (a_{0i}r_{0i}/w_i)\exp(-r^2/w_i^2)$. Assuming $r_{0i} \gg \lambda_i$, the axial field a_z is given by $\nabla \cdot \mathbf{a} = 0$,

$$a_z = \sum_{i=1,2} \frac{2\hat{a}_i}{k_i w_i^2}\left[(x - \hat{z}_i y)\sin\psi_i - (\hat{z}_i x + y)\cos\psi_i\right], \tag{15}$$

where $\hat{z}_i = z/Z_{Ri}$. The electron energy is obtained from the equation

$$\frac{d\gamma}{dz} = \frac{\hat{a}_1\hat{a}_2}{\gamma\beta_z}\Delta k \sin(\psi_2 - \psi_1) - \sum_{i=1,2}\frac{2\hat{a}_i}{w_i^2}\left[(x - \hat{z}_i y)\cos\psi_i + (\hat{z}_i x + y)\sin\psi_i\right], \tag{16}$$

where $\omega_2 - \omega_1 = \Delta\omega = c\Delta k > 0$ and nonresonant (slowly varying) terms proportional to a_{0i}^2 are neglected. The effective accelerating gradient is inversely proportional to the electron energy. The transverse orbits are calculated by

$$\frac{dx}{dz} \simeq \frac{1}{\gamma\beta_z} \sum_{i=1,2} \hat{a}_i \cos\psi_i, \quad \frac{dy}{dz} \simeq \frac{1}{\gamma\beta_z} \sum_{i=1,2} \hat{a}_i \sin\psi_i, \tag{17}$$

where $\gamma = \gamma_z\gamma_\perp$, $\gamma_z = (1 - \beta_z^2)^{-1/2}$, $\beta_z = v_z/c$, $\gamma_\perp = (1 + a_\perp^2)^{1/2}$, and $t = \int dz/v_z$.

Along the axis $r = 0$ and near the focus $|z| < Z_{Ri}$, the phase velocity of the accelerating field is

$$v_{ph}/c = (1 + 1/\Delta kZ_{R1} - 1/\Delta kZ_{R2})^{-1} \simeq 1 - (1 - Z_{R1}/Z_{R2})/(\Delta kZ_{R1}), \tag{18}$$

which is less than c for $Z_{R2} > Z_{R1}$ and can be controlled by choosing the laser spot sizes. Therefore the acceleration distance is limited not by the slippage distance but by the diffraction range to be approximately two Rayleigh lengths. Setting $x = y = 0$, $t = z/c$ and $Z_R = Z_{R1} = Z_{R2}$, the electron energy can be found by

$$\gamma_F^2 - \gamma_I^2 = 2a_{01}a_{02}\Delta kZ_R \sin(\phi_{02} - \phi_{01})(\tan^{-1}\hat{z}_F - \tan^{-1}\hat{z}_I), \tag{19}$$

where $\gamma_{I,F} = \gamma(z = z_{I,F})$ is the electron initial (final) energy. In an infinite interaction region $z_I = -\infty$ and $z_F = -\infty$, it is $\gamma_F^2 - \gamma_I^2 = 2\pi a_{01}a_{02}\Delta kZ_R$, assuming $\sin(\phi_{02} - \phi_{01}) = 1$. In terms of $P_1(TW) = 0.043(a_{01}r_{01}/\lambda_1)^2$ for $a_{01} = a_{02}$, the electron energy is given by

$$W_F[\text{MeV}] = \left[W_I^2[\text{MeV}] + 750(\lambda_1/\lambda_2 - 1)P_1[\text{TW}]\right]^{1/2}, \tag{20}$$

where $W_F(W_I)$ is the final (initial) electron energy and $W_I \gg m_ec^2$. As an example, for $W_I = 20$ MeV, $\lambda_1 = 2\lambda_2 = 1\mu m$, and $P = 10$ TW, the final electron energy is $W_F = 89$ MeV.

In the laser beat wave accelerator, an attainable maximum energy is limited by radiative losses caused by transverse quiver oscillations when electrons interact with laser fields. The power radiated by a single electron is given by the relativistic Larmor formula as

$$P_R \simeq (2/3)r_e m_ec^3\gamma^2 a_0^2 k^2 [1/(2\gamma_z^2) + 1/(kZ_R)]^2, \tag{21}$$

where $r_e = e^2/m_ec^2$ is the classical electron radius. Including this term, the electron energy equation is

$$d\gamma/dz \simeq a_{01}a_{02}\Delta k/\gamma - (2/3)r_e\gamma^2(a_{01}^2/Z_{R1}^2 + a_{02}^2/Z_{R2}^2), \tag{22}$$

where $\gamma_z^2 \gg k_iZ_{Ri}/2$ is assumed. The maximum value of γ is given by setting $d\gamma/dz = 0$. Assuming $\Delta k = k_1$, $a_{01} = a_{02}$, and $Z_{R1} = Z_{R2}$, the maximum electron energy is given by

$$W_{max} \simeq m_ec^2(\pi r_{01}/\lambda_1)(3r_{01}/2r_e)^{1/3}. \tag{23}$$

In terms of the practical unit, $W_{max} \simeq 1.3(r_{01}^{4/3}/\lambda_1)$, where W_{max} is in GeV and r_{01} and λ_1 are in μm. As an example, for $r_{01} \approx \lambda_1 = 1\mu m$, $W_{max} \simeq 1.3$ GeV.

LASER ACCELERATION IN PLASMAS

Plasma wave excitation by laser-plasma interaction

Plasmas provide some advantages as an accelerating medium in laser-driven accelerators. Plasmas can sustain ultrahigh electric fields, and can optically guide the laser beam and the particle beam as well under appropriate conditions. For a nonrelativistic plasma wave, the acceleration gradients are limited to the order of the wave-breaking field given by

$$eE_0[\text{eV/cm}] = m_e c \omega_p \simeq 0.96 n_0^{1/2} [\text{cm}^{-3}], \tag{24}$$

where $\omega_p = (4\pi n_0 e^2/m_e)^{1/2}$ is the electron plasma frequency and n_0 is the ambient electron plasma density. It means that the plasma density of $n_e = 10^{18}$ cm^{-3} can sustain the acceleration gradient of 100 GeV/m.

The dispersion relation of the transverse electromagnetic wave in a plasma is given by

$$\omega^2 = \omega_p^2 + c^2 k^2. \tag{25}$$

The phase velocity and the group velocity of the electromagnetic wave in a plasma are

$$v_{ph} = \omega/k = c/\sqrt{1 - \omega_p^2/\omega^2}, \tag{26}$$

$$v_g = \partial\omega/\partial k = c^2/v_{ph} = c\sqrt{1 - \omega_p^2/\omega^2}. \tag{27}$$

An index of refraction is defined as $\eta \equiv ck/\omega = \sqrt{1 - \omega_p^2/\omega^2}$. If $\omega > \omega_p$, an electromagnetic wave packet (a laser pulse) can propagate in a plasma with the group velocity v_g. This wave, however, can not efficiently accelerate particles because of the phase velocity $v_{ph} > c$. When $\omega < \omega_p$, the electromagnetic wave becomes evanescent with an imaginary refractive index.

The dispersion relation for the electron plasma wave is $\omega = \pm\omega_p$, i.e. dispersionless for a cold plasma wave. Assuming the wave numbers $k \ll k_D = \omega_p/v_T$, where k_D is the Debye wave number and $v_T = KT/m_e$ is the thermal velocity, the phase and group velocities of the plasma wave are:

$$v_{ph} = \omega_p/k \gg v_T, \ v_g = \partial\omega/\partial k = 0. \tag{28}$$

This plasma wave will not propagate, but merely oscillates at the frequency $\omega = \omega_p$. In the plasma waves, it is possible to make the phase velocity lower than c.

In laser-driven plasma-based accelerator, plasma waves are driven by the ponderomotive force arising from the $-e(\mathbf{v} \times \mathbf{B})/c$. In a cold fluid limit of a plasma, the momentum equation of electron fluid is

$$\frac{\partial \mathbf{p}}{\partial t} + \mathbf{v} \cdot \nabla \mathbf{p} = -e\left(\mathbf{E} + \frac{1}{c}\mathbf{v} \times \mathbf{B}\right) \tag{29}$$

The electric $\mathbf{E}$ and magnetic $\mathbf{B}$ fields of the laser can be written as

$$\mathbf{E} = -\frac{1}{c}\frac{\partial \mathbf{A}}{\partial t}, \ \mathbf{B} = \nabla \times \mathbf{A}, \tag{30}$$

where **A** is the vector potential of the laser. For the linearly polarized laser, the vector potential is written as

$$\mathbf{A} = A_0 \cos(kz - \omega t)\mathbf{e}_x \tag{31}$$

In the linear limit $|a| = eA_0/mc^2 \ll 1$, letting $\mathbf{v} = \mathbf{v}_0 + \mathbf{v}_1$, the leading order equation of motion $m\partial\mathbf{v}_0/\partial t = -e\mathbf{E}$ gives the quiver velocity, $\mathbf{v}_0 = e\mathbf{A}/m_e c = c\mathbf{a}$. The second-order motion is given by

$$m_e \frac{\partial \mathbf{v}_1}{\partial t} = -m_e[(\mathbf{v}_0 \cdot \nabla)\mathbf{v}_0 + c\mathbf{v}_0 \times (\nabla \times \mathbf{a})] = -m_e c^2 \nabla(a^2/2). \tag{32}$$

$\mathbf{F}_p = -m_e c^2 \nabla(a^2/2)$ is the ponderomotive force defined by averaging the nonlinear force over $2\pi/\omega_0$, exerted on plasma electrons by a laser field with frequency ω_0, and the ponderomotive potential $\phi_p = m_e c^2(a^2/2)$ can be defined. Using the linearlized Poisson's equation and the continuity equation, i.e.

$$\nabla \cdot \mathbf{E} = -4\pi en, \quad \frac{\partial n_1}{\partial t} + n_0 \nabla \cdot \mathbf{v} = 0, \tag{33}$$

excitation of the plasma wave is described by

$$\left(\frac{\partial^2}{\partial t^2} + \omega_p^2\right)\frac{n_1}{n_0} = \frac{c^2}{2}\nabla^2 a^2, \tag{34}$$

where $a^2 \ll 1$ is the normalized intensity of the driving laser beam, and $n_1/n_0 \ll 1$ is the perturbed density of the plasma wave. If the electric field generated by the plasma wave is written by $\mathbf{E} = -\nabla\phi$, where ϕ is the electrostatic potential, Eq. (34) leads to

$$\left(\frac{\partial^2}{\partial t^2} + \omega_p^2\right)\phi = \omega_p^2 \frac{m_e c^2}{2e}a^2. \tag{35}$$

Laser Wake-Field Accelerator

As an intense laser pulse propagates through an underdense plasma, $\omega_p^2/\omega_0^2 \ll 1$, the ponderomotive force associated with the laser pulse envelope expels electrons from the region of the laser pulse. This effect excites a large amplitude plasma wave (wakefield) with phase velocity approximately equal to the group velocity of laser pulse. Assuming that all of the axial and time dependencies can be expressed as a function of a single variable $\zeta = z - v_p t$ with a phase velocity v_p of the plasma wave, a simple-harmonic osillator equation for the wake potential ϕ is

$$\frac{\partial^2 \phi}{\partial \zeta^2} + k_p^2 \phi = k_p^2 \frac{m_e c^2}{2e}a^2(r, \zeta), \tag{36}$$

where $k_p = \omega_p/v_p$ is the plasma wave number. The phase velocity of the plasma wave is equal to the group velocity of the laser pulse in a plasma, given by $v_p = c(1 - \omega_p^2/\omega_0^2)^{1/2}$.

Consider laser wakefields driven by a circularly polarized laser pulse with a normalized intensity profile given by $|\mathbf{a}(r,\zeta)| = a_0 \exp(-r^2/r_0^2 - \zeta^2/2\sigma_z^2)$, where σ_z is the temporal $1/e$ half-width of the pulse and r_0 is the spot size. The axial and radial wakefields are given by[11]

$$eE_z(r,\zeta) = \frac{\sqrt{\pi}}{4}m_e c^2 k_p^2 \sigma_z a_0^2 \exp(-\frac{2r^2}{r_0^2} - \frac{k_p^2\sigma_z^2}{4})[C(\zeta)\cos k_p\zeta + S(\zeta)\sin k_p\zeta], \quad (37)$$

$$eE_r(r,\zeta) = -\sqrt{\pi}m_e c^2 k_p \sigma_z a_0^2 \frac{r}{r_0^2}\exp(-\frac{2r^2}{r_0^2} - \frac{k_p^2\sigma_z^2}{4})[C(\zeta)\sin k_p\zeta - S(\zeta)\cos k_p\zeta], (38)$$

where $C(\zeta) = 1 - \mathrm{Re}[\mathrm{erf}(\zeta/\sigma_z - ik_p\sigma_z/2)]$, and $S(\zeta) = -\mathrm{Im}[\mathrm{erf}(\zeta/\sigma_z - ik_p\sigma_z/2)]$. For $\zeta \ll \sigma_z$, i.e. the region away behind the laser pulse, $C(\zeta) \to 2$ and $S(\zeta) \to 0$. Hence the radial wakefield is zero along the axis and its phase is shifted by $\pi/2$ from that of the axial wakefield. An electron displaced from the axis will experience simultaneous axial accelerating and radial focusing forces in a phase region of $k_p\Delta\zeta = \pi/4$. The maximum accelerating gradient is achieved at the plasma wavelength $\lambda_p = \pi\sigma_z$ as $(eE_z)_{\max} = 2\sqrt{\pi}e^{-1}m_e c^2 a_0^2/\sigma_z$. For linear polarization, a_0^2 is replaced with $a_0^2/2$ in all equations describing wakefields. In practical units, the maximum axial wakefield occurs at the plasma wavelength, $\lambda_p[\mu m] \simeq 0.57\tau$ in a plasma with the resonant electron density, $n_0[\mathrm{cm}^{-3}] = 1/(\pi r_e \sigma_z^2) \simeq 3.5 \times 10^{21}/\tau^2$ in terms of a FWHM pulse duration τ [fs], where a FWHM pulse width is given by $c\tau = 2\sqrt{\ln 2}\sigma_z$. When a Gaussian driving laser pulse with the peak power P [TW] is focused on the spot size r_0 [μm], the maximum axial wakefield yields $(eE_z)_{\max}[\mathrm{GeV/m}] \simeq 8.6 \times 10^4 P\lambda_0^2/(\tau r_0^2\gamma_L)$, where $\gamma_L = (1 + a_0^2/2)^{1/2}$ takes account of nonlinear relativistic effects, and $a_0 = 6.8\lambda_0 P^{1/2}/r_0$ for the linear polarization.

Acceleration Energy Gain

Assuming a Gaussian beam propagation of the laser pulse with the peak power P in an underdense plasma ($\omega \gg \omega_p$), the effective acceleration length can be limited to a diffraction length $L_{dif} = \pi Z_R$. For a properly phased electron, the maximum energy gain is given by

$$\Delta W_{dif}[\mathrm{GeV}] \simeq 0.85 P[\mathrm{TW}]\lambda_0[\mu m]/(\gamma_L \tau[\mathrm{fs}]). \quad (39)$$

Note that the maximum energy gain is independent of the focusing property of the laser beam due to diffraction effects in the limit of $a_0^2 \ll 1$. For example, the maximum energy gain of the LWFA driven by the laser pulse with $\lambda_0 = 0.8\mu$m, $P = 2$ TW, $\tau = 100$ fs, and $r_0 = 10\mu$m is limited to $\Delta W_{dif} = 12$ MeV by the diffraction length $L_{dif} = 1.2$ mm.

An electron can be accelerated along the z-axis by an electrostatic plasma wave of the form $E_z = E_{z\max}\sin\omega_p(z/v_{ph} - t)$. As the electron is accelerated, its velocity v_z will increase and approach the speed of light, $v_z \to c$. If the phase velocity of the plasma wave is constant with $v_{ph} < c$, the electrons will eventually outrun the accelerating phase and move into the decelerating phase. This dephasing effect in the plasma wave limits the energy gain of the electron. The dephasing length is defined as the length the electron must travel before its phase slips by one-half of a period with respect to the plasma wave. For a highly relativistic electron with $v_z \simeq c$, the dephasing length is given by

$$L_d = (\lambda_p/2)/(1 - v_{ph}/c) \simeq \lambda_p(\omega_0^2/\omega_p^2) = \lambda_p \gamma_p^2, \tag{40}$$

where $\gamma_p = (1 - v_{ph}^2/c^2)^{-1/2}$ is the relativistic factor associated with the phase velocity of the plasma wave. The energy gain of a highly relativistic electron is obtained from integrating eE_z over the acceleration length L_d, The maximum energy gain is given by $\Delta W_d = (2/\pi)eE_{zmax}L_d \simeq (2/\pi)eE_{zmax}\lambda_p\gamma_p^2$. For the optimum plasma condition, $\lambda_p = \pi\sigma_z$, the dephasing length is given by

$$L_d[\text{cm}] = 0.18 \times 10^{-4}\tau^3[\text{fs}]\gamma_L/\lambda_0^2[\mu\text{m}], \tag{41}$$

and the maximum energy gain is

$$\Delta W_d[\text{GeV}] = 0.01P[\text{TW}]\tau^2[\text{fs}]/r_0^2[\mu\text{m}]. \tag{42}$$

For example, the maximum energy gain of the LWFA driven by the laser pulse with $\lambda_0 = 0.8\mu\text{m}$, $P = 2$ TW, $\tau = 100$ fs, and $r_0 = 10\mu\text{m}$ is limited to $\Delta W_d = 2$ GeV by the dephasing length $L_d = 32$ cm.

In the laser-plasma accelerator, as the laser driver excites a plasma wave, it loses energy. The pump depletion length L_{pd}, in which the laser pulse loses a half of its total energy to excite plasma waves, is estimated by equating the laser pulse energy to the energy left behind in the wakefield, $E_z^2 L_{pd} = (1/2)E_L^2 L$, where E_L is the laser field. For a Gaussian laser pulse, the pump depletion length is given by

$$L_{pd} = \frac{8}{\sqrt{\pi}}\frac{\gamma_p^2}{a_0^2 k_p^2 \sigma_z}\exp\left(\frac{k_p^2\sigma_z^2}{2}\right). \tag{43}$$

For the optimum condition, $\lambda_p = \pi\sigma_z$, the pump depletion length is $L_{pd} \simeq 2.65\lambda_p\gamma_p^2 a_0^{-2}$. Note that if the laser strength parameter is $a_0 < 1.6$, the pump depletion length is larger than the dephasing length, $L_d < L_{pd}$. The energy gain limited by the pump depletion is given by

$$\Delta W_{pd} = eE_z L_{pd} = 2m_e c^2 \gamma_p^2 \exp(k_p^2\sigma_z^2/4). \tag{44}$$

For the optimum plasma condition, $\lambda_p = \pi\sigma_z$, the pump depletion length is given by

$$L_{pd}[\text{m}] = 1.06 \times 10^{-8}\tau^3[\text{fs}]r_0^2[\mu\text{m}]\gamma_L^3/(\lambda_0^4[\mu\text{m}]P[\text{TW}]), \tag{45}$$

and the maximum energy gain is

$$\Delta W_{pd}[\text{GeV}] = 0.91 \times 10^{-3}\tau^2[\text{fs}]\gamma_L^2/\lambda_0^2[\mu\text{m}]. \tag{46}$$

For example, the maximum energy gain of the LWFA driven by the laser pulse with $\lambda_0 = 0.8\mu\text{m}$, $P = 2$ TW, $\tau = 100$ fs, and $r_0 = 10\mu\text{m}$ is limited to $\Delta W_{pd} = 18.5$ GeV by the pump depletion length $L_d = 1.9$ m.

TABLE 1. The design parameters of the capillary-guided laser wakefield accelerators.

Energy gain ΔW [GeV]	0.5	1	5
Pulse duratuon τ [fs]	20	50	100
Peak power P [TW]	100	40	20
Spot radius r_0 [μm]	30	20	10
Laser strength a_0	1.8	1.7	2.4
Plasma density n_e [cm^{-3}]	8.8×10^{18}	1.4×10^{18}	3.5×10^{17}
Accelerating gradient eE_{zmax} [GeV/cm]	1.9	0.7	0.55
Diffraction length L_{dif} [cm]	1.1	0.5	0.12
Dephasing length L_d [cm]	0.4	5.5	56
Capillary length L_c [cm]	No	1.5	10
N_{max} [10^9]	7	1.1	0.2

In order to achieve the acceleration energy gains higher than 1 GeV in a single stage of cm-scale, it is necessary to extend the acceleration length limited by diffraction effects of laser beams. We propose the capillary-guided laser wakefield accelerators in which both the driving laser pulses and particle beams can be guided through the capillary discharge plasmas of cm-scale[12]. The design parameters to test electron acceleration of GeV energies are shown in Table 1. The design of the laser wakefield accelerators is based on availability of the 10 Hz table-top ultrashort, ultrahigh peak power Ti:Sapphire laser with 20 fs and 100 TW developed by JAERI-KANSAI[13]. The parameters are optimized according to the criteria described in ref. [14] so as to maximize the energy gain in the single-stage acceleration under the conditions that the wakefield amplitudes are less than the relativistic wave-breaking field and that the longitudinal wakefield should be larger than the transverse field. The energy gain of 5 GeV will be achieved with a 10 cm long capillary plasma waveguide driven by a 20 TW, 100 fs laser pulse in the plasma density of $n_e = 3.5 \times 10^{17}$ cm^{-3} without wave-breaking. The maximum number of electrons capable of accelerating with 100% energy spread is estimated to be $N_{max} \sim 3.55 \times 10^9 P \lambda_0^2 / (\tau \gamma_L)$[15].

PONDEROMOTIVE ACCELERATION IN PLASMAS

Ponderomotive potential scattering

Let us consider the ultraintense ultrashort laser pulse propagating in a plasma with the electron density n_e and the plasma frequency ω_p. In the reference frame that is moving with the group velocity $v_g = c(1 - \omega_p^2/\omega^2)^{1/2} < c$ of the laser pulse along the z axis. In the reference frame, since the laser ponderomotive potential becomes static, the interaction can be considered as the elastic potential scattering. Let the initial energy and velocity of the particle γ_0 and β_0, respectively, and the initial angle of collision is θ_0. The Lorentz transformation gives the energy of the particle in the reference frame:

$$\gamma^* = (1 - \beta_g \beta_0 \cos \theta_0) \gamma_0 \gamma_g, \tag{47}$$

where $\gamma_g = (1-\beta_g^2)^{-1/2}$, and $\beta_g = v_g/c$. According to the classical scattering theory, if this energy is less than the field peak ponderomotive potential, i.e.

$$\gamma^* \le \gamma_L = (1+a_0^2/2)^{1/2}, \tag{48}$$

the particle must be reflected. As an effective Hamiltonian is time-independent in the reference frame, the final energy γ of the particle does not change after scattering, and after the Lorentz transformation, we have in the laboratory frame

$$\gamma_g \gamma (1 - \beta_g \beta \cos\theta) = \gamma^* \tag{49}$$

In this relation, β is the particle velocity, and θ is the final scattering angle, where $\theta = 0$ corresponds to the co-propagation case. From these equations, the following conservation law is given by

$$\gamma(1 - \beta_g \beta_z) = \gamma_0 (1 - \beta_g \beta_{0z}) = \text{const}. \tag{50}$$

This result of simple kinematics approach coincides with that of electrodynamics consideration[16, 17]. The maximum final energy of particle is given for scattering at $\theta = 0$;

$$\gamma = [(\beta_g - \beta_0)^2 + \gamma_0^{-2}]\gamma_0 \gamma_g^2. \tag{51}$$

In a quantum mechanical analysis, the value of ponderomotive potential determines the probability of scattering. To estimate its value, one can start with Klein - Gordon equation, as the spin effects are evidently not important for the scattering. It is convenient to consider the problem in the reference frame, where the potential is time-independent. We are interested in the backscattering, i.e. scattering at $\theta = 0$, and, consequently, one can use the 1D approximation. In the Lorentz gauge and paraxial ray approximation, the vector potential $A(z)$ is the transverse one, and the Klein - Gordon equation is reduced to the following equation

$$\frac{\partial^2 \psi}{\partial z^2} - \frac{1}{c^2}\frac{\partial^2 \psi}{\partial t^2} - \frac{m^2 c^2}{\hbar^2}\left(1 + \frac{e^2 A^2(z)}{m^2 c^4}\right)\psi = 0. \tag{52}$$

Here and below we omit (*) in the coordinates of the reference frame. As the potential is time-independent, the energy of the particle is conserved. One can obtain the solution in the form

$$\psi(z,t) = R(z)\exp(-i\varepsilon t/\hbar), \tag{53}$$

where ε is the energy of a particle with the mass m in the reference frame, $\varepsilon = mc^2\gamma^*$. Assuming that the stationary flow of particles collides with the potential, we obtain reflected and transmitted parts of this flow. The correspondent asymptotic behavior of solution must be the following

$$R(z \to \infty) = \exp(-ik_0 z) + a\exp(ik_0 z), \quad R(z \to -\infty) = b\exp(-ik_0 z), \tag{54}$$

where a and b are the reflection and transmission amplitudes, normalized to the amplitude of the incident flow. As the Klein - Gordon current is given by

$$j = -i\frac{e\hbar}{2mc^2}(\psi\nabla\psi^* - \psi^*\nabla\psi), \tag{55}$$

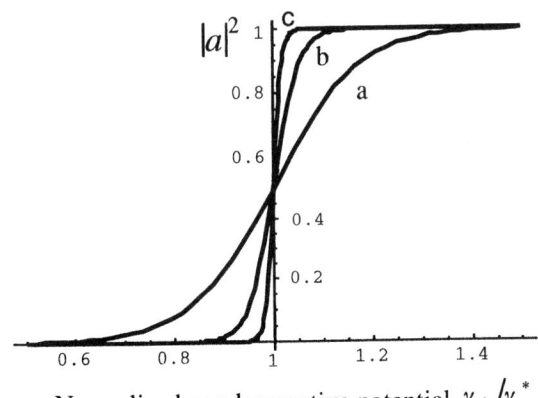

FIGURE 1. The scattering probability on the normalized ponderomotive potential γ_L/γ^* for (a) $k_0 l^* = 3$, (b) $k_0 l^* = 10$, (c) $k_0 l^* = 20$

and asymptotic wavenumbers are the same for all three waves, the absolute values $|a|^2$ and $|b|^2$ are the reflected and transmitted fractions of the incident flow, respectively. The asymptotic wavenumber $k_0 = \sqrt{\varepsilon^2 - m^2 c^4}/\hbar c$ corresponds to the field-free momentum of an electron of a given energy. The spatial part of the wave function is governed by

$$\frac{d^2 R}{dz^2} + \left(k_0^2 - \frac{e^2}{\hbar^2 c^2} A^2(z) \right) R(z) = 0 \tag{56}$$

The solution is obtained by considering a specific intensity profile of laser pulse, such as the modefied Pöschl-Teller potential[18], $A^2(z) = A_0^2 \cosh^{-2}(z/l^*)$, where $l^* = \gamma_g \beta_g c \tau_p$ is the characteristic pulse length in the reference frame for the pulse duration τ_p in the laboratory frame. Then the reflection and transmitted probabilities are calculated by

$$|a|^2 = \frac{1}{1 + \Delta^2}, \quad |b|^2 = \frac{\Delta^2}{1 + \Delta^2}, \quad \Delta = \frac{\tanh \delta_+ + \tanh \delta_-}{1 - \tanh \delta_+ \tanh \delta_-}, \tag{57}$$

where the characterestic phases $\delta_{\pm}$ are determined by the relation between the particle momentum and the ponderomotive potential as

$$\delta_{\pm} = \frac{\pi}{2} \left[k_0 l^* \pm \left(\frac{e^2 A_0^2 l^{*2}}{\hbar^2 c^2} - \frac{1}{4} \right)^{1/2} \right]. \tag{58}$$

For an incident particle energy higher than the laser ponderomotive potential barrier, these large positive phases result in a large Δ so that the reflection probability vanishes, whereas the trasmission probability is close to 1. When the ponderomotive potential exceeds the particle energy, δ_- becomes negative to lead to a small Δ as the laser intensity increases. Then the reflection probability rapidly reaches unity as shown in Fig. 1. In this regime the ponderomotive scattering off the laser pulse works as a particle acceleration mechanism which we call as "Dirac Accelerator".

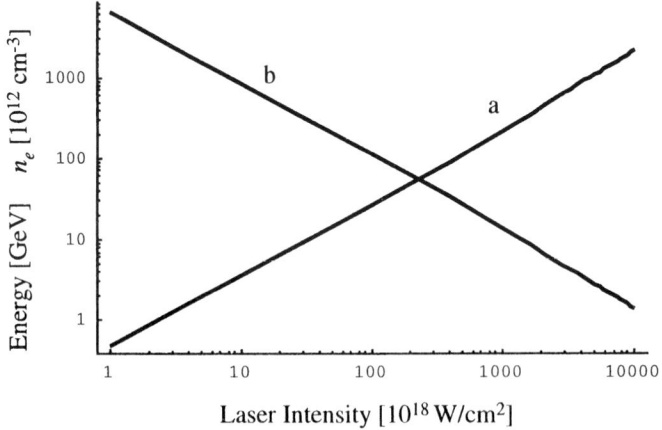

FIGURE 2. The maximum final energy (a) of an electron accelerated from the injection energy $E_{inj} =$ 150 MeV and the required plasma density (b) as a function of the laser intensity.

High energy "Dirac Accelerator"

The ponderomotive potential scattering results in acceleration of electrons via a point-like interaction with the strong laser fields. This acceleration length will be at most a half of the laser pulse length when the reflection condition Eq. (48) is satisfied. Let us consider acceleration of an electron from the injection energy γ_0 to the accelerated final energy γ via the ponderomotive interaction with the laser field in a plasma with the density n_e. The required group velocity and the corresponding Lorentz factor of the laser pulse are given by

$$\beta_g = \frac{\gamma_0 \beta_0 + \gamma \beta}{\gamma_0 + \gamma}, \quad \gamma_g = \frac{\gamma_0 + \gamma}{\sqrt{2}[1 + \gamma \gamma_0 (1 - \beta \beta_0)]^{1/2}}. \tag{59}$$

The plasma density can be determined as $n_e[\mathrm{cm}^{-3}] \approx 1.115 \times 10^{21} \gamma_g^{-2} \lambda_0^{-2}$. For this condition of "Dirac accelerator", the minimum laser field is given by

$$a_0 = \sqrt{2}[\gamma_g^2 \gamma_0^2 (1 - \beta_g \beta_0)^2 - 1]^{1/2} \tag{60}$$

The corresponding laser intensity is $I[\mathrm{W/cm}^2] = 1.37 \times 10^{18} \lambda_0^{-2} [\mu\mathrm{m}] a_0^2$. The laser pulse with the ponderomotive energy γ_L can accelerate the electron with the initial energy γ_0 up to the maximum final energy $\gamma_{\max}$ given by

$$\gamma_{\max} = 2\gamma_L^2 \gamma_0 (1 + \beta_0 \beta_L) - \gamma_0, \tag{61}$$

where $\beta_L = (\gamma_L^2 - 1)^{1/2}/\gamma_L$. Fig. 2 shows the maximum final energy for the electron acceleration from the injection energy $E_{inj} = 150\,\mathrm{MeV}$ as a function of the laser intensity.

The "Dirac accelerator" makes it possible to accelerate electrons up to the super high energy, such as 1 PeV ($\gamma \approx 2 \times 10^9$), provided with the injection energy 250 MeV

$(\gamma_0 \approx 500)$, the plasma density $n_e \approx 1.1 \times 10^9$ cm^{-3} $(\gamma_g \approx 10^6)$, and the laser intensity 2.7×10^{24} W/cm^2 $(\gamma_L \approx 1000, a_0 \approx 1414)$ for the wavelength $\lambda_L = 1 \mu$m. This laser intensity can be produced by focusing ~ 400 PW on the spot radius $r_0 = 3\mu$m.

CONCLUSIONS

The particle acceleration with strong laser fields is reviewed on interactions in vacuum and plasmas. In vacuum, the laser beat wave scheme can produce efficient particle acceleration in the energy range less than 1 GeV. In plasmas, wakefields excited by ultrashort intense laser pulses can produce ultrahigh gradient particle acceleration of the order of 100 GeV/m, which is 3 magnitudes higher than the conventional accelerators. The high energy gain exceeding 1 GeV will be achieved by using the optical guiding technique, such as a capillary discharge plasma. In the ultra-relativistic regime of the laser intensity higher than 10^{20} W/cm^2, the laser ponderomotive scattering results in particle acceleration in plasmas, which is called as "Dirac accelerator" based on the strong field-particle interaction different from classical acceleration mechanism such as RF accelerators and the laser wakefield accelerators. The "Dirac accelerator" may accelerate electrons up to PeV range energies with the laser intensity of $> 10^{24}$ W/cm^2.

REFERENCES

1. R. B. Palmer, AIP Conf. Proc. **91**, 179 (1982).
2. T. Tajima, J. M. Dawson, Phys. Rev. Lett., **43**, 267 (1979).
3. K. Nakajima et al., Phys. Rev. Lett. **74** , 4428 (1995).
4. A. Modena et al., Nature **377**, 806 (1995).
5. H. Dewa et al., Nucl. Inst. and Meth. **A410** 357 (1998); M. Kando et al., Jpn. J. Appl. Phys. **38** L967 (1999).
6. P. L. Kapitza and P. A. M. Dirac, Proc. Cambridge Philos. Soc. **29**, 297 (1933)
7. E. Esarey, P. Sprangle, and J. Krall, Phys. Rev. E, **52**, 5443 (1995).
8. A. Yariv, Quantum Electronics, New York: John Wiley & Sons, 1975, p. 118.
9. J. D. Lawson, IEEE Trans. Nucl. Sci. **NS-26**, 4217 (1979).
10. E. Esarey et al., Phys. Plasmas **2**, 1432 (1995).
11. K. Nakajima, Phys. Plasmas **3**, 2169 (1996).
12. T. Hosokai et al., Optics Letters, **25**,10-12 (2000).
13. K. Yamakawa, M. Aoyama, S. Matsuoka, T. Kase, Y. Akahane, and H. Takuma, Opt. Lett. **23**, 1468 (1998).
14. K. Nakajima, Nucl. Instr. and Meth. **A410**, 514 (1998).
15. K. Nakajima, Nucl. Instr. and Meth. **A455**, 140-147 (2000).
16. F. V. Hartemann, S. N. Fochs, G. P. LeSage, N. C. Luhmann, Jr., J. G. Woodworth, M. D. Perry, Y. J. Chen, A. K. Kerman, Phys. Rev. E, **51**, 4833 (1995).
17. C. J. McInstrie and E. A. Startsev, Phys. Rev. E **54**, R1070 (1996); Phys. Rev. E **56**, 2130 (1997).
18. S. Flügge, Practical Quantum Mechanics, New York: Springer-Verlag, 1974.

Searching for the Origin of Masses

Hiroyuki Iwasaki

KEK, High Energy Accelerator Research Organization
1-1 Oho, Tsukuba-shi, Ibaraki-ken, 305-0801 Japan

Abstract. The origin of masses of weak gauge bosons as well as quarks and leptons is one of the most mysterious themes in high energy physics. In the standard model, it is explained as a result of a spontaneous symmetry breaking of the vacuum, which is caused by self-interaction of an unknown complex scalar field. Three weak gauge bosons acquire their masses by "eating" the three components of the scalar field. The remaining one component survives and called the "Higgs boson." Masses of quarks and leptons are generated by an interaction between those particles and the scalar field. Searching for the Higgs boson is a key step toward deeper understanding of nature. In the lecture, a brief introduction of theoretical aspects and an experimental approach based on an ongoing project will be given.

1. INTRODUCTION

There are four fundamental interactions: gravitaional, electromagnetic, weak, and strong interaction. Indeed classical gravitational interaction is well understood in the framework of general theory of relativity, but there is not a satisfactory quantum theory treating gravity. The other three interactions are well formulated in a consistent manner called "the standard model." Although it is called model, all the experimental results are consistent with this model with good accuracy.[1] Since quantum effect of gravity becomes significant only when the energy scale is order of 10^{19} GeV, it can be ignored as long as we consider phenomena at least below 1 TeV energy scale.

In the modern picture, an interaction is caused through exchanging particles between subject particles. The mediating particles are called "gauge bosons," which are photon for the electromagnetic interaction, weak bosons (W, Z) for the weak interaction, and gluons for the strong interaction. Elementary particles in the standard model are categorized into matter fermions, gauge bosons, and a yet-discovered Higgs boson. The matter fermions are six quarks, three charged leptons, and three neutrinos.

In order to find out a dynamics of an interaction between the elementary particles, one needs a guiding principle. A requirement of the "local gauge invariance" of the Lagrangian is such a principle in the standard model. We can deduce the interaction by demanding the gauge invariance on the Lagrangian for free particles. Its direct

---

[1] Only one exception is neutrino masses observed recently. But it can be incorporated into the framework of the model. In fact, neutrino masses are very small and have been assumed massless in the standard model.

CP634, *Science of Superstrong Field Interactions*, edited by K. Nakajima and M. Deguchi
© 2002 American Institute of Physics 0-7354-0089-X/02/$19.00

consequence is that the gauge bosons have to be massless. Indeed photon and gluons are massless, but weak gauge bosons are quite heavy. In addition, the weak interaction violates parity: the W boson couple to left-handed fermions only. In such a case, fermions have to be also massless, which also contradicts with the real world. A key solution of the mass problem is so called the "Higgs mechanism." And the model predicts a massive spin-0 boson called the "Higgs boson."

In the following section, an overview of the Higgs mechanism is given. In section 3, a big international project aiming at discovery of the Higgs boson is introduced. It is called "LHC." A pp-collider and two detectors under construction are briefly reported. In section 4, it is explained how the experiments reveal the Higgs boson. Discovery potential of the Higgs mass is studied. A summary of the lecture is given in the last section.

2. HIGGS MECHANISM

Here we will see only flavor of a basic idea of the Higgs mechanism and not get into the electroweak theory,[2] which is beyond our scope of this short course [1]. We consider $U(1)$ gauge invariant Lagrangian for a complex scalar field $\phi = (\phi_1 + i\phi_2)/\sqrt{2}$ described by

$$L = (D^\mu \phi)^* (D_\mu \phi) - V(\phi) - \frac{1}{4} F_{\mu\nu} F^{\mu\nu}, \tag{2.1}$$

where D_μ is a covariant derivative,

$$D_\mu = \partial_\mu + ig A_\mu, \tag{2.2}$$

where g is the coupling constant, A_μ is the $U(1)$ gauge field and the last term is the kinetic energy of A_μ, and

$$F_{\mu\nu} = \partial_\mu A_\nu - \partial_\nu A_\mu. \tag{2.3}$$

The potential is

$$V(\phi) = \mu^2 \phi \phi^* + \lambda (\phi \phi^*)^2 = \frac{1}{2}\mu^2(\phi_1^2 + \phi_2^2) + \frac{1}{4}\lambda(\phi_1^2 + \phi_2^2)^2. \tag{2.4}$$

One can see that the Lagrangian is invariant under $U(1)$ local gauge transformation,

$$\phi \to e^{i\alpha(x)}\phi, \tag{2.5}$$

$$A_\mu \to A_\mu - \frac{1}{g}\partial_\mu \alpha. \tag{2.6}$$

If $\mu^2 > 0$ and $\lambda = 0$, the potential $V(\phi)$ simply describes the mass term of the scalar fields. If, however, somehow $\mu^2 < 0$ and $\lambda > 0$, the potential looks like a bottom of a wine bottle as shown in Fig.1. In this case, the point $(\phi_1, \phi_2) = (0,0)$ is no

---

[2] In the electroweak theory, we have to deal with $SU(2) \times U(1)$, instead of $U(1)$. We will briefly comment on the electroweak theory at the end of this section.

longer the minimum. It has a circle of minima[3] of the potential in the ϕ_1-ϕ_2 plane of radius v, such that

$$\phi_1^2 + \phi_2^2 = v^2 \quad \text{with} \quad v^2 = -\frac{\mu^2}{\lambda}. \tag{2.7}$$

We can take any point in this circle as an absolute minimum of energy, which corresponds to a "vacuum." We choose a point $(\phi_1, \phi_2) = (v, 0)$ as the vacuum and introduce new real fields, η and ξ as

$$\phi(x) = \sqrt{\frac{1}{2}} e^{i\xi(x)/v} [v + \eta(x)] \approx \sqrt{\frac{1}{2}} [v + \eta(x) + i\xi(x)]. \tag{2.8}$$

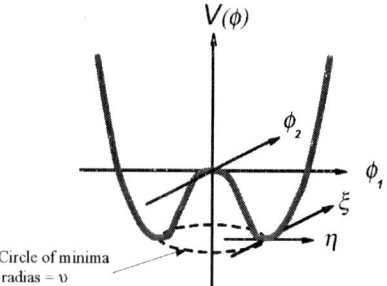

FIGURE 1. The potential V(ϕ) for a complex scalar field.

Once a certain vacuum is chosen, the vacuum no longer possesses a global gauge symmetry. It is called the "spontaneous symmetry braking." Now if we fix the gage as

$$\alpha(x) = -\xi(x)/v, \tag{2.9}$$

we get

$$\phi'(x) = e^{i\alpha(x)}\phi(x) = \sqrt{\frac{1}{2}} [v + \eta(x)]. \tag{2.10}$$

By substituting $\phi'(x)$ into the Lagrangian (2.1) and neglecting a constant, we can rewrite it as,

$$
\begin{aligned}
L' ={}& \frac{1}{2}(\partial_\mu \eta)^2 - \lambda v^2 \eta^2 + \frac{1}{2} g^2 v^2 A_\mu^2 - \lambda v \eta^3 - \frac{1}{4}\lambda \eta^4 \\
&+ \frac{1}{2} g^2 A_\mu^2 \eta^2 + v g^2 A_\mu^2 \eta - \frac{1}{4} F_{\mu\nu} F^{\mu\nu}.
\end{aligned} \tag{2.11}
$$

There are only two fields left, a scalar η and a vector gauge boson A_μ. The third term shows that the gauge boson A_μ has acquired a mass,

---

[3] The value v is called the "vacuum expectation value."

414

$$m_A = gv. \tag{2.12}$$

The spurious ξ-field has disappeared[4] and its freedom is turned into a longitudinally polarized (helicity zero) state of the massive gauge boson A_μ. This is called the "Higgs mechanism." The second term is a mass term of η with,

$$m_\eta = \sqrt{2\lambda v^2}. \tag{2.13}$$

The massive η is called the Higgs boson. We can deduce v from the coupling constant g and the mass of the gauge boson m_A by using (2.12). Nevertheless, since we don't know the parameters λ, we cannot estimate the Higgs mass m_η. On the other hand, we know how the Higgs boson couples to the gauge boson from the sixth (two Higgs bosons and two gauge bosons) and the seventh term (a Higgs boson and two gauge bosons) of (2.11). Once we discover the Higgs boson and measure its mass, we can deduce the parameter λ. Then the self-couplings of the Higgs boson are known from the fourth and fifth terms of (2.11).

This is the mechanism of how the gauge boson acquires the mass. Then one may ask why the photon and the gluons are still massless. The reason for the gluon is simple. The gluons couple to "color charges" which represent three degrees of freedom.[5] Since the scalar field ϕ does not have a color charge, and hence does not couple to gluons. The gluons are free from the spontaneous symmetry breaking of the vacuum and it can be massless.

In case of the photon, it is not so simple. In the standard model, the electromagnetic interaction and the weak interaction are unified in a single framework called the "electroweak theory," where the gauge group is $SU(2)_L \times U(1)_Y$. There are three gauge fields, W_1, W_2, W_3 for $SU(2)_L$ and one gauge field, B for $U(1)_Y$. These four fields couple to four scalar fields ϕ_i ($i = 1,2,3,4$) which belong to $SU(2)_L$. Three scalar fields out of the four disappear when $SU(2)_L$ gauge is fixed. After the spontaneous symmetry breaking, W_1 and W_2 become massive charged $W^\pm$ (80 GeV); properly chosen linear combinations of W_3 and B make massive neutral Z (91 GeV) and massless photon; one real scalar field remains, which is the Higgs boson.

Masses of the fermions are generated through the couplings of fermions with the scalar field ϕ. The mass of a fermion is simply proportional to the coupling. If a fermion is heavy, it is because its coupling to the scalar field is strong. Due to the same reason, the Higgs boson couples to a fermion in proportional to its mass. Now that we know how the Higgs boson couples to all the other particles (as well as self-couplings), we can calculate the production cross sections of the Higgs boson and its decay modes with good precision as a parameter of its mass.

---

[4] If the gauge is not fixed and (2.8) is simply substituted into (2.1), there appear additional terms which contain the kinetic term of the ξ-field without mass term, $m_\xi^2 \xi^2 / 2$. It means there appears additionally an unwanted massless scalar boson, called a "Goldstone boson."

[5] The gauge group of the strong interaction is $SU(3)$.

3. LHC PROJECT

The Large Hadron Collider (LHC) is an accelerator under construction at European Organization for Nuclear Research (CERN) near Geneva in Switzerland [2]. It is a circular machine assembled in a tunnel approximately 100m deep underground, circumference of which is 27 km. Protons of 7 TeV collide on protons of the same energy (pp collider), namely 14 TeV in proton-proton center-of-mass energy (E_{CM}).[6] Since the Higgs mass is not predicted from the theory, the machine have to be capable to produce Higgs particles over the wide range up to 1 TeV mass scale. Protons are accelerated with several steps to 7 TeV: up to 0.05 GeV with a linear accelerator (Linac), up to 1.4 GeV with Proton Synchrotron Booster (PSB), up to 26 GeV with Proton Synchrotron (PS), up to 450 GeV with Super-Proton Synchrotron (SPS), then up to 7 TeV with LHC. There are two general-purpose detectors called ATLAS and CMS. A conceptual figure of the LHC complex is shown Fig.2.

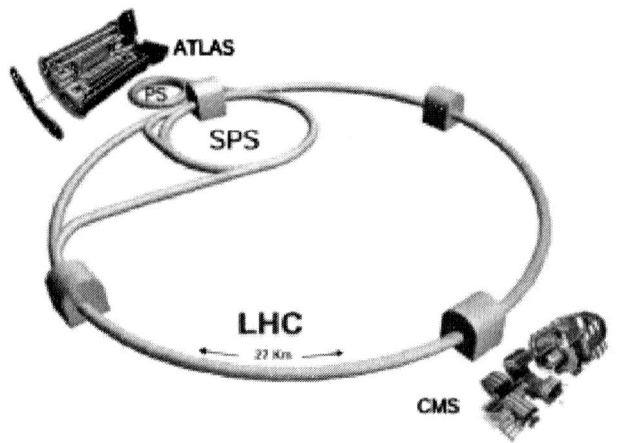

FIGURE 2. An image of the LHC complex [6].

In addition to the beam energy, another important parameter of a collider is "luminosity." It is defined as

$$N = \mathcal{L}\sigma, \qquad (3.1)$$

where, N is a number of events, $\mathcal{L}$ is the luminosity, and σ is a cross section. Cross sections for interesting processes are usually small. Therefore, the luminosity has to be large enough to get statistically significant events. The luminosity is proportional to

---

[6] Accelerators being used for energy frontier particle physics are collider type machines. The present world highest record of E_{CM} is 2 TeV. The beam energy is most effectively used in collision processes. In the fixed target case, on the other hand, most of the beam energy is just spent to boost the whole system. If we need E_{CM} of 14 TeV in a pp fixed target experiment, the beam energy has to be 100,000TeV!

each beam current and a collision frequency, and inversely proportional to the beam cross sections. Protons in a beam are grouped into many clusters called bunches, each of which contains 10^{11} protons. Bunch spacing is 7.5m and beam collision occurs every 25 ns (40 MHz). The beam shape at the interaction point is round and its transverse size (1 sigma) is 16 µm. The design luminosity is 10^{34} cm^{-2}s^{-1} at each collision point. The main machine parameters are summarized in Table1.

TABLE 1.   Main parameters of LHC.

Parameter		Unit
Circumference	27	km
Collision type	Proton on proton	
Beam energy	7.0	TeV
Design luminosity	10^{34}	cm^{-2}s^{-1}
Circulating current/beam	0.54	A
Bunch spacing	25	ns
Bending radius	2.8	km
Dipole field	8.4	T
Number of dipole magnets	1232	

Although the LHC ring is huge, a magnetic field has to be 8.4 Tesla in order to bend the 7 TeV protons. Such a high field can be achieved with a NbTi superconducting magnet operated at 1.9 K. Since the colliding beams are both protons, two beam-channels are necessary. These two are embedded in a single cryostat. Each coil aperture is 56 mm, and the magnet length is 14.2m.

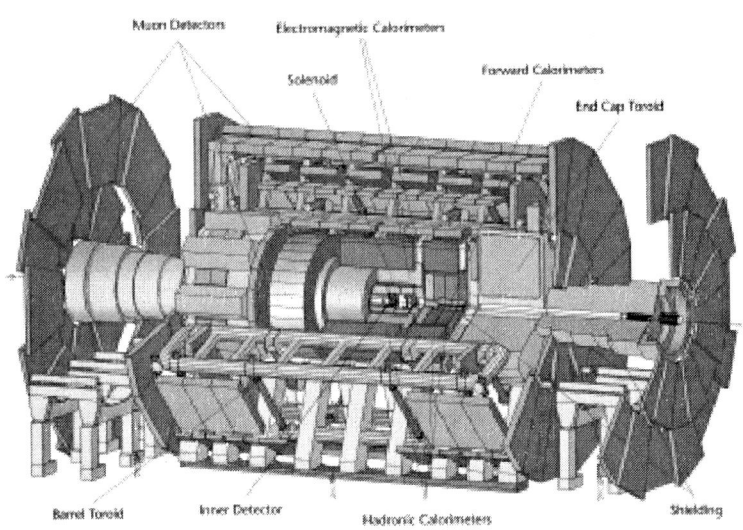

FIGURE 3. Conceptual bird's-eye view of the ATLAS detector. Approximate dimensions are 24m in diameter, 42m in total length. The overall weight is about 7,000 tons. Air-core toroid magnets are used in barrel and endcap regions for the muon spectrometer.

There are two general-purpose detectors for the LHC experiment, ATLAS (A Toroidal LHC ApparatuS, Fig.3) [3] and CMS (Compact Muon Solenoid, Fig.4) [4]. Both are gigantic detectors having cylindrical structure[7] and are hermetic to capture almost all the emerged particles.

The innermost detector surrounding the interaction region is a tracking system with light materials to measure momenta of charged particles. Outside the tracker, there exists an electromagnetic calorimeter system with high-atomic-number materials to measure energy of electrons and photons. Then a hadron calorimeter system comes next. It is made of heavy materials to measure energy of hadronic partcles, such as pions, kaons, protons, neutrons. The outermost device is a muon spectrometer to detect muons and measure their momenta. Since the muon interacts with materials electromagnetically and much heavier than the electron, it can easily penetrate heavy materials. A neutrino escapes whole detector system and cannot be detected directly,

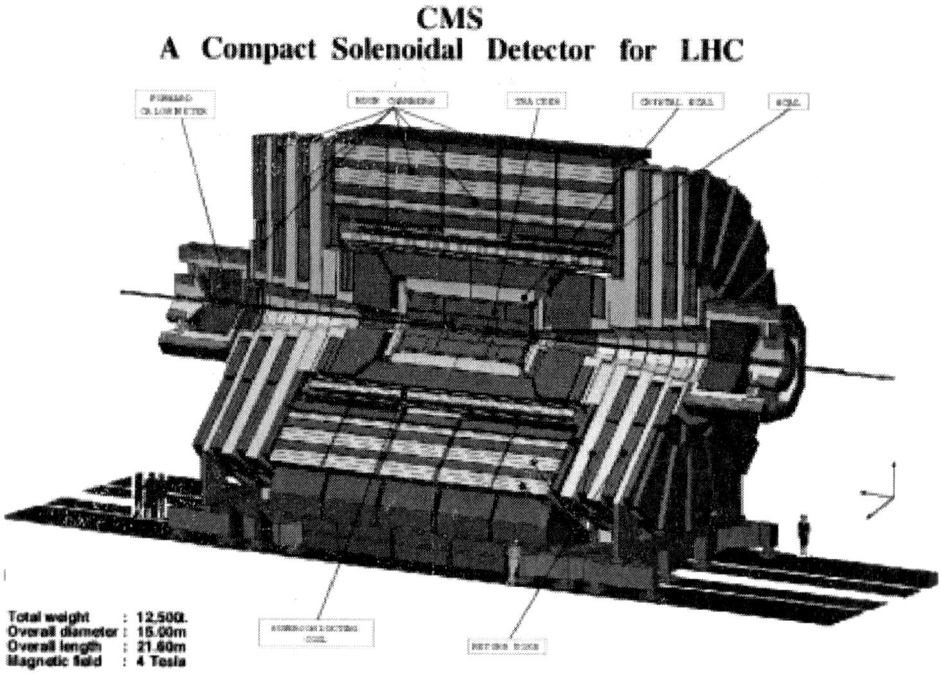

CMS
A Compact Solenoidal Detector for LHC

Total weight   : 12.500.
Overall diameter : 15.00m
Overall length   : 21.60m
Magnetic field   : 4 Tesla

FIGURE 4. Conceptual bird's-eye view of the CMS detector. Approximate dimensions are 15m in diameter, 22m in total length. The overall weight is about 12,500 tons. It is much compact compared with ATLAS, but is heavier because of using iron yoke outside of a solenoid magnet. The magnet is superconducting one with inner diameter of 5.9m and length of 13m. It generates a 4 Tesla magnetic field inside the coil.

---

[7] In a collider type experiment, particles produced at the collision emerge in all directions. In a fixed target experiment, on the other hand, the particles are boosted to the forward direction.

but energy carried by the neutrino causes energy imbalance in the transverse plane. Since the detector is hermetic, total sum of the transverse energy has to be zero. In this way, we can identify its existence and even measure its transverse energy.[8] A high energy quark or gluon is observed as a bundle of particles, called "jet."

While cross sections for interesting processes are small, typically 1 pb or less,[9] proton-proton total cross section is 100 mb. It means non-interesting underling events occur with a rate of 1 GHz at the normal luminosity of 10^{34} cm^{-2}s^{-1}. In addition, about 100 particles are emerged from each of such an event. Therefore radiation environment is harsh and the detectors, including readout electronics, have to be radiation hard. In order to select only interesting events, a sophisticated event triggering and data-taking system is necessary. Furthermore, the large volume data flow (500 Gbits/s) requires a huge mass storage (1 PB/year) system and gigantic computing power (5TIPS).

4. HIGGS SEARCH

A proton is not an elementary particle, but composed of quarks and gluons. Its static picture is a system made of three valence quarks, namely two up-quarks and one

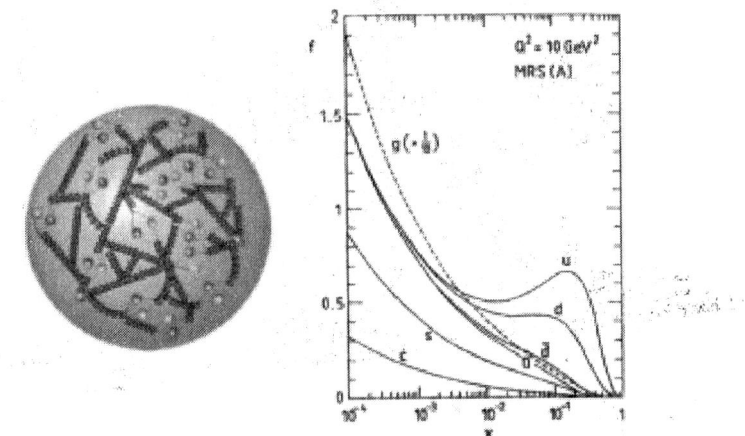

FIGURE 5. **Left:** Conceptual picture inside a proton. Quarks are shown with small balls, gluons with springs. Sea quarks as well as valence quarks exist inside a proton. These quarks and gluons are not stable, but are crated at a certain time then annihilate soon after. **Right:** Parton distribution function at $Q^2=10$ GeV^2. It shows probability density of a certain parton carrying momentum fraction x of proton, $x = p_{parton} / p_{proton}$.

---

[8] Why only the transverse energy? Along the beam direction, most of the energy is carried by proton remnants and these particles escape into the beam pipe. Therefore, the longitudinal energy generally does not balance when we use detected particles even in an event without neutrino.

[9] The unit "pb" means pico (10^{-12}) barn (10^{-24} cm^2), namely 10^{-36} cm^2. Similarly, "mb" is mili (10^{-3}) barn and "fb" is femto (10^{-15}) barn.

down-quark. However, other types of quark-antiquark pairs, as well as gluons, are also crated and annihilate continuously inside the proton. The quark and gluon are called "parton" as a collective name. Each type of parton carries momentum fraction of the proton. It is not constant but depends on Q^2, which is the momentum-transfer squared. This dynamical image of the proton and a parton distribution function at certain Q^2 is shown in Fig.5. One should note that an elementary process is a parton-parton collision, and usually only a small fraction of the center-of-mass energy (14 TeV) is used.

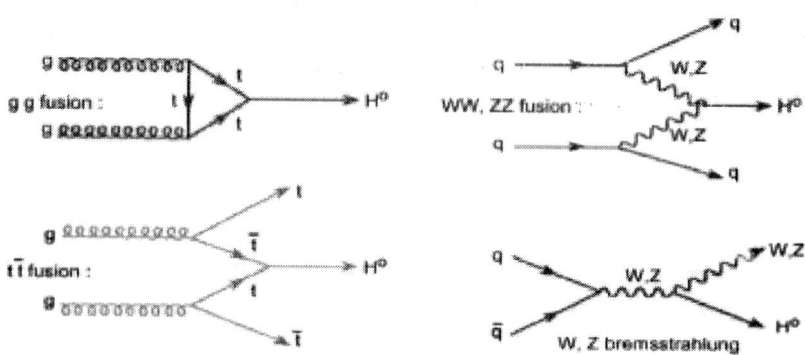

FIGURE 6. Main diagrams of Higgs production.

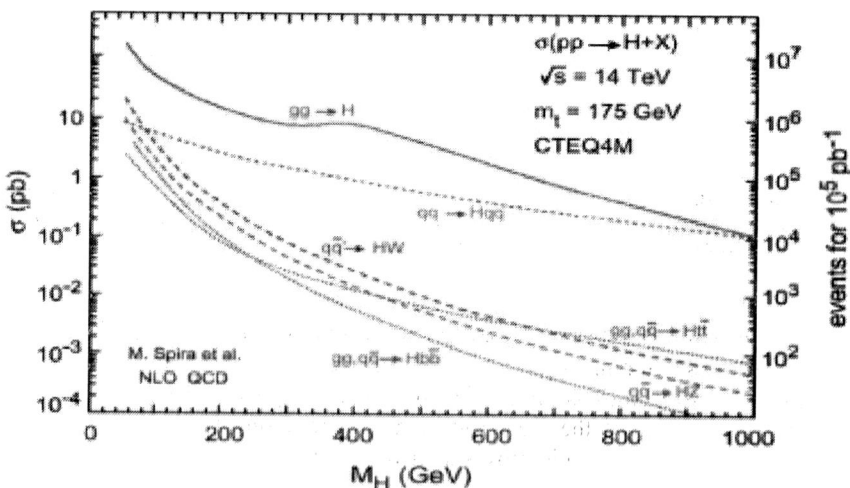

FIGURE 7. Higgs production cross section as a function of its mass for various processes. The unit of 1pb (left vertical scale) is 10^{-36} cm^2. The number of events for 10^5 pb^{-1} (right vertical scale) corresponds to the data accumulated for about 120 net days with the nominal luminosity, 10^{34} cm^{-2}s^{-1}.

Since Higgs boson couples strongly with the heavier particles, it is produced through top-quark (174 GeV) or weak gage bosons, W (80 GeV) or Z (91 GeV). Main diagrams for the Higgs production are shown in Fig.6. The production cross section through each process is shown in Fig.7 as a function of Higgs mass. For any of the Higgs mass, "gluon-gluon fusion" is the largest cross section, and "WW and ZZ fusion" is the next except for the Higgs mass of below 100 GeV. If the Higgs mass is around 400 GeV, millions of events will be produced within a year.

Branching ratios of the Higgs boson is shown in Fig.8 as a function of its mass. Above 180 GeV where real WW or ZZ decay is kinematically allowed, these decay modes are dominant. Below 100 GeV, it predominantly decays to a b-quark pair, and about 8% to a tau pair. Around 100 to 150 GeV, it decays to a gamma pair with about 0.1% level. Although it is a small branching ratio, this decay mode is promising because of its clean signature.

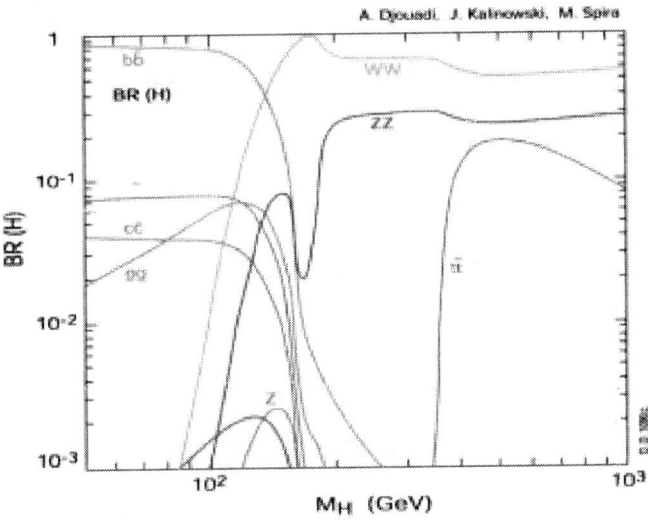

FIGURE 8. Branching ratio of the Higgs particle as a function of its mass. Above 180 GeV, it dominantly decays to a gauge-boson pair. The gorge at 160 GeV is due to the fact that the real WW pair becomes possible, while the real ZZ is still not open.

We can study whether a signal process stands out or is hidden by background processes by simulating both processes. Since background events usually dominate over signal events, various event-selection cuts are applied in the event analysis which are effective to reject the backgrounds while accepting the signals. Although the signal-to-background ratio improves after the event selection, number of signal events is also inevitably reduced. As a measure of observability of a signal process, we define "signal significance" as $N_S / \sqrt{N_B}$, where N_S is the number of the signal events and N_B is that of the background events. If it exceeds five, the signal process is expected to be visible. Since the significance is approximately proportional to the square-root of an integrated luminosity, we can estimate how much integrated luminosity is necessary and how long it takes to accumulate.

Here we show some examples of how the Higgs boson can be seen for several processes [5]. An expected invariant mass distribution of two gammas for H $\rightarrow$ 2γ is shown in Fig.9 in a case that the Higgs mass is 120 GeV and an integrated luminosity is 100 fb^{-1}. It takes 120 net days and corresponds to about one calendar year to accumulate 100 fb^{-1} with the nominal luminosity. A clear peak can be seen on the smooth background slope. The signal significance becomes 6.5. For the heavier Higgs, the decay mode, H $\rightarrow$ ZZ $\rightarrow$ 4ℓ, namely Higgs decays to a ZZ pair and each

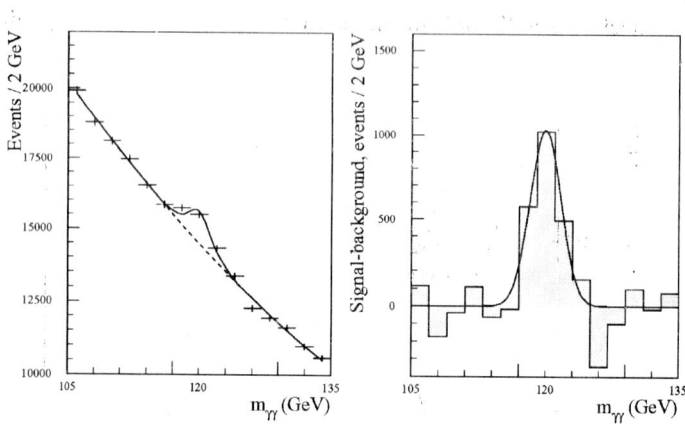

FIGURE 9. Expected invariant mass distribution of two gammas for M_H=120 GeV, and for an integrated luminosity of 100 fb^{-1}. A clear peak can be seen on the smooth background slope (left). The peak after subtraction of the background (right).

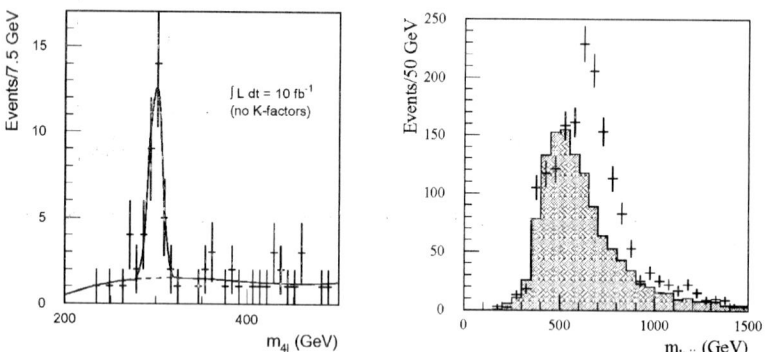

FIGURE 10. **Left:** Expected invariant mass distribution of four leptons coming from H $\rightarrow$ ZZ $\rightarrow$ 4ℓ for M_H=300 GeV and an integrated luminosity of 10 fb-1. **Right:** Expected invariant mass distribution of one lepton, one neutrino, and two jets coming from H $\rightarrow$ WW $\rightarrow$ $\ell\nu$jj for M_H=600 GeV and an integrated luminosity of 100 fb-1.

Z decays to a charged-lepton pair, is the most promising channel. Such a case is shown in Fig.10 (Left) for M_H=300 GeV and an integrated luminosity of 10 fb^{-1}. If the Higgs mass is much heavier, we can use H $\rightarrow$ WW $\rightarrow$ $\ell\nu$jj. Distribution of invariant mass for this mode is shown in Fig.10 (Right) for M_H=600 GeV and an integrated luminosity of 100 fb^{-1}.

A summary plot of the signal significance for various decay channels is shown in Fig.11 as a function of Higgs mass for an integrated luminosity of 100 fb^{-1}. After one year of LHC running with the nominal luminosity, we can discover the Higgs boson of any mass range below 1 TeV with a high significance. We can then measure the decay width of Higgs boson, as well as its mass, and its couplings to fermions. Through the consistency check, we will confirm the particle as the Higgs boson predicted in the standard model.

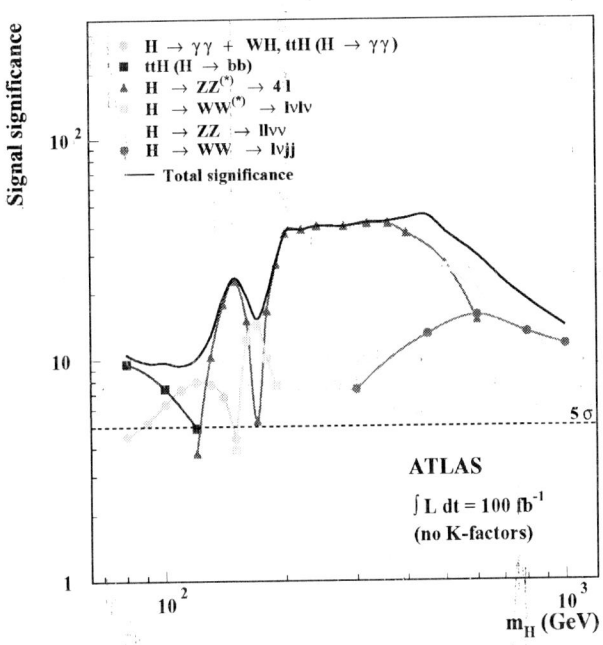

FIGURE 11. Statistical significance for various decay modes with the ATLAS detector for an integrated luminosity of 100 fb^{-1}.

5. SUMMARY

The origin of masses of weak gauge bosons as well as quarks and leptons is one of the most mysterious themes in high energy physics. In the standard model, it is explained as a result of a spontaneous symmetry breaking of the vacuum, which is caused by a self-interaction of unknown complex scalar field. Three degrees of freedom of the scalar field out of the four disappear when $SU(2)_L$ gauge is fixed, but appear as the longitudinally polarized states of the three massive gauge bosons. The remaining one freedom survives, which is the Higgs boson. Masses of quarks and leptons are generated by an interaction between those fermions and the scalar field.

The LHC project is to discover the Higgs boson. Since the Higgs mass is not predicted, the machine and detectors were designed capable of covering full rage of the Higgs mass below 1 TeV. The machine is a pp-collider operated at $E_{CM} = 14$ TeV with a nominal luminosity of 10^{34} cm^{-2}s^{-1}. The energy is approximately 7 times larger than the present world record. There are two general-purpose detectors, ATLAS and CMS. The machine and the detectors are under construction at CERN, and its first collision is scheduled in April 2007.

Once the machine is operational as designed, the Higgs boson will be discovered within a year. Then its property will be studied and checked whether it is consistent with the standard model in the following years. If the Higgs is not discovered, by any chance, it also opens a new interesting scenario, "beyond the standard model." In either case, the LHC project will bring us deeper understanding of nature.

ACKNOWLEDGMENTS

The author greatly appreciates Prof. Nakajima for giving him an opportunity to have a short lecture to young and motivated students. Prof. Nakajima also kindly invited the author to attend the symposium, "Science of Super-Strong Field Interactions" being held just after the "Shonan lectures." It covered variety of fields from basic theoretical aspects to high-energy cosmological phenomena. Most of those talks were new to him and were very interesting. The author appreciates Prof. Nakajima's effort to lead the symposium as well as the Shonan lectures so successful.

REFERENCES

1. If one wants to know more details, see for example: Abers. E., and Lee, B. W., "Gauge Theories," Phys. Rep. 9C, 1(1973); Halzen, F., and Martin. A. D., "Quarks & Leptons: An Introductory Course in Modern Particle Physics," John Wiley & Sons, Inc. New York, 1984, pp. 311-354.
2. LHC study group, "The Large Hadron Collider," CERN/AC/95-05(LHC), 20 October 1995.
3. ATLAS collaboration, "ATLAS, Technical Proposal for a General-Purpose pp Experiment at the Large Hadron Collider at CERN," CERN/LHCC/94-43, LHCC/P2, 15 December 1994.
4. CMS collaboration, "Technical Proposal," CERN/LHCC/94-38, LHCC/P1, 15 December 1994.
5. ATLAS collaboration, "ATLAS Detector and Physics Performance, Technical Design Report Volume II" CERN/LHCC/99-15, ATLAS TDR 15, 25 May 1999.
6. One can get information on the LHC project via CERN Web site, http://welcome.cern.ch/welcome/gateway.html.